DIGITAL IMAGE
PROCESSING

DIGITAL IMAGE PROCESSING

PIKS Inside

Third Edition

WILLIAM K. PRATT
PixelSoft, Inc.
Los Altos, California

A Wiley-Interscience Publication

JOHN WILEY & SONS, INC.

New York • Chichester • Weinheim • Brisbane • Singapore • Toronto

About the cover:

The upper left corner image is the color `peppers_gamma` original image.

The upper center image is the horizontal Sobel gradient of the luma component of the original image.

The lower left image is the color `dolls_gamma` original image.

The lower center image is a blurred version of the original image obtained by convolution of the red, green, and blue components of the original image with a 5 × 5 uniform rectangular impulse response array.

The lower right image image is a sharpened version of the original image obtained by subtracting an amplitude weighted version of the blurred image from a weighted version of the original image. The processing technique is called unsharp masking.

Library of Congress Cataloging-in-Publication Data Is Available

ISBN 0-471-37407-5

Printed in the United States of America

To my wife, Shelly
whose image needs no enhancement

CONTENTS

PREFACE

In January 1978, I began the preface to the first edition of *Digital Image Processing* with the following statement:

The field of image processing has grown considerably during the past decade with the increased utilization of imagery in myriad applications coupled with improvements in the size, speed, and cost effectiveness of digital computers and related signal processing technologies. Image processing has found a significant role in scientific, industrial, space, and government applications.

In January 1991, in the preface to the second edition, I stated:

Thirteen years later as I write this preface to the second edition, I find the quoted statement still to be valid. The 1980s have been a decade of significant growth and maturity in this field. At the beginning of that decade, many image processing techniques were of academic interest only; their execution was too slow and too costly. Today, thanks to algorithmic and implementation advances, image processing has become a vital cost-effective technology in a host of applications.

Now, in this beginning of the twenty-first century, image processing has become a mature engineering discipline. But advances in the theoretical basis of image processing continue. Some of the reasons for this third edition of the book are to correct defects in the second edition, delete content of marginal interest, and add discussion of new, important topics. Another motivating factor is the inclusion of interactive, computer display imaging examples to illustrate image processing concepts. Finally, this third edition includes computer programming exercises to bolster its theoretical content. These exercises can be implemented using the Programmer's Imaging Kernel System (PIKS) application program interface (API). PIKS is an International

Standards Organization (ISO) standard library of image processing operators and associated utilities. The PIKS Core version is included on a CD affixed to the back cover of this book.

The book is intended to be an "industrial strength" introduction to digital image processing to be used as a text for an electrical engineering or computer science course in the subject. Also, it can be used as a reference manual for scientists who are engaged in image processing research, developers of image processing hardware and software systems, and practicing engineers and scientists who use image processing as a tool in their applications. Mathematical derivations are provided for most algorithms. The reader is assumed to have a basic background in linear system theory, vector space algebra, and random processes. Proficiency in C language programming is necessary for execution of the image processing programming exercises using PIKS.

The book is divided into six parts. The first three parts cover the basic technologies that are needed to support image processing applications. Part 1 contains three chapters concerned with the characterization of continuous images. Topics include the mathematical representation of continuous images, the psychophysical properties of human vision, and photometry and colorimetry. In Part 2, image sampling and quantization techniques are explored along with the mathematical representation of discrete images. Part 3 discusses two-dimensional signal processing techniques, including general linear operators and unitary transforms such as the Fourier, Hadamard, and Karhunen–Loeve transforms. The final chapter in Part 3 analyzes and compares linear processing techniques implemented by direct convolution and Fourier domain filtering.

The next two parts of the book cover the two principal application areas of image processing. Part 4 presents a discussion of image enhancement and restoration techniques, including restoration models, point and spatial restoration, and geometrical image modification. Part 5, entitled "Image Analysis," concentrates on the extraction of information from an image. Specific topics include morphological image processing, edge detection, image feature extraction, image segmentation, object shape analysis, and object detection.

Part 6 discusses the software implementation of image processing applications. This part describes the PIKS API and explains its use as a means of implementing image processing algorithms. Image processing programming exercises are included in Part 6.

This third edition represents a major revision of the second edition. In addition to Part 6, new topics include an expanded description of color spaces, the Hartley and Daubechies transforms, wavelet filtering, watershed and snake image segmentation, and Mellin transform matched filtering. Many of the photographic examples in the book are supplemented by executable programs for which readers can adjust algorithm parameters and even substitute their own source images.

Although readers should find this book reasonably comprehensive, many important topics allied to the field of digital image processing have been omitted to limit the size and cost of the book. Among the most prominent omissions are the topics of pattern recognition, image reconstruction from projections, image understanding, image coding, scientific visualization, and computer graphics. References to some of these topics are provided in the bibliography.

WILLIAM K. PRATT

Los Altos, California
August 2000

ACKNOWLEDGMENTS

The first edition of this book was written while I was a professor of electrical engineering at the University of Southern California (USC). Image processing research at USC began in 1962 on a very modest scale, but the program increased in size and scope with the attendant international interest in the field. In 1971, Dr. Zohrab Kaprielian, then dean of engineering and vice president of academic research and administration, announced the establishment of the USC Image Processing Institute. This environment contributed significantly to the preparation of the first edition. I am deeply grateful to Professor Kaprielian for his role in providing university support of image processing and for his personal interest in my career.

Also, I wish to thank the following past and present members of the Institute's scientific staff who rendered invaluable assistance in the preparation of the first-edition manuscript: Jean-François Abramatic, Harry C. Andrews, Lee D. Davisson, Olivier Faugeras, Werner Frei, Ali Habibi, Anil K. Jain, Richard P. Kruger, Nasser E. Nahi, Ramakant Nevatia, Keith Price, Guner S. Robinson, Alexander A. Sawchuk, and Lloyd R. Welsh.

In addition, I sincerely acknowledge the technical help of my graduate students at USC during preparation of the first edition: Ikram Abdou, Behnam Ashjari, Wen-Hsiung Chen, Faramarz Davarian, Michael N. Huhns, Kenneth I. Laws, Sang Uk Lee, Clanton Mancill, Nelson Mascarenhas, Clifford Reader, John Roese, and Robert H. Wallis.

The first edition was the outgrowth of notes developed for the USC course "Image Processing." I wish to thank the many students who suffered through the early versions of the notes for their valuable comments. Also, I appreciate the reviews of the notes provided by Harry C. Andrews, Werner Frei, Ali Habibi, and Ernest L. Hall, who taught the course.

With regard to the first edition, I wish to offer words of appreciation to the Information Processing Techniques Office of the Advanced Research Projects Agency, directed by Larry G. Roberts, which provided partial financial support of my research at USC.

During the academic year 1977–1978, I performed sabbatical research at the Institut de Recherche d'Informatique et Automatique in LeChesney, France and at the Université de Paris. My research was partially supported by these institutions, USC, and a Guggenheim Foundation fellowship. For this support, I am indebted.

I left USC in 1979 with the intention of forming a company that would put some of my research ideas into practice. Toward that end, I joined a startup company, Compression Labs, Inc., of San Jose, California. There I worked on the development of facsimile and video coding products with Dr., Wen-Hsiung Chen and Dr. Robert H. Wallis. Concurrently, I directed a design team that developed a digital image processor called VICOM. The early contributors to its hardware and software design were William Bryant, Howard Halverson, Stephen K. Howell, Jeffrey Shaw, and William Zech. In 1981, I formed Vicom Systems, Inc., of San Jose, California, to manufacture and market the VICOM image processor. Many of the photographic examples in this book were processed on a VICOM.

Work on the second edition began in 1986. In 1988, I joined Sun Microsystems, of Mountain View, California. At Sun, I collaborated with Stephen A. Howell and Ihtisham Kabir on the development of image processing software. During my time at Sun, I participated in the specification of the Programmers Imaging Kernel application program interface which was made an International Standards Organization standard in 1994. Much of the PIKS content is present in this book. Some of the principal contributors to PIKS include Timothy Butler, Adrian Clark, Patrick Krolak, and Gerard A. Paquette.

In 1993, I formed PixelSoft, Inc., of Los Altos, California, to commercialize the PIKS standard. The PIKS Core version of the PixelSoft implementation is affixed to the back cover of this edition. Contributors to its development include Timothy Butler, Larry R. Hubble, and Gerard A. Paquette.

In 1996, I joined Photon Dynamics, Inc., of San Jose, California, a manufacturer of machine vision equipment for the inspection of electronics displays and printed circuit boards. There, I collaborated with Larry R. Hubble, Sunil S. Sawkar, and Gerard A. Paquette on the development of several hardware and software products based on PIKS.

I wish to thank all those previously cited, and many others too numerous to mention, for their assistance in this industrial phase of my career. Having participated in the design of hardware and software products has been an arduous but intellectually rewarding task. This industrial experience, I believe, has significantly enriched this third edition.

I offer my appreciation to Ray Schmidt, who was responsible for many photographic reproductions in the book, and to Kris Pendelton, who created much of the line art. Also, thanks are given to readers of the first two editions who reported errors both typographical and mental.

Most of all, I wish to thank my wife, Shelly, for her support in the writing of the third edition.

W. K. P.

PART 1

CONTINUOUS IMAGE CHARACTERIZATION

Although this book is concerned primarily with digital, as opposed to analog, image processing techniques. It should be remembered that most digital images represent continuous natural images. Exceptions are artificial digital images such as test patterns that are numerically created in the computer and images constructed by tomographic systems. Thus, it is important to understand the "physics" of image formation by sensors and optical systems including human visual perception. Another important consideration is the measurement of light in order quantitatively to describe images. Finally, it is useful to establish spatial and temporal characteristics of continuous image fields which provide the basis for the interrelationship of digital image samples. These topics are covered in the following chapters.

1

CONTINUOUS IMAGE MATHEMATICAL CHARACTERIZATION

In the design and analysis of image processing systems, it is convenient and often necessary mathematically to characterize the image to be processed. There are two basic mathematical characterizations of interest: deterministic and statistical. In *deterministic image representation*, a mathematical image function is defined and point properties of the image are considered. For a *statistical image representation*, the image is specified by average properties. The following sections develop the deterministic and statistical characterization of continuous images. Although the analysis is presented in the context of visual images, many of the results can be extended to general two-dimensional time-varying signals and fields.

1.1. IMAGE REPRESENTATION

Let $C(x, y, t, \lambda)$ represent the spatial energy distribution of an image source of radiant energy at spatial coordinates (x, y), at time t and wavelength λ. Because light intensity is a real positive quantity, that is, because intensity is proportional to the modulus squared of the electric field, the image light function is real and nonnegative. Furthermore, in all practical imaging systems, a small amount of background light is always present. The physical imaging system also imposes some restriction on the maximum intensity of an image, for example, film saturation and cathode ray tube (CRT) phosphor heating. Hence it is assumed that

$$0 < C(x, y, t, \lambda) \le A \tag{1.1-1}$$

where A is the maximum image intensity. A physical image is necessarily limited in extent by the imaging system and image recording media. For mathematical simplicity, all images are assumed to be nonzero only over a rectangular region for which

$$-L_x \leq x \leq L_x \qquad (1.1\text{-}2a)$$

$$-L_y \leq y \leq L_y \qquad (1.1\text{-}2b)$$

The physical image is, of course, observable only over some finite time interval. Thus let

$$-T \leq t \leq T \qquad (1.1\text{-}2c)$$

The image light function $C(x, y, t, \lambda)$ is, therefore, a bounded four-dimensional function with bounded independent variables. As a final restriction, it is assumed that the image function is continuous over its domain of definition.

The intensity response of a standard human observer to an image light function is commonly measured in terms of the instantaneous luminance of the light field as defined by

$$Y(x, y, t) = \int_0^\infty C(x, y, t, \lambda) V(\lambda) \, d\lambda \qquad (1.1\text{-}3)$$

where $V(\lambda)$ represents the *relative luminous efficiency function*, that is, the spectral response of human vision. Similarly, the color response of a standard observer is commonly measured in terms of a set of tristimulus values that are linearly proportional to the amounts of red, green, and blue light needed to match a colored light. For an arbitrary red–green–blue coordinate system, the instantaneous tristimulus values are

$$R(x, y, t) = \int_0^\infty C(x, y, t, \lambda) R_S(\lambda) \, d\lambda \qquad (1.1\text{-}4a)$$

$$G(x, y, t) = \int_0^\infty C(x, y, t, \lambda) G_S(\lambda) \, d\lambda \qquad (1.1\text{-}4b)$$

$$B(x, y, t) = \int_0^\infty C(x, y, t, \lambda) B_S(\lambda) \, d\lambda \qquad (1.1\text{-}4c)$$

where $R_S(\lambda)$, $G_S(\lambda)$, $B_S(\lambda)$ are spectral tristimulus values for the set of red, green, and blue primaries. The spectral tristimulus values are, in effect, the tristimulus values required to match a unit amount of narrowband light at wavelength λ. In a multispectral imaging system, the image field observed is modeled as a spectrally weighted integral of the image light function. The ith spectral image field is then given as

$$F_i(x, y, t) = \int_0^\infty C(x, y, t, \lambda) S_i(\lambda) \, d\lambda \qquad (1.1\text{-}5)$$

where $S_i(\lambda)$ is the spectral response of the ith sensor.

For notational simplicity, a single image function $F(x, y, t)$ is selected to represent an image field in a physical imaging system. For a monochrome imaging system, the image function $F(x, y, t)$ nominally denotes the image luminance, or some converted or corrupted physical representation of the luminance, whereas in a color imaging system, $F(x, y, t)$ signifies one of the tristimulus values, or some function of the tristimulus value. The image function $F(x, y, t)$ is also used to denote general three-dimensional fields, such as the time-varying noise of an image scanner.

In correspondence with the standard definition for one-dimensional time signals, the time average of an image function at a given point (x, y) is defined as

$$\langle F(x, y, t) \rangle_T = \lim_{T \to \infty} \left[\frac{1}{2T} \int_{-T}^{T} F(x, y, t) L(t) \, dt \right] \qquad (1.1\text{-}6)$$

where $L(t)$ is a time-weighting function. Similarly, the average image brightness at a given time is given by the spatial average,

$$\langle F(x, y, t) \rangle_S = \lim_{\substack{L_x \to \infty \\ L_y \to \infty}} \left[\frac{1}{4 L_x L_y} \int_{-L_x}^{L_x} \int_{-L_y}^{L_y} F(x, y, t) \, dx \, dy \right] \qquad (1.1\text{-}7)$$

In many imaging systems, such as image projection devices, the image does not change with time, and the time variable may be dropped from the image function. For other types of systems, such as movie pictures, the image function is time sampled. It is also possible to convert the spatial variation into time variation, as in television, by an image scanning process. In the subsequent discussion, the time variable is dropped from the image field notation unless specifically required.

1.2. TWO-DIMENSIONAL SYSTEMS

A *two-dimensional system*, in its most general form, is simply a mapping of some input set of two-dimensional functions $F_1(x, y)$, $F_2(x, y)$,..., $F_N(x, y)$ to a set of output two-dimensional functions $G_1(x, y)$, $G_2(x, y)$,..., $G_M(x, y)$, where $(-\infty < x, y < \infty)$ denotes the independent, continuous spatial variables of the functions. This mapping may be represented by the operators $O\{\cdot\}$ for $m = 1, 2,..., M$, which relate the input to output set of functions by the set of equations

$$\begin{bmatrix} G_1(x, y) = O_1\{F_1(x, y), F_2(x, y), ..., F_N(x, y)\} \\ \vdots \\ G_m(x, y) = O_m\{F_1(x, y), F_2(x, y), ..., F_N(x, y)\} \\ \vdots \\ G_M(x, y) = O_M\{F_1(x, y), F_2(x, y), ..., F_N(x, y)\} \end{bmatrix} \qquad (1.2\text{-}1)$$

In specific cases, the mapping may be many-to-few, few-to-many, or one-to-one. The *one-to-one mapping* is defined as

$$G(x, y) = O\{F(x, y)\} \qquad (1.2\text{-}2)$$

To proceed further with a discussion of the properties of two-dimensional systems, it is necessary to direct the discourse toward specific types of operators.

1.2.1. Singularity Operators

Singularity operators are widely employed in the analysis of two-dimensional systems, especially systems that involve sampling of continuous functions. The two-dimensional *Dirac delta function* is a singularity operator that possesses the following properties:

$$\int_{-\varepsilon}^{\varepsilon} \int_{-\varepsilon}^{\varepsilon} \delta(x, y)\, dx\, dy = 1 \qquad \text{for } \varepsilon > 0 \qquad (1.2\text{-}3a)$$

$$\int_{-\infty}^{\infty} \int_{-\infty}^{\infty} F(\xi, \eta)\delta(x - \xi, y - \eta)\, d\xi\, d\eta = F(x, y) \qquad (1.2\text{-}3b)$$

In Eq. 1.2-3a, ε is an infinitesimally small limit of integration; Eq. 1.2-3b is called the *sifting property* of the Dirac delta function.

The two-dimensional delta function can be decomposed into the product of two one-dimensional delta functions defined along orthonormal coordinates. Thus

$$\delta(x, y) = \delta(x)\delta(y) \qquad (1.2\text{-}4)$$

where the one-dimensional delta function satisfies one-dimensional versions of Eq. 1.2-3. The delta function also can be defined as a limit on a family of functions. General examples are given in References 1 and 2.

1.2.2. Additive Linear Operators

A two-dimensional system is said to be an *additive linear system* if the system obeys the law of additive superposition. In the special case of one-to-one mappings, the additive superposition property requires that

$$O\{a_1F_1(x, y) + a_2F_2(x, y)\} = a_1O\{F_1(x, y)\} + a_2O\{F_2(x, y)\} \qquad (1.2\text{-}5)$$

where a_1 and a_2 are constants that are possibly complex numbers. This additive superposition property can easily be extended to the general mapping of Eq. 1.2-1.

A system input function $F(x, y)$ can be represented as a sum of amplitude-weighted Dirac delta functions by the sifting integral,

$$F(x, y) = \int_{-\infty}^{\infty} \int_{-\infty}^{\infty} F(\xi, \eta)\delta(x - \xi, y - \eta) \, d\xi \, d\eta \qquad (1.2\text{-}6)$$

where $F(\xi, \eta)$ is the weighting factor of the impulse located at coordinates (ξ, η) in the x–y plane, as shown in Figure 1.2-1. If the output of a general linear one-to-one system is defined to be

$$G(x, y) = O\{F(x, y)\} \qquad (1.2\text{-}7)$$

then

$$G(x, y) = O\left\{ \int_{-\infty}^{\infty} \int_{-\infty}^{\infty} F(\xi, \eta)\delta(x - \xi, y - \eta) \, d\xi \, d\eta \right\} \qquad (1.2\text{-}8a)$$

or

$$G(x, y) = \int_{-\infty}^{\infty} \int_{-\infty}^{\infty} F(\xi, \eta)O\{\delta(x - \xi, y - \eta)\} \, d\xi \, d\eta \qquad (1.2\text{-}8b)$$

In moving from Eq. 1.2-8a to Eq. 1.2-8b, the application order of the general linear operator $O\{ \cdot \}$ and the integral operator have been reversed. Also, the linear operator has been applied only to the term in the integrand that is dependent on the

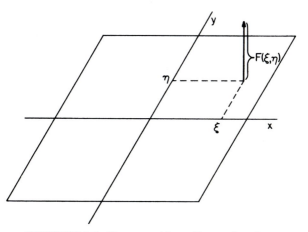

FIGURE 1.2-1. Decomposition of image function.

spatial variables (x, y). The second term in the integrand of Eq. 1.2-8b, which is redefined as

$$H(x, y; \xi, \eta) \equiv O\{\delta(x - \xi, y - \eta)\} \qquad (1.2\text{-}9)$$

is called the *impulse response* of the two-dimensional system. In optical systems, the impulse response is often called the *point spread function* of the system. Substitution of the impulse response function into Eq. 1.2-8b yields the additive *superposition integral*

$$G(x, y) = \int_{-\infty}^{\infty}\int_{-\infty}^{\infty} F(\xi, \eta)H(x, y; \xi, \eta)\,d\xi\,d\eta \qquad (1.2\text{-}10)$$

An additive linear two-dimensional system is called *space invariant* (isoplanatic) if its impulse response depends only on the factors $x - \xi$ and $y - \eta$. In an optical system, as shown in Figure 1.2-2, this implies that the image of a point source in the focal plane will change only in location, not in functional form, as the placement of the point source moves in the object plane. For a space-invariant system

$$H(x, y; \xi, \eta) = H(x - \xi, y - \eta) \qquad (1.2\text{-}11)$$

and the superposition integral reduces to the special case called the *convolution integral*, given by

$$G(x, y) = \int_{-\infty}^{\infty}\int_{-\infty}^{\infty} F(\xi, \eta)H(x - \xi, y - \eta)\,d\xi\,d\eta \qquad (1.2\text{-}12a)$$

Symbolically,

$$G(x, y) = F(x, y) \circledast H(x, y) \qquad (1.2\text{-}12b)$$

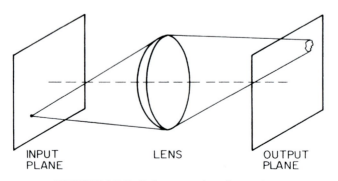

FIGURE 1.2-2. Point-source imaging system.

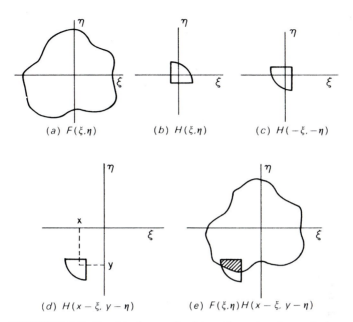

FIGURE 1.2-3. Graphical example of two-dimensional convolution.

denotes the *convolution operation*. The convolution integral is symmetric in the sense that

$$G(x, y) = \int_{-\infty}^{\infty} \int_{-\infty}^{\infty} F(x - \xi, y - \eta) H(\xi, \eta) \, d\xi \, d\eta \qquad (1.2\text{-}13)$$

Figure 1.2-3 provides a visualization of the convolution process. In Figure 1.2-3*a* and *b*, the input function $F(x, y)$ and impulse response are plotted in the dummy coordinate system (ξ, η). Next, in Figures 1.2-3*c* and *d* the coordinates of the impulse response are reversed, and the impulse response is offset by the spatial values (x, y). In Figure 1.2-3*e*, the integrand product of the convolution integral of Eq. 1.2-12 is shown as a crosshatched region. The integral over this region is the value of $G(x, y)$ at the offset coordinate (x, y). The complete function $F(x, y)$ could, in effect, be computed by sequentially scanning the reversed, offset impulse response across the input function and simultaneously integrating the overlapped region.

1.2.3. Differential Operators

Edge detection in images is commonly accomplished by performing a spatial differentiation of the image field followed by a thresholding operation to determine points of steep amplitude change. Horizontal and vertical spatial derivatives are defined as

$$d_x = \frac{\partial F(x, y)}{\partial x} \tag{1.2-14a}$$

$$d_y = \frac{\partial F(x, y)}{\partial y} \tag{1.2-14b}$$

The directional derivative of the image field along a vector direction z subtending an angle ϕ with respect to the horizontal axis is given by (3, p. 106)

$$\nabla\{F(x, y)\} = \frac{\partial F(x, y)}{\partial z} = d_x \cos \phi + d_y \sin \phi \tag{1.2-15}$$

The gradient magnitude is then

$$|\nabla\{F(x, y)\}| = \sqrt{d_x^2 + d_y^2} \tag{1.2-16}$$

Spatial second derivatives in the horizontal and vertical directions are defined as

$$d_{xx} = \frac{\partial^2 F(x, y)}{\partial x^2} \tag{1.2-17a}$$

$$d_{yy} = \frac{\partial^2 F(x, y)}{\partial y^2} \tag{1.2-17b}$$

The sum of these two spatial derivatives is called the *Laplacian operator*:

$$\nabla^2\{F(x, y)\} = \frac{\partial^2 F(x, y)}{\partial x^2} + \frac{\partial^2 F(x, y)}{\partial y^2} \tag{1.2-18}$$

1.3. TWO-DIMENSIONAL FOURIER TRANSFORM

The two-dimensional *Fourier transform* of the image function $F(x, y)$ is defined as (1,2)

$$\mathcal{F}(\omega_x, \omega_y) = \int_{-\infty}^{\infty}\int_{-\infty}^{\infty} F(x, y) \exp\{-i(\omega_x x + \omega_y y)\}\, dx\, dy \tag{1.3-1}$$

where ω_x and ω_y are *spatial frequencies* and $i = \sqrt{-1}$. Notationally, the Fourier transform is written as

$$\mathcal{H}(\omega_x, \omega_y) = O_{\mathcal{F}}\{F(x, y)\} \tag{1.3-2}$$

In general, the Fourier coefficient $\mathcal{H}(\omega_x, \omega_y)$ is a complex number that may be represented in real and imaginary form,

$$\mathcal{H}(\omega_x, \omega_y) = \mathcal{R}(\omega_x, \omega_y) + iI(\omega_x, \omega_y) \tag{1.3-3a}$$

or in magnitude and phase-angle form,

$$\mathcal{H}(\omega_x, \omega_y) = \mathcal{M}(\omega_x, \omega_y) \exp\{i\phi(\omega_x, \omega_y)\} \tag{1.3-3b}$$

where

$$\mathcal{M}(\omega_x, \omega_y) = [\mathcal{R}^2(\omega_x, \omega_y) + I^2(\omega_x, \omega_y)]^{1/2} \tag{1.3-4a}$$

$$\phi(\omega_x, \omega_y) = \arctan\left\{\frac{I(\omega_x, \omega_y)}{\mathcal{R}(\omega_x, \omega_y)}\right\} \tag{1.3-4b}$$

A sufficient condition for the existence of the Fourier transform of $F(x, y)$ is that the function be absolutely integrable. That is,

$$\int_{-\infty}^{\infty}\int_{-\infty}^{\infty} |F(x, y)|\, dx\, dy < \infty \tag{1.3-5}$$

The input function $F(x, y)$ can be recovered from its Fourier transform by the inversion formula

$$F(x, y) = \frac{1}{4\pi^2}\int_{-\infty}^{\infty}\int_{-\infty}^{\infty} \mathcal{H}(\omega_x, \omega_y) \exp\{i(\omega_x x + \omega_y y)\}\, d\omega_x\, d\omega_y \tag{1.3-6a}$$

or in operator form

$$F(x, y) = O_{\mathcal{F}}^{-1}\{\mathcal{H}(\omega_x, \omega_y)\} \tag{1.3-6b}$$

The functions $F(x, y)$ and $\mathcal{H}(\omega_x, \omega_y)$ are called *Fourier transform pairs*.

The two-dimensional Fourier transform can be computed in two steps as a result of the separability of the kernel. Thus, let

$$\mathcal{F}_y(\omega_x, y) = \int_{-\infty}^{\infty} F(x, y) \exp\{-i(\omega_x x)\}\, dx \qquad (1.3\text{-}7)$$

then

$$\mathcal{F}(\omega_x, \omega_y) = \int_{-\infty}^{\infty} \mathcal{F}_y(\omega_x, y) \exp\{-i(\omega_y y)\}\, dy \qquad (1.3\text{-}8)$$

Several useful properties of the two-dimensional Fourier transform are stated below. Proofs are given in References 1 and 2.

Separability. If the image function is spatially separable such that

$$F(x, y) = f_x(x) f_y(y) \qquad (1.3\text{-}9)$$

then

$$\mathcal{F}_y(\omega_x, \omega_y) = f_x(\omega_x) f_y(\omega_y) \qquad (1.3\text{-}10)$$

where $f_x(\omega_x)$ and $f_y(\omega_y)$ are one-dimensional Fourier transforms of $f_x(x)$ and $f_y(y)$, respectively. Also, if $F(x, y)$ and $\mathcal{F}(\omega_x, \omega_y)$ are two-dimensional Fourier transform pairs, the Fourier transform of $F^*(x, y)$ is $\mathcal{F}^*(-\omega_x, -\omega_y)$. An asterisk[*] used as a superscript denotes complex conjugation of a variable (i.e. if $F = A + iB$, then $F^* = A - iB$). Finally, if $F(x, y)$ is symmetric such that $F(x, y) = F(-x, -y)$, then $\mathcal{F}(\omega_x, \omega_y) = \mathcal{F}(-\omega_x, -\omega_y)$.

Linearity. The Fourier transform is a linear operator. Thus

$$O_{\mathcal{F}}\{aF_1(x, y) + bF_2(x, y)\} = a\mathcal{F}_1(\omega_x, \omega_y) + b\mathcal{F}_2(\omega_x, \omega_y) \qquad (1.3\text{-}11)$$

where a and b are constants.

Scaling. A linear scaling of the spatial variables results in an inverse scaling of the spatial frequencies as given by

$$O_{\mathcal{F}}\{F(ax, by)\} = \frac{1}{|ab|}\, \mathcal{F}\!\left(\frac{\omega_x}{a}, \frac{\omega_y}{b}\right) \qquad (1.3\text{-}12)$$

Hence, stretching of an axis in one domain results in a contraction of the corresponding axis in the other domain plus an amplitude change.

Shift. A positional shift in the input plane results in a phase shift in the output plane:

$$O_{\mathcal{F}}\{F(x-a, y-b)\} = \mathcal{F}(\omega_x, \omega_y)\exp\{-i(\omega_x a + \omega_y b)\} \qquad (1.3\text{-}13\text{a})$$

Alternatively, a frequency shift in the Fourier plane results in the equivalence

$$O_{\mathcal{F}}^{-1}\{\mathcal{F}(\omega_x - a, \omega_y - b)\} = F(x, y)\exp\{i(ax + by)\} \qquad (1.3\text{-}13\text{b})$$

Convolution. The two-dimensional Fourier transform of two convolved functions is equal to the products of the transforms of the functions. Thus

$$O_{\mathcal{F}}\{F(x, y) \circledast H(x, y)\} = \mathcal{F}(\omega_x, \omega_y)\mathcal{H}(\omega_x, \omega_y) \qquad (1.3\text{-}14)$$

The inverse theorem states that

$$O_{\mathcal{F}}\{F(x, y)H(x, y)\} = \frac{1}{4\pi^2}\mathcal{F}(\omega_x, \omega_y) \circledast \mathcal{H}(\omega_x, \omega_y) \qquad (1.3\text{-}15)$$

Parseval 's Theorem. The energy in the spatial and Fourier transform domains is related by

$$\int_{-\infty}^{\infty}\int_{-\infty}^{\infty}|F(x, y)|^2\,dx\,dy = \frac{1}{4\pi^2}\int_{-\infty}^{\infty}\int_{-\infty}^{\infty}|\mathcal{F}(\omega_x, \omega_y)|^2\,d\omega_x\,d\omega_y \qquad (1.3\text{-}16)$$

Autocorrelation Theorem. The Fourier transform of the spatial autocorrelation of a function is equal to the magnitude squared of its Fourier transform. Hence

$$O_{\mathcal{F}}\left\{\int_{-\infty}^{\infty}\int_{-\infty}^{\infty}F(\alpha, \beta)F^*(\alpha-x, \beta-y)\,d\alpha\,d\beta\right\} = |\mathcal{F}(\omega_x, \omega_y)|^2 \qquad (1.3\text{-}17)$$

Spatial Differentials. The Fourier transform of the directional derivative of an image function is related to the Fourier transform by

$$O_{\mathcal{F}}\left\{\frac{\partial F(x, y)}{\partial x}\right\} = -i\omega_x \mathcal{F}(\omega_x, \omega_y) \qquad (1.3\text{-}18\text{a})$$

$$O_{\mathcal{F}}\left\{\frac{\partial F(x, y)}{\partial y}\right\} = -i\omega_y \, \mathcal{F}(\omega_x, \omega_y) \qquad (1.3\text{-}18b)$$

Consequently, the Fourier transform of the Laplacian of an image function is equal to

$$O_{\mathcal{F}}\left\{\frac{\partial^2 F(x, y)}{\partial x^2} + \frac{\partial^2 F(x, y)}{\partial y^2}\right\} = -(\omega_x^2 + \omega_y^2) \, \mathcal{F}(\omega_x, \omega_y) \qquad (1.3\text{-}19)$$

The Fourier transform convolution theorem stated by Eq. 1.3-14 is an extremely useful tool for the analysis of additive linear systems. Consider an image function $F(x, y)$ that is the input to an additive linear system with an impulse response $H(x, y)$. The output image function is given by the convolution integral

$$G(x, y) = \int_{-\infty}^{\infty} \int_{-\infty}^{\infty} F(\alpha, \beta) H(x - \alpha, y - \beta) \, d\alpha \, d\beta \qquad (1.3\text{-}20)$$

Taking the Fourier transform of both sides of Eq. 1.3-20 and reversing the order of integration on the right-hand side results in

$$\mathcal{G}(\omega_x, \omega_y) = \int_{-\infty}^{\infty} \int_{-\infty}^{\infty} F(\alpha, \beta) \left[\int_{-\infty}^{\infty} \int_{-\infty}^{\infty} H(x - \alpha, y - \beta) \exp\{-i(\omega_x x + \omega_y y)\} dx \, dy\right] d\alpha \, d\beta$$

$$(1.3\text{-}21)$$

By the Fourier transform shift theorem of Eq. 1.3-13, the inner integral is equal to the Fourier transform of $H(x, y)$ multiplied by an exponential phase-shift factor. Thus

$$\mathcal{G}(\omega_x, \omega_y) = \int_{-\infty}^{\infty} \int_{-\infty}^{\infty} F(\alpha, \beta) \mathcal{H}(\omega_x, \omega_y) \exp\{-i(\omega_x \alpha + \omega_y \beta)\} \, d\alpha \, d\beta \qquad (1.3\text{-}22)$$

Performing the indicated Fourier transformation gives

$$\mathcal{G}(\omega_x, \omega_y) = \mathcal{H}(\omega_x, \omega_y) \, \mathcal{F}(\omega_x, \omega_y) \qquad (1.3\text{-}23)$$

Then an inverse transformation of Eq. 1.3-23 provides the output image function

$$G(x, y) = \frac{1}{4\pi^2} \int_{-\infty}^{\infty} \int_{-\infty}^{\infty} \mathcal{H}(\omega_x, \omega_y) \, \mathcal{F}(\omega_x, \omega_y) \exp\{i(\omega_x x + \omega_y y)\} \, d\omega_x \, d\omega_y \qquad (1.3\text{-}24)$$

Equations 1.3-20 and 1.3-24 represent two alternative means of determining the output image response of an additive, linear, space-invariant system. The analytic or operational choice between the two approaches, convolution or Fourier processing, is usually problem dependent.

1.4. IMAGE STOCHASTIC CHARACTERIZATION

The following presentation on the statistical characterization of images assumes general familiarity with probability theory, random variables, and stochastic process. References 2 and 4 to 7 can provide suitable background. The primary purpose of the discussion here is to introduce notation and develop stochastic image models.

It is often convenient to regard an image as a sample of a stochastic process. For continuous images, the image function $F(x, y, t)$ is assumed to be a member of a continuous three-dimensional stochastic process with space variables (x, y) and time variable (t).

The stochastic process $F(x, y, t)$ can be described completely by knowledge of its *joint probability density*

$$p\{F_1, F_2 ..., F_J; x_1, y_1, t_1, x_2, y_2, t_2, ..., x_J, y_J, t_J\}$$

for all sample points J, where (x_j, y_j, t_j) represent space and time samples of image function $F_j(x_j, y_j, t_j)$. In general, high-order joint probability densities of images are usually not known, nor are they easily modeled. The first-order probability density $p(F; x, y, t)$ can sometimes be modeled successfully on the basis of the physics of the process or histogram measurements. For example, the first-order probability density of random noise from an electronic sensor is usually well modeled by a *Gaussian density* of the form

$$p\{F; x, y, t\} = [2\pi\sigma_F^2(x, y, t)]^{-1/2} \exp\left\{-\frac{[F(x, y, t) - \eta_F(x, y, t)]^2}{2\sigma_F^2(x, y, t)}\right\} \quad (1.4-1)$$

where the parameters $\eta_F(x, y, t)$ and $\sigma_F^2(x, y, t)$ denote the mean and variance of the process. The Gaussian density is also a reasonably accurate model for the probability density of the amplitude of unitary transform coefficients of an image. The probability density of the luminance function must be a one-sided density because the luminance measure is positive. Models that have found application include the *Rayleigh density*,

$$p\{F; x, y, t\} = \frac{F(x, y, t)}{\alpha^2} \exp\left\{-\frac{[F(x, y, t)]^2}{2\alpha^2}\right\} \quad (1.4-2a)$$

the *log-normal density*,

$$p\{F; x, y, t\} = [2\pi F^2(x, y, t)\sigma_F^2(x, y, t)]^{-1/2} \exp\left\{-\frac{[\log\{F(x, y, t)\} - \eta_F(x, y, t)]^2}{2\sigma_F^2(x, y, t)}\right\}$$

(1.4-2b)

and the *exponential density*,

$$p\{F; x, y, t\} = \alpha \exp\{-\alpha|F(x, y, t)|\}$$

(1.4-2c)

all defined for $F \geq 0$, where α is a constant. The *two-sided*, or *Laplacian density*,

$$p\{F; x, y, t\} = \frac{\alpha}{2} \exp\{-\alpha|F(x, y, t)|\}$$

(1.4-3)

where α is a constant, is often selected as a model for the probability density of the difference of image samples. Finally, the *uniform density*

$$p\{F; x, y, t\} = \frac{1}{2\pi}$$

(1.4-4)

for $-\pi \leq F \leq \pi$ is a common model for phase fluctuations of a random process. Conditional probability densities are also useful in characterizing a stochastic process. The *conditional density* of an image function evaluated at (x_1, y_1, t_1) given knowledge of the image function at (x_2, y_2, t_2) is defined as

$$p\{F_1; x_1, y_1, t_1 | F_2; x_2, y_2, t_2\} = \frac{p\{F_1, F_2; x_1, y_1, t_1, x_2, y_2, t_2\}}{p\{F_2; x_2, y_2, t_2\}}$$

(1.4-5)

Higher-order conditional densities are defined in a similar manner.

Another means of describing a stochastic process is through computation of its ensemble averages. The *first moment* or *mean* of the image function is defined as

$$\eta_F(x, y, t) = E\{F(x, y, t)\} = \int_{-\infty}^{\infty} F(x, y, t)p\{F; x, y, t\}\, dF$$

(1.4-6)

where $E\{\cdot\}$ is the *expectation operator*, as defined by the right-hand side of Eq. 1.4-6.

The *second moment* or *autocorrelation function* is given by

$$R(x_1, y_1, t_1; x_2, y_2, t_2) = E\{F(x_1, y_1, t_1)F^*(x_2, y_2, t_2)\}$$

(1.4-7a)

or in explicit form

$$R(x_1, y_1, t_1; x_2, y_2, t_2) = \int_{-\infty}^{\infty} \int_{-\infty}^{\infty} F(x_1, x_1, y_1) F^*(x_2, y_2, t_2)$$

$$\times p\{F_1, F_2; x_1, y_1, t_1, x_2, y_2, t_2\} dF_1 dF_2 \qquad (1.4\text{-}7b)$$

The autocovariance of the image process is the autocorrelation about the mean, defined as

$$K(x_1, y_1, t_1; x_2, y_2, t_2) = E\{[F(x_1, y_1, t_1) - \eta_F(x_1, y_1, t_1)][F^*(x_2, y_2, t_2) - \eta_F^*(x_2, y_2, t_2)]\}$$

$$(1.4\text{-}8a)$$

or

$$K(x_1, y_1, t_1; x_2, y_2, t_2) = R(x_1, y_1, t_1; x_2, y_2, t_2) - \eta_F(x_1, y_1, t_1) \eta_F^*(x_2, y_2, t_2) \quad (1.4\text{-}8b)$$

Finally, the *variance* of an image process is

$$\sigma_F^2(x, y, t) = K(x, y, t; x, y, t) \qquad (1.4\text{-}9)$$

An image process is called *stationary in the strict sense* if its moments are unaffected by shifts in the space and time origins. The image process is said to be *stationary in the wide sense* if its mean is constant and its autocorrelation is dependent on the differences in the image coordinates, $x_1 - x_2$, $y_1 - y_2$, $t_1 - t_2$, and not on their individual values. In other words, the image autocorrelation is not a function of position or time. For stationary image processes,

$$E\{F(x, y, t)\} = \eta_F \qquad (1.4\text{-}10a)$$

$$R(x_1, y_1, t_1; x_2, y_2, t_2) = R(x_1 - x_2, y_1 - y_2, t_1 - t_2) \qquad (1.4\text{-}10b)$$

The autocorrelation expression may then be written as

$$R(\tau_x, \tau_y, \tau_t) = E\{F(x + \tau_x, y + \tau_y, t + \tau_t) F^*(x, y, t)\} \qquad (1.4\text{-}11)$$

Because

$$R(-\tau_x, -\tau_y, -\tau_t) = R^*(\tau_x, \tau_y, \tau_t) \tag{1.4-12}$$

then for an image function with F real, the autocorrelation is real and an even function of τ_x, τ_y, τ_t. The *power spectral density*, also called the *power spectrum*, of a stationary image process is defined as the three-dimensional Fourier transform of its autocorrelation function as given by

$$\mathcal{W}(\omega_x, \omega_y, \omega_t) = \int_{-\infty}^{\infty}\int_{-\infty}^{\infty}\int_{-\infty}^{\infty} R(\tau_x, \tau_y, \tau_t)\exp\{-i(\omega_x\tau_x + \omega_y\tau_y + \omega_t\tau_t)\}\, d\tau_x\, d\tau_y\, d\tau_t$$

$$\tag{1.4-13}$$

In many imaging systems, the spatial and time image processes are separable so that the stationary correlation function may be written as

$$R(\tau_x, \tau_y, \tau_t) = R_{xy}(\tau_x, \tau_y)R_t(\tau_t) \tag{1.4-14}$$

Furthermore, the spatial autocorrelation function is often considered as the product of x and y axis autocorrelation functions,

$$R_{xy}(\tau_x, \tau_y) = R_x(\tau_x)R_y(\tau_y) \tag{1.4-15}$$

for computational simplicity. For scenes of manufactured objects, there is often a large amount of horizontal and vertical image structure, and the spatial separation approximation may be quite good. In natural scenes, there usually is no preferential direction of correlation; the spatial autocorrelation function tends to be rotationally symmetric and not separable.

An image field is often modeled as a sample of a first-order Markov process for which the correlation between points on the image field is proportional to their geometric separation. The *autocovariance* function for the two-dimensional Markov process is

$$R_{xy}(\tau_x, \tau_y) = C \exp\left\{-\sqrt{\alpha_x^2\tau_x^2 + \alpha_y^2\tau_y^2}\right\} \tag{1.4-16}$$

where C is an energy scaling constant and α_x and α_y are spatial scaling constants. The corresponding power spectrum is

$$\mathcal{W}(\omega_x, \omega_y) = \frac{1}{\sqrt{\alpha_x\alpha_y}}\frac{2C}{1 + [\omega_x^2/\alpha_x^2 + \omega_y^2/\alpha_y^2]} \tag{1.4-17}$$

As a simplifying assumption, the Markov process is often assumed to be of separable form with an autocovariance function

$$K_{xy}(\tau_x, \tau_y) = C \exp\{-\alpha_x|\tau_x| - \alpha_y|\tau_y|\} \tag{1.4-18}$$

The power spectrum of this process is

$$\mathcal{W}(\omega_x, \omega_y) = \frac{4\alpha_x\alpha_y C}{(\alpha_x^2 + \omega_x^2)(\alpha_y^2 + \omega_y^2)} \tag{1.4-19}$$

In the discussion of the deterministic characteristics of an image, both time and space averages of the image function have been defined. An ensemble average has also been defined for the statistical image characterization. A question of interest is: What is the relationship between the spatial-time averages and the ensemble averages? The answer is that for certain stochastic processes, which are called *ergodic processes*, the spatial-time averages and the ensemble averages are equal. Proof of the ergodicity of a process in the general case is often difficult; it usually suffices to determine second-order ergodicity in which the first- and second-order space-time averages are equal to the first- and second-order ensemble averages.

Often, the probability density or moments of a stochastic image field are known at the input to a system, and it is desired to determine the corresponding information at the system output. If the system transfer function is algebraic in nature, the output probability density can be determined in terms of the input probability density by a probability density transformation. For example, let the system output be related to the system input by

$$G(x, y, t) = O_F\{F(x, y, t)\} \tag{1.4-20}$$

where $O_F\{\cdot\}$ is a monotonic operator on $F(x, y)$. The probability density of the output field is then

$$p\{G; x, y, t\} = \frac{p\{F; x, y, t\}}{|dO_F\{F(x, y, t)\}/dF|} \tag{1.4-21}$$

The extension to higher-order probability densities is straightforward, but often cumbersome.

The moments of the output of a system can be obtained directly from knowledge of the output probability density, or in certain cases, indirectly in terms of the system operator. For example, if the system operator is additive linear, the mean of the system output is

$$E\{G(x, y, t)\} = E\{O_F\{F(x, y, t)\}\} = O_F\{E\{F(x, y, t)\}\} \tag{1.4-22}$$

It can be shown that if a system operator is additive linear, and if the system input image field is stationary in the strict sense, the system output is also stationary in the strict sense. Furthermore, if the input is stationary in the wide sense, the output is also wide-sense stationary.

Consider an additive linear space-invariant system whose output is described by the three-dimensional convolution integral

$$G(x, y, t) = \int_{-\infty}^{\infty} \int_{-\infty}^{\infty} \int_{-\infty}^{\infty} F(x - \alpha, y - \beta, t - \gamma) H(\alpha, \beta, \gamma) \, d\alpha \, d\beta \, d\gamma \tag{1.4-23}$$

where $H(x, y, t)$ is the system impulse response. The mean of the output is then

$$E\{G(x, y, t)\} = \int_{-\infty}^{\infty} \int_{-\infty}^{\infty} \int_{-\infty}^{\infty} E\{F(x - \alpha, y - \beta, t - \gamma)\} H(\alpha, \beta, \gamma) \, d\alpha \, d\beta \, d\gamma \tag{1.4-24}$$

If the input image field is stationary, its mean η_F is a constant that may be brought outside the integral. As a result,

$$E\{G(x, y, t)\} = \eta_F \int_{-\infty}^{\infty} \int_{-\infty}^{\infty} \int_{-\infty}^{\infty} H(\alpha, \beta, \gamma) \, d\alpha \, d\beta \, d\gamma = \eta_F \, \mathcal{H}(0, 0, 0) \tag{1.4-25}$$

where $\mathcal{H}(0, 0, 0)$ is the transfer function of the linear system evaluated at the origin in the spatial-time frequency domain. Following the same techniques, it can easily be shown that the autocorrelation functions of the system input and output are related by

$$R_G(\tau_x, \tau_y, \tau_t) = R_F(\tau_x, \tau_y, \tau_t) \circledast H(\tau_x, \tau_y, \tau_t) \circledast H^*(-\tau_x, -\tau_y, -\tau_t) \tag{1.4-26}$$

Taking Fourier transforms on both sides of Eq. 1.4-26 and invoking the Fourier transform convolution theorem, one obtains the relationship between the power spectra of the input and output image,

$$\mathcal{W}_G(\omega_x, \omega_y, \omega_t) = \mathcal{W}_F(\omega_x, \omega_y, \omega_t) \mathcal{H}(\omega_x, \omega_y, \omega_t) \mathcal{H}^*(\omega_x, \omega_y, \omega_t) \tag{1.4-27a}$$

or

$$\mathcal{W}_G(\omega_x, \omega_y, \omega_t) = \mathcal{W}_F(\omega_x, \omega_y, \omega_t) |\mathcal{H}(\omega_x, \omega_y, \omega_t)|^2 \tag{1.4-27b}$$

This result is found useful in analyzing the effect of noise in imaging systems.

REFERENCES

1. J. W. Goodman, *Introduction to Fourier Optics*, 2nd Ed., McGraw-Hill, New York, 1996.

2. A. Papoulis, *Systems and Transforms with Applications in Optics*, McGraw-Hill, New York, 1968.

3. J. M. S. Prewitt, "Object Enhancement and Extraction," in *Picture Processing and Psychopictorics*, B. S. Lipkin and A. Rosenfeld, Eds., Academic Press, New York, 1970.

4. A. Papoulis, *Probability, Random Variables, and Stochastic Processes*, 3rd ed., McGraw-Hill, New York, 1991.

5. J. B. Thomas, *An Introduction to Applied Probability Theory and Random Processes*, Wiley, New York, 1971.

6. J. W. Goodman, *Statistical Optics*, Wiley, New York, 1985.

7. E. R. Dougherty, *Random Processes for Image and Signal Processing*, Vol. PM44, SPIE Press, Bellingham, Wash., 1998.

2

PSYCHOPHYSICAL VISION PROPERTIES

For efficient design of imaging systems for which the output is a photograph or display to be viewed by a human observer, it is obviously beneficial to have an understanding of the mechanism of human vision. Such knowledge can be utilized to develop conceptual models of the human visual process. These models are vital in the design of image processing systems and in the construction of measures of image fidelity and intelligibility.

2.1. LIGHT PERCEPTION

Light, according to *Webster's Dictionary* (1), is "radiant energy which, by its action on the organs of vision, enables them to perform their function of sight." Much is known about the physical properties of light, but the mechanisms by which light interacts with the *organs of vision* is not as well understood. Light is known to be a form of electromagnetic radiation lying in a relatively narrow region of the electromagnetic spectrum over a wavelength band of about 350 to 780 nanometers (nm). A physical light source may be characterized by the rate of radiant energy (radiant intensity) that it emits at a particular spectral wavelength. Light entering the human visual system originates either from a self-luminous source or from light reflected from some object or from light transmitted through some translucent object. Let $E(\lambda)$ represent the *spectral energy distribution* of light emitted from some primary light source, and also let $t(\lambda)$ and $r(\lambda)$ denote the wavelength-dependent transmissivity and reflectivity, respectively, of an object. Then, for a *transmissive object,* the observed light spectral energy distribution is

$$C(\lambda) = t(\lambda)E(\lambda) \qquad (2.1\text{-}1)$$

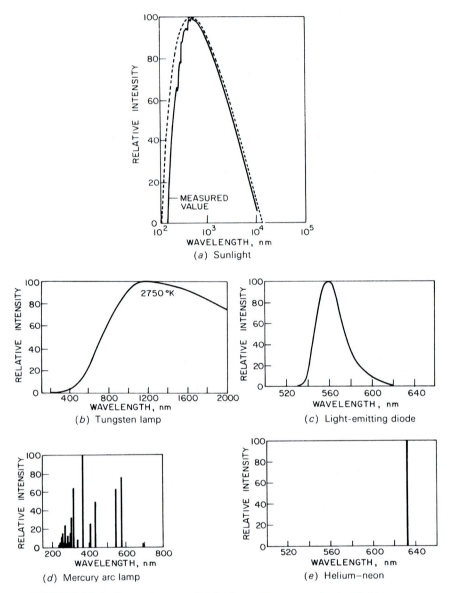

FIGURE 2.1-1. Spectral energy distributions of common physical light sources.

and for a *reflective object*

$$C(\lambda) = r(\lambda)E(\lambda) \tag{2.1-2}$$

Figure 2.1-1 shows plots of the spectral energy distribution of several common sources of light encountered in imaging systems: sunlight, a tungsten lamp, a light-emitting diode, a mercury arc lamp, and a helium–neon laser (2). A human being viewing each of the light sources will perceive the sources differently. Sunlight appears as an extremely bright yellowish-white light, while the tungsten light bulb appears less bright and somewhat yellowish. The light-emitting diode appears to be a dim green; the mercury arc light is a highly bright bluish-white light; and the laser produces an extremely bright and pure red beam. These observations provoke many questions. What are the attributes of the light sources that cause them to be perceived differently? Is the spectral energy distribution sufficient to explain the differences in perception? If not, what are adequate descriptors of visual perception? As will be seen, answers to these questions are only partially available.

There are three common perceptual descriptors of a light sensation: brightness, hue, and saturation. The characteristics of these descriptors are considered below.

If two light sources with the same spectral shape are observed, the source of greater physical intensity will generally appear to be perceptually *brighter*. However, there are numerous examples in which an object of uniform intensity appears not to be of uniform brightness. Therefore, intensity is not an adequate quantitative measure of brightness.

The attribute of light that distinguishes a red light from a green light or a yellow light, for example, is called the *hue* of the light. A prism and slit arrangement (Figure 2.1-2) can produce narrowband wavelength light of varying color. However, it is clear that the light wavelength is not an adequate measure of color because some colored lights encountered in nature are not contained in the rainbow of light produced by a prism. For example, purple light is absent. Purple light can be produced by combining equal amounts of red and blue narrowband lights. Other counterexamples exist. If two light sources with the same spectral energy distribution are observed under identical conditions, they will appear to possess the same hue. However, it is possible to have two light sources with different spectral energy distributions that are perceived identically. Such lights are called *metameric pairs.*

The third perceptual descriptor of a colored light is its *saturation*, the attribute that distinguishes a spectral light from a pastel light of the same hue. In effect, saturation describes the whiteness of a light source. Although it is possible to speak of the percentage saturation of a color referenced to a spectral color on a chromaticity diagram of the type shown in Figure 3.3-3, saturation is not usually considered to be a quantitative measure.

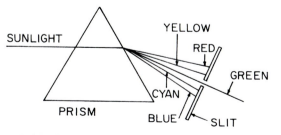

FIGURE 2.1-2. Refraction of light from a prism.

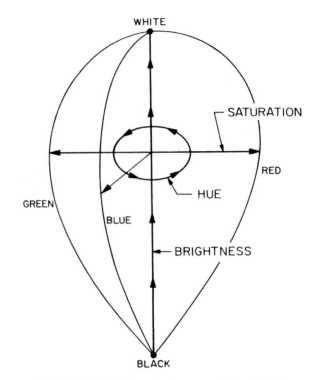

FIGURE 2.1-3. Perceptual representation of light.

As an aid to classifying colors, it is convenient to regard colors as being points in some color solid, as shown in Figure 2.1-3. The Munsell system of color classification actually has a form similar in shape to this figure (3). However, to be quantitatively useful, a color solid should possess metric significance. That is, a unit distance within the color solid should represent a constant perceptual color difference regardless of the particular pair of colors considered. The subject of perceptually significant color solids is considered later.

2.2. EYE PHYSIOLOGY

A conceptual technique for the establishment of a model of the human visual system would be to perform a physiological analysis of the eye, the nerve paths to the brain, and those parts of the brain involved in visual perception. Such a task, of course, is presently beyond human abilities because of the large number of infinitesimally small elements in the visual chain. However, much has been learned from physiological studies of the eye that is helpful in the development of visual models (4–7).

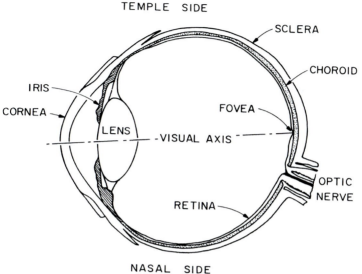

FIGURE 2.2-1. Eye cross section.

Figure 2.2-1 shows the horizontal cross section of a human eyeball. The front of the eye is covered by a transparent surface called the *cornea*. The remaining outer cover, called the *sclera*, is composed of a fibrous coat that surrounds the *choroid*, a layer containing blood capillaries. Inside the choroid is the *retina*, which is composed of two types of receptors: *rods* and *cones*. Nerves connecting to the retina leave the eyeball through the *optic nerve bundle*. Light entering the cornea is focused on the retina surface by a *lens* that changes shape under muscular control to

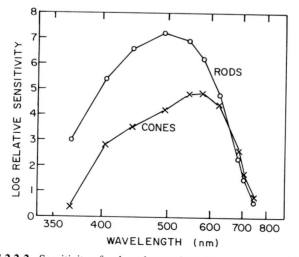

FIGURE 2.2-2. Sensitivity of rods and cones based on measurements by Wald.

perform proper focusing of near and distant objects. An *iris* acts as a diaphram to control the amount of light entering the eye.

The rods in the retina are long slender receptors; the cones are generally shorter and thicker in structure. There are also important operational distinctions. The rods are more sensitive than the cones to light. At low levels of illumination, the rods provide a visual response called *scotopic vision*. Cones respond to higher levels of illumination; their response is called *photopic vision*. Figure 2.2-2 illustrates the relative sensitivities of rods and cones as a function of illumination wavelength (7,8). An eye contains about 6.5 million cones and 100 million cones distributed over the retina (4). Figure 2.2-3 shows the distribution of rods and cones over a horizontal line on the retina (4). At a point near the optic nerve called the *fovea*, the density of cones is greatest. This is the region of sharpest photopic vision. There are no rods or cones in the vicinity of the optic nerve, and hence the eye has a blind spot in this region.

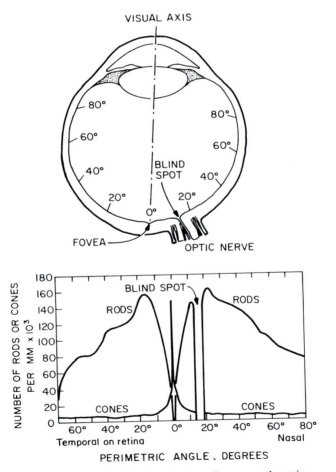

FIGURE 2.2-3. Distribution of rods and cones on the retina.

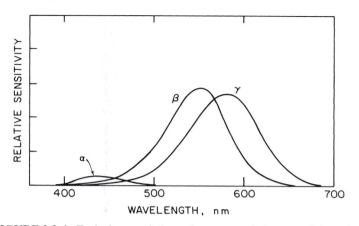

FIGURE 2.2-4. Typical spectral absorption curves of pigments of the retina.

In recent years, it has been determined experimentally that there are three basic types of cones in the retina (9, 10). These cones have different absorption characteristics as a function of wavelength with peak absorptions in the red, green, and blue regions of the optical spectrum. Figure 2.2-4 shows curves of the measured spectral absorption of pigments in the retina for a particular subject (10). Two major points of note regarding the curves are that the α cones, which are primarily responsible for blue light perception, have relatively low sensitivity, and the absorption curves overlap considerably. The existence of the three types of cones provides a physiological basis for the trichromatic theory of color vision.

When a light stimulus activates a rod or cone, a photochemical transition occurs, producing a nerve impulse. The manner in which nerve impulses propagate through the visual system is presently not well established. It is known that the optic nerve bundle contains on the order of 800,000 nerve fibers. Because there are over 100,000,000 receptors in the retina, it is obvious that in many regions of the retina, the rods and cones must be interconnected to nerve fibers on a many-to-one basis. Because neither the photochemistry of the retina nor the propagation of nerve impulses within the eye is well understood, a deterministic characterization of the visual process is unavailable. One must be satisfied with the establishment of models that characterize, and hopefully predict, human visual response. The following section describes several visual phenomena that should be considered in the modeling of the human visual process.

2.3. VISUAL PHENOMENA

The visual phenomena described below are interrelated, in some cases only minimally, but in others, to a very large extent. For simplification in presentation and, in some instances, lack of knowledge, the phenomena are considered disjoint.

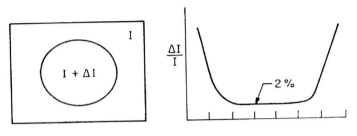

(a) No background

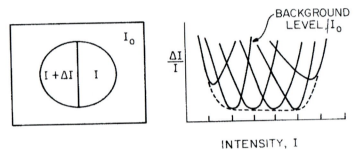

(b) With background

FIGURE 2.3-1. Contrast sensitivity measurements.

Contrast Sensitivity. The response of the eye to changes in the intensity of illumination is known to be nonlinear. Consider a patch of light of intensity $I + \Delta I$ surrounded by a background of intensity I (Figure 2.3-1a). The just noticeable difference ΔI is to be determined as a function of I. Over a wide range of intensities, it is found that the ratio $\Delta I / I$, called the *Weber fraction*, is nearly constant at a value of about 0.02 (11; 12, p. 62). This result does not hold at very low or very high intensities, as illustrated by Figure 2.3-1a (13). Furthermore, contrast sensitivity is dependent on the intensity of the surround. Consider the experiment of Figure 2.3-1b, in which two patches of light, one of intensity I and the other of intensity $I + \Delta I$, are surrounded by light of intensity I_o. The Weber fraction $\Delta I / I$ for this experiment is plotted in Figure 2.3-1b as a function of the intensity of the background. In this situation it is found that the range over which the Weber fraction remains constant is reduced considerably compared to the experiment of Figure 2.3-1a. The envelope of the lower limits of the curves of Figure 2.3-1b is equivalent to the curve of Figure 2.3-1a. However, the range over which $\Delta I / I$ is approximately constant for a fixed background intensity I_o is still comparable to the dynamic range of most electronic imaging systems.

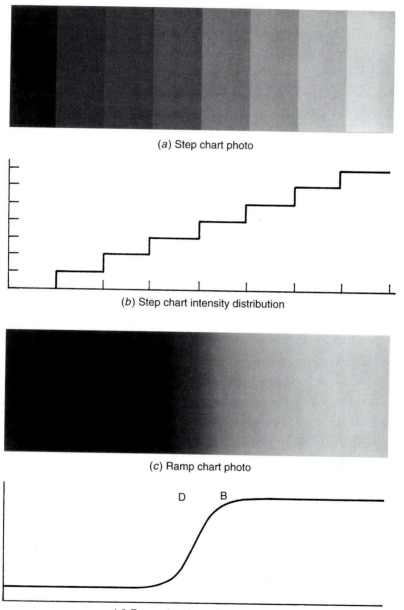

(*a*) Step chart photo

(*b*) Step chart intensity distribution

(*c*) Ramp chart photo

(*d*) Ramp chart intensity distribution

FIGURE 2.3-2. Mach band effect.

Because the differential of the logarithm of intensity is

$$d(\log I) = \frac{dI}{I} \tag{2.3-1}$$

equal changes in the logarithm of the intensity of a light can be related to equal just noticeable changes in its intensity over the region of intensities, for which the Weber fraction is constant. For this reason, in many image processing systems, operations are performed on the logarithm of the intensity of an image point rather than the intensity.

Mach Band. Consider the set of gray scale strips shown in of Figure 2.3-2*a*. The reflected light intensity from each strip is uniform over its width and differs from its neighbors by a constant amount; nevertheless, the visual appearance is that each strip is darker at its right side than at its left. This is called the *Mach band effect* (14). Figure 2.3-2*c* is a photograph of the Mach band pattern of Figure 2.3-2*d*. In the photograph, a bright bar appears at position *B* and a dark bar appears at *D*. Neither bar would be predicted purely on the basis of the intensity distribution. The apparent Mach band overshoot in brightness is a consequence of the spatial frequency response of the eye. As will be seen shortly, the eye possesses a lower sensitivity to high and low spatial frequencies than to midfrequencies. The implication for the designer of image processing systems is that perfect fidelity of edge contours can be sacrificed to some extent because the eye has imperfect response to high-spatial-frequency brightness transitions.

Simultaneous Contrast. The simultaneous contrast phenomenon (7) is illustrated in Figure 2.3-3. Each small square is actually the same intensity, but because of the different intensities of the surrounds, the small squares do not appear equally bright. The hue of a patch of light is also dependent on the wavelength composition of surrounding light. A white patch on a black background will appear to be yellowish if the surround is a blue light.

Chromatic Adaption. The hue of a perceived color depends on the adaption of a viewer (15). For example, the American flag will not immediately appear red, white, and blue if the viewer has been subjected to high-intensity red light before viewing the flag. The colors of the flag will appear to shift in hue toward the red complement, cyan.

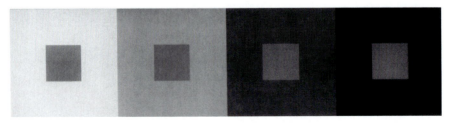

FIGURE 2.3-3. Simultaneous contrast.

Color Blindness. Approximately 8% of the males and 1% of the females in the world population are subject to some form of color blindness (16, p. 405). There are various degrees of color blindness. Some people, called *monochromats*, possess only rods or rods plus one type of cone, and therefore are only capable of monochromatic vision. *Dichromats* are people who possess two of the three types of cones. Both monochromats and dichromats can distinguish colors insofar as they have learned to associate particular colors with particular objects. For example, dark roses are assumed to be red, and light roses are assumed to be yellow. But if a red rose were painted yellow such that its reflectivity was maintained at the same value, a monochromat might still call the rose red. Similar examples illustrate the inability of dichromats to distinguish hue accurately.

2.4. MONOCHROME VISION MODEL

One of the modern techniques of optical system design entails the treatment of an optical system as a two-dimensional linear system that is linear in intensity and can be characterized by a two-dimensional transfer function (17). Consider the linear optical system of Figure 2.4-1. The system input is a spatial light distribution obtained by passing a constant-intensity light beam through a transparency with a spatial sine-wave transmittance. Because the system is linear, the spatial output intensity distribution will also exhibit sine-wave intensity variations with possible changes in the amplitude and phase of the output intensity compared to the input intensity. By varying the spatial frequency (number of intensity cycles per linear dimension) of the input transparency, and recording the output intensity level and phase, it is possible, in principle, to obtain the *optical transfer function* (OTF) of the optical system.

Let $\mathcal{H}(\omega_x, \omega_y)$ represent the optical transfer function of a two-dimensional linear system where $\omega_x = 2\pi/T_x$ and $\omega_y = 2\pi/T_y$ are angular spatial frequencies with spatial periods T_x and T_y in the x and y coordinate directions, respectively. Then, with $I_I(x, y)$ denoting the input intensity distribution of the object and $I_o(x, y)$

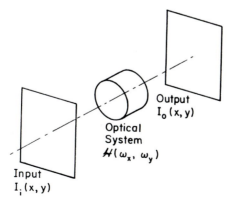

FIGURE 2.4-1. Linear systems analysis of an optical system.

representing the output intensity distribution of the image, the frequency spectra of the input and output signals are defined as

$$\mathcal{I}_I(\omega_x, \omega_y) = \int_{-\infty}^{\infty} \int_{-\infty}^{\infty} I_I(x, y) \exp\{-i(\omega_x x + \omega_y y)\}\, dx\, dy \qquad (2.4\text{-}1)$$

$$\mathcal{I}_O(\omega_x, \omega_y) = \int_{-\infty}^{\infty} \int_{-\infty}^{\infty} I_O(x, y) \exp\{-i(\omega_x x + \omega_y y)\}\, dx\, dy \qquad (2.4\text{-}2)$$

The input and output intensity spectra are related by

$$\mathcal{I}_O(\omega_x, \omega_y) = \mathcal{H}(\omega_x, \omega_y)\mathcal{I}_I(\omega_x, \omega_y) \qquad (2.4\text{-}3)$$

The spatial distribution of the image intensity can be obtained by an inverse Fourier transformation of Eq. 2.4-2, yielding

$$I_O(x, y) = \frac{1}{4\pi^2} \int_{-\infty}^{\infty} \int_{-\infty}^{\infty} \mathcal{I}_O(\omega_x, \omega_y) \exp\{i(\omega_x x + \omega_y y)\}\, d\omega_x\, d\omega_y \qquad (2.4\text{-}4)$$

In many systems, the designer is interested only in the magnitude variations of the output intensity with respect to the magnitude variations of the input intensity, not the phase variations. The ratio of the magnitudes of the Fourier transforms of the input and output signals,

$$\frac{|\mathcal{I}_O(\omega_x, \omega_y)|}{|\mathcal{I}_I(\omega_x, \omega_y)|} = |\mathcal{H}(\omega_x, \omega_y)| \qquad (2.4\text{-}5)$$

is called the *modulation transfer function* (MTF) of the optical system.

Much effort has been given to application of the linear systems concept to the human visual system (18–24). A typical experiment to test the validity of the linear systems model is as follows. An observer is shown two sine-wave grating transparencies, a reference grating of constant contrast and spatial frequency and a variable-contrast test grating whose spatial frequency is set at a value different from that of the reference. *Contrast* is defined as the ratio

$$\frac{\text{max} - \text{min}}{\text{max} + \text{min}}$$

where max and min are the maximum and minimum of the grating intensity, respectively. The contrast of the test grating is varied until the brightnesses of the bright and dark regions of the two transparencies appear identical. In this manner it is possible to develop a plot of the MTF of the human visual system. Figure 2.4-2a is a

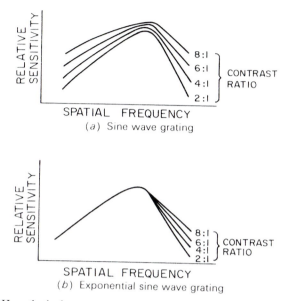

(a) Sine wave grating

(b) Exponential sine wave grating

FIGURE 2.4-2. Hypothetical measurements of the spatial frequency response of the human visual system.

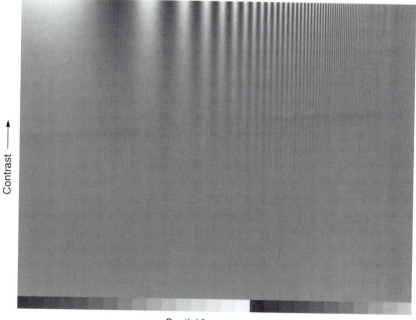

FIGURE 2.4-3. MTF measurements of the human visual system by modulated sine-wave grating.

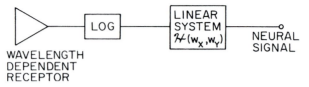

FIGURE 2.4-4. Logarithmic model of monochrome vision.

hypothetical plot of the MTF as a function of the input signal contrast. Another indication of the form of the MTF can be obtained by observation of the composite sine-wave grating of Figure 2.4-3, in which spatial frequency increases in one coordinate direction and contrast increases in the other direction. The envelope of the visible bars generally follows the MTF curves of Figure 2.4-2a (23).

Referring to Figure 2.4-2a, it is observed that the MTF measurement depends on the input contrast level. Furthermore, if the input sine-wave grating is rotated relative to the optic axis of the eye, the shape of the MTF is altered somewhat. Thus, it can be concluded that the human visual system, as measured by this experiment, is nonlinear and anisotropic (rotationally variant).

It has been postulated that the nonlinear response of the eye to intensity variations is logarithmic in nature and occurs near the beginning of the visual information processing system, that is, near the rods and cones, before spatial interaction occurs between visual signals from individual rods and cones. Figure 2.4-4 shows a simple logarithmic eye model for monochromatic vision. If the eye exhibits a logarithmic response to input intensity, then if a signal grating contains a recording of an exponential sine wave, that is, $\exp\{\sin\{I_I(x, y)\}\}$, the human visual system can be linearized. A hypothetical MTF obtained by measuring an observer's response to an exponential sine-wave grating (Figure 2.4-2b) can be fitted reasonably well by a single curve for low-and mid-spatial frequencies. Figure 2.4-5 is a plot of the measured MTF of the human visual system obtained by Davidson (25) for an exponential

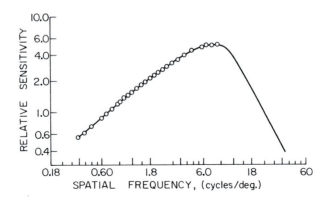

FIGURE 2.4-5. MTF measurements with exponential sine-wave grating.

sine-wave test signal. The high-spatial-frequency portion of the curve has been extrapolated for an average input contrast.

The logarithmic/linear system eye model of Figure 2.4-4 has proved to provide a reasonable prediction of visual response over a wide range of intensities. However, at high spatial frequencies and at very low or very high intensities, observed responses depart from responses predicted by the model. To establish a more accurate model, it is necessary to consider the physical mechanisms of the human visual system.

The nonlinear response of rods and cones to intensity variations is still a subject of active research. Hypotheses have been introduced suggesting that the nonlinearity is based on chemical activity, electrical effects, and neural feedback. The basic logarithmic model assumes the form

$$I_O(x, y) = K_1 \log\{K_2 + K_3 I_I(x, y)\} \tag{2.4-6}$$

where the K_i are constants and $I_I(x, y)$ denotes the input field to the nonlinearity and $I_O(x, y)$ is its output. Another model that has been suggested (7, p. 253) follows the fractional response

$$I_O(x, y) = \frac{K_1 I_I(x, y)}{K_2 + I_I(x, y)} \tag{2.4-7}$$

where K_1 and K_2 are constants. Mannos and Sakrison (26) have studied the effect of various nonlinearities employed in an analytical visual fidelity measure. Their results, which are discussed in greater detail in Chapter 3, establish that a power law nonlinearity of the form

$$I_O(x, y) = [I_I(x, y)]^s \tag{2.4-8}$$

where s is a constant, typically 1/3 or 1/2, provides good agreement between the visual fidelity measure and subjective assessment. The three models for the nonlinear response of rods and cones defined by Eqs. 2.4-6 to 2.4-8 can be forced to a reasonably close agreement over some midintensity range by an appropriate choice of scaling constants.

The physical mechanisms accounting for the spatial frequency response of the eye are partially optical and partially neural. As an optical instrument, the eye has limited resolution because of the finite size of the lens aperture, optical aberrations, and the finite dimensions of the rods and cones. These effects can be modeled by a low-pass transfer function inserted between the receptor and the nonlinear response element. The most significant contributor to the frequency response of the eye is the lateral inhibition process (27). The basic mechanism of *lateral inhibition* is illustrated in

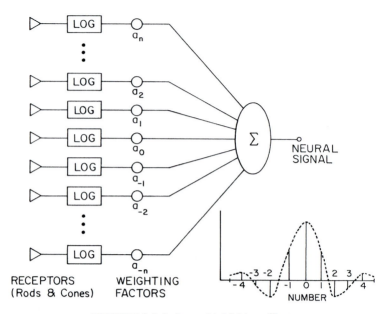

FIGURE 2.4-6. Lateral inhibition effect.

Figure 2.4-6. A neural signal is assumed to be generated by a weighted contribution of many spatially adjacent rods and cones. Some receptors actually exert an inhibitory influence on the neural response. The weighting values are, in effect, the impulse response of the human visual system beyond the retina. The two-dimensional Fourier transform of this impulse response is the postretina transfer function.

When a light pulse is presented to a human viewer, there is a measurable delay in its perception. Also, perception continues beyond the termination of the pulse for a short period of time. This delay and lag effect arising from neural temporal response limitations in the human visual system can be modeled by a linear temporal transfer function.

Figure 2.4-7 shows a model for monochromatic vision based on results of the preceding discussion. In the model, the output of the wavelength-sensitive receptor is fed to a low-pass type of linear system that represents the optics of the eye. Next follows a general monotonic nonlinearity that represents the nonlinear intensity response of rods or cones. Then the lateral inhibition process is characterized by a linear system with a bandpass response. Temporal filtering effects are modeled by the following linear system. Hall and Hall (28) have investigated this model extensively and have found transfer functions for the various elements that accurately model the total system response. The monochromatic vision model of Figure 2.4-7, with appropriately scaled parameters, seems to be sufficiently detailed for most image processing applications. In fact, the simpler logarithmic model of Figure 2.4-4 is probably adequate for the bulk of applications.

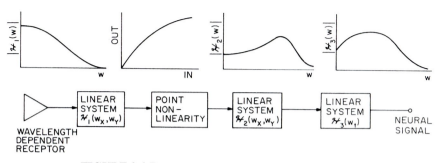

FIGURE 2.4-7. Extended model of monochrome vision.

2.5. COLOR VISION MODEL

There have been many theories postulated to explain human color vision, beginning with the experiments of Newton and Maxwell (29–32). The classical model of human color vision, postulated by Thomas Young in 1802 (31), is the trichromatic model in which it is assumed that the eye possesses three types of sensors, each sensitive over a different wavelength band. It is interesting to note that there was no direct physiological evidence of the existence of three distinct types of sensors until about 1960 (9,10).

Figure 2.5-1 shows a color vision model proposed by Frei (33). In this model, three receptors with spectral sensitivities $s_1(\lambda), s_2(\lambda), s_3(\lambda)$, which represent the absorption pigments of the retina, produce signals

$$e_1 = \int C(\lambda)s_1(\lambda) \, d\lambda \qquad\qquad (2.5\text{-}1a)$$

$$e_2 = \int C(\lambda)s_2(\lambda) \, d\lambda \qquad\qquad (2.5\text{-}1b)$$

$$e_3 = \int C(\lambda)s_3(\lambda) \, d\lambda \qquad\qquad (2.5\text{-}1c)$$

where $C(\lambda)$ is the spectral energy distribution of the incident light source. The three signals e_1, e_2, e_3 are then subjected to a logarithmic transfer function and combined to produce the outputs

$$d_1 = \log e_1 \qquad\qquad (2.5\text{-}2a)$$

$$d_2 = \log e_2 - \log e_1 = \log \frac{e_2}{e_1} \qquad\qquad (2.5\text{-}2b)$$

$$d_3 = \log e_3 - \log e_1 = \log \frac{e_3}{e_1} \qquad\qquad (2.5\text{-}2c)$$

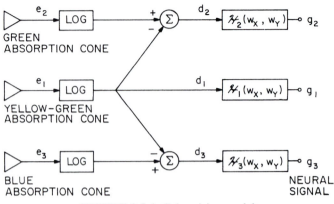

FIGURE 2.5-1 Color vision model.

Finally, the signals d_1, d_2, d_3 pass through linear systems with transfer functions $\mathcal{H}_1(\omega_x, \omega_y)$, $\mathcal{H}_2(\omega_x, \omega_y)$, $\mathcal{H}_3(\omega_x, \omega_y)$ to produce output signals g_1, g_2, g_3 that provide the basis for perception of color by the brain.

In the model of Figure 2.5-1, the signals d_2 and d_3 are related to the chromaticity of a colored light while signal d_1 is proportional to its luminance. This model has been found to predict many color vision phenomena quite accurately, and also to satisfy the basic laws of colorimetry. For example, it is known that if the spectral energy of a colored light changes by a constant multiplicative factor, the hue and saturation of the light, as described quantitatively by its chromaticity coordinates, remain invariant over a wide dynamic range. Examination of Eqs. 2.5-1 and 2.5-2 indicates that the chrominance signals d_2 and d_3 are unchanged in this case, and that the luminance signal d_1 increases in a logarithmic manner. Other, more subtle evaluations of the model are described by Frei (33).

As shown in Figure 2.2-4, some indication of the spectral sensitivities $s_i(\lambda)$ of the three types of retinal cones has been obtained by spectral absorption measurements of cone pigments. However, direct physiological measurements are difficult to perform accurately. Indirect estimates of cone spectral sensitivities have been obtained from measurements of the color response of color-blind peoples by Konig and Brodhun (34). Judd (35) has used these data to produce a linear transformation relating the spectral sensitivity functions $s_i(\lambda)$ to spectral tristimulus values obtained by colorimetric testing. The resulting sensitivity curves, shown in Figure 2.5-2, are unimodal and strictly positive, as expected from physiological considerations (34).

The logarithmic color vision model of Figure 2.5-1 may easily be extended, in analogy with the monochromatic vision model of Figure 2.4-7, by inserting a linear transfer function after each cone receptor to account for the optical response of the eye. Also, a general nonlinearity may be substituted for the logarithmic transfer function. It should be noted that the order of the receptor summation and the transfer function operations can be reversed without affecting the output, because both are

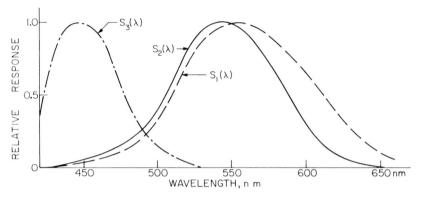

FIGURE 2.5-2. Spectral sensitivity functions of retinal cones based on Konig's data.

linear operations. Figure 2.5-3 shows the extended model for color vision. It is expected that the spatial frequency response of the g_1 neural signal through the color vision model should be similar to the luminance spatial frequency response discussed in Section 2.4. Sine-wave response measurements for colored lights obtained by van der Horst et al. (36), shown in Figure 2.5-4, indicate that the chromatic response is shifted toward low spatial frequencies relative to the luminance response. Lateral inhibition effects should produce a low spatial frequency rolloff below the measured response.

Color perception is relative; the perceived color of a given spectral energy distribution is dependent on the viewing surround and state of adaption of the viewer. A human viewer can adapt remarkably well to the surround or viewing illuminant of a scene and essentially normalize perception to some reference white or overall color balance of the scene. This property is known as *chromatic adaption.*

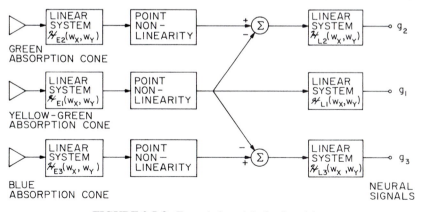

FIGURE 2.5-3. Extended model of color vision.

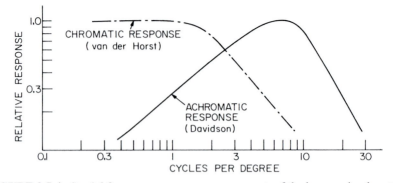

FIGURE 2.5-4. Spatial frequency response measurements of the human visual system.

The simplest visual model for chromatic adaption, proposed by von Kries (37, 16, p. 435), involves the insertion of automatic gain controls between the cones and first linear system of Figure 2.5-2. These gains

$$a_i = \left[\int W(\lambda) s_i(\lambda) \, d\lambda \right]^{-1} \tag{2.5-3}$$

for $i = 1, 2, 3$ are adjusted such that the modified cone response is unity when viewing a reference white with spectral energy distribution $W(\lambda)$. Von Kries's model is attractive because of its qualitative reasonableness and simplicity, but chromatic testing (16, p. 438) has shown that the model does not completely predict the chromatic adaptation effect. Wallis (38) has suggested that chromatic adaption may, in part, result from a post-retinal neural inhibition mechanism that linearly attenuates slowly varying visual field components. The mechanism could be modeled by the low-spatial-frequency attenuation associated with the post-retinal transfer functions $\mathcal{H}_{Li}(\omega_x, \omega_y)$ of Figure 2.5-3. Undoubtedly, both retinal and post-retinal mechanisms are responsible for the chromatic adaption effect. Further analysis and testing are required to model the effect adequately.

REFERENCES

1. *Webster's New Collegiate Dictionary*, G. & C. Merriam Co. (The Riverside Press), Springfield, MA, 1960.
2. H. H. Malitson, "The Solar Energy Spectrum," *Sky and Telescope*, **29**, 4, March 1965, 162–165.
3. *Munsell Book of Color*, Munsell Color Co., Baltimore.
4. M. H. Pirenne, *Vision and the Eye*, 2nd ed., Associated Book Publishers, London, 1967.
5. S. L. Polyak, *The Retina*, University of Chicago Press, Chicago, 1941.

6. L. H. Davson, *The Physiology of the Eye*, McGraw-Hill (Blakiston), New York, 1949.

7. T. N. Cornsweet, *Visual Perception*, Academic Press, New York, 1970.

8. G. Wald, "Human Vision and the Spectrum," *Science*, **101**, 2635, June 29, 1945, 653–658.

9. P. K. Brown and G. Wald, "Visual Pigment in Single Rods and Cones of the Human Retina," *Science*, **144**, 3614, April 3, 1964, 45–52.

10. G. Wald, "The Receptors for Human Color Vision," *Science*, **145**, 3636, September 4, 1964, 1007–1017.

11. S. Hecht, "The Visual Discrimination of Intensity and the Weber–Fechner Law," *J. General. Physiology.*, **7**, 1924, 241.

12. W. F. Schreiber, *Fundamentals of Electronic Imaging Systems,* Springer-Verlag, Berlin, 1986.

13. S. S. Stevens, *Handbook of Experimental Psychology*, Wiley, New York, 1951.

14. F. Ratliff, *Mach Bands: Quantitative Studies on Neural Networks in the Retina*, Holden-Day, San Francisco, 1965.

15. G. S. Brindley, "Afterimages," *Scientific American*, **209**, 4, October 1963, 84–93.

16. G. Wyszecki and W. S. Stiles, *Color Science*, 2nd ed., Wiley, New York, 1982.

17. J. W. Goodman, *Introduction to Fourier Optics*, 2nd ed., McGraw-Hill, New York, 1996.

18. F. W. Campbell, "The Human Eye as an Optical Filter," *Proc. IEEE*, **56**, 6, June 1968, 1009–1014.

19. O. Bryngdahl, "Characteristics of the Visual System: Psychophysical Measurement of the Response to Spatial Sine-Wave Stimuli in the Mesopic Region," *J. Optical. Society of America*, **54**, 9, September 1964, 1152–1160.

20. E. M. Lowry and J. J. DePalma, "Sine Wave Response of the Visual System, I. The Mach Phenomenon," *J. Optical Society of America*, **51**, 7, July 1961, 740–746.

21. E. M. Lowry and J. J. DePalma, "Sine Wave Response of the Visual System, II. Sine Wave and Square Wave Contrast Sensitivity," *J. Optical Society of America*, **52**, 3, March 1962, 328–335.

22. M. B. Sachs, J. Nachmias, and J. G. Robson, "Spatial Frequency Channels in Human Vision," *J. Optical Society of America*, **61**, 9, September 1971, 1176–1186.

23. T. G. Stockham, Jr., "Image Processing in the Context of a Visual Model," *Proc. IEEE*, **60**, 7, July 1972, 828–842.

24. D. E. Pearson, "A Realistic Model for Visual Communication Systems," *Proc. IEEE*, **55**, 3, March 1967, 380–389.

25. M. L. Davidson, "Perturbation Approach to Spatial Brightness Interaction in Human Vision," *J. Optical Society of America*, **58**, 9, September 1968, 1300–1308.

26. J. L. Mannos and D. J. Sakrison, "The Effects of a Visual Fidelity Criterion on the Encoding of Images," *IEEE Trans. Information. Theory*, **IT-20**, 4, July 1974, 525–536.

27. F. Ratliff, H. K. Hartline, and W. H. Miller, "Spatial and Temporal Aspects of Retinal Inhibitory Interaction," *J. Optical Society of America*, **53**, 1, January 1963, 110–120.

28. C. F. Hall and E. L. Hall, "A Nonlinear Model for the Spatial Characteristics of the Human Visual System," *IEEE Trans, Systems, Man and Cybernetics*, **SMC-7**, 3, March 1977, 161–170.

29. J. J. McCann, "Human Color Perception," in *Color: Theory and Imaging Systems*, R. A. Enyard, Ed., Society of Photographic Scientists and Engineers, Washington, DC, 1973, 1–23.

30. I. Newton, *Optiks*, 4th ed., 1730; Dover Publications, New York, 1952.

31. T. Young, *Philosophical Trans*, **92**, 1802, 12–48.

32. J. C. Maxwell, *Scientific Papers of James Clerk Maxwell*, W. D. Nevern, Ed., Dover Publications, New York, 1965.

33. W. Frei, "A New Model of Color Vision and Some Practical Limitations," USCEE Report 530, University of Southern California, Image Processing Institute, Los Angeles March 1974, 128–143.

34. A. Konig and E. Brodhun, "Experimentell Untersuchungen uber die Psycho-physische fundamental in Bezug auf den Gesichtssinn," *Zweite Mittlg. S.B. Preuss Akademic der Wissenschaften*, 1889, 641.

35. D. B. Judd, "Standard Response Functions for Protanopic and Deuteranopic Vision," *J. Optical Society of America*, **35**, 3, March 1945, 199–221.

36. C. J. C. van der Horst, C. M. de Weert, and M. A. Bouman, "Transfer of Spatial Chromaticity Contrast at Threshold in the Human Eye," *J. Optical Society of America*, **57**, 10, October 1967, 1260–1266.

37. J. von Kries, "Die Gesichtsempfindungen," *Nagel's Handbuch der. Physiologie der Menschen*, Vol. 3, 1904, 211.

38. R. H. Wallis, "Film Recording of Digital Color Images," USCEE Report 570, University of Southern California, Image Processing Institute, Los Angeles, June 1975.

3

PHOTOMETRY AND COLORIMETRY

Chapter 2 dealt with human vision from a qualitative viewpoint in an attempt to establish models for monochrome and color vision. These models may be made quantitative by specifying measures of human light perception. Luminance measures are the subject of the science of photometry, while color measures are treated by the science of colorimetry.

3.1. PHOTOMETRY

A source of radiative energy may be characterized by its spectral energy distribution $C(\lambda)$, which specifies the time rate of energy the source emits per unit wavelength interval. The total power emitted by a radiant source, given by the integral of the spectral energy distribution,

$$P = \int_0^\infty C(\lambda)\, d\lambda \tag{3.1-1}$$

is called the *radiant flux* of the source and is normally expressed in watts (W).

A body that exists at an elevated temperature radiates electromagnetic energy proportional in amount to its temperature. A *blackbody* is an idealized type of heat radiator whose radiant flux is the maximum obtainable at any wavelength for a body at a fixed temperature. The spectral energy distribution of a blackbody is given by *Planck's law* (1):

$$C(\lambda) = \frac{C_1}{\lambda^5 [\exp\{C_2/\lambda T\} - 1]} \tag{3.1-2}$$

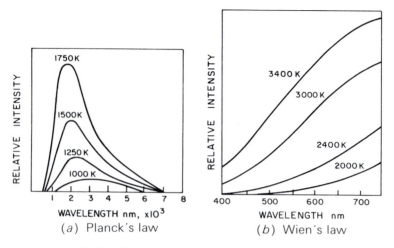

FIGURE 3.1-1. Blackbody radiation functions.

where λ is the radiation wavelength, T is the temperature of the body, and C_1 and C_2 are constants. Figure 3.1-1a is a plot of the spectral energy of a blackbody as a function of temperature and wavelength. In the visible region of the electromagnetic spectrum, the blackbody spectral energy distribution function of Eq. 3.1-2 can be approximated by *Wien's radiation law* (1):

$$C(\lambda) = \frac{C_1}{\lambda^5 \exp\{C_2/\lambda T\}} \tag{3.1-3}$$

Wien's radiation function is plotted in Figure 3.1-1b over the visible spectrum.

The most basic physical light source, of course, is the sun. Figure 2.1-1a shows a plot of the measured spectral energy distribution of sunlight (2). The dashed line in

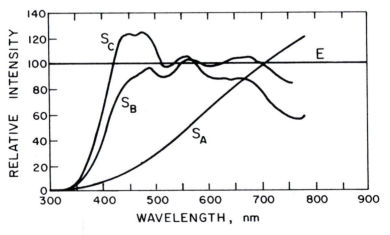

FIGURE 3.1-2. CIE standard illumination sources.

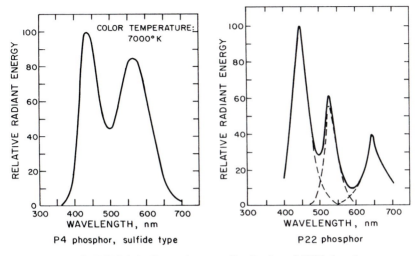

FIGURE 3.1-3. Spectral energy distribution of CRT phosphors.

this figure, approximating the measured data, is a 6000 kelvin (K) blackbody curve. Incandescent lamps are often approximated as blackbody radiators of a given temperature in the range 1500 to 3500 K (3).

The Commission Internationale de l'Eclairage (CIE), which is an international body concerned with standards for light and color, has established several standard sources of light, as illustrated in Figure 3.1-2 (4). Source S_A is a tungsten filament lamp. Over the wavelength band 400 to 700 nm, source S_B approximates direct sunlight, and source S_C approximates light from an overcast sky. A hypothetical source, called *Illuminant E*, is often employed in colorimetric calculations. Illuminant E is assumed to emit constant radiant energy at all wavelengths.

Cathode ray tube (CRT) phosphors are often utilized as light sources in image processing systems. Figure 3.1-3 describes the spectral energy distributions of common phosphors (5). Monochrome television receivers generally use a P4 phosphor, which provides a relatively bright blue-white display. Color television displays utilize cathode ray tubes with red, green, and blue emitting phosphors arranged in triad dots or strips. The P22 phosphor is typical of the spectral energy distribution of commercial phosphor mixtures. Liquid crystal displays (LCDs) typically project a white light through red, green and blue vertical strip pixels. Figure 3.1-4 is a plot of typical color filter transmissivities (6).

Photometric measurements seek to describe quantitatively the perceptual brightness of visible electromagnetic energy (7,8). The link between photometric measurements and radiometric measurements (physical intensity measurements) is the photopic luminosity function, as shown in Figure 3.1-5a (9). This curve, which is a CIE standard, specifies the spectral sensitivity of the human visual system to optical radiation as a function of wavelength for a typical person referred to as the *standard*

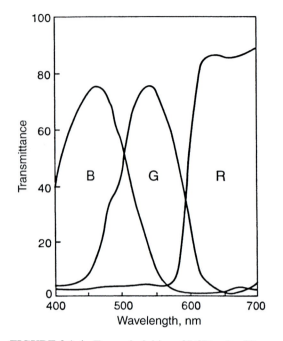

FIGURE 3.1-4. Transmissivities of LCD color filters.

observer. In essence, the curve is a standardized version of the measurement of cone sensitivity given in Figure 2.2-2 for photopic vision at relatively high levels of illumination. The standard luminosity function for scotopic vision at relatively low levels of illumination is illustrated in Figure 3.1-5*b*. Most imaging system designs are based on the photopic luminosity function, commonly called the *relative luminous efficiency.*

The perceptual brightness sensation evoked by a light source with spectral energy distribution $C(\lambda)$ is specified by its *luminous flux*, as defined by

$$F = K_m \int_0^\infty C(\lambda) V(\lambda) \, d\lambda \tag{3.1-4}$$

where $V(\lambda)$ represents the relative luminous efficiency and K_m is a scaling constant. The modern unit of luminous flux is the lumen (lm), and the corresponding value for the scaling constant is $K_m = 685$ lm/W. An infinitesimally narrowband source of 1 W of light at the peak wavelength of 555 nm of the relative luminous efficiency curve therefore results in a luminous flux of 685 lm.

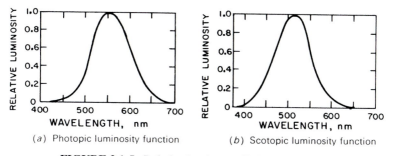

(a) Photopic luminosity function

(a) Photopic luminosity function   (b) Scotopic luminosity function

FIGURE 3.1-5. Relative luminous efficiency functions.

3.2. COLOR MATCHING

The basis of the trichromatic theory of color vision is that it is possible to match an arbitrary color by superimposing appropriate amounts of three primary colors (10–14). In an *additive color reproduction system* such as color television, the three primaries are individual red, green, and blue light sources that are projected onto a common region of space to reproduce a colored light. In a *subtractive color system*, which is the basis of most color photography and color printing, a white light sequentially passes through cyan, magenta, and yellow filters to reproduce a colored light.

3.2.1. Additive Color Matching

An additive color-matching experiment is illustrated in Figure 3.2-1. In Figure 3.2-1a, a patch of light (C) of arbitrary spectral energy distribution $C(\lambda)$, as shown in Figure 3.2-2a, is assumed to be imaged onto the surface of an ideal diffuse reflector (a surface that reflects uniformly over all directions and all wavelengths). A reference white light (W) with an energy distribution, as in Figure 3.2-2b, is imaged onto the surface along with three primary lights (P_1), (P_2), (P_3) whose spectral energy distributions are sketched in Figure 3.2-2c to e. The three primary lights are first overlapped and their intensities are adjusted until the overlapping region of the three primary lights perceptually matches the reference white in terms of brightness, hue, and saturation. The amounts of the three primaries $A_1(W)$, $A_2(W)$, $A_3(W)$ are then recorded in some physical units, such as watts. These are the matching values of the reference white. Next, the intensities of the primaries are adjusted until a match is achieved with the colored light (C), if a match is possible. The procedure to be followed if a match cannot be achieved is considered later. The intensities of the primaries

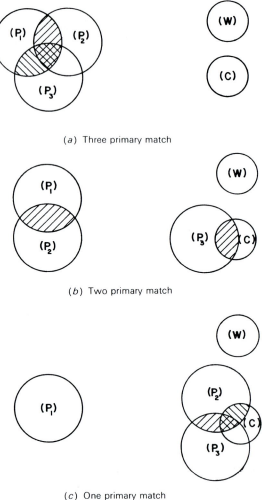

(*a*) Three primary match

(*b*) Two primary match

(*c*) One primary match

FIGURE 3.2-1. Color matching.

$A_1(C)$, $A_2(C)$, $A_3(C)$ when a match is obtained are recorded, and normalized matching values $T_1(C)$, $T_2(C)$, $T_3(C)$, called *tristimulus values*, are computed as

$$T_1(C) = \frac{A_1(C)}{A_1(W)} \qquad T_2(C) = \frac{A_2(C)}{A_2(W)} \qquad T_3(C) = \frac{A_3(C)}{A_3(W)}$$

$$(3.2\text{-}1)$$

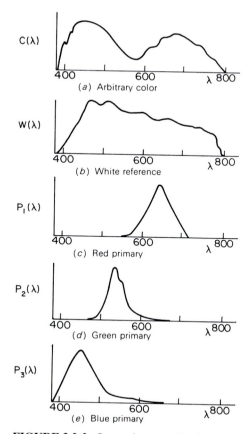

FIGURE 3.2-2. Spectral energy distributions.

If a match cannot be achieved by the procedure illustrated in Figure 3.2-1a, it is often possible to perform the color matching outlined in Figure 3.2-1b. One of the primaries, say (P_3), is superimposed with the light (C), and the intensities of all three primaries are adjusted until a match is achieved between the overlapping region of primaries (P_1) and (P_2) with the overlapping region of (P_3) and (C). If such a match is obtained, the tristimulus values are

$$T_1(C) = \frac{A_1(C)}{A_1(W)} \qquad T_2(C) = \frac{A_2(C)}{A_2(W)} \qquad T_3(C) = \frac{-A_3(C)}{A_3(W)} \qquad (3.2\text{-}2)$$

In this case, the tristimulus value $T_3(C)$ is negative. If a match cannot be achieved with this geometry, a match is attempted between (P_1) plus (P_3) and (P_2) plus (C). If a match is achieved by this configuration, tristimulus value $T_2(C)$ will be negative. If this configuration fails, a match is attempted between (P_2) plus (P_3) and (P_1) plus (C). A correct match is denoted with a negative value for $T_1(C)$.

Finally, in the rare instance in which a match cannot be achieved by either of the configurations of Figure 3.2-1a or b, two of the primaries are superimposed with (C) and an attempt is made to match the overlapped region with the remaining primary. In the case illustrated in Figure 3.2-1c, if a match is achieved, the tristimulus values become

$$T_1(C) = \frac{A_1(C)}{A_1(W)} \qquad T_2(C) = \frac{-A_2(C)}{A_2(W)} \qquad T_3(C) = \frac{-A_3(C)}{A_3(W)} \qquad (3.2-3)$$

If a match is not obtained by this configuration, one of the other two possibilities will yield a match.

The process described above is a direct method for specifying a color quantitatively. It has two drawbacks: The method is cumbersome and it depends on the perceptual variations of a single observer. In Section 3.3 we consider standardized quantitative color measurement in detail.

3.2.2. Subtractive Color Matching

A subtractive color-matching experiment is shown in Figure 3.2-3. An illumination source with spectral energy distribution $E(\lambda)$ passes sequentially through three dye filters that are nominally cyan, magenta, and yellow. The spectral absorption of the dye filters is a function of the dye concentration. It should be noted that the spectral transmissivities of practical dyes change shape in a nonlinear manner with dye concentration.

In the first step of the subtractive color-matching process, the dye concentrations of the three spectral filters are varied until a perceptual match is obtained with a reference white (W). The dye concentrations are the matching values of the color match $A_1(W), A_2(W), A_3(W)$. Next, the three dye concentrations are varied until a match is obtained with a desired color (C). These matching values $A_1(C), A_2(C), A_3(C)$, are then used to compute the tristimulus values $T_1(C), T_2(C), T_3(C)$, as in Eq. 3.2-1.

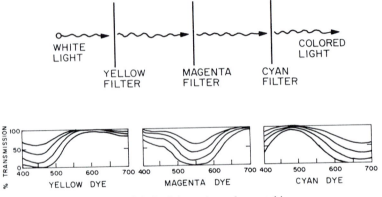

FIGURE 3.2-3. Subtractive color matching.

It should be apparent that there is no fundamental theoretical difference between color matching by an additive or a subtractive system. In a subtractive   system, the yellow dye acts as a variable absorber of blue light, and with ideal dyes, the yellow dye effectively forms a blue primary light. In a similar manner, the magenta filter ideally forms the green primary, and the cyan filter ideally forms the red primary. Subtractive color systems ordinarily utilize cyan, magenta, and yellow dye spectral filters rather than red, green, and blue dye filters because the cyan, magenta, and yellow filters are notch filters which permit a greater transmission of light energy than do narrowband red, green, and blue bandpass filters. In color printing, a fourth filter layer of variable gray level density is often introduced to achieve a higher contrast in reproduction because common dyes do not possess a wide density range.

3.2.3. Axioms of Color Matching

The color-matching experiments described for additive and subtractive color matching have been performed quite accurately by a number of researchers. It has been found that perfect color matches sometimes cannot be obtained at either very high or very low levels of illumination. Also, the color matching results do depend to some extent on the spectral composition of the surrounding light. Nevertheless, the simple color matching experiments have been found to hold over a wide range of conditions.

Grassman (15) has developed a set of eight axioms that define trichromatic color matching and that serve as a basis for quantitative color measurements. In the following presentation of these axioms, the symbol $\Diamond$ indicates a color match; the symbol $\oplus$ indicates an additive color mixture; the symbol $\bullet$ indicates units of a color. These axioms are:

1. Any color can be matched by a mixture of no more than three colored lights.
2. A color match at one radiance level holds over a wide range of levels.
3. Components of a mixture of colored lights cannot be resolved by the human eye.
4. The luminance of a color mixture is equal to the sum of the luminance of its components.
5. *Law of addition.* If color (M) matches color (N) and color (P) matches color (Q), then color (M) mixed with color (P) matches color (N) mixed with color (Q):

$$(M) \Diamond (N) \cap (P) \Diamond (Q) \Rightarrow [(M) \oplus (P)] \Diamond [(N) \oplus (Q)] \qquad (3.2\text{-}4)$$

6. *Law of subtraction.* If the mixture of (M) plus (P) matches the mixture of (N) plus (Q) and if (P) matches (Q), then (M) matches (N):

$$[(M) \oplus (P)] \Diamond [(N) \oplus (Q)] \cap [(P) \Diamond (Q)] \Rightarrow (M) \Diamond (N) \qquad (3.2\text{-}5)$$

7. *Transitive law.* If (M) matches (N) and if (N) matches (P), then (M) matches (P):

$$[(M) \lozenge (N)] \cap [(N) \lozenge (P)] \Rightarrow (M) \lozenge (P) \qquad (3.2\text{-}6)$$

8. *Color matching.* (a) c units of (C) matches the mixture of m units of (M) plus n units of (N) plus p units of (P):

$$c \bullet C \lozenge [m \bullet (M)] \oplus [n \bullet (N)] \oplus [p \bullet (P)] \qquad (3.2\text{-}7)$$

or (b) a mixture of c units of C plus m units of M matches the mixture of n units of N plus p units of P:

$$[c \bullet (C)] \oplus [m \bullet (M)] \lozenge [n \bullet (N)] \oplus [p \bullet (P)] \qquad (3.2\text{-}8)$$

or (c) a mixture of c units of (C) plus m units of (M) plus n units of (N) matches p units of P:

$$[c \bullet (C)] \oplus [m \bullet (M)] \oplus [n \bullet (N)] \lozenge [p \bullet (P)] \qquad (3.2\text{-}9)$$

With Grassman's laws now specified, consideration is given to the development of a quantitative theory for color matching.

3.3. COLORIMETRY CONCEPTS

Colorimetry is the science of quantitatively measuring color. In the trichromatic color system, color measurements are in terms of the tristimulus values of a color or a mathematical function of the tristimulus values.

Referring to Section 3.2.3, the axioms of color matching state that a color C can be matched by three primary colors P_1, P_2, P_3. The qualitative match is expressed as

$$(C) \lozenge [A_1(C) \bullet (P_1)] \oplus [A_2(C) \bullet (P_2)] \oplus [A_3(C) \bullet (P_3)] \qquad (3.3\text{-}1)$$

where $A_1(C)$, $A_2(C)$, $A_3(C)$ are the matching values of the color (C). Because the intensities of incoherent light sources add linearly, the spectral energy distribution of a color mixture is equal to the sum of the spectral energy distributions of its components. As a consequence of this fact and Eq. 3.3-1, the spectral energy distribution $C(\lambda)$ can be replaced by its color-matching equivalent according to the relation

$$C(\lambda) \lozenge A_1(C)P_1(\lambda) + A_2(C)P_2(\lambda) + A_3(C)P_3(\lambda) = \sum_{j=1}^{3} A_j(C)P_j(\lambda) \qquad (3.3\text{-}2)$$

Equation 3.3-2 simply means that the spectral energy distributions on both sides of the equivalence operator ◊ evoke the same color sensation. Color matching is usually specified in terms of tristimulus values, which are normalized matching values, as defined by

$$T_j(C) = \frac{A_j(C)}{A_j(W)} \tag{3.3-3}$$

where $A_j(W)$ represents the matching value of the reference white. By this substitution, Eq. 3.3-2 assumes the form

$$C(\lambda) \, \Diamond \, \sum_{j=1}^{3} T_j(C)A_j(W)P_j(\lambda) \tag{3.3-4}$$

From Grassman's fourth law, the luminance of a color mixture $Y(C)$ is equal to the luminance of its primary components. Hence

$$Y(C) = \int C(\lambda)V(\lambda)\,d\lambda = \sum_{j=1}^{3} \int A_j(C)P_j(\lambda)V(\lambda)\,d\lambda \tag{3.3-5a}$$

or

$$Y(C) = \sum_{j=1}^{3} \int T_j(C)A_j(W)P_j(\lambda)V(\lambda)\,d\lambda \tag{3.3-5b}$$

where $V(\lambda)$ is the relative luminous efficiency and $P_j(\lambda)$ represents the spectral energy distribution of a primary. Equations 3.3-4 and 3.3-5 represent the quantitative foundation for colorimetry.

3.3.1. Color Vision Model Verification

Before proceeding further with quantitative descriptions of the color-matching process, it is instructive to determine whether the matching experiments and the axioms of color matching are satisfied by the color vision model presented in Section 2.5. In that model, the responses of the three types of receptors with sensitivities $s_1(\lambda)$, $s_2(\lambda)$, $s_3(\lambda)$ are modeled as

$$e_1(C) = \int C(\lambda)s_1(\lambda)\,d\lambda \tag{3.3-6a}$$

$$e_2(C) = \int C(\lambda)s_2(\lambda)\,d\lambda \tag{3.3-6b}$$

$$e_3(C) = \int C(\lambda)s_3(\lambda)\,d\lambda \tag{3.3-6c}$$

If a viewer observes the primary mixture instead of C, then from Eq. 3.3-4, substitution for $C(\lambda)$ should result in the same cone signals $e_i(C)$. Thus

$$e_1(C) = \sum_{j=1}^{3} T_j(C)A_j(W) \int P_j(\lambda)s_1(\lambda)\,d\lambda \qquad (3.3\text{-}7a)$$

$$e_2(C) = \sum_{j=1}^{3} T_j(C)A_j(W) \int P_j(\lambda)s_2(\lambda)\,d\lambda \qquad (3.3\text{-}7b)$$

$$e_3(C) = \sum_{j=1}^{3} T_j(C)A_j(W) \int P_j(\lambda)s_3(\lambda)\,d\lambda \qquad (3.3\text{-}7c)$$

Equation 3.3-7 can be written more compactly in matrix form by defining

$$k_{ij} = \int P_j(\lambda)s_i(\lambda)\,d\lambda \qquad (3.3\text{-}8)$$

Then

$$\begin{bmatrix} e_1(C) \\ e_2(C) \\ e_3(C) \end{bmatrix} = \begin{bmatrix} k_{11} & k_{12} & k_{13} \\ k_{21} & k_{22} & k_{23} \\ k_{31} & k_{32} & k_{33} \end{bmatrix} \begin{bmatrix} A_1(W) & 0 & 0 \\ 0 & A_2(W) & 0 \\ 0 & 0 & A_3(W) \end{bmatrix} \begin{bmatrix} T_1(C) \\ T_2(C) \\ T_3(C) \end{bmatrix} \qquad (3.3\text{-}9)$$

or in yet more abbreviated form,

$$\mathbf{e}(C) = \mathbf{KAt}(C) \qquad (3.3\text{-}10)$$

where the vectors and matrices of Eq. 3.3-10 are defined in correspondence with Eqs. 3.3-7 to 3.3-9. The vector space notation used in this section is consistent with the notation formally introduced in Appendix 1. Matrices are denoted as boldface uppercase symbols, and vectors are denoted as boldface lowercase symbols. It should be noted that for a given set of primaries, the matrix $\mathbf{K}$ is constant valued, and for a given reference white, the white matching values of the matrix $\mathbf{A}$ are constant. Hence, if a set of cone signals $e_i(C)$ were known for a color (C), the corresponding tristimulus values $T_j(C)$ could in theory be obtained from

$$\mathbf{t}(C) = [\mathbf{KA}]^{-1}\mathbf{e}(C) \qquad (3.3\text{-}11)$$

provided that the matrix inverse of [**KA**] exists. Thus, it has been shown that with proper selection of the tristimulus signals $T_j(C)$, any color can be matched in the sense that the cone signals will be the same for the primary mixture as for the actual color C. Unfortunately, the cone signals $e_i(C)$ are not easily measured physical quantities, and therefore, Eq. 3.3-11 cannot be used directly to compute the tristimulus values of a color. However, this has not been the intention of the derivation. Rather, Eq. 3.3-11 has been developed to show the consistency of the color-matching experiment with the color vision model.

3.3.2. Tristimulus Value Calculation

It is possible indirectly to compute the tristimulus values of an arbitrary color for a particular set of primaries if the tristimulus values of the spectral colors (narrow-band light) are known for that set of primaries. Figure 3.3-1 is a typical sketch of the tristimulus values required to match a unit energy spectral color with three arbitrary primaries. These tristimulus values, which are fundamental to the definition of a primary system, are denoted as $T_{s_1}(\lambda)$, $T_{s_2}(\lambda)$, $T_{s_3}(\lambda)$, where λ is a particular wavelength in the visible region. A unit energy spectral light (C_ψ) at wavelength ψ with energy distribution $\delta(\lambda - \psi)$ is matched according to the equation

$$e_i(C_\psi) = \int \delta(\lambda - \psi) s_i(\lambda)\, d\lambda = \sum_{j=1}^{3} \int A_j(W) P_j(\lambda) T_{s_j}(\psi) s_i(\lambda)\, d\lambda \qquad (3.3\text{-}12)$$

Now, consider an arbitrary color [C] with spectral energy distribution $C(\lambda)$. At wavelength ψ, $C(\psi)$ units of the color are matched by $C(\psi)T_{s_1}(\psi)$, $C(\psi)T_{s_2}(\psi)$, $C(\psi)T_{s_3}(\psi)$ tristimulus units of the primaries as governed by

$$\int C(\psi)\delta(\lambda - \psi) s_i(\lambda)\, d\lambda = \sum_{j=1}^{3} \int A_j(W) P_j(\lambda) C(\psi) T_{s_j}(\psi) s_i(\lambda)\, d\lambda \qquad (3.3\text{-}13)$$

Integrating each side of Eq. 3.3-13 over ψ and invoking the sifting integral gives the cone signal for the color (C). Thus

$$\iint C(\psi)\delta(\lambda - \psi) s_i(\lambda)\, d\lambda\, d\psi = e_i(C) = \sum_{j=1}^{3} \iint A_j(W) P_j(\lambda) C(\psi) T_{s_j}(\psi) s_i(\lambda)\, d\psi\, d\lambda$$

$$(3.3\text{-}14)$$

By correspondence with Eq. 3.3-7, the tristimulus values of (C) must be equivalent to the second integral on the right of Eq. 3.3-14. Hence

$$T_j(C) = \int C(\psi) T_{s_j}(\psi)\, d\psi \qquad (3.3\text{-}15)$$

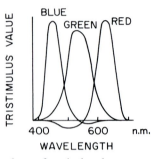

FIGURE 3.3-1. Tristimulus values of typical red, green, and blue primaries required to match unit energy throughout the spectrum.

From Figure 3.3-1 it is seen that the tristimulus values obtained from solution of Eq. 3.3-11 may be negative at some wavelengths. Because the tristimulus values represent units of energy, the physical interpretation of this mathematical result is that a color match can be obtained by adding the primary with negative tristimulus value to the original color and then matching this resultant color with the remaining primary. In this sense, any color can be matched by any set of primaries. However, from a practical viewpoint, negative tristimulus values are not physically realizable, and hence there are certain colors that cannot be matched in a practical color display (e.g., a color television receiver) with fixed primaries. Fortunately, it is possible to choose primaries so that most commonly occurring natural colors can be matched.

The three tristimulus values T_1, T_2, T_3 can be considered to form the three axes of a color space as illustrated in Figure 3.3-2. A particular color may be described as a a vector in the color space, but it must be remembered that it is the coordinates of the vectors (tristimulus values), rather than the vector length, that specify the color. In Figure 3.3-2, a triangle, called a *Maxwell triangle*, has been drawn between the three primaries. The intersection point of a color vector with the triangle gives an indication of the hue and saturation of the color in terms of the distances of the point from the vertices of the triangle.

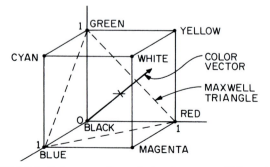

FIGURE 3.3-2 Color space for typical red, green, and blue primaries.

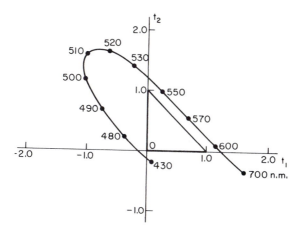

FIGURE 3.3-3. Chromaticity diagram for typical red, green, and blue primaries.

Often the luminance of a color is not of interest in a color match. In such situations, the hue and saturation of color (*C*) can be described in terms of *chromaticity coordinates*, which are normalized tristimulus values, as defined by

$$t_1 \equiv \frac{T_1}{T_1 + T_2 + T_3} \tag{3.3-16a}$$

$$t_2 \equiv \frac{T_2}{T_1 + T_2 + T_3} \tag{3.3-16b}$$

$$t_3 \equiv \frac{T_3}{T_1 + T_2 + T_3} \tag{3.3-16c}$$

Clearly, $t_3 = 1 - t_1 - t_2$, and hence only two coordinates are necessary to describe a color match. Figure 3.3-3 is a plot of the chromaticity coordinates of the spectral colors for typical primaries. Only those colors within the triangle defined by the three primaries are realizable by physical primary light sources.

3.3.3. Luminance Calculation

The tristimulus values of a color specify the amounts of the three primaries required to match a color where the units are measured relative to a match of a reference white. Often, it is necessary to determine the absolute rather than the relative amount of light from each primary needed to reproduce a color match. This information is found from luminance measurements of calculations of a color match.

From Eq. 3.3-5 it is noted that the *luminance* of a matched color $Y(C)$ is equal to the sum of the luminances of its primary components according to the relation

$$Y(C) = \sum_{j=1}^{3} T_j(C) \int A_j(C) P_j(\lambda) V(\lambda) \, d\lambda \qquad (3.3\text{-}17)$$

The integrals of Eq. 3.3-17,

$$Y(P_j) = \int A_j(C) P_j(\lambda) V(\lambda) \, d\lambda \qquad (3.3\text{-}18)$$

are called *luminosity coefficients* of the primaries. These coefficients represent the luminances of unit amounts of the three primaries for a match to a specific reference white. Hence the luminance of a matched color can be written as

$$Y(C) = T_1(C) Y(P_1) + T_2(C) Y(P_2) + T_3(C) Y(P_3) \qquad (3.3\text{-}19)$$

Multiplying the right and left sides of Eq. 3.3-19 by the right and left sides, respectively, of the definition of the chromaticity coordinate

$$t_1(C) = \frac{T_1(C)}{T_1(C) + T_2(C) + T_3(C)} \qquad (3.3\text{-}20)$$

and rearranging gives

$$T_1(C) = \frac{t_1(C) Y(C)}{t_1(C) Y(P_1) + t_2(C) Y(P_2) + t_3(C) Y(P_3)} \qquad (3.3\text{-}21a)$$

Similarly,

$$T_2(C) = \frac{t_2(C) Y(C)}{t_1(C) Y(P_1) + t_2(C) Y(P_2) + t_3(C) Y(P_3)} \qquad (3.3\text{-}21b)$$

$$T_3(C) = \frac{t_3(C) Y(C)}{t_1(C) Y(P_1) + t_2(C) Y(P_2) + t_3(C) Y(P_3)} \qquad (3.3\text{-}21c)$$

Thus the tristimulus values of a color can be expressed in terms of the luminance and chromaticity coordinates of the color.

3.4. TRISTIMULUS VALUE TRANSFORMATION

From Eq. 3.3-7 it is clear that there is no unique set of primaries for matching colors. If the tristimulus values of a color are known for one set of primaries, a simple coordinate conversion can be performed to determine the tristimulus values for another set of primaries (16). Let (P_1), (P_2), (P_3) be the original set of primaries with spectral energy distributions $P_1(\lambda)$, $P_2(\lambda)$, $P_3(\lambda)$, with the units of a match determined by a white reference (W) with matching values $A_1(W)$, $A_2(W)$, $A_3(W)$. Now, consider a new set of primaries $(\tilde{P}_1)$, $(\tilde{P}_2)$, $(\tilde{P}_3)$ with spectral energy distributions $\tilde{P}_1(\lambda)$, $\tilde{P}_2(\lambda)$, $\tilde{P}_3(\lambda)$. Matches are made to a reference white $(\tilde{W})$, which may be different than the reference white of the original set of primaries, by matching values $\tilde{A}_1(W)$, $\tilde{A}_2(W)$, $\tilde{A}_3(W)$. Referring to Eq. 3.3-10, an arbitrary color (C) can be matched by the tristimulus values $T_1(C)$, $T_2(C)$, $T_3(C)$ with the original set of primaries or by the tristimulus values $\tilde{T}_1(C)$, $\tilde{T}_2(C)$, $\tilde{T}_3(C)$ with the new set of primaries, according to the matching matrix relations

$$\mathbf{e}(C) = \mathbf{KA}(W)\mathbf{t}(C) = \mathbf{\tilde{K}\tilde{A}}(\tilde{W})\mathbf{\tilde{t}}(C) \tag{3.4-1}$$

The tristimulus value units of the new set of primaries, with respect to the original set of primaries, must now be found. This can be accomplished by determining the color signals of the reference white for the second set of primaries in terms of both sets of primaries. The color signal equations for the reference white $\tilde{W}$ become

$$\mathbf{e}(\tilde{W}) = \mathbf{KA}(W)\mathbf{t}(\tilde{W}) = \mathbf{\tilde{K}\tilde{A}}(\tilde{W})\mathbf{\tilde{t}}(\tilde{W}) \tag{3.4-2}$$

where $\tilde{T}_1(\tilde{W}) = \tilde{T}_2(\tilde{W}) = \tilde{T}_3(\tilde{W}) = 1$. Finally, it is necessary to relate the two sets of primaries by determining the color signals of each of the new primary colors $(\tilde{P}_1)$, $(\tilde{P}_2)$, $(\tilde{P}_3)$ in terms of both primary systems. These color signal equations are

$$\mathbf{e}(\tilde{P}_1) = \mathbf{KA}(W)\mathbf{t}(\tilde{P}_1) = \mathbf{\tilde{K}\tilde{A}}(\tilde{W})\mathbf{\tilde{t}}(\tilde{P}_1) \tag{3.4-3a}$$

$$\mathbf{e}(\tilde{P}_2) = \mathbf{KA}(W)\mathbf{t}(\tilde{P}_2) = \mathbf{\tilde{K}\tilde{A}}(\tilde{W})\mathbf{\tilde{t}}(\tilde{P}_2) \tag{3.4-3b}$$

$$\mathbf{e}(\tilde{P}_3) = \mathbf{KA}(W)\mathbf{t}(\tilde{P}_3) = \mathbf{\tilde{K}\tilde{A}}(\tilde{W})\mathbf{\tilde{t}}(\tilde{P}_3) \tag{3.4-3c}$$

where

$$\mathbf{\tilde{t}}(\tilde{P}_1) = \begin{bmatrix} \dfrac{1}{\tilde{A}_1(\tilde{W})} \\ 0 \\ 0 \end{bmatrix} \qquad \mathbf{\tilde{t}}(\tilde{P}_2) = \begin{bmatrix} 0 \\ \dfrac{1}{A_2(\tilde{W})} \\ 0 \end{bmatrix} \qquad \mathbf{\tilde{t}}(\tilde{P}_2) = \begin{bmatrix} 0 \\ 0 \\ \dfrac{1}{A_3(\tilde{W})} \end{bmatrix}$$

Matrix equations 3.4-1 to 3.4-3 may be solved jointly to obtain a relationship between the tristimulus values of the original and new primary system:

$$
\tilde{T}_1(C) = \frac{\begin{vmatrix} T_1(C) & T_1(\tilde{P}_2) & T_1(\tilde{P}_3) \\ T_2(C) & T_2(\tilde{P}_2) & T_2(\tilde{P}_3) \\ T_3(C) & T_3(\tilde{P}_2) & T_3(\tilde{P}_3) \end{vmatrix}}{\begin{vmatrix} T_1(\tilde{W}) & T_1(\tilde{P}_2) & T_1(\tilde{P}_3) \\ T_2(\tilde{W}) & T_2(\tilde{P}_2) & T_2(\tilde{P}_3) \\ T_3(\tilde{W}) & T_3(\tilde{P}_2) & T_3(\tilde{P}_3) \end{vmatrix}}
\tag{3.4-4a}
$$

$$
\tilde{T}_2(C) = \frac{\begin{vmatrix} T_1(\tilde{P}_1) & T_1(C) & T_1(\tilde{P}_3) \\ T_2(\tilde{P}_1) & T_2(C) & T_2(\tilde{P}_3) \\ T_3(\tilde{P}_1) & T_3(C) & T_3(\tilde{P}_3) \end{vmatrix}}{\begin{vmatrix} T_1(\tilde{P}_1) & T_1(\tilde{W}) & T_1(\tilde{P}_3) \\ T_2(\tilde{P}_1) & T_2(\tilde{W}) & T_2(\tilde{P}_3) \\ T_3(\tilde{P}_1) & T_3(\tilde{W}) & T_3(\tilde{P}_3) \end{vmatrix}}
\tag{3.4-4b}
$$

$$
\tilde{T}_3(C) = \frac{\begin{vmatrix} T_1(\tilde{P}_1) & T_1(\tilde{P}_2) & T_1(C) \\ T_2(\tilde{P}_1) & T_2(\tilde{P}_2) & T_2(C) \\ T_3(\tilde{P}_1) & T_3(\tilde{P}_2) & T_3(C) \end{vmatrix}}{\begin{vmatrix} T_1(\tilde{P}_1) & T_1(\tilde{P}_2) & T_1(\tilde{W}) \\ T_2(\tilde{P}_1) & T_2(\tilde{P}_2) & T_2(\tilde{W}) \\ T_3(\tilde{P}_1) & T_3(\tilde{P}_2) & T_3(\tilde{W}) \end{vmatrix}}
\tag{3.4-4c}
$$

where $|\mathbf{T}|$ denotes the determinant of matrix $\mathbf{T}$. Equations 3.4-4 then may be written in terms of the chromaticity coordinates $t_i(\tilde{P}_1)$, $t_i(\tilde{P}_2)$, $t_i(\tilde{P}_3)$ of the new set of primaries referenced to the original primary coordinate system.

With this revision,

$$
\begin{bmatrix} \tilde{T}_1(C) \\ \tilde{T}_2(C) \\ \tilde{T}_3(C) \end{bmatrix} = \begin{bmatrix} m_{11} & m_{12} & m_{13} \\ m_{21} & m_{22} & m_{31} \\ m_{31} & m_{32} & m_{33} \end{bmatrix} \begin{bmatrix} T_1(C) \\ T_2(C) \\ T_3(C) \end{bmatrix}
\tag{3.4-5}
$$

where

$$m_{ij} = \frac{\Delta_{ij}}{\Delta_i}$$

and

$$\Delta_1 = T_1(\tilde{W})\Delta_{11} + T_2(\tilde{W})\Delta_{12} + T_3(\tilde{W})\Delta_{13}$$

$$\Delta_2 = T_1(\tilde{W})\Delta_{21} + T_2(\tilde{W})\Delta_{22} + T_3(\tilde{W})\Delta_{23}$$

$$\Delta_3 = T_1(\tilde{W})\Delta_{31} + T_2(\tilde{W})\Delta_{32} + T_3(\tilde{W})\Delta_{33}$$

$$\Delta_{11} = t_2(\tilde{P}_2)t_3(\tilde{P}_3) - t_3(\tilde{P}_2)t_2(\tilde{P}_3)$$

$$\Delta_{12} = t_3(\tilde{P}_2)t_1(\tilde{P}_3) - t_1(\tilde{P}_2)t_3(\tilde{P}_3)$$

$$\Delta_{13} = t_1(\tilde{P}_2)t_2(\tilde{P}_3) - t_2(\tilde{P}_2)t_1(\tilde{P}_3)$$

$$\Delta_{21} = t_3(\tilde{P}_1)t_2(\tilde{P}_3) - t_2(\tilde{P}_1)t_3(\tilde{P}_3)$$

$$\Delta_{22} = t_1(\tilde{P}_1)t_3(\tilde{P}_3) - t_3(\tilde{P}_1)t_1(\tilde{P}_3)$$

$$\Delta_{23} = t_2(\tilde{P}_1)t_1(\tilde{P}_3) - t_1(\tilde{P}_1)t_2(\tilde{P}_3)$$

$$\Delta_{31} = t_2(\tilde{P}_1)t_3(\tilde{P}_2) - t_3(\tilde{P}_1)t_2(\tilde{P}_2)$$

$$\Delta_{32} = t_3(\tilde{P}_1)t_1(\tilde{P}_2) - t_1(\tilde{P}_1)t_3(\tilde{P}_2)$$

$$\Delta_{33} = t_1(\tilde{P}_1)t_2(\tilde{P}_2) - t_2(\tilde{P}_1)t_1(\tilde{P}_2)$$

Thus, if the tristimulus values are known for a given set of primaries, conversion to another set of primaries merely entails a simple linear transformation of coordinates.

3.5. COLOR SPACES

It has been shown that a color (C) can be matched by its tristimulus values $T_1(C)$, $T_2(C)$, $T_3(C)$ for a given set of primaries. Alternatively, the color may be specified by its chromaticity values $t_1(C)$, $t_2(C)$ and its luminance $Y(C)$. Appendix 2 presents formulas for color coordinate conversion between tristimulus values and chromaticity coordinates for various representational combinations. A third approach in specifying a color is to represent the color by a linear or nonlinear invertible function of its tristimulus or chromaticity values.

In this section we describe several standard and nonstandard color spaces for the representation of color images. They are categorized as colorimetric, subtractive, video, or nonstandard. Figure 3.5-1 illustrates the relationship between these color spaces. The figure also lists several example color spaces.

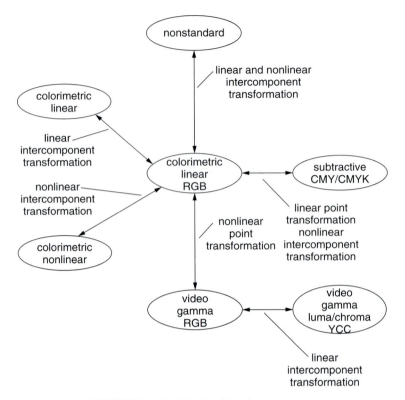

FIGURE 3.5-1. Relationship of color spaces.

Natural color images, as opposed to computer-generated images, usually originate from a color scanner or a color video camera. These devices incorporate three
sensors that are spectrally sensitive to the red, green, and blue portions of the light
spectrum. The color sensors typically generate red, green, and blue color signals that
are linearly proportional to the amount of red, green, and blue light detected by each
sensor. These signals are linearly proportional to the tristimulus values of a color at
each pixel. As indicated in Figure 3.5-1, linear *RGB* images are the basis for the generation of the various color space image representations.

3.5.1. Colorimetric Color Spaces

The class of colorimetric color spaces includes all linear *RGB* images and the standard colorimetric images derived from them by linear and nonlinear intercomponent
transformations.

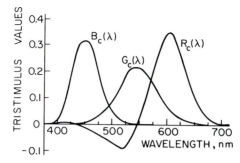

FIGURE 3.5-2. Tristimulus values of CIE spectral primaries required to match unit energy throughout the spectrum. Red = 700 nm, green = 546.1 nm, and blue = 435.8 nm.

$R_C G_C B_C$ *Spectral Primary Color Coordinate System.* In 1931, the CIE developed a standard primary reference system with three monochromatic primaries at wavelengths: red = 700 nm; green = 546.1 nm; blue = 435.8 nm (11). The units of the tristimulus values are such that the tristimulus values R_C, G_C, B_C are equal when matching an equal-energy white, called *Illuminant E*, throughout the visible spectrum. The primary system is defined by tristimulus curves of the spectral colors, as shown in Figure 3.5-2. These curves have been obtained indirectly by experimental color-matching experiments performed by a number of observers. The collective color-matching response of these observers has been called the *CIE Standard Observer.* Figure 3.5-3 is a chromaticity diagram for the CIE spectral coordinate system.

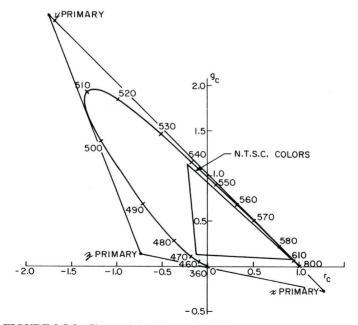

FIGURE 3.5-3. Chromaticity diagram for CIE spectral primary system.

$R_N G_N B_N$ NTSC Receiver Primary Color Coordinate System. Commercial television receivers employ a cathode ray tube with three phosphors that glow in the red,
green, and blue regions of the visible spectrum. Although the phosphors of
commercial television receivers differ from manufacturer to manufacturer, it is
common practice to reference them to the National Television Systems Committee
(NTSC) receiver phosphor standard (14). The standard observer data for the CIE
spectral primary system is related to the NTSC primary system by a pair of linear
coordinate conversions.

Figure 3.5-4 is a chromaticity diagram for the NTSC primary system. In this
system, the units of the tristimulus values are normalized so that the tristimulus
values are equal when matching the *Illuminant C* white reference. The NTSC
phosphors are not pure monochromatic sources of radiation, and hence the gamut of
colors producible by the NTSC phosphors is smaller than that available from the
spectral primaries. This fact is clearly illustrated by Figure 3.5-3, in which the gamut
of NTSC reproducible colors is plotted in the spectral primary chromaticity diagram
(11). In modern practice, the NTSC chromaticities are combined with *Illuminant
D65*.

$R_E G_E B_E$ EBU Receiver Primary Color Coordinate System. The European Broadcast Union (EBU) has established a receiver primary system whose chromaticities
are close in value to the CIE chromaticity coordinates, and the reference white is
Illuminant C (17). The EBU chromaticities are also combined with the D65 illuminant.

$R_R G_R B_R$ CCIR Receiver Primary Color Coordinate Systems. In 1990, the International Telecommunications Union (ITU) issued its Recommendation 601, which

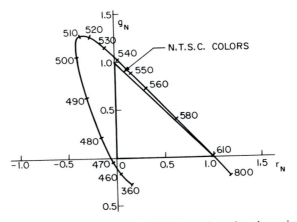

FIGURE 3.5-4. Chromaticity diagram for NTSC receiver phosphor primary system.

specified the receiver primaries for standard resolution digital television (18). Also, in 1990 the ITU published its Recommendation 709 for digital high-definition television systems (19). Both standards are popularly referenced as CCIR Rec. 601 and CCIR Rec. 709, abbreviations of the former name of the standards committee, Comité Consultatif International des Radiocommunications.

$R_S G_S B_S$ SMPTE Receiver Primary Color Coordinate System. The Society of Motion Picture and Television Engineers (SMPTE) has established a standard receiver primary color coordinate system with primaries that match modern receiver phosphors better than did the older NTSC primary system (20). In this coordinate system, the reference white is Illuminant D65.

XYZ Color Coordinate System. In the CIE spectral primary system, the tristimulus values required to achieve a color match are sometimes negative. The CIE has developed a standard artificial primary coordinate system in which all tristimulus values required to match colors are positive (4). These artificial primaries are shown in the CIE primary chromaticity diagram of Figure 3.5-3 (11). The *XYZ* system primaries have been chosen so that the *Y* tristimulus value is equivalent to the luminance of the color to be matched. Figure 3.5-5 is the chromaticity diagram for the CIE *XYZ* primary system referenced to equal-energy white (4). The linear transformations between $R_C G_C B_C$ and *XYZ* are given by

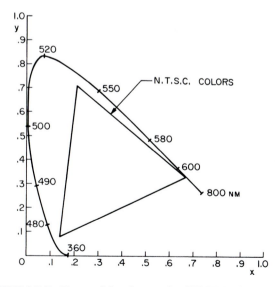

FIGURE 3.5-5. Chromaticity diagram for CIE *XYZ* primary system.

$$
\begin{bmatrix} X \\ Y \\ Z \end{bmatrix} = \begin{bmatrix} 0.49018626 & 0.30987954 & 0.19993420 \\ 0.17701522 & 0.81232418 & 0.01066060 \\ 0.00000000 & 0.01007720 & 0.98992280 \end{bmatrix} \begin{bmatrix} R_C \\ G_C \\ B_C \end{bmatrix} \qquad (3.5\text{-}1a)
$$

$$
\begin{bmatrix} R_C \\ G_C \\ B_C \end{bmatrix} = \begin{bmatrix} 2.36353918 & -0.89582361 & -0.46771557 \\ -0.51511248 & 1.42643694 & 0.08867553 \\ 0.00524373 & -0.01452082 & 1.00927709 \end{bmatrix} \begin{bmatrix} X \\ Y \\ Z \end{bmatrix} \qquad (3.5\text{-}1b)
$$

The color conversion matrices of Eq. 3.5-1 and those color conversion matrices defined later are quoted to eight decimal places (21,22). In many instances, this quotation is to a greater number of places than the original specification. The number of places has been increased to reduce computational errors when concatenating transformations between color representations.

The color conversion matrix between XYZ and any other linear RGB color space can be computed by the following algorithm.

1. Compute the colorimetric weighting coefficients $a(1)$, $a(2)$, $a(3)$ from

$$
\begin{bmatrix} a(1) \\ a(2) \\ a(3) \end{bmatrix} = \begin{bmatrix} x_R & x_G & x_B \\ y_R & y_G & y_B \\ z_R & z_G & z_B \end{bmatrix}^{-1} \begin{bmatrix} x_W/y_W \\ 1 \\ z_W/y_W \end{bmatrix} \qquad (3.5\text{-}2a)
$$

where x_k, y_k, z_k are the chromaticity coordinates of the RGB primary set.

2. Compute the RGB-to-XYZ conversion matrix.

$$
\begin{bmatrix} M(1,1) & M(1,2) & M(1,3) \\ M(2,1) & M(2,2) & M(2,3) \\ M(3,1) & M(3,2) & M(3,3) \end{bmatrix} = \begin{bmatrix} x_R & x_G & x_B \\ y_R & y_G & y_B \\ z_R & z_G & z_B \end{bmatrix} \begin{bmatrix} a(1) & 0 & 0 \\ 0 & a(2) & 0 \\ 0 & 0 & a(3) \end{bmatrix} \qquad (3.5\text{-}2b)
$$

The XYZ-to-RGB conversion matrix is, of course, the matrix inverse of **M**. Table 3.5-1 lists the XYZ tristimulus values of several standard illuminants. The XYZ chromaticity coordinates of the standard linear RGB color systems are presented in Table 3.5-2.

From Eqs. 3.5-1 and 3.5-2 it is possible to derive a matrix transformation between $R_C G_C B_C$ and any linear colorimetric RGB color space. The book CD contains a file that lists the transformation matrices (22) between the standard RGB color coordinate systems and XYZ and UVW, defined below.

TABLE 3.5-1. *XYZ* **Tristimulus Values of Standard Illuminants**

Illuminant	X_0	Y_0	Z_0
A	1.098700	1.000000	0.355900
C	0.980708	1.000000	1.182163
D50	0.964296	1.000000	0.825105
D65	0.950456	1.000000	1.089058
E	1.000000	1.000000	1.000000

TABLE 3.5-2. *XYZ* **Chromaticity Coordinates of Standard Primaries**

Standard		x	y	z
CIE	R_C	0.640000	0.330000	0.030000
	G_C	0.300000	0.600000	0.100000
	B_C	0.150000	0.06000	0.790000
NTSC	R_N	0.670000	0.330000	0.000000
	G_N	0.210000	0.710000	0.080000
	B_N	0.140000	0.080000	0.780000
SMPTE	R_S	0.630000	0.340000	0.030000
	G_S	0.310000	0.595000	0.095000
	B_S	0.155000	0.070000	0.775000
EBU	R_E	0.640000	0.330000	0.030000
	G_E	0.290000	0.60000	0.110000
	B_E	0.150000	0.060000	0.790000
CCIR	R_R	0.640000	0.330000	0.030000
	G_R	0.30000	0.600000	0.100000
	B_R	0.150000	0.060000	0.790000

UVW Uniform Chromaticity Scale Color Coordinate System. In 1960, the CIE. adopted a coordinate system, called the *Uniform Chromaticity Scale* (UCS), in which, to a good approximation, equal changes in the chromaticity coordinates result in equal, just noticeable changes in the perceived hue and saturation of a color. The V component of the UCS coordinate system represents luminance. The u, v chromaticity coordinates are related to the x, y chromaticity coordinates by the relations (23)

$$u = \frac{4x}{-2x + 12y + 3} \tag{3.5-3a}$$

$$v = \frac{6y}{-2x + 12y + 3} \tag{3.5-3b}$$

$$x = \frac{3u}{2u - 8v - 4} \tag{3.5-3c}$$

$$y = \frac{2v}{2u - 8v - 4} \tag{3.5-3d}$$

Figure 3.5-6 is a UCS chromaticity diagram.

The tristimulus values of the uniform chromaticity scale coordinate system *UVW* are related to the tristimulus values of the spectral coordinate primary system by

$$\begin{bmatrix} U \\ V \\ W \end{bmatrix} = \begin{bmatrix} 0.32679084 & 0.20658636 & 0.13328947 \\ 0.17701522 & 0.81232418 & 0.01066060 \\ 0.02042971 & 1.06858510 & 0.41098519 \end{bmatrix} \begin{bmatrix} R_C \\ G_C \\ B_C \end{bmatrix} \tag{3.5-4a}$$

$$\begin{bmatrix} R_C \\ G_C \\ B_C \end{bmatrix} = \begin{bmatrix} 2.84373542 & 0.50732308 & -0.93543113 \\ -0.63965541 & 1.16041034 & 0.17735107 \\ 1.52178123 & -3.04235208 & 2.01855417 \end{bmatrix} \begin{bmatrix} U \\ V \\ W \end{bmatrix} \tag{3.5-4b}$$

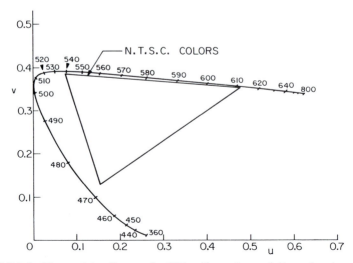

FIGURE 3.5-6. Chromaticity diagram for CIE uniform chromaticity scale primary system.

U*V*W* Color Coordinate System. The $U^*V^*W^*$ color coordinate system, adopted by the CIE in 1964, is an extension of the UVW coordinate system in an attempt to obtain a color solid for which unit shifts in luminance and chrominance are uniformly perceptible. The $U^*V^*W^*$ coordinates are defined as (24)

$$U^* = 13W^*(u - u_o) \tag{3.5-5a}$$

$$V^* = 13W^*(v - v_o) \tag{3.5-5b}$$

$$W^* = 25(100Y)^{1/3} - 17 \tag{3.5-5c}$$

where the luminance Y is measured over a scale of 0.0 to 1.0 and u_o and v_o are the chromaticity coordinates of the reference illuminant.

The UVW and $U^*V^*W^*$ coordinate systems were rendered obsolete in 1976 by the introduction by the CIE of the more accurate $L^*a^*b^*$ and $L^*u^*v^*$ color coordinate systems. Although depreciated by the CIE, much valuable data has been collected in the UVW and $U^*V^*W^*$ color systems.

L*a*b* Color Coordinate System. The $L^*a^*b^*$ cube root color coordinate system was developed to provide a computationally simple measure of color in agreement with the Munsell color system (25). The color coordinates are

$$L^* = \begin{cases} 116\left(\dfrac{Y}{Y_o}\right)^{1/3} - 16 & \text{for } \dfrac{Y}{Y_o} > 0.008856 \qquad (3.5\text{-}6a) \\[4mm] 903.3\,\dfrac{Y}{Y_o} & \text{for } 0.0 \le \dfrac{Y}{Y_o} \le 0.008856 \quad (3.5\text{-}6b) \end{cases}$$

$$a^* = 500\left[f\left\{\frac{X}{X_o}\right\} - f\left\{\frac{Y}{Y_o}\right\}\right] \tag{3.5-6c}$$

$$b^* = 200\left[f\left\{\frac{X}{X_o}\right\} - f\left\{\frac{Z}{Z_o}\right\}\right] \tag{3.5-6d}$$

where

$$f(w) = \begin{cases} w^{1/3} & \text{for } w > 0.008856 \qquad\qquad (3.6\text{-}6e) \\[3mm] 7.787(w) + 0.1379 & \text{for } 0.0 \le w \le 0.008856 \quad (3.6\text{-}6f) \end{cases}$$

The terms X_o, Y_o, Z_o are the tristimulus values for the reference white. Basically, L^* is correlated with brightness, a^* with redness-greenness, and b^* with yellowness-blueness. The inverse relationship between $L^*a^*b^*$ and XYZ is

$$X = X_o \left[g \left\{ \frac{L^* + 16}{25} \right\} \right] \tag{3.5-7a}$$

$$Y = Y_o \left[g \left\{ f \left\{ \frac{Y}{Y_o} \right\} + \frac{a^*}{500} \right\} \right] \tag{3.5-7b}$$

$$Z = Z_o \left[g \left\{ f \left\{ \frac{Y}{Y_o} \right\} - \frac{b^*}{200} \right\} \right] \tag{3.5-7c}$$

where

$$g(w) = \begin{cases} w^3 & \text{for } w > 0.20681 \tag{3.6-7d} \\ \\ 0.1284(w - 0.1379) & \text{if } 0.0 \le w \le 0.20689 \tag{3.6-7e} \end{cases}$$

L*u*v* Color Coordinate System. The $L^*u^*v^*$ coordinate system (26), which has evolved from the $L^*a^*b^*$ and the $U^*V^*W^*$ coordinate systems, became a CIE standard in 1976. It is defined as

$$L^* = \begin{cases} 25 \left(100 \frac{Y}{Y_o} \right)^{1/3} - 16 & \text{for } \frac{Y}{Y_o} \ge 0.008856 \tag{3.5-8a} \\ \\ 903.3 \frac{Y}{Y_o} & \text{for } \frac{Y}{Y_o} < 0.008856 \tag{3.5-8b} \end{cases}$$

$$u^* = 13L^*(u' - u'_o) \tag{3.5-8c}$$

$$v^* = 13L^*(v' - v'_o) \tag{3.5-8d}$$

where

$$u' = \frac{4X}{X + 15Y + 3Z} \tag{3.5-8e}$$

$$v' = \frac{9Y}{X + 15Y + 3Z} \tag{3.5-8f}$$

and u'_o and v'_o are obtained by substitution of the tristimulus values X_o, Y_o, Z_o for the reference white. The inverse relationship is given by

$$X = \frac{9u'}{4v'}Y \tag{3.5-9a}$$

$$Y = Y_o \left(\frac{L^* + 16}{25}\right)^3 \tag{3.5-9b}$$

$$Z = Y\frac{12 - 3u' - 20v'}{4v'} \tag{3.5-9c}$$

where

$$u' = \frac{u^*}{13L^*} + u'_o \tag{3.5-9d}$$

$$v' = \frac{v^*}{13L^*} + u'_o \tag{3.5-9e}$$

Figure 3.5-7 shows the linear *RGB* components of an NTSC receiver primary color image. This color image is printed in the color insert. If printed properly, the color image and its monochromatic component images will appear to be of "normal" brightness. When displayed electronically, the linear images will appear too dark. Section 3.5.3 discusses the proper display of electronic images. Figures 3.5-8 to 3.5-10 show the *XYZ*, *Yxy*, and *L*a*b** components of Figure 3.5-7. Section 10.1.1 describes amplitude-scaling methods for the display of image components outside the unit amplitude range. The amplitude range of each component is printed below each photograph.

3.5.2. Subtractive Color Spaces

The color printing and color photographic processes (see Section 11-3) are based on a subtractive color representation. In color printing, the linear *RGB* color components are transformed to cyan (*C*), magenta (*M*), and yellow (*Y*) inks, which are overlaid at each pixel on a, usually, white paper. The simplest transformation relationship is

$$C = 1.0 - R \tag{3.5-10a}$$

$$M = 1.0 - G \tag{3.5-10b}$$

$$Y = 1.0 - B \tag{3.5-10c}$$

(*a*) Linear *R*, 0.000 to 0.965

(*b*) Linear *G*, 0.000 to 1.000   (*c*) Linear *B*, 0.000 to 0.965

FIGURE 3.5-7. Linear *RGB* components of the `dolls_linear` color image. See insert for a color representation of this figure.

where the linear *RGB* components are tristimulus values over [0.0, 1.0]. The inverse relations are

$$R = 1.0 - C \tag{3.5-11a}$$

$$G = 1.0 - M \tag{3.5-11b}$$

$$B = 1.0 - Y \tag{3.5-11c}$$

In high-quality printing systems, the *RGB*-to-*CMY* transformations, which are usually proprietary, involve color component cross-coupling and point nonlinearities.

(*a*) *X*, 0.000 to 0.952

(*b*) *Y*, 0.000 to 0.985

(*c*) *Z*, 0.000 to 1,143

FIGURE 3.5-8. *XYZ* components of the `dolls_linear` color image.

To achieve dark black printing without using excessive amounts of *CMY* inks, it is common to add a fourth component, a black ink, called the *key (K)* or *black component*. The black component is set proportional to the smallest of the *CMY* components as computed by Eq. 3.5-10. The common *RGB*-to-*CMYK* transformation, which is based on the *undercolor removal algorithm* (27), is

$$C = 1.0 - R - uK_b \tag{3.5-12a}$$

$$M = 1.0 - G - uK_b \tag{3.5-12b}$$

$$Y = 1.0 - B - uK_b \tag{3.5-12c}$$

$$K = bK_b \tag{3.5-12d}$$

(a) Y, 0.000 to 0.965

(b) x, 0.140 to 0.670   (c) y, 0.080 to 0.710

FIGURE 3.5-9. *Yxy* components of the `dolls_linear` color image.

where

$$K_b = \text{MIN}\{1.0 - R, 1.0 - G, 1.0 - B\} \tag{3.5-12e}$$

and $0.0 \leq u \leq 1.0$ is the undercolor removal factor and $0.0 \leq b \leq 1.0$ is the blackness factor. Figure 3.5-11 presents the *CMY* components of the color image of Figure 3.5-7.

3.5.3 Video Color Spaces

The red, green, and blue signals from video camera sensors typically are linearly proportional to the light striking each sensor. However, the light generated by cathode tube displays is approximately proportional to the display amplitude drive signals

(*a*) *L**, −16.000 to 99.434

(*b*) *a**, −55.928 to 69.291   (*c*) *b**, −65.224 to 90.171

FIGURE 3.5-10. *L*a*b** components of the `dolls_linear` color image.

raised to a power in the range 2.0 to 3.0 (28). To obtain a good-quality display, it is necessary to compensate for this point nonlinearity. The compensation process, called *gamma correction*, involves passing the camera sensor signals through a point nonlinearity with a power, typically, of about 0.45. In television systems, to reduce receiver cost, gamma correction is performed at the television camera rather than at the receiver. A linear *RGB* image that has been gamma corrected is called a *gamma RGB image*. Liquid crystal displays are reasonably linear in the sense that the light generated is approximately proportional to the display amplitude drive signal. But because LCDs are used in lieu of CRTs in many applications, they usually employ circuitry to compensate for the gamma correction at the sensor.

(*a*) *C*, 0.0035 to 1.000

(*b*) *M*, 0.000 to 1.000   (*c*) *Y*, 0.0035 to 1.000

FIGURE 3.5-11. *CMY* components of the `dolls_linear` color image.

In high-precision applications, gamma correction follows a linear law for low-amplitude components and a power law for high-amplitude components according to the relations (22)

$$
\tilde{K} = \begin{cases} c_1 K^{c_2} + c_3 & \text{for } K \geq b \qquad (3.5\text{-}13\text{a}) \\[2em] c_4 K & \text{for } 0.0 \leq K < b \qquad (3.5\text{-}13\text{b}) \end{cases}
$$

where K denotes a linear RGB component and $\tilde{K}$ is the gamma-corrected component. The constants c_k and the breakpoint b are specified in Table 3.5-3 for the general case and for conversion to the SMPTE, CCIR and CIE lightness components. Figure 3.5-12 is a plot of the gamma correction curve for the CCIR Rec. 709 primaries.

TABLE 3.5-3. Gamma Correction Constants

	General	SMPTE	CCIR	CIE L*
c_1	1.00	1.1115	1.099	116.0
c_2	0.45	0.45	0.45	0.3333
c_3	0.00	-0.1115	-0.099	-16.0
c_4	0.00	4.0	4.5	903.3
b	0.00	0.0228	0.018	0.008856

The inverse gamma correction relation is

$$k = \begin{cases} \left[\left(\dfrac{\tilde{K} - c_3}{c_1} \right)^{1/c_2} \right] & \text{for } \tilde{K} \geq c_4 b \qquad (3.5\text{-}14a) \\[2em] \dfrac{\tilde{K}}{c_4} & \text{for } 0.0 \leq \tilde{K} < c_4 b \qquad (3.5\text{-}14b) \end{cases}$$

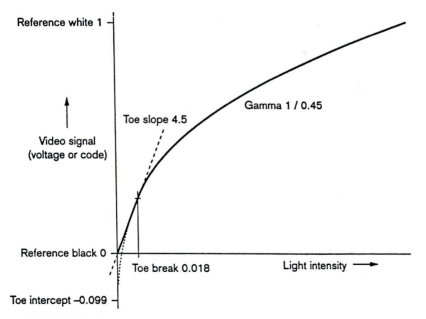

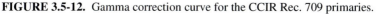

FIGURE 3.5-12. Gamma correction curve for the CCIR Rec. 709 primaries.

(*a*) Gamma *R*, 0.000 to 0.984

(*b*) Gamma *G*, 0.000 to 1.000   (*c*) Gamma *B*, 0.000 to 0.984

FIGURE 3.5-13. Gamma *RGB* components of the `dolls_gamma` color image. See insert for a color representation of this figure.

Figure 3.5-13 shows the gamma *RGB* components of the color image of Figure 3.5-7. The gamma color image is printed in the color insert. The gamma components have been printed as if they were linear components to illustrate the effects of the point transformation. When viewed on an electronic display, the gamma *RGB* color image will appear to be of "normal" brightness.

YIQ NTSC Transmission Color Coordinate System. In the development of the color television system in the United States, NTSC formulated a color coordinate system for transmission composed of three values, *Y, I, Q* (14). The *Y* value, called *luma*, is proportional to the gamma-corrected luminance of a color. The other two components, *I* and *Q*, called *chroma*, jointly describe the hue and saturation attributes of an image. The reasons for transmitting the *YIQ* components rather than the gamma-corrected $\tilde{R}_N \tilde{G}_N \tilde{B}_N$ components directly from a color camera were two fold: The *Y* signal alone could be used with existing monochrome receivers to display monochrome images; and it was found possible to limit the spatial bandwidth of the *I* and *Q* signals without noticeable image degradation. As a result of the latter property, a clever analog modulation scheme was developed such that the bandwidth of a color television carrier could be restricted to the same bandwidth as a monochrome carrier.

The *YIQ* transformations for an Illuminant C reference white are given by

$$
\begin{bmatrix} Y \\ I \\ Q \end{bmatrix} = \begin{bmatrix} 0.29889531 & 0.58662247 & 0.11448223 \\ 0.59597799 & -0.27417610 & -0.32180189 \\ 0.21147017 & -0.52261711 & 0.31114694 \end{bmatrix} \begin{bmatrix} \tilde{R}_N \\ \tilde{G}_N \\ \tilde{B}_N \end{bmatrix} \tag{3.5-15a}
$$

$$
\begin{bmatrix} \tilde{R}_N \\ \tilde{G}_N \\ \tilde{B}_N \end{bmatrix} = \begin{bmatrix} 1.00000000 & 0.95608445 & 0.62088850 \\ 1.00000000 & -0.27137664 & -0.64860590 \\ 1.00000000 & -1.10561724 & 1.70250126 \end{bmatrix} \begin{bmatrix} Y \\ I \\ Q \end{bmatrix} \tag{3.5-15b}
$$

where the tilde denotes that the component has been gamma corrected.

Figure 3.5-14 presents the *YIQ* components of the gamma color image of Figure 3.5-13.

YUV EBU Transmission Color Coordinate System. In the PAL and SECAM color television systems (29) used in many countries, the luma *Y* and two color differences,

$$
U = \frac{\tilde{B}_E - Y}{2.03} \tag{3.5-16a}
$$

$$
V = \frac{\tilde{R}_E - Y}{1.14} \tag{3.5-16b}
$$

are used as transmission coordinates, where $\tilde{R}_E$ and $\tilde{B}_E$ are the gamma-corrected EBU red and blue components, respectively. The *YUV* coordinate system was initially proposed as the NTSC transmission standard but was later replaced by the *YIQ* system because it was found (4) that the *I* and *Q* signals could be reduced in bandwidth to a greater degree than the *U* and *V* signals for an equal level of visual quality. The *I* and *Q* signals are related to the *U* and *V* signals by a simple rotation of coordinates in color space:

(a) Y, 0.000 to 0.994

(b) I, −0.276 to 0.347   *(c) Q,* = 0.147 to 0.169

FIGURE 3.5-14. *YIQ* components of the gamma corrected `dolls_gamma` color image.

$$I = -U\sin 33° + V\cos 33° \qquad (3.5\text{-}17a)$$

$$Q = U\cos 33° + V\sin 33° \qquad (3.5\text{-}17b)$$

It should be noted that the *U* and *V* components of the *YUV* video color space are not equivalent to the *U* and V components of the UVW uniform chromaticity system.

YCbCr CCIR Rec. 601 Transmission Color Coordinate System. The CCIR Rec. 601 color coordinate system *YCbCr* is defined for the transmission of luma and chroma components coded in the integer range 0 to 255. The *YCbCr* transformations for unit range components are defined as (28)

$$\begin{bmatrix} Y \\ Cb \\ Cr \end{bmatrix} = \begin{bmatrix} 0.29900000 & 0.58700000 & 0.11400000 \\ -0.16873600 & -0.33126400 & 0.50000000 \\ 0.50000000 & -0.4186680 & -0.08131200 \end{bmatrix} \begin{bmatrix} \tilde{R}_S \\ \tilde{G}_S \\ \tilde{B}_S \end{bmatrix} \qquad \text{(3.5-18a)}$$

$$\begin{bmatrix} \tilde{R}_S \\ \tilde{G}_S \\ \tilde{B}_S \end{bmatrix} = \begin{bmatrix} 1.00000000 & -0.0009264 & 1.40168676 \\ 1.00000000 & -0.34369538 & -0.71416904 \\ 1.00000000 & 1.77216042 & 0.00099022 \end{bmatrix} \begin{bmatrix} Y \\ Cb \\ Cr \end{bmatrix} \qquad \text{(3.5-18b)}$$

where the tilde denotes that the component has been gamma corrected.

Photo YCC Color Coordinate System. Eastman Kodak company has developed an image storage system, called *PhotoCD*, in which a photographic negative is scanned, converted to a luma/chroma format similar to Rec. 601*YCbCr*, and recorded in a proprietary compressed form on a compact disk. The PhotoYCC format and its associated RGB display format have become defacto standards. PhotoYCC employs the CCIR Rec. 709 primaries for scanning. The conversion to YCC is defined as (27,28,30)

$$\begin{bmatrix} Y \\ C_1 \\ C_2 \end{bmatrix} = \begin{bmatrix} 0.299 & 0.587 & 0.114 \\ -0.299 & -0.587 & 0.500 \\ 0.500 & -0.587 & 0.114 \end{bmatrix} \begin{bmatrix} \tilde{R}_{709} \\ \tilde{G}_{709} \\ \tilde{B}_{709} \end{bmatrix} \qquad \text{(3.5-19a)}$$

Transformation from PhotoCD components for display is not an exact inverse of Eq. 3.5-19a, in order to preserve the extended dynamic range of film images. The YC_1C_2-to-$R_DG_DB_D$ display components is given by

$$\begin{bmatrix} R_D \\ G_D \\ B_D \end{bmatrix} = \begin{bmatrix} 0.969 & 0.000 & 1.000 \\ 0.969 & -0.194 & -0.509 \\ 0.969 & 1.000 & 0.000 \end{bmatrix} \begin{bmatrix} Y \\ C_1 \\ C_2 \end{bmatrix} \qquad \text{(3.5-19b)}$$

3.5.4. Nonstandard Color Spaces

Several nonstandard color spaces used for image processing applications are described in this section.

IHS Color Coordinate System. The IHS coordinate system (31) has been used within the image processing community as a quantitative means of specifying the intensity, hue, and saturation of a color. It is defined by the relations

$$
\begin{bmatrix} I \\ V_1 \\ V_2 \end{bmatrix} = \begin{bmatrix} \dfrac{1}{3} & \dfrac{1}{3} & \dfrac{1}{3} \\ \dfrac{-1}{\sqrt{6}} & \dfrac{-1}{\sqrt{6}} & \dfrac{2}{\sqrt{6}} \\ \dfrac{1}{\sqrt{6}} & \dfrac{-1}{\sqrt{6}} & 0 \end{bmatrix} \begin{bmatrix} R \\ G \\ B \end{bmatrix} \tag{3.5-20a}
$$

$$
H = \arctan\left\{ \frac{V_2}{V_1} \right\} \tag{3.5-20b}
$$

$$
S = (V_1^2 + V_2^2)^{1/2} \tag{3.5-20c}
$$

By this definition, the color blue is the zero reference for hue. The inverse relationship is

$$
V_1 = S\cos\{H\} \tag{3.5-21a}
$$

$$
V_2 = S\sin\{H\} \tag{3.5-21b}
$$

$$
\begin{bmatrix} R \\ G \\ B \end{bmatrix} = \begin{bmatrix} 1 & \dfrac{-\sqrt{6}}{6} & \dfrac{\sqrt{6}}{2} \\ 1 & \dfrac{\sqrt{6}}{6} & \dfrac{-\sqrt{6}}{2} \\ 1 & \dfrac{\sqrt{6}}{3} & 0 \end{bmatrix} \begin{bmatrix} I \\ V_1 \\ V_2 \end{bmatrix} \tag{3.5-21c}
$$

Figure 3.5-15 shows the *IHS* components of the gamma *RGB* image of Figure 3.5-13.

Karhunen–Loeve Color Coordinate System. Typically, the *R*, *G*, and *B* tristimulus values of a color image are highly correlated with one another (32). In the development of efficient quantization, coding, and processing techniques for color images, it is often desirable to work with components that are uncorrelated. If the second-order moments of the *RGB* tristimulus values are known, or at least estimable, it is

(*a*) *I*, 0.000 to 0.989

(*b*) *H*, −3.136 to 3.142   (*c*) *S*, 0.000 to 0.476

FIGURE 3.5-15. *IHS* components of the `dolls_gamma` color image.

possible to derive an orthogonal coordinate system, in which the components are uncorrelated, by a Karhunen–Loeve (K–L) transformation of the *RGB* tristimulus values. The K-L color transform is defined as

$$
\begin{bmatrix} K_1 \\ K_2 \\ K_3 \end{bmatrix} = \begin{bmatrix} m_{11} & m_{12} & m_{13} \\ m_{21} & m_{22} & m_{23} \\ m_{31} & m_{32} & m_{33} \end{bmatrix} \begin{bmatrix} R \\ G \\ B \end{bmatrix}
\tag{3.5-22a}
$$

$$
\begin{bmatrix} R \\ G \\ B \end{bmatrix} = \begin{bmatrix} m_{11} & m_{21} & m_{31} \\ m_{12} & m_{22} & m_{32} \\ m_{13} & m_{23} & m_{33} \end{bmatrix} \begin{bmatrix} K_1 \\ K_2 \\ K_3 \end{bmatrix} \tag{3.5-22b}
$$

where the transformation matrix with general term m_{ij} composed of the eigenvectors of the *RGB* covariance matrix with general term u_{ij}. The transformation matrix satisfies the relation

$$
\begin{bmatrix} m_{11} & m_{12} & m_{13} \\ m_{21} & m_{22} & m_{23} \\ m_{31} & m_{32} & m_{33} \end{bmatrix} \begin{bmatrix} u_{11} & u_{12} & u_{13} \\ u_{12} & u_{22} & u_{23} \\ u_{13} & u_{23} & u_{33} \end{bmatrix} \begin{bmatrix} m_{11} & m_{21} & m_{31} \\ m_{12} & m_{22} & m_{32} \\ m_{13} & m_{23} & m_{33} \end{bmatrix} = \begin{bmatrix} \lambda_1 & 0 & 0 \\ 0 & \lambda_2 & 0 \\ 0 & 0 & \lambda_3 \end{bmatrix}
$$

$$\tag{3.5-23}$$

where λ_1, λ_2, λ_3 are the eigenvalues of the covariance matrix and

$$
u_{11} = E\{(R - \bar{R})^2\} \tag{3.5-24a}
$$

$$
u_{22} = E\{(G - \bar{G})^2\} \tag{3.5-24b}
$$

$$
u_{33} = E\{(B - \bar{B})^2\} \tag{3.5-24c}
$$

$$
u_{12} = E\{(R - \bar{R})(G - \bar{G})\} \tag{3.5-24d}
$$

$$
u_{13} = E\{(R - \bar{R})(B - \bar{B})\} \tag{3.5-24e}
$$

$$
u_{23} = E\{(G - \bar{G})(B - \bar{B})\} \tag{3.5-24f}
$$

In Eq. 3.5-23, $E\{\cdot\}$ is the expectation operator and the overbar denotes the mean value of a random variable.

Retinal Cone Color Coordinate System. As indicated in Chapter 2, in the discussion of models of the human visual system for color vision, indirect measurements of the spectral sensitivities $s_1(\lambda)$, $s_2(\lambda)$, $s_3(\lambda)$ have been made for the three types of retinal cones. It has been found that these spectral sensitivity functions can be linearly related to spectral tristimulus values established by colorimetric experimentation. Hence a set of cone signals T_1, T_2, T_3 may be regarded as tristimulus values in a retinal cone color coordinate system. The tristimulus values of the retinal cone color coordinate system are related to the *XYZ* system by the coordinate conversion matrix (33)

$$\begin{bmatrix} T_1 \\ T_2 \\ T_3 \end{bmatrix} = \begin{bmatrix} 0.000000 & 1.000000 & 0.000000 \\ -0.460000 & 1.359000 & 0.101000 \\ 0.000000 & 0.000000 & 1.000000 \end{bmatrix} \begin{bmatrix} X \\ Y \\ Z \end{bmatrix} \qquad (3.5\text{-}25)$$

REFERENCES

1. T. P. Merrit and F. F. Hall, Jr., "Blackbody Radiation," *Proc. IRE*, **47**, 9, September 1959, 1435–1442.

2. H. H. Malitson, "The Solar Energy Spectrum," *Sky and Telescope,* **29**, 4, March 1965, 162–165.

3. R. D. Larabee, "Spectral Emissivity of Tungsten," *J. Optical of Society America*, **49**, 6, June 1959, 619–625.

4. *The Science of Color*, Crowell, New York, 1973.

5. D. G. Fink, Ed., *Television Engineering Handbook*, McGraw-Hill, New York, 1957.

6. Toray Industries, Inc. *LCD* Color Filter Specification.

7. J. W. T. Walsh, *Photometry*, Constable, London, 1953.

8. M. Born and E. Wolf, *Principles of Optics*, 6th ed., Pergamon Press, New York, 1981.

9. K. S. Weaver, "The Visibility of Radiation at Low Intensities," *J. Optical Society of America*, **27**, 1, January 1937, 39–43.

10. G. Wyszecki and W. S. Stiles, *Color Science*, 2nd ed., Wiley, New York, 1982.

11. R. W. G. Hunt, *The Reproduction of Colour*, 5th ed., Wiley, New York, 1957.

12. W. D. Wright, *The Measurement of Color*, Adam Hilger, London, 1944, 204–205.

13. R. A. Enyord, Ed., *Color: Theory and Imaging Systems*, Society of Photographic Scientists and Engineers, Washington, DC, 1973.

14. F. J. Bingley, "Color Vision and Colorimetry," *in Television Engineering Handbook*, D. G. Fink, ed., McGraw–Hill, New York, 1957.

15. H. Grassman, "On the Theory of Compound Colours," *Philosophical Magazine*, Ser. 4, **7**, April 1854, 254–264.

16. W. T. Wintringham, "Color Television and Colorimetry," *Proc. IRE*, **39**, 10, October 1951, 1135–1172.

17. "EBU Standard for Chromaticity Tolerances for Studio Monitors," Technical Report 3213-E, European Broadcast Union, Brussels, 1975.

18. "Encoding Parameters of Digital Television for Studios", Recommendation ITU-R BT.601-4, (International Telecommunications Union, Geneva; 1990).

19 "Basic Parameter Values for the HDTV Standard for the Studio and for International Programme Exchange," Recommendation ITU-R BT 709, International Telecommunications Unions, Geneva; 1990.

20. L. E. DeMarsh, "Colorimetric Standards in U.S. Color Television. A Report to the Subcommittee on Systems Colorimetry of the SMPTE Television Committee," *J. Society of Motion Picture and Television Engineers*, **83**, 1974.

21. "Information Technology, Computer Graphics and Image Processing, Image Processing and Interchange, Part 1: Common Architecture for Imaging," ISO/IEC 12087-1:1995(E).

22. "Information Technology, Computer Graphics and Image Processing, Image Processing and Interchange, Part 2: Programmer's Imaging Kernel System Application Program Interface," ISO/IEC 12087-2:1995(E).

23. D. L. MacAdam, "Projective Transformations of ICI Color Specifications," *J. Optical Society of America*, **27**, 8, August 1937, 294–299.

24. G. Wyszecki, "Proposal for a New Color-Difference Formula," *J. Optical Society of America*, **53**, 11, November 1963, 1318–1319.

25. "CIE Colorimetry Committee Proposal for Study of Color Spaces," Technical, Note, *J. Optical Society of America*, **64**, 6, June 1974, 896–897.

26. *Colorimetry*, 2nd ed., Publication 15.2, Central Bureau, Commission Internationale de l'Eclairage, Vienna, 1986.

27. W. K. Pratt, *Developing Visual Applications, XIL: An Imaging Foundation Library*, Sun Microsystems Press, Mountain View, CA, 1997.

28. C. A. Poynton, *A Technical Introduction to Digital Video*, Wiley, New York, 1996.

29. P. S. Carnt and G. B. Townsend, *Color Television* Vol. 2; *PAL, SECAM, and Other Systems*, Iliffe, London, 1969.

30. I. Kabir, *High Performance Computer Imaging*, Manning Publications, Greenwich, CT, 1996.

31. W. Niblack, *An Introduction to Digital Image Processing*, Prentice Hall, Englewood Cliffs, NJ, 1985.

32. W. K. Pratt, "Spatial Transform Coding of Color Images," *IEEE Trans. Communication Technology*, **COM-19**, 12, December 1971, 980–992.

33. D. B. Judd, "Standard Response Functions for Protanopic and Deuteranopic Vision," *J. Optical Society of America*, **35**, 3, March 1945, 199–221.

PART 2

DIGITAL IMAGE CHARACTERIZATION

Digital image processing is based on the conversion of a continuous image field to equivalent digital form. This part of the book considers the image sampling and quantization processes that perform the analog image to digital image conversion. The inverse operation of producing continuous image displays from digital image arrays is also analyzed. Vector-space methods of image representation are developed for deterministic and stochastic image arrays.

4

IMAGE SAMPLING AND RECONSTRUCTION

In digital image processing systems, one usually deals with arrays of numbers obtained by spatially sampling points of a physical image. After processing, another array of numbers is produced, and these numbers are then used to reconstruct a continuous image for viewing. Image samples nominally represent some physical measurements of a continuous image field, for example, measurements of the image intensity or photographic density. Measurement uncertainties exist in any physical measurement apparatus. It is important to be able to model these measurement errors in order to specify the validity of the measurements and to design processes for compensation of the measurement errors. Also, it is often not possible to measure an image field directly. Instead, measurements are made of some function related to the desired image field, and this function is then inverted to obtain the desired image field. Inversion operations of this nature are discussed in the sections on image restoration. In this chapter the image sampling and reconstruction process is considered for both theoretically exact and practical systems.

4.1. IMAGE SAMPLING AND RECONSTRUCTION CONCEPTS

In the design and analysis of image sampling and reconstruction systems, input images are usually regarded as deterministic fields (1–5). However, in some situations it is advantageous to consider the input to an image processing system, especially a noise input, as a sample of a two-dimensional random process (5–7). Both viewpoints are developed here for the analysis of image sampling and reconstruction methods.

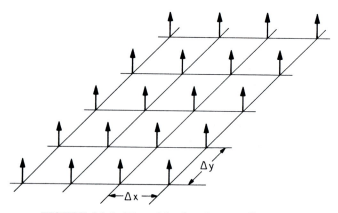

FIGURE 4.1-1. Dirac delta function sampling array.

4.1.1. Sampling Deterministic Fields

Let $F_I(x, y)$ denote a continuous, infinite-extent, ideal image field representing the luminance, photographic density, or some desired parameter of a physical image. In a perfect image sampling system, spatial samples of the ideal image would, in effect, be obtained by multiplying the ideal image by a spatial sampling function

$$S(x, y) = \sum_{j = -\infty}^{\infty} \sum_{k = -\infty}^{\infty} \delta(x - j\,\Delta x, y - k\,\Delta y) \qquad (4.1\text{-}1)$$

composed of an infinite array of Dirac delta functions arranged in a grid of spacing $(\Delta x, \Delta y)$ as shown in Figure 4.1-1. The sampled image is then represented as

$$F_P(x, y) = F_I(x, y)S(x, y) = \sum_{j = -\infty}^{\infty} \sum_{k = -\infty}^{\infty} F_I(j\,\Delta x, k\,\Delta y)\delta(x - j\,\Delta x, y - k\,\Delta y) \qquad (4.1\text{-}2)$$

where it is observed that $F_I(x, y)$ may be brought inside the summation and evaluated only at the sample points $(j\,\Delta x, k\,\Delta y)$. It is convenient, for purposes of analysis, to consider the spatial frequency domain representation $\mathcal{F}_P(\omega_x, \omega_y)$ of the sampled image obtained by taking the continuous two-dimensional Fourier transform of the sampled image. Thus

$$\mathcal{F}_P(\omega_x, \omega_y) = \int_{-\infty}^{\infty}\int_{-\infty}^{\infty} F_P(x, y)\exp\{-i(\omega_x x + \omega_y y)\}\,dx\,dy \qquad (4.1\text{-}3)$$

By the Fourier transform convolution theorem, the Fourier transform of the sampled image can be expressed as the convolution of the Fourier transforms of the ideal image $\mathcal{F}_I(\omega_x, \omega_y)$ and the sampling function $S(\omega_x, \omega_y)$ as expressed by

$$\mathcal{F}_P(\omega_x, \omega_y) = \frac{1}{4\pi^2} \mathcal{F}_I(\omega_x, \omega_y) \circledast S(\omega_x, \omega_y) \qquad (4.1\text{-}4)$$

The two-dimensional Fourier transform of the spatial sampling function is an infinite array of Dirac delta functions in the spatial frequency domain as given by (4, p. 22)

$$S(\omega_x, \omega_y) = \frac{4\pi^2}{\Delta x \, \Delta y} \sum_{j=-\infty}^{\infty} \sum_{k=-\infty}^{\infty} \delta(\omega_x - j\,\omega_{xs}, \omega_y - k\,\omega_{ys}) \qquad (4.1\text{-}5)$$

where $\omega_{xs} = 2\pi/\Delta x$ and $\omega_{ys} = 2\pi/\Delta y$ represent the Fourier domain sampling frequencies. It will be assumed that the spectrum of the ideal image is bandlimited to some bounds such that $\mathcal{F}_I(\omega_x, \omega_y) = 0$ for $|\omega_x| > \omega_{xc}$ and $|\omega_y| > \omega_{yc}$. Performing the convolution of Eq. 4.1-4 yields

$$\mathcal{F}_P(\omega_x, \omega_y) = \frac{1}{\Delta x \, \Delta y} \int_{-\infty}^{\infty} \int_{-\infty}^{\infty} \mathcal{F}_I(\omega_x - \alpha, \omega_y - \beta)$$

$$\times \sum_{j=-\infty}^{\infty} \sum_{k=-\infty}^{\infty} \delta(\omega_x - j\,\omega_{xs}, \omega_y - k\,\omega_{ys})\, d\alpha \, d\beta \qquad (4.1\text{-}6)$$

Upon changing the order of summation and integration and invoking the sifting property of the delta function, the sampled image spectrum becomes

$$\mathcal{F}_P(\omega_x, \omega_y) = \frac{1}{\Delta x \, \Delta y} \sum_{j=-\infty}^{\infty} \sum_{k=-\infty}^{\infty} \mathcal{F}_I(\omega_x - j\,\omega_{xs}, \omega_y - k\,\omega_{ys}) \qquad (4.1\text{-}7)$$

As can be seen from Figure 4.1-2, the spectrum of the sampled image consists of the spectrum of the ideal image infinitely repeated over the frequency plane in a grid of resolution $(2\pi/\Delta x, 2\pi/\Delta y)$. It should be noted that if Δx and Δy are chosen too large with respect to the spatial frequency limits of $\mathcal{F}_I(\omega_x, \omega_y)$, the individual spectra will overlap.

A continuous image field may be obtained from the image samples of $F_P(x, y)$ by linear spatial interpolation or by linear spatial filtering of the sampled image. Let $R(x, y)$ denote the continuous domain impulse response of an interpolation filter and $\mathcal{R}(\omega_x, \omega_y)$ represent its transfer function. Then the reconstructed image is obtained

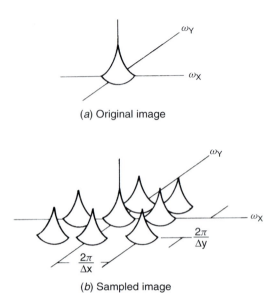

(a) Original image

(b) Sampled image

FIGURE 4.1-2. Typical sampled image spectra.

by a convolution of the samples with the reconstruction filter impulse response. The reconstructed image then becomes

$$F_R(x, y) = F_P(x, y) \circledast R(x, y) \tag{4.1-8}$$

Upon substituting for $F_P(x, y)$ from Eq. 4.1-2 and performing the convolution, one obtains

$$F_R(x, y) = \sum_{j=-\infty}^{\infty} \sum_{k=-\infty}^{\infty} F_I(j\,\Delta x, k\,\Delta y) R(x - j\,\Delta x, y - k\,\Delta y) \tag{4.1-9}$$

Thus it is seen that the impulse response function $R(x, y)$ acts as a two-dimensional interpolation waveform for the image samples. The spatial frequency spectrum of the reconstructed image obtained from Eq. 4.1-8 is equal to the product of the reconstruction filter transform and the spectrum of the sampled image,

$$\mathcal{F}_R(\omega_x, \omega_y) = \mathcal{F}_P(\omega_x, \omega_y) \mathcal{R}(\omega_x, \omega_y) \tag{4.1-10}$$

or, from Eq. 4.1-7,

$$\mathcal{F}_R(\omega_x, \omega_y) = \frac{1}{\Delta x\,\Delta y} \mathcal{R}(\omega_x, \omega_y) \sum_{j=-\infty}^{\infty} \sum_{k=-\infty}^{\infty} \mathcal{F}_I(\omega_x - j\,\omega_{xs}, \omega_y - k\,\omega_{ys}) \tag{4.1-11}$$

It is clear from Eq. 4.1-11 that if there is no spectrum overlap and if $\mathcal{R}(\omega_x, \omega_y)$ filters out all spectra for $j, k \neq 0$, the spectrum of the reconstructed image can be made equal to the spectrum of the ideal image, and therefore the images themselves can be made identical. The first condition is met for a bandlimited image if the sampling period is chosen such that the rectangular region bounded by the image cutoff frequencies $(\omega_{xc}, \omega_{yc})$ lies within a rectangular region defined by one-half the sampling frequency. Hence

$$\omega_{xc} \leq \frac{\omega_{xs}}{2} \qquad\qquad \omega_{yc} \leq \frac{\omega_{ys}}{2} \qquad\qquad (4.1\text{-}12a)$$

or, equivalently,

$$\Delta x \leq \frac{\pi}{\omega_{xc}} \qquad\qquad \Delta y \leq \frac{\pi}{\omega_{yc}} \qquad\qquad (4.1\text{-}12b)$$

In physical terms, the sampling period must be equal to or smaller than one-half the period of the finest detail within the image. This sampling condition is equivalent to the one-dimensional sampling theorem constraint for time-varying signals that requires a time-varying signal to be sampled at a rate of at least twice its highest-frequency component. If equality holds in Eq. 4.1-12, the image is said to be sampled at its *Nyquist rate*; if Δx and Δy are smaller than required by the *Nyquist criterion*, the image is called *oversampled*; and if the opposite case holds, the image is *undersampled*.

If the original image is sampled at a spatial rate sufficient to prevent spectral overlap in the sampled image, exact reconstruction of the ideal image can be achieved by spatial filtering the samples with an appropriate filter. For example, as shown in Figure 4.1-3, a filter with a transfer function of the form

$$\mathcal{R}(\omega_x, \omega_y) = \begin{cases} K & \text{for } |\omega_x| \leq \omega_{xL} \text{ and } |\omega_y| \leq \omega_{yL} & (4.1\text{-}13a) \\[2ex] 0 & \text{otherwise} & (4.1\text{-}13b) \end{cases}$$

where K is a scaling constant, satisfies the condition of exact reconstruction if $\omega_{xL} > \omega_{xc}$ and $\omega_{yL} > \omega_{yc}$. The point-spread function or impulse response of this reconstruction filter is

$$R(x, y) = \frac{K\omega_{xL}\omega_{yL}}{\pi^2} \frac{\sin\{\omega_{xL}x\}}{\omega_{xL}x} \frac{\sin\{\omega_{yL}y\}}{\omega_{yL}y} \qquad\qquad (4.1\text{-}14)$$

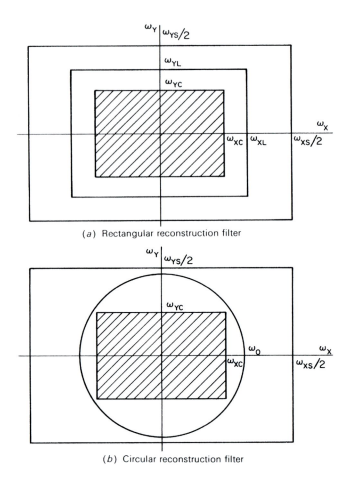

(a) Rectangular reconstruction filter

(b) Circular reconstruction filter

FIGURE 4.1-3. Sampled image reconstruction filters.

With this filter, an image is reconstructed with an infinite sum of $(\sin\theta)/\theta$ functions, called *sinc functions*. Another type of reconstruction filter that could be employed is the cylindrical filter with a transfer function

$$
R(\omega_x, \omega_y) = \begin{cases} K & \text{for } \sqrt{\omega_x^2 + \omega_y^2} \le \omega_0 & (4.1\text{-}15a) \\[2em] 0 & \text{otherwise} & (4.1\text{-}15b) \end{cases}
$$

provided that $\omega_0^2 > \omega_{xc}^2 + \omega_{yc}^2$. The impulse response for this filter is

$$R(x, y) = 2\pi\omega_0 K \frac{J_1\left\{\omega_0\sqrt{x^2 + y^2}\right\}}{\sqrt{x^2 + y^2}} \qquad (4.1\text{-}16)$$

where $J_1\{\cdot\}$ is a first-order *Bessel function*. There are a number of reconstruction filters, or equivalently, interpolation waveforms, that could be employed to provide perfect image reconstruction. In practice, however, it is often difficult to implement optimum reconstruction filters for imaging systems.

4.1.2. Sampling Random Image Fields

In the previous discussion of image sampling and reconstruction, the ideal input image field has been considered to be a deterministic function. It has been shown that if the Fourier transform of the ideal image is bandlimited, then discrete image samples taken at the Nyquist rate are sufficient to reconstruct an exact replica of the ideal image with proper sample interpolation. It will now be shown that similar results hold for sampling two-dimensional random fields.

Let $F_I(x, y)$ denote a continuous two-dimensional stationary random process with known mean η_{F_I} and autocorrelation function

$$R_{F_I}(\tau_x, \tau_y) = E\{F_I(x_1, y_1)F_I^*(x_2, y_2)\} \qquad (4.1\text{-}17)$$

where $\tau_x = x_1 - x_2$ and $\tau_y = y_1 - y_2$. This process is spatially sampled by a Dirac sampling array yielding

$$F_P(x, y) = F_I(x, y)S(x, y) = F_I(x, y) \sum_{j=-\infty}^{\infty} \sum_{k=-\infty}^{\infty} \delta(x - j \, \Delta x, y - k \, \Delta y) \qquad (4.1\text{-}18)$$

The autocorrelation of the sampled process is then

$$R_{F_P}(\tau_x, \tau_y) = E\{F_P(x_1, y_1) F_P^*(x_2, y_2)\} \qquad (4.1\text{-}19)$$

$$= E\{F_I(x_1, y_1) F_I^*(x_2, y_2)\}S(x_1, y_1)S(x_2, y_2)$$

The first term on the right-hand side of Eq. 4.1-19 is the autocorrelation of the stationary ideal image field. It should be observed that the product of the two Dirac sampling functions on the right-hand side of Eq. 4.1-19 is itself a Dirac sampling function of the form

$$S(x_1, y_1)S(x_2, y_2) = S(x_1 - x_2, y_1 - y_2) = S(\tau_x, \tau_y) \tag{4.1-20}$$

Hence the sampled random field is also stationary with an autocorrelation function

$$R_{F_P}(\tau_x, \tau_y) = R_{F_I}(\tau_x, \tau_y)S(\tau_x, \tau_y) \tag{4.1-21}$$

Taking the two-dimensional Fourier transform of Eq. 4.1-21 yields the power spectrum of the sampled random field. By the Fourier transform convolution theorem

$$\mathcal{W}_{F_P}(\omega_x, \omega_y) = \frac{1}{4\pi^2} \mathcal{W}_{F_I}(\omega_x, \omega_y) \circledast S(\omega_x, \omega_y) \tag{4.1-22}$$

where $\mathcal{W}_{F_I}(\omega_x, \omega_y)$ and $\mathcal{W}_{F_P}(\omega_x, \omega_y)$ represent the power spectral densities of the ideal image and sampled ideal image, respectively, and $S(\omega_x, \omega_y)$ is the Fourier transform of the Dirac sampling array. Then, by the derivation leading to Eq. 4.1-7, it is found that the spectrum of the sampled field can be written as

$$\mathcal{W}_{F_P}(\omega_x, \omega_y) = \frac{1}{\Delta x \, \Delta y} \sum_{j=-\infty}^{\infty} \sum_{k=-\infty}^{\infty} \mathcal{W}_{F_I}(\omega_x - j \, \omega_{xs}, \omega_y - k \, \omega_{ys}) \tag{4.1-23}$$

Thus the sampled image power spectrum is composed of the power spectrum of the continuous ideal image field replicated over the spatial frequency domain at integer multiples of the sampling spatial frequency $(2\pi/\Delta x, 2\pi/\Delta y)$. If the power spectrum of the continuous ideal image field is bandlimited such that $\mathcal{W}_{F_I}(\omega_x, \omega_y) = 0$ for $|\omega_x| > \omega_{xc}$ and $|\omega_y| > \omega_{yc}$, where ω_{xc} and are ω_{yc} cutoff frequencies, the individual spectra of Eq. 4.1-23 will not overlap if the spatial sampling periods are chosen such that $\Delta x < \pi/\omega_{xc}$ and $\Delta y < \pi/\omega_{yc}$. A continuous random field $F_R(x, y)$ may be reconstructed from samples of the random ideal image field by the interpolation formula

$$F_R(x, y) = \sum_{j=-\infty}^{\infty} \sum_{k=-\infty}^{\infty} F_I(j \, \Delta x, k \, \Delta y)R(x - j \, \Delta x, y - k \, \Delta y) \tag{4.1-24}$$

where $R(x, y)$ is the deterministic interpolation function. The reconstructed field and the ideal image field can be made equivalent in the mean-square sense (5, p. 284), that is,

$$E\{|F_I(x, y) - F_R(x, y)|^2\} = 0 \tag{4.1-25}$$

if the Nyquist sampling criteria are met and if suitable interpolation functions, such as the sinc function or Bessel function of Eqs. 4.1-14 and 4.1-16, are utilized.

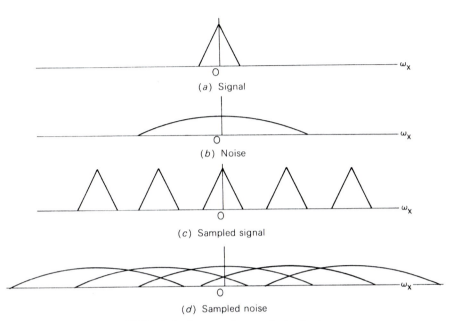

(a) Signal

(b) Noise

(c) Sampled signal

(d) Sampled noise

FIGURE 4.1-4. Spectra of a sampled noisy image.

The preceding results are directly applicable to the practical problem of sampling a deterministic image field plus additive noise, which is modeled as a random field. Figure 4.1-4 shows the spectrum of a sampled noisy image. This sketch indicates a significant potential problem. The spectrum of the noise may be wider than the ideal image spectrum, and if the noise process is undersampled, its tails will overlap into the passband of the image reconstruction filter, leading to additional noise artifacts. A solution to this problem is to prefilter the noisy image before sampling to reduce the noise bandwidth.

4.2. IMAGE SAMPLING SYSTEMS

In a physical image sampling system, the sampling array will be of finite extent, the sampling pulses will be of finite width, and the image may be undersampled. The consequences of nonideal sampling are explored next.

As a basis for the discussion, Figure 4.2-1 illustrates a common image scanning system. In operation, a narrow light beam is scanned directly across a positive photographic transparency of an ideal image. The light passing through the transparency is collected by a condenser lens and is directed toward the surface of a photodetector. The electrical output of the photodetector is integrated over the time period during which the light beam strikes a resolution cell. In the analysis it will be assumed that the sampling is noise-free. The results developed in Section 4.1 for

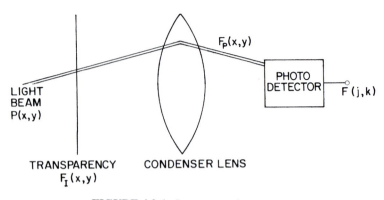

FIGURE 4.2-1. Image scanning system.

sampling noisy images can be combined with the results developed in this section quite readily. Also, it should be noted that the analysis is easily extended to a wide class of physical image sampling systems.

4.2.1. Sampling Pulse Effects

Under the assumptions stated above, the sampled image function is given by

$$F_P(x, y) = F_I(x, y)S(x, y) \tag{4.2-1}$$

where the sampling array

$$S(x, y) = \sum_{j=-J}^{J} \sum_{k=-K}^{K} P(x - j\,\Delta x, y - k\,\Delta y) \tag{4.2-2}$$

is composed of $(2J + 1)(2K + 1)$ identical pulses $P(x, y)$ arranged in a grid of spacing $\Delta x, \Delta y$. The symmetrical limits on the summation are chosen for notational simplicity. The sampling pulses are assumed scaled such that

$$\int_{-\infty}^{\infty} \int_{-\infty}^{\infty} P(x, y)\, dx\, dy = 1 \tag{4.2-3}$$

For purposes of analysis, the sampling function may be assumed to be generated by a finite array of Dirac delta functions $D_T(x, y)$ passing through a linear filter with impulse response $P(x, y)$. Thus

$$S(x, y) = D_T(x, y) \circledast P(x, y) \tag{4.2-4}$$

where

$$D_T(x, y) = \sum_{j=-J}^{J} \sum_{k=-K}^{K} \delta(x - j\,\Delta x, y - k\,\Delta y) \tag{4.2-5}$$

Combining Eqs. 4.2-1 and 4.2-2 results in an expression for the sampled image function,

$$F_P(x, y) = \sum_{j=-J}^{J} \sum_{k=-K}^{K} F_I(j\,\Delta x, k\,\Delta y)P(x - j\,\Delta x, y - k\,\Delta y) \tag{4.2-6}$$

The spectrum of the sampled image function is given by

$$\mathcal{F}_P(\omega_x, \omega_y) = \frac{1}{4\pi^2} \mathcal{F}_I(\omega_x, \omega_y) \circledast [\mathcal{D}_T(\omega_x, \omega_y)\mathcal{P}(\omega_x, \omega_y)] \tag{4.2-7}$$

where $\mathcal{P}(\omega_x, \omega_y)$ is the Fourier transform of $P(x, y)$. The Fourier transform of the truncated sampling array is found to be (5, p. 105)

$$\mathcal{D}_T(\omega_x, \omega_y) = \frac{\sin\left\{\omega_x(J + \tfrac{1}{2})\Delta x\right\}}{\sin\left\{\omega_x\,\Delta x/2\right\}} \frac{\sin\left\{\omega_y(K + \tfrac{1}{2})\Delta y\right\}}{\sin\left\{\omega_y\,\Delta y/2\right\}} \tag{4.2-8}$$

Figure 4.2-2 depicts $\mathcal{D}_T(\omega_x, \omega_y)$. In the limit as J and K become large, the right-hand side of Eq. 4.2-7 becomes an array of Dirac delta functions.

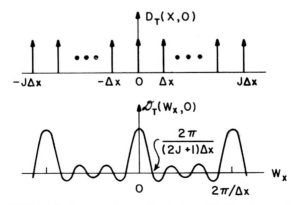

FIGURE 4.2-2. Truncated sampling train and its Fourier spectrum.

In an image reconstruction system, an image is reconstructed by interpolation of its samples. Ideal interpolation waveforms such as the sinc function of Eq. 4.1-14 or the Bessel function of Eq. 4.1-16 generally extend over the entire image field. If the sampling array is truncated, the reconstructed image will be in error near its boundary because the tails of the interpolation waveforms will be truncated in the vicinity of the boundary (8,9). However, the error is usually negligibly small at distances of about 8 to 10 Nyquist samples or greater from the boundary.

The actual numerical samples of an image are obtained by a spatial integration of $F_S(x, y)$ over some finite resolution cell. In the scanning system of Figure 4.2-1, the integration is inherently performed on the photodetector surface. The image sample value of the resolution cell (j, k) may then be expressed as

$$F_S(j\,\Delta x, k\,\Delta y) = \int_{j\Delta x - A_x}^{j\Delta x + A_x} \int_{k\Delta y - A_y}^{k\Delta y + A_y} F_I(x, y)P(x - j\,\Delta x, y - k\,\Delta y)\,dx\,dy \qquad (4.2\text{-}9)$$

where A_x and A_y denote the maximum dimensions of the resolution cell. It is assumed that only one sample pulse exists during the integration time of the detector. If this assumption is not valid, consideration must be given to the difficult problem of sample crosstalk. In the sampling system under discussion, the width of the resolution cell may be larger than the sample spacing. Thus the model provides for sequentially overlapped samples in time.

By a simple change of variables, Eq. 4.2-9 may be rewritten as

$$F_S(j\,\Delta x, k\,\Delta y) = \int_{-A_x}^{A_x} \int_{-A_y}^{A_y} F_I(j\,\Delta x - \alpha, k\,\Delta y - \beta)P(-\alpha, -\beta)\,dx\,dy \qquad (4.2\text{-}10)$$

Because only a single sampling pulse is assumed to occur during the integration period, the limits of Eq. 4.2-10 can be extended infinitely . In this formulation, Eq. 4.2-10 is recognized to be equivalent to a convolution of the ideal continuous image $F_I(x, y)$ with an impulse response function $P(-x, -y)$ with reversed coordinates, followed by sampling over a finite area with Dirac delta functions. Thus, neglecting the effects of the finite size of the sampling array, the model for finite extent pulse sampling becomes

$$F_S(j\,\Delta x, k\,\Delta y) = [F_I(x, y) \circledast P(-x, -y)]\delta(x - j\,\Delta x, y - k\,\Delta y) \qquad (4.2\text{-}11)$$

In most sampling systems, the sampling pulse is symmetric, so that $P(-x, -y) = P(x, y)$.

Equation 4.2-11 provides a simple relation that is useful in assessing the effect of finite extent pulse sampling. If the ideal image is bandlimited and A_x and A_y satisfy the Nyquist criterion, the finite extent of the sample pulse represents an equivalent linear spatial degradation (an image blur) that occurs before ideal sampling. Part 4 considers methods of compensating for this degradation. A finite-extent sampling pulse is not always a detriment, however. Consider the situation in which the ideal image is insufficiently bandlimited so that it is undersampled. The finite-extent pulse, in effect, provides a low-pass filtering of the ideal image, which, in turn, serves to limit its spatial frequency content, and hence to minimize aliasing error.

4.2.2. Aliasing Effects

To achieve perfect image reconstruction in a sampled imaging system, it is necessary to bandlimit the image to be sampled, spatially sample the image at the Nyquist or higher rate, and properly interpolate the image samples. Sample interpolation is considered in the next section; an analysis is presented here of the effect of undersampling an image.

If there is spectral overlap resulting from undersampling, as indicated by the shaded regions in Figure 4.2-3, spurious spatial frequency components will be introduced into the reconstruction. The effect is called an *aliasing error* (10,11). Aliasing effects in an actual image are shown in Figure 4.2-4. Spatial undersampling of the image creates artificial low-spatial-frequency components in the reconstruction. In the field of optics, aliasing errors are called *moiré patterns*.

From Eq. 4.1-7 the spectrum of a sampled image can be written in the form

$$\mathcal{F}_P(\omega_x, \omega_y) = \frac{1}{\Delta x \, \Delta y}[\,\mathcal{F}_I(\omega_x, \omega_y) + \mathcal{F}_Q(\omega_x, \omega_y)] \qquad (4.2\text{-}12)$$

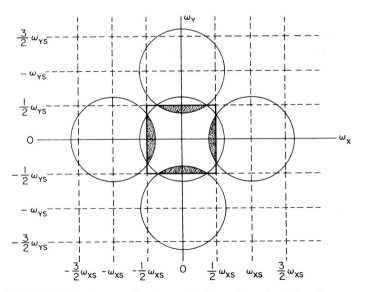

FIGURE 4.2-3. Spectra of undersampled two-dimensional function.

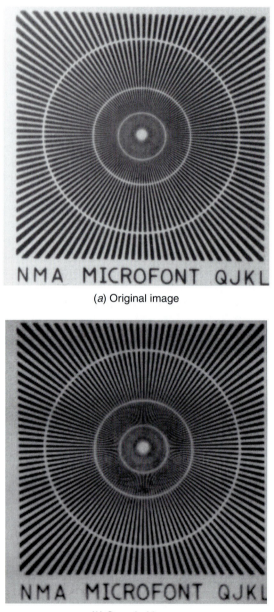

(*a*) Original image

(*b*) Sampled image

FIGURE 4.2-4. Example of aliasing error in a sampled image.

where $\mathcal{F}_I(\omega_x, \omega_y)$ represents the spectrum of the original image sampled at period $(\Delta x, \Delta y)$. The term

$$\mathcal{F}_Q(\omega_x, \omega_y) = \frac{1}{\Delta x \, \Delta y} \sum_{j=-\infty}^{\infty} \sum_{k=-\infty}^{\infty} \mathcal{F}_I(\omega_x - j \, \omega_{xs}, \omega_y - k \, \omega_{ys}) \qquad (4.2\text{-}13)$$

for $j \neq 0$ and $k \neq 0$ describes the spectrum of the higher-order components of the sampled image repeated over spatial frequencies $\omega_{xs} = 2\pi/\Delta x$ and $\omega_{ys} = 2\pi/\Delta y$. If there were no spectral foldover, optimal interpolation of the sampled image components could be obtained by passing the sampled image through a zonal low-pass filter defined by

$$\mathcal{R}(\omega_x, \omega_y) = \begin{cases} K & \text{for } |\omega_x| \leq \omega_{xs}/2 \text{ and } |\omega_y| \leq \omega_{ys}/2 \quad (4.2\text{-}14a) \\ 0 & \text{otherwise} \hspace{4.5cm} (4.2\text{-}14b) \end{cases}$$

where K is a scaling constant. Applying this interpolation strategy to an undersampled image yields a reconstructed image field

$$F_R(x, y) = F_I(x, y) + A(x, y) \qquad (4.2\text{-}15)$$

where

$$A(x, y) = \frac{1}{4\pi^2} \int_{-\omega_{xs}/2}^{\omega_{xs}/2} \int_{-\omega_{ys}/2}^{\omega_{ys}/2} \mathcal{F}_Q(\omega_x, \omega_y) \exp\{i(\omega_x x + \omega_y y)\} d\omega_x \, d\omega_y \quad (4.2\text{-}16)$$

represents the aliasing error artifact in the reconstructed image. The factor K has absorbed the amplitude scaling factors. Figure 4.2-5 shows the reconstructed image

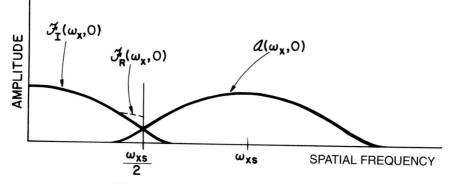

FIGURE 4.2-5. Reconstructed image spectrum.

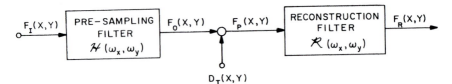

FIGURE 4.2-6. Model for analysis of aliasing effect.

spectrum that illustrates the spectral foldover in the zonal low-pass filter passband. The aliasing error component of Eq. 4.2-16 can be reduced substantially by low-pass filtering before sampling to attenuate the spectral foldover.

Figure 4.2-6 shows a model for the quantitative analysis of aliasing effects. In this model, the ideal image $F_I(x, y)$ is assumed to be a sample of a two-dimensional random process with known power-spectral density $W_{F_I}(\omega_x, \omega_y)$. The ideal image is linearly filtered by a presampling spatial filter with a transfer function $\mathcal{H}(\omega_x, \omega_y)$. This filter is assumed to be a low-pass type of filter with a smooth attenuation of high spatial frequencies (i.e., not a zonal low-pass filter with a sharp cutoff). The filtered image is then spatially sampled by an ideal Dirac delta function sampler at a resolution $\Delta x, \Delta y$. Next, a reconstruction filter interpolates the image samples to produce a replica of the ideal image. From Eq. 1.4-27, the power spectral density at the presampling filter output is found to be

$$W_{F_O}(\omega_x, \omega_y) = |\mathcal{H}(\omega_x, \omega_y)|^2 W_{F_I}(\omega_x, \omega_y) \tag{4.2-17}$$

and the Fourier spectrum of the sampled image field is

$$W_{F_P}(\omega_x, \omega_y) = \frac{1}{\Delta x \, \Delta y} \sum_{j = -\infty}^{\infty} \sum_{k = -\infty}^{\infty} W_{F_O}(\omega_x - j\,\omega_{xs}, \omega_y - k\,\omega_{ys}) \tag{4.2-18}$$

Figure 4.2-7 shows the sampled image power spectral density and the foldover aliasing spectral density from the first sideband with and without presampling low-pass filtering.

It is desirable to isolate the undersampling effect from the effect of improper reconstruction. Therefore, assume for this analysis that the reconstruction filter $\mathcal{R}(\omega_x, \omega_y)$ is an optimal filter of the form given in Eq. 4.2-14. The energy passing through the reconstruction filter for $j = k = 0$ is then

$$E_R = \int_{-\omega_{xs}/2}^{\omega_{xs}/2} \int_{-\omega_{ys}/2}^{\omega_{ys}/2} W_{F_I}(\omega_x, \omega_y) |\mathcal{H}(\omega_x, \omega_y)|^2 \, d\omega_x \, d\omega_y \tag{4.2-19}$$

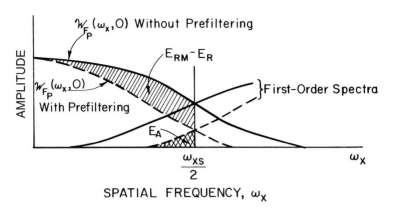

FIGURE 4.2-7. Effect of presampling filtering on a sampled image.

Ideally, the presampling filter should be a low-pass zonal filter with a transfer function identical to that of the reconstruction filter as given by Eq. 4.2-14. In this case, the sampled image energy would assume the maximum value

$$E_{RM} = \int_{-\omega_{xs}/2}^{\omega_{xs}/2} \int_{-\omega_{ys}/2}^{\omega_{ys}/2} W_{F_I}(\omega_x, \omega_y) \, d\omega_x \, d\omega_y \qquad (4.2\text{-}20)$$

Image resolution degradation resulting from the presampling filter may then be measured by the ratio

$$\mathcal{E}_R = \frac{E_{RM} - E_R}{E_{RM}} \qquad (4.2\text{-}21)$$

The aliasing error in a sampled image system is generally measured in terms of the energy, from higher-order sidebands, that folds over into the passband of the reconstruction filter. Assume, for simplicity, that the sampling rate is sufficient so that the spectral foldover from spectra centered at $(\pm j\,\omega_{xs}/2, \pm k\,\omega_{ys}/2)$ is negligible for $j \geq 2$ and $k \geq 2$. The total aliasing error energy, as indicated by the doubly cross-hatched region of Figure 4.2-7, is then

$$E_A = E_O - E_R \qquad (4.2\text{-}22)$$

where

$$E_O = \int_{-\infty}^{\infty} \int_{-\infty}^{\infty} W_{F_I}(\omega_x, \omega_y) |\mathcal{H}(\omega_x, \omega_y)|^2 \, d\omega_x \, d\omega_y \qquad (4.2\text{-}23)$$

denotes the energy of the output of the presampling filter. The aliasing error is defined as (10)

$$\mathcal{E}_A = \frac{E_A}{E_O} \tag{4.2-24}$$

Aliasing error can be reduced by attenuating high spatial frequencies of $F_I(x, y)$ with the presampling filter. However, any attenuation within the passband of the reconstruction filter represents a loss of resolution of the sampled image. As a result, there is a trade-off between sampled image resolution and aliasing error.

Consideration is now given to the aliasing error versus resolution performance of several practical types of presampling filters. Perhaps the simplest means of spatially filtering an image formed by incoherent light is to pass the image through a lens with a restricted aperture. Spatial filtering can then be achieved by controlling the degree of lens misfocus. Figure 11.2-2 is a plot of the optical transfer function of a circular lens as a function of the degree of lens misfocus. Even a perfectly focused lens produces some blurring because of the diffraction limit of its aperture. The transfer function of a diffraction-limited circular lens of diameter d is given by (12, p. 83)

$$\mathcal{H}(\omega) = \begin{cases} \dfrac{2}{\pi}\left[a \cos\left\{\dfrac{\omega}{\omega_0}\right\} - \dfrac{\omega}{\omega_0}\sqrt{1 - \left(\dfrac{\omega}{\omega_0}\right)^2}\, \right] & \text{for } 0 \le \omega \le \omega_0 \tag{4.2-25a} \\[20pt] 0 & \text{for } |\omega| > \omega_0 \tag{4.2-25b} \end{cases}$$

where $\omega_0 = \pi d / R$ and R is the distance from the lens to the focal plane. In Section 4.2.1, it was noted that sampling with a finite-extent sampling pulse is equivalent to ideal sampling of an image that has been passed through a spatial filter whose impulse response is equal to the pulse shape of the sampling pulse with reversed coordinates. Thus the sampling pulse may be utilized to perform presampling filtering. A common pulse shape is the rectangular pulse

$$P(x, y) = \begin{cases} \dfrac{1}{T^2} & \text{for } |x, y| \le \dfrac{T}{2} \tag{4.2-26a} \\[16pt] 0 & \text{for } |x, y| > \dfrac{T}{2} \tag{4.2-26b} \end{cases}$$

obtained with an incoherent light imaging system of a scanning microdensitometer. The transfer function for a square scanning spot is

$$\mathcal{P}(\omega_x, \omega_y) = \frac{\sin\{\omega_x T/2\}}{\omega_x T/2} \; \frac{\sin\{\omega_y T/2\}}{\omega_y T/2} \qquad (4.2\text{-}27)$$

Cathode ray tube displays produce display spots with a two-dimensional Gaussian shape of the form

$$P(x, y) = \frac{1}{2\pi\sigma_w^2} \; \exp\left\{ -\frac{x^2 + y^2}{2\sigma_w^2} \right\} \qquad (4.2\text{-}28)$$

where σ_w is a measure of the spot spread. The equivalent transfer function of the Gaussian-shaped scanning spot

$$\mathcal{P}(\omega_x, \omega_y) = \exp\left\{ -\frac{(\omega_x^2 + \omega_y^2)\sigma_w^2}{2} \right\} \qquad (4.2\text{-}29)$$

Examples of the aliasing error-resolution trade-offs for a diffraction-limited aperture, a square sampling spot, and a Gaussian-shaped spot are presented in Figure 4.2-8 as a function of the parameter ω_0. The square pulse width is set at $T = 2\pi/\omega_0$, so that the first zero of the sinc function coincides with the lens cutoff frequency. The spread of the Gaussian spot is set at $\sigma_w = 2/\omega_0$, corresponding to two standard deviation units in crosssection. In this example, the input image spectrum is modeled as

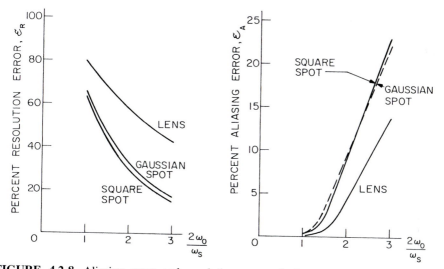

FIGURE 4.2-8. Aliasing error and resolution error obtained with different types of prefiltering.

$$W_{F_I}(\omega_x, \omega_y) = \frac{A}{1 + (\omega/\omega_c)^{2m}} \qquad (4.2\text{-}30)$$

where A is an amplitude constant, m is an integer governing the rate of falloff of the Fourier spectrum, and ω_c is the spatial frequency at the half-amplitude point. The curves of Figure 4.2-8 indicate that the Gaussian spot and square spot scanning prefilters provide about the same results, while the diffraction-limited lens yields a somewhat greater loss in resolution for the same aliasing error level. A defocused lens would give even poorer results.

4.3. IMAGE RECONSTRUCTION SYSTEMS

In Section 4.1 the conditions for exact image reconstruction were stated: The original image must be spatially sampled at a rate of at least twice its highest spatial frequency, and the reconstruction filter, or equivalent interpolator, must be designed to pass the spectral component at $j = 0$, $k = 0$ without distortion and reject all spectra for which $j, k \neq 0$. With physical image reconstruction systems, these conditions are impossible to achieve exactly. Consideration is now given to the effects of using imperfect reconstruction functions.

4.3.1. Implementation Techniques

In most digital image processing systems, electrical image samples are sequentially output from the processor in a normal raster scan fashion. A continuous image is generated from these electrical samples by driving an optical display such as a cathode ray tube (CRT) with the intensity of each point set proportional to the image sample amplitude. The light array on the CRT can then be imaged onto a ground-glass screen for viewing or onto photographic film for recording with a light projection system incorporating an incoherent spatial filter possessing a desired optical transfer function. Optimal transfer functions with a perfectly flat passband over the image spectrum and a sharp cutoff to zero outside the spectrum cannot be physically implemented.

The most common means of image reconstruction is by use of electro-optical techniques. For example, image reconstruction can be performed quite simply by electrically defocusing the writing spot of a CRT display. The drawback of this technique is the difficulty of accurately controlling the spot shape over the image field. In a scanning microdensitometer, image reconstruction is usually accomplished by projecting a rectangularly shaped spot of light onto photographic film. Generally, the spot size is set at the same size as the sample spacing to fill the image field completely. The resulting interpolation is simple to perform, but not optimal. If a small writing spot can be achieved with a CRT display or a projected light display, it is possible approximately to synthesize any desired interpolation by subscanning a resolution cell, as shown in Figure 4.3-1.

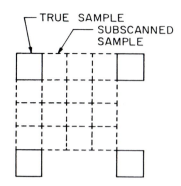

FIGURE 4.3-1. Image reconstruction by subscanning.

The following subsections introduce several one- and two-dimensional interpolation functions and discuss their theoretical performance. Chapter 13 presents methods of digitally implementing image reconstruction systems.

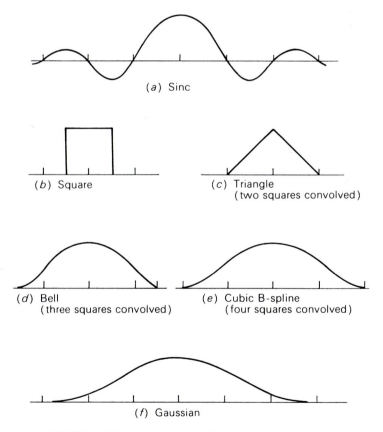

FIGURE 4.3-2. One-dimensional interpolation waveforms.

4.3.2. Interpolation Functions

Figure 4.3-2 illustrates several one-dimensional interpolation functions. As stated previously, the sinc function, provides an exact reconstruction, but it cannot be physically generated by an incoherent optical filtering system. It is possible to approximate the sinc function by truncating it and then performing subscanning (Figure 4.3-1). The simplest interpolation waveform is the square pulse function, which results in a zero-order interpolation of the samples. It is defined mathematically as

$$R_0(x) = 1 \qquad \text{for } -\tfrac{1}{2} \le x \le \tfrac{1}{2} \qquad (4.3\text{-}1)$$

and zero otherwise, where for notational simplicity, the sample spacing is assumed to be of unit dimension. A triangle function, defined as

$$R_1(x) = \begin{cases} x + 1 & \text{for } -1 \le x \le 0 \qquad (4.3\text{-}2a) \\ \\ 1 - x & \text{for } 0 < x \le 1 \qquad (4.3\text{-}2b) \end{cases}$$

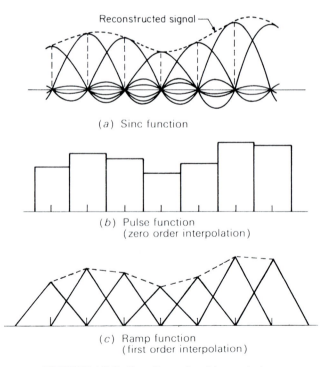

(a) Sinc function

(b) Pulse function
(zero order interpolation)

(c) Ramp function
(first order interpolation)

FIGURE 4.3-3. One-dimensional interpolation.

provides the first-order linear sample interpolation with trianglar interpolation waveforms. Figure 4.3-3 illustrates one-dimensional interpolation using sinc, square, and triangle functions.

The triangle function may be considered to be the result of convolving a square function with itself. Convolution of the triangle function with the square function yields a bell-shaped interpolation waveform (in Figure 4.3-2d). It is defined as

$$
R_2(x) = \begin{cases} \frac{1}{2}(x + \frac{3}{2})^2 & \text{for } -\frac{3}{2} \le x \le -\frac{1}{2} & (4.3\text{-}3a) \\[2mm] \frac{3}{4} - (x)^2 & \text{for } -\frac{1}{2} < x \le \frac{1}{2} & (4.3\text{-}3b) \\[2mm] \frac{1}{2}(x - \frac{3}{2})^2 & \text{for } \frac{1}{2} < x \le \frac{3}{2} & (4.3\text{-}3c) \end{cases}
$$

This process quickly converges to the Gaussian-shaped waveform of Figure 4.3-2f. Convolving the bell-shaped waveform with the square function results in a third-order polynomial function called a *cubic B-spline* (13,14). It is defined mathematically as

$$
R_3(x) = \begin{cases} \frac{2}{3} + \frac{1}{2}|x|^3 - (x)^2 & \text{for } 0 \le |x| \le 1 & (4.3\text{-}4a) \\[2mm] \frac{1}{6}(2 - |x|)^3 & \text{for } 1 < |x| \le 2 & (4.3\text{-}4b) \end{cases}
$$

The cubic B-spline is a particularly attractive candidate for image interpolation because of its properties of continuity and smoothness at the sample points. It can be shown by direct differentiation of Eq. 4.3-4, that $R_3(x)$ is continuous in its first and second derivatives at the sample points.

As mentioned earlier, the sinc function can be approximated by truncating its tails. Typically, this is done over a four-sample interval. The problem with this approach is that the slope discontinuity at the ends of the waveform leads to amplitude ripples in a reconstructed function. This problem can be eliminated by generating a cubic convolution function (15,16), which forces the slope of the ends of the interpolation to be zero. The *cubic convolution* interpolation function can be expressed in the following general form:

$$
R_c(x) = \begin{cases} A_1|x|^3 + B_1|x|^2 + C_1|x| + D_1 & \text{for } 0 \le |x| \le 1 & (4.3\text{-}5a) \\[2mm] A_2|x|^3 + B_2|x|^2 + C_2|x| + D_2 & \text{for } 1 < |x| \le 2 & (4.3\text{-}5b) \end{cases}
$$

where A_i, B_i, C_i, D_i are weighting factors. The weighting factors are determined by satisfying two sets of extraneous conditions:

1. $R_c(x) = 1$ at $x = 0$, and $R_c(x) = 0$ at $x = 1, 2$.

2.   The first-order derivative $R'_c(x) = 0$ at $x = 0, 1, 2$.

These conditions results in seven equations for the eight unknowns and lead to the parametric expression

$$R_c(x) = \begin{cases} (a+2)|x|^3 - (a+3)|x|^2 + 1 & \text{for } 0 \le |x| \le 1 \quad (4.3\text{-}6a) \\ a|x|^3 - 5a|x|^2 + 8a|x| - 4a & \text{for } 1 < |x| \le 2 \quad (4.3\text{-}6b) \end{cases}$$

where $a \equiv A_2$ of Eq. 4.3-5 is the remaining unknown weighting factor. Rifman (15) and Bernstein (16) have set $a = -1$, which causes $R_c(x)$ to have the same slope, - 1, at $x = 1$ as the sinc function. Keys (17) has proposed setting $a = -1/2$, which provides an interpolation function that approximates the original unsampled image to as high a degree as possible in the sense of a power series expansion. The factor a in Eq. 4.3-6 can be used as a tuning parameter to obtain a best visual interpolation (18,19).

Table 4.3-1 defines several orthogonally separable two-dimensional interpolation functions for which $R(x, y) = R(x)R(y)$. The separable square function has a square peg shape. The separable triangle function has the shape of a pyramid. Using a triangle interpolation function for one-dimensional interpolation is equivalent to linearly connecting adjacent sample peaks as shown in Figure 4.3-3c. The extension to two dimensions does not hold because, in general, it is not possible to fit a plane to four adjacent samples. One approach, illustrated in Figure 4.3-4a, is to perform a planar fit in a piecewise fashion. In region I of Figure 4.3-4a, points are linearly interpolated in the plane defined by pixels A, B, C, while in region II, interpolation is performed in the plane defined by pixels B, C, D. A computationally simpler method, called *bilinear interpolation*, is described in Figure 4.3-4b. Bilinear interpolation is performed by linearly interpolating points along separable orthogonal coordinates of the continuous image field. The resultant interpolated surface of Figure 4.3-4b, connecting pixels A, B, C, D, is generally nonplanar. Chapter 13 shows that bilinear interpolation is equivalent to interpolation with a pyramid function.

TABLE 4.3-1. Two-Dimensional Interpolation Functions

Function	Definition
Separable sinc	$R(x, y) = \dfrac{4}{T_x T_y} \dfrac{\sin\{2\pi x/T_x\}}{2\pi x/T_x} \dfrac{\sin\{2\pi y/T_y\}}{2\pi y/T_y}$ $\qquad T_x = \dfrac{2\pi}{\omega_{xs}}$ $\qquad\qquad\qquad\qquad\qquad\qquad\qquad\qquad T_y = \dfrac{2\pi}{\omega_{ys}}$ $\mathcal{R}(\omega_x, \omega_y) = \begin{cases} 1 & \lvert\omega_x\rvert \le \omega_{xs}, \quad \lvert\omega_y\rvert \le \omega_{ys} \\ 0 & \text{otherwise} \end{cases}$
Separable square	$R_0(x, y) = \begin{cases} \dfrac{1}{T_x T_y} & \lvert x\rvert \le \dfrac{T_x}{2}, \quad \lvert y\rvert \le \dfrac{T_y}{2} \\ 0 & \text{otherwise} \end{cases}$ $\mathcal{R}_0(\omega_x, \omega_y) = \dfrac{\sin\{\omega_x T_x/2\}\sin\{\omega_y T_y/2\}}{(\omega_x T_x/2)(\omega_y T_y/2)}$
Separable triangle	$R_1(x, y) = R_0(x, y) \circledast R_0(x, y)$ $\mathcal{R}_1(\omega_x, \omega_y) = \mathcal{R}_0^2(\omega_x, \omega_y)$
Separable bell	$R_2(x, y) = R_0(x, y) \circledast R_1(x, y)$ $\mathcal{R}_2(\omega_x, \omega_y) = \mathcal{R}_0^3(\omega_x, \omega_y)$
Separable cubic B-spline	$R_3(x, y) = R_0(x, y) \circledast R_2(x, y)$ $\mathcal{R}_3(\omega_x, \omega_y) = \mathcal{R}_0^4(\omega_x, \omega_y)$
Gaussian	$R(x, y) = [2\pi\sigma_w^2]^{-1}\exp\left\{-\dfrac{x^2 + y^2}{2\sigma_w^3}\right\}$ $\mathcal{R}(\omega_x, \omega_y) = \exp\left\{-\dfrac{\sigma_w^2(\omega_x^2 + \omega_y^2)}{2}\right\}$

4.3.3. Effect of Imperfect Reconstruction Filters

The performance of practical image reconstruction systems will now be analyzed. It will be assumed that the input to the image reconstruction system is composed of samples of an ideal image obtained by sampling with a finite array of Dirac samples at the Nyquist rate. From Eq. 4.1-9 the reconstructed image is found to be

$$F_R(x, y) = \sum_{j=-\infty}^{\infty} \sum_{k=-\infty}^{\infty} F_I(j\,\Delta x, k\,\Delta y)R(x - j\,\Delta x, y - k\,\Delta y) \qquad (4.3\text{-}7)$$

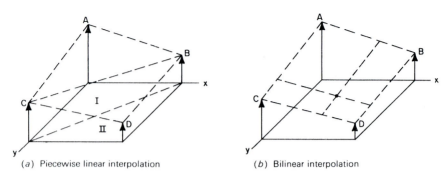

(a) Piecewise linear interpolation   (b) Bilinear interpolation

FIGURE 4.3-4. Two-dimensional linear interpolation.

where $R(x, y)$ is the two-dimensional interpolation function of the image reconstruction system. Ideally, the reconstructed image would be the exact replica of the ideal image as obtained from Eq. 4.1-9. That is,

$$\hat{F}_R(x, y) = \sum_{j=-\infty}^{\infty} \sum_{k=-\infty}^{\infty} F_I(j\,\Delta x, k\,\Delta y) R_I(x - j\,\Delta x, y - k\,\Delta y) \qquad (4.3-8)$$

where $R_I(x, y)$ represents an optimum interpolation function such as given by Eq. 4.1-14 or 4.1-16. The reconstruction error over the bounds of the sampled image is then

$$E_D(x, y) = \sum_{j=-\infty}^{\infty} \sum_{k=-\infty}^{\infty} F_I(j\Delta x, k\Delta y)[R(x - j\Delta x, y - k\Delta y) - R_I(x - j\Delta x, y - k\Delta y)] \quad (4.3-9)$$

There are two contributors to the reconstruction error: (1) the physical system interpolation function $R(x, y)$ may differ from the ideal interpolation function $R_I(x, y)$, and (2) the finite bounds of the reconstruction, which cause truncation of the interpolation functions at the boundary. In most sampled imaging systems, the boundary reconstruction error is ignored because the error generally becomes negligible at distances of a few samples from the boundary. The utilization of nonideal interpolation functions leads to a potential loss of image resolution and to the introduction of high-spatial-frequency artifacts.

The effect of an imperfect reconstruction filter may be analyzed conveniently by examination of the frequency spectrum of a reconstructed image, as derived in Eq. 4.1-11:

$$\mathcal{F}_R(\omega_x, \omega_y) = \frac{1}{\Delta x\,\Delta y} \mathcal{R}(\omega_x, \omega_y) \sum_{j=-\infty}^{\infty} \sum_{k=-\infty}^{\infty} \mathcal{F}_I(\omega_x - j\,\omega_{xs}, \omega_y - k\,\omega_{ys}) \quad (4.3-10)$$

Ideally, $\mathcal{R}(\omega_x, \omega_y)$ should select the spectral component for $j = 0, k = 0$ with uniform attenuation at all spatial frequencies and should reject all other spectral components. An imperfect filter may attenuate the frequency components of the zero-order spectra, causing a loss of image resolution, and may also permit higher-order spectral modes to contribute to the restoration, and therefore introduce distortion in the restoration. Figure 4.3-5 provides a graphic example of the effect of an imperfect image reconstruction filter. A typical cross section of a sampled image is shown in Figure 4.3-5a. With an ideal reconstruction filter employing sinc functions for interpolation, the central image spectrum is extracted and all sidebands are rejected, as shown in Figure 4.3-5c. Figure 4.3-5d is a plot of the transfer function for a zero-order interpolation reconstruction filter in which the reconstructed pixel amplitudes over the pixel sample area are set at the sample value. The resulting spectrum shown in Figure 4.3-5e exhibits distortion from attenuation of the central spectral mode and spurious high-frequency signal components.

Following the analysis leading to Eq. 4.2-21, the resolution loss resulting from the use of a nonideal reconstruction function $R(x, y)$ may be specified quantitatively as

$$\mathcal{E}_R = \frac{E_{RM} - E_R}{E_{RM}} \tag{4.3-11}$$

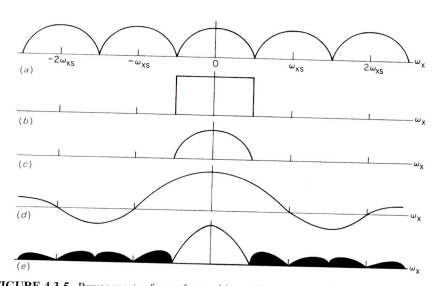

FIGURE 4.3-5. Power spectra for perfect and imperfect reconstruction: (*a*) Sampled image input $\mathcal{W}_{F_J}(\omega_x, 0)$; (*b*) sinc function reconstruction filter transfer function $\mathcal{R}(\omega_x, 0)$; (*c*) sinc function interpolator output $\mathcal{W}_{F_O}(\omega_x, 0)$; (*d*) zero-order interpolation reconstruction filter transfer function $\mathcal{R}(\omega_x, 0)$; (*e*) zero-order interpolator output $\mathcal{W}_{F_O}(\omega_x, 0)$.

where

$$E_R = \int_{-\omega_{xs}/2}^{\omega_{xs}/2} \int_{-\omega_{ys}/2}^{\omega_{ys}/2} W_{F_I}(\omega_x, \omega_y) |\mathcal{H}(\omega_x, w_y)|^2 \, d\omega_x \, d\omega_y \qquad (4.3\text{-}12)$$

represents the actual interpolated image energy in the Nyquist sampling band limits, and

$$E_{RM} = \int_{-\omega_{xs}/2}^{\omega_{xs}/2} \int_{-\omega_{ys}/2}^{\omega_{ys}/2} W_{F_I}(\omega_x, \omega_y) \, d\omega_x \, d\omega_y \qquad (4.3\text{-}13)$$

is the ideal interpolated image energy. The interpolation error attributable to high-spatial-frequency artifacts may be defined as

$$\mathcal{E}_H = \frac{E_H}{E_T} \qquad (4.3\text{-}14)$$

where

$$E_T = \int_{-\infty}^{\infty} \int_{-\infty}^{\infty} W_{F_I}(\omega_x, \omega_y) |\mathcal{H}(\omega_x, \omega_y)|^2 \, d\omega_x \, d\omega_y \qquad (4.3\text{-}15)$$

denotes the total energy of the interpolated image and

$$E_H = E_T - E_R \qquad (4.3\text{-}16)$$

is that portion of the interpolated image energy lying outside the Nyquist band limits.

Table 4.3-2 lists the resolution error and interpolation error obtained with several separable two-dimensional interpolation functions. In this example, the power spectral density of the ideal image is assumed to be of the form

$$W_{F_I}(\omega_x, \omega_y) = \sqrt{\left(\frac{\omega_s}{2}\right)^2 - \omega^2} \qquad \text{for } \omega^2 \le \left(\frac{\omega_s}{2}\right)^2 \qquad (4.3\text{-}17)$$

and zero elsewhere. The interpolation error contribution of highest-order components, $j_1, j_2 > 2$, is assumed negligible. The table indicates that zero-order

TABLE 4.3-2. Interpolation Error and Resolution Error for Various Separable Two-Dimensional Interpolation Functions

Function	Percent Resoluton Error E_R	Percent Interpolation Error E_H
Sinc	0.0	0.0
Square	26.9	15.7
Triangle	44.0	3.7
Bell	55.4	1.1
Cubic B-spline	63.2	0.3
Gaussian $\sigma_w = \dfrac{3T}{8}$	38.6	10.3
Gaussian $\sigma_w = \dfrac{T}{2}$	54.6	2.0
Gaussian $\sigma_w = \dfrac{5T}{8}$	66.7	0.3

interpolation with a square interpolation function results in a significant amount of resolution error. Interpolation error reduces significantly for higher-order convolutional interpolation functions, but at the expense of resolution error.

REFERENCES

1. F. T. Whittaker, "On the Functions Which Are Represented by the Expansions of the Interpolation Theory," *Proc. Royal Society of Edinburgh*, **A35**, 1915, 181–194.

2. C. E. Shannon, "Communication in the Presence of Noise," *Proc. IRE*, **37**, 1, January 1949, 10–21.

3. H. J. Landa, "Sampling, Data Transmission, and the Nyquist Rate," *Proc. IEEE*, **55**, 10, October 1967, 1701–1706.

4. J. W. Goodman, *Introduction to Fourier Optics*, 2nd ed., McGraw-Hill, New York, 1996.

5. A. Papoulis, *Systems and Transforms with Applications in Optics*, McGraw-Hill, New York, 1966.

6. S. P. Lloyd, "A Sampling Theorem for Stationary (Wide Sense) Stochastic Processes," *Trans. American Mathematical Society*, **92**, 1, July 1959, 1–12.

7. H. S. Shapiro and R. A. Silverman, "Alias-Free Sampling of Random Noise," *J. SIAM*, **8**, 2, June 1960, 225–248.

8. J. L. Brown, Jr., "Bounds for Truncation Error in Sampling Expansions of Band-Limited Signals," *IEEE Trans. Information Theory*, **IT-15**, 4, July 1969, 440–444.

9. H. D. Helms and J. B. Thomas, "Truncation Error of Sampling Theory Expansions," *Proc. IRE*, **50**, 2, February 1962, 179–184.

10. J. J. Downing, "Data Sampling and Pulse Amplitude Modulation," in *Aerospace Telemetry*, H. L. Stiltz, Ed., Prentice Hall, Englewood Cliffs, NJ, 1961.

11. D. G. Childers, "Study and Experimental Investigation on Sampling Rate and Aliasing in Time Division Telemetry Systems," *IRE Trans. Space Electronics and Telemetry*, **SET-8**, December 1962, 267–283.

12. E. L. O'Neill, *Introduction to Statistical Optics*, Addison-Wesley, Reading, MA, 1963.

13. H. S. Hou and H. C. Andrews, "Cubic Splines for Image Interpolation and Digital Filtering," *IEEE Trans. Acoustics, Speech, and Signal Processing*, **ASSP-26**, 6, December 1978, 508–517.

14. T. N. E. Greville, "Introduction to Spline Functions," in *Theory and Applications of Spline Functions*, T. N. E. Greville, Ed., Academic Press, New York, 1969.

15. S. S. Rifman, "Digital Rectification of ERTS Multispectral Imagery," *Proc. Symposium on Significant Results Obtained from ERTS-1 (NASA SP-327)*, **I**, Sec. B, 1973, 1131–1142.

16. R. Bernstein, "Digital Image Processing of Earth Observation Sensor Data," *IBM J. Research and Development*, **20**, 1976, 40–57.

17. R. G. Keys, "Cubic Convolution Interpolation for Digital Image Processing," *IEEE Trans. Acoustics, Speech, and Signal Processing*, **AASP-29**, 6, December 1981, 1153–1160.

18. K. W. Simon, "Digital Image Reconstruction and Resampling of Landsat Imagery," *Proc. Symposium on Machine Processing of Remotely Sensed Data*, Purdue University, Lafayette, IN, IEEE 75, CH 1009-0-C, June 1975, 3A-1–3A-11.

19. S. K. Park and R. A. Schowengerdt, "Image Reconstruction by Parametric Cubic Convolution," *Computer Vision, Graphics, and Image Processing*, **23**, 3, September 1983, 258–272.

5

DISCRETE IMAGE MATHEMATICAL CHARACTERIZATION

Chapter 1 presented a mathematical characterization of continuous image fields. This chapter develops a vector-space algebra formalism for representing discrete image fields from a deterministic and statistical viewpoint. Appendix 1 presents a summary of vector-space algebra concepts.

5.1. VECTOR-SPACE IMAGE REPRESENTATION

In Chapter 1 a generalized continuous image function $F(x, y, t)$ was selected to represent the luminance, tristimulus value, or some other appropriate measure of a physical imaging system. Image sampling techniques, discussed in Chapter 4, indicated means by which a discrete array $F(j, k)$ could be extracted from the continuous image field at some time instant over some rectangular area $-J \leq j \leq J$, $-K \leq k \leq K$. It is often helpful to regard this sampled image array as a $N_1 \times N_2$ element matrix

$$\mathbf{F} = [F(n_1, n_2)] \tag{5.1-1}$$

for $1 \leq n_i \leq N_i$ where the indices of the sampled array are reindexed for consistency with standard vector-space notation. Figure 5.1-1 illustrates the geometric relationship between the Cartesian coordinate system of a continuous image and its array of samples. Each image sample is called a *pixel*.

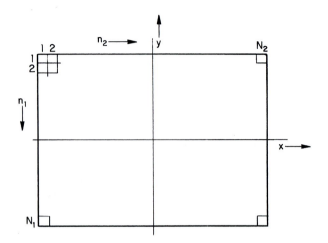

FIGURE 5.1-1. Geometric relationship between a continuous image and its array of samples.

For purposes of analysis, it is often convenient to convert the image matrix to vector form by column (or row) scanning $\mathbf{F}$, and then stringing the elements together in a long vector (1). An equivalent scanning operation can be expressed in quantitative form by the use of a $N_2 \times 1$ operational vector $\mathbf{v}_n$ and a $N_1 \cdot N_2 \times N_2$ matrix $\mathbf{N}_n$ defined as

$$
\mathbf{v}_n =
\begin{bmatrix}
0 \\
\vdots \\
0 \\
1 \\
0 \\
\vdots \\
0
\end{bmatrix}
\begin{matrix}
1 \\
\vdots \\
n-1 \\
n \\
n+1 \\
\vdots \\
N_2
\end{matrix}
\qquad
\mathbf{N}_n =
\begin{bmatrix}
\mathbf{0} \\
\vdots \\
\mathbf{0} \\
\mathbf{1} \\
\mathbf{0} \\
\vdots \\
\mathbf{0}
\end{bmatrix}
\begin{matrix}
1 \\
\vdots \\
n-1 \\
n \\
n+1 \\
\vdots \\
N_2
\end{matrix}
\qquad (5.1\text{-}2)
$$

Then the vector representation of the image matrix $\mathbf{F}$ is given by the stacking operation

$$
\mathbf{f} = \sum_{n=1}^{N_2} \mathbf{N}_n \mathbf{F} \mathbf{v}_n \qquad (5.1\text{-}3)
$$

In essence, the vector $\mathbf{v}_n$ extracts the nth column from $\mathbf{F}$ and the matrix $\mathbf{N}_n$ places this column into the nth segment of the vector $\mathbf{f}$. Thus, $\mathbf{f}$ contains the column-scanned elements of **F**. The inverse relation of casting the vector **f** into matrix form is obtained from

$$\mathbf{F} = \sum_{n=1}^{N_2} \mathbf{N}_n^T \mathbf{f} \mathbf{v}_n^T \qquad (5.1\text{-}4)$$

With the matrix-to-vector operator of Eq. 5.1-3 and the vector-to-matrix operator of Eq. 5.1-4, it is now possible easily to convert between vector and matrix representations of a two-dimensional array. The advantages of dealing with images in vector form are a more compact notation and the ability to apply results derived previously for one-dimensional signal processing applications. It should be recognized that Eqs 5.1-3 and 5.1-4 represent more than a lexicographic ordering between an array and a vector; these equations define mathematical operators that may be manipulated analytically. Numerous examples of the applications of the stacking operators are given in subsequent sections.

5.2. GENERALIZED TWO-DIMENSIONAL LINEAR OPERATOR

A large class of image processing operations are linear in nature; an output image field is formed from linear combinations of pixels of an input image field. Such operations include superposition, convolution, unitary transformation, and discrete linear filtering.

Consider the $N_1 \times N_2$ element input image array $F(n_1, n_2)$. A generalized linear operation on this image field results in a $M_1 \times M_2$ output image array $P(m_1, m_2)$ as defined by

$$P(m_1, m_2) = \sum_{n_1=1}^{N_1} \sum_{n_2=1}^{N_2} F(n_1, n_2) O(n_1, n_2; m_1, m_2) \qquad (5.2\text{-}1)$$

where the operator kernel $O(n_1, n_2; m_1, m_2)$ represents a weighting constant, which, in general, is a function of both input and output image coordinates (1).

For the analysis of linear image processing operations, it is convenient to adopt the vector-space formulation developed in Section 5.1. Thus, let the input image array $F(n_1, n_2)$ be represented as matrix **F** or alternatively, as a vector **f** obtained by column scanning **F**. Similarly, let the output image array $P(m_1, m_2)$ be represented by the matrix **P** or the column-scanned vector **p**. For notational simplicity, in the subsequent discussions, the input and output image arrays are assumed to be square and of dimensions $N_1 = N_2 = N$ and $M_1 = M_2 = M$, respectively. Now, let **T** denote the $M^2 \times N^2$ matrix performing a linear transformation on the $N^2 \times 1$ input image vector **f** yielding the $M^2 \times 1$ output image vector

$$\mathbf{p} = \mathbf{Tf} \qquad (5.2\text{-}2)$$

The matrix $\mathbf{T}$ may be partitioned into $M \times N$ submatrices $\mathbf{T}_{mn}$ and written as

$$
\mathbf{T} = \begin{bmatrix} \mathbf{T}_{11} & \mathbf{T}_{12} & \cdots & \mathbf{T}_{1N} \\ \mathbf{T}_{21} & \mathbf{T}_{22} & \cdots & \mathbf{T}_{2N} \\ \vdots & \vdots & & \vdots \\ \mathbf{T}_{M1} & \mathbf{T}_{M2} & \cdots & \mathbf{T}_{MN} \end{bmatrix}
\tag{5.2-3}
$$

From Eq. 5.1-3, it is possible to relate the output image vector $\mathbf{p}$ to the input image matrix $\mathbf{F}$ by the equation

$$
\mathbf{p} = \sum_{n=1}^{N} \mathbf{TN}_n \mathbf{Fv}_n
\tag{5.2-4}
$$

Furthermore, from Eq. 5.1-4, the output image matrix $\mathbf{P}$ is related to the input image vector $\mathbf{p}$ by

$$
\mathbf{P} = \sum_{m=1}^{M} \mathbf{M}_m^T \mathbf{p} \mathbf{u}_m^T
\tag{5.2-5}
$$

Combining the above yields the relation between the input and output image matrices,

$$
\mathbf{P} = \sum_{m=1}^{M} \sum_{n=1}^{N} (\mathbf{M}_m^T \mathbf{TN}_n) \mathbf{F}(\mathbf{v}_n \mathbf{u}_m^T)
\tag{5.2-6}
$$

where it is observed that the operators $\mathbf{M}_m$ and $\mathbf{N}_n$ simply extract the partition $\mathbf{T}_{mn}$ from $\mathbf{T}$. Hence

$$
\mathbf{P} = \sum_{m=1}^{M} \sum_{n=1}^{N} \mathbf{T}_{mn} \mathbf{F}(\mathbf{v}_n \mathbf{u}_m^T)
\tag{5.2-7}
$$

If the linear transformation is separable such that $\mathbf{T}$ may be expressed in the direct product form

$$
\mathbf{T} = \mathbf{T}_C \otimes \mathbf{T}_R
\tag{5.2-8}
$$

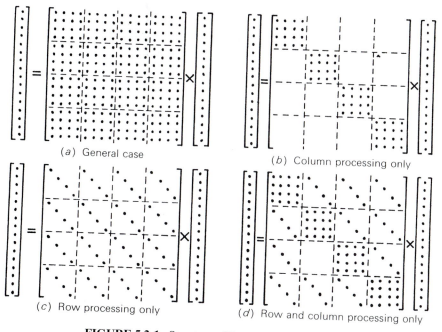

(a) General case

(b) Column processing only

(c) Row processing only

(d) Row and column processing only

FIGURE 5.2-1. Structure of linear operator matrices.

where $\mathbf{T}_R$ and $\mathbf{T}_C$ are row and column operators on $\mathbf{F}$, then

$$\mathbf{T}_{mn} = T_R(m, n)\mathbf{T}_C \qquad (5.2\text{-}9)$$

As a consequence,

$$\mathbf{P} = \mathbf{T}_C \mathbf{F} \sum_{m=1}^{M} \sum_{n=1}^{N} T_R(m, n)\mathbf{v}_n \mathbf{u}_m^T = \mathbf{T}_C \mathbf{F} \mathbf{T}_R^T \qquad (5.2\text{-}10)$$

Hence the output image matrix $\mathbf{P}$ can be produced by sequential row and column operations.

In many image processing applications, the linear transformations operator $\mathbf{T}$ is highly structured, and computational simplifications are possible. Special cases of interest are listed below and illustrated in Figure 5.2-1 for the case in which the input and output images are of the same dimension, $M = N$.

1. *Column processing of* **F**:

$$\mathbf{T} = \text{diag}[\mathbf{T}_{C1}, \mathbf{T}_{C2}, ..., \mathbf{T}_{CN}] \qquad (5.2\text{-}11)$$

where $\mathbf{T}_{Cj}$ is the transformation matrix for the *j*th column.

2. *Identical column processing of* **F**:

$$\mathbf{T} = \text{diag}[\mathbf{T}_C, \mathbf{T}_C, ..., \mathbf{T}_C] = \mathbf{T}_C \otimes \mathbf{I}_N \qquad (5.2\text{-}12)$$

3. *Row processing of* **F**:

$$\mathbf{T}_{mn} = \text{diag}[T_{R1}(m, n), T_{R2}(m, n), ..., T_{RN}(m, n)] \qquad (5.2\text{-}13)$$

where $\mathbf{T}_{Rj}$ is the transformation matrix for the *j*th row.

4. *Identical row processing of* **F**:

$$\mathbf{T}_{mn} = \text{diag}[T_R(m, n), T_R(m, n), ..., T_R(m, n)] \qquad (5.2\text{-}14a)$$

and

$$\mathbf{T} = \mathbf{I}_N \otimes \mathbf{T}_R \qquad (5.2\text{-}14b)$$

5. *Identical row and identical column processing of* **F**:

$$\mathbf{T} = \mathbf{T}_C \otimes \mathbf{I}_N + \mathbf{I}_N \otimes \mathbf{T}_R \qquad (5.2\text{-}15)$$

The number of computational operations for each of these cases is tabulated in Table 5.2-1.

Equation 5.2-10 indicates that separable two-dimensional linear transforms can be computed by sequential one-dimensional row and column operations on a data array. As indicated by Table 5.2-1, a considerable savings in computation is possible for such transforms: computation by Eq 5.2-2 in the general case requires M^2N^2 operations; computation by Eq. 5.2-10, when it applies, requires only $MN^2 + M^2N$ operations. Furthermore, **F** may be stored in a serial memory and fetched line by line. With this technique, however, it is necessary to transpose the result of the column transforms in order to perform the row transforms. References 2 and 3 describe algorithms for line storage matrix transposition.

TABLE 5.2-1. Computational Requirements for Linear Transform Operator

Case	Operations (Multiply and Add)
General	N^4
Column processing	N^3
Row processing	N^3
Row and column processing	$2N^3 - N^2$
Separable row and column processing matrix form	$2N^3$

5.3. IMAGE STATISTICAL CHARACTERIZATION

The statistical descriptors of continuous images presented in Chapter 1 can be applied directly to characterize discrete images. In this section, expressions are developed for the statistical moments of discrete image arrays. Joint probability density models for discrete image fields are described in the following section. Reference 4 provides background information for this subject.

The moments of a discrete image process may be expressed conveniently in vector-space form. The mean value of the discrete image function is a matrix of the form

$$E\{\mathbf{F}\} = [E\{F(n_1, n_2)\}] \qquad (5.3\text{-}1)$$

If the image array is written as a column-scanned vector, the mean of the image vector is

$$\boldsymbol{\eta_f} = E\{\mathbf{f}\} = \sum_{n=1}^{N_2} \mathbf{N}_n E\{\mathbf{F}\} \mathbf{v}_n \qquad (5.3\text{-}2)$$

The correlation function of the image array is given by

$$R(n_1, n_2 ; n_3, n_4) = E\{F(n_1, n_2)F^*(n_3, n_4)\} \qquad (5.3\text{-}3)$$

where the n_i represent points of the image array. Similarly, the covariance function of the image array is

$$K(n_1, n_2; n_3, n_4) = E\{[F(n_1, n_2) - E\{F(n_1, n_2)\}][F^*(n_3, n_4) - E\{F^*(n_3, n_4)\}]\}$$
$$(5.3\text{-}4)$$

Finally, the variance function of the image array is obtained directly from the covariance function as

$$\sigma^2(n_1, n_2) = K(n_1, n_2; n_1, n_2) \tag{5.3-5}$$

If the image array is represented in vector form, the correlation matrix of $\mathbf{f}$ can be written in terms of the correlation of elements of $\mathbf{F}$ as

$$\mathbf{R_f} = E\{\mathbf{ff}^{*T}\} = E\left\{ \left(\sum_{m=1}^{N_2} \mathbf{N}_m \mathbf{F} \mathbf{v}_m \right) \left(\sum_{n=1}^{N_2} \mathbf{v}_n^T \mathbf{F}^{*T} \mathbf{N}_n^T \right) \right\} \tag{5.3-6a}$$

or

$$\mathbf{R_f} = \sum_{m=1}^{N_2} \sum_{n=1}^{N_2} \mathbf{N}_m E\left\{ \mathbf{F} \mathbf{v}_m \mathbf{v}_n^T \mathbf{F}^{*T} \right\} \mathbf{N}_n^T \tag{5.3-6b}$$

The term

$$E\left\{ \mathbf{F} \mathbf{v}_m \mathbf{v}_n^T \mathbf{F}^{*T} \right\} = \mathbf{R}_{mn} \tag{5.3-7}$$

is the $N_1 \times N_1$ correlation matrix of the mth and nth columns of $\mathbf{F}$. Hence it is possible to express $\mathbf{R_f}$ in partitioned form as

$$\mathbf{R_f} = \begin{bmatrix} \mathbf{R}_{11} & \mathbf{R}_{12} & \cdots & \mathbf{R}_{1N_2} \\ \mathbf{R}_{21} & \mathbf{R}_{22} & \cdots & \mathbf{R}_{2N_2} \\ \vdots & \vdots & & \vdots \\ \mathbf{R}_{N_2 1} & \mathbf{R}_{N_2 2} & \cdots & \mathbf{R}_{N_2 N_2} \end{bmatrix} \tag{5.3-8}$$

The covariance matrix of $\mathbf{f}$ can be found from its correlation matrix and mean vector by the relation

$$\mathbf{K_f} = \mathbf{R_f} - \boldsymbol{\eta}_f \boldsymbol{\eta}_f^{*T} \tag{5.3-9}$$

A variance matrix $\mathbf{V_F}$ of the array $F(n_1, n_2)$ is defined as a matrix whose elements represent the variances of the corresponding elements of the array. The elements of this matrix may be extracted directly from the covariance matrix partitions of $\mathbf{K_f}$. That is,

$$\mathbf{V_F}(n_1, n_2) = \mathbf{K}_{n_2, n_2}(n_1, n_1) \tag{5.3-10}$$

If the image matrix $\mathbf{F}$ is wide-sense stationary, the correlation function can be expressed as

$$R(n_1, n_2; n_3, n_4) = R(n_1 - n_3, n_2 - n_4) = R(j, k) \tag{5.3-11}$$

where $j = n_1 - n_3$ and $k = n_2 - n_4$. Correspondingly, the covariance matrix partitions of Eq. 5.3-9 are related by

$$\mathbf{K}_{mn} = \mathbf{K}_k \qquad m \geq n \tag{5.3-12a}$$

$$\mathbf{K}^*_{mn} = \mathbf{K}^*_k \qquad m < n \tag{5.3-12b}$$

where $k = |m - n| + 1$. Hence, for a wide-sense-stationary image array

$$\mathbf{K_f} = \begin{bmatrix} \mathbf{K}_1 & \mathbf{K}_2 & \cdots & \mathbf{K}_{N_2} \\ \mathbf{K}^*_2 & \mathbf{K}_1 & \cdots & \mathbf{K}_{N_2-1} \\ \vdots & \vdots & & \vdots \\ \mathbf{K}^*_{N_2} & \mathbf{K}^*_{N_2-1} & \cdots & \mathbf{K}_1 \end{bmatrix} \tag{5.3-13}$$

The matrix of Eq. 5.3-13 is of block Toeplitz form (5). Finally, if the covariance between elements is separable into the product of row and column covariance functions, then the covariance matrix of the image vector can be expressed as the direct product of row and column covariance matrices. Under this condition

$$\mathbf{K_f} = \mathbf{K}_C \otimes \mathbf{K}_R = \begin{bmatrix} \mathbf{K}_R(1, 1)\mathbf{K}_C & \mathbf{K}_R(1, 2)\mathbf{K}_C & \cdots & \mathbf{K}_R(1, N_2)\mathbf{K}_C \\ \mathbf{K}_R(2, 1)\mathbf{K}_C & \mathbf{K}_R(2, 2)\mathbf{K}_C & \cdots & \mathbf{K}_R(2, N_2)\mathbf{K}_C \\ \vdots & \vdots & & \vdots \\ \mathbf{K}_R(N_2, 1)\mathbf{K}_C & \mathbf{K}_R(N_2, 2)\mathbf{K}_C & \cdots & \mathbf{K}_R(N_2, N_2)\mathbf{K}_C \end{bmatrix} \tag{5.3-14}$$

where $\mathbf{K}_C$ is a $N_1 \times N_1$ covariance matrix of each column of $\mathbf{F}$ and $\mathbf{K}_R$ is a $N_2 \times N_2$ covariance matrix of the rows of $\mathbf{F}$.

As a special case, consider the situation in which adjacent pixels along an image row have a correlation of $(0.0 \leq \rho_R \leq 1.0)$ and a self-correlation of unity. Then the covariance matrix reduces to

$$\mathbf{K}_R = \sigma_R^2 \begin{bmatrix} 1 & \rho_R & \cdots & \rho_R^{N_2-1} \\ \rho_R & 1 & \cdots & \rho_R^{N_2-2} \\ \vdots & \vdots & & \vdots \\ \rho_R^{N_2-1} & \rho_R^{N_2-2} & \cdots & 1 \end{bmatrix} \tag{5.3-15}$$

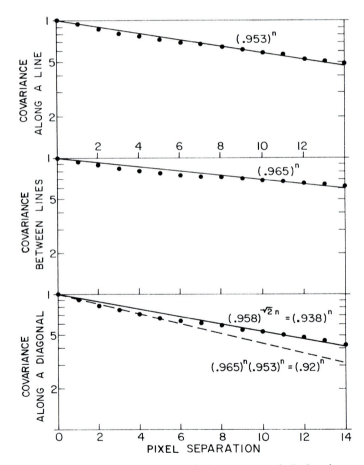

FIGURE 5.3-1. Covariance measurements of the `smpte_girl_luminance` monochrome image.

FIGURE 5.3-2. Photograph of `smpte_girl_luminance` image.

where σ_R^2 denotes the variance of pixels along a row. This is an example of the covariance matrix of a *Markov process*, analogous to the continuous autocovariance function $\exp(-\alpha|x|)$. Figure 5.3-1 contains a plot by Davisson (6) of the measured covariance of pixels along an image line of the monochrome image of Figure 5.3-2. The data points can be fit quite well with a Markov covariance function with $\rho = 0.953$. Similarly, the covariance between lines can be modeled well with a Markov covariance function with $\rho = 0.965$. If the horizontal and vertical covariances were exactly separable, the covariance function for pixels along the image diagonal would be equal to the product of the horizontal and vertical axis covariance functions. In this example, the approximation was found to be reasonably accurate for up to five pixel separations.

The discrete power-spectral density of a discrete image random process may be defined, in analogy with the continuous power spectrum of Eq. 1.4-13, as the two-dimensional discrete Fourier transform of its stationary autocorrelation function. Thus, from Eq. 5.3-11

$$\mathcal{W}(u, v) = \sum_{j=0}^{N_1-1} \sum_{k=0}^{N_2-1} R(j, k) \exp\left\{-2\pi i\left(\frac{ju}{N_1} + \frac{kv}{N_2}\right)\right\} \qquad (5.3\text{-}16)$$

Figure 5.3-3 shows perspective plots of the power-spectral densities for separable and circularly symmetric Markov processes.

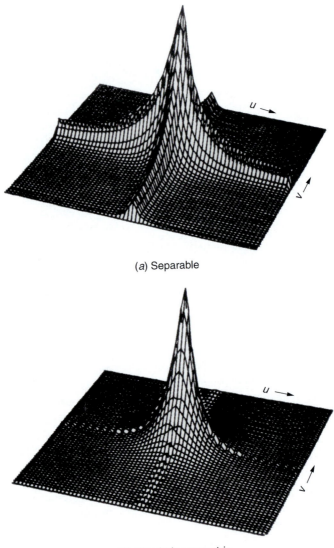

(*a*) Separable

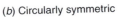

(*b*) Circularly symmetric

FIGURE 5.3-3. Power spectral densities of Markov process sources; $N = 256$, log magnitude displays.

5.4. IMAGE PROBABILITY DENSITY MODELS

A discrete image array $F(n_1, n_2)$ can be completely characterized statistically by its joint probability density, written in matrix form as

$$p(\mathbf{F}) \equiv p\{F(1, 1), F(2, 1), ..., F(N_1, N_2)\} \tag{5.4-1a}$$

or in corresponding vector form as

$$p(\mathbf{f}) \equiv p\{f(1), f(2), ..., f(Q)\} \tag{5.4-1b}$$

where $Q = N_1 \cdot N_2$ is the order of the joint density. If all pixel values are statistically independent, the joint density factors into the product

$$p(\mathbf{f}) \equiv p\{f(1)\}p\{f(2)\}...p\{f(Q)\} \tag{5.4-2}$$

of its first-order marginal densities.

The most common joint probability density is the joint Gaussian, which may be expressed as

$$p(\mathbf{f}) = (2\pi)^{-Q/2}|\mathbf{K_f}|^{-1/2}\exp\left\{-\frac{1}{2}(\mathbf{f} - \boldsymbol{\eta_f})^T \mathbf{K_f}^{-1}(\mathbf{f} - \boldsymbol{\eta_f})\right\} \tag{5.4-3}$$

where $\mathbf{K_f}$ is the covariance matrix of $\mathbf{f}$, $\boldsymbol{\eta_f}$ is the mean of $\mathbf{f}$ and $|\mathbf{K_f}|$ denotes the determinant of $\mathbf{K_f}$. The joint Gaussian density is useful as a model for the density of unitary transform coefficients of an image. However, the Gaussian density is not an adequate model for the luminance values of an image because luminance is a positive quantity and the Gaussian variables are bipolar.

Expressions for joint densities, other than the Gaussian density, are rarely found in the literature. Huhns (7) has developed a technique of generating high-order densities in terms of specified first-order marginal densities and a specified covariance matrix between the ensemble elements.

In Chapter 6, techniques are developed for quantizing variables to some discrete set of values called *reconstruction levels*. Let $r_{j_q}(q)$ denote the reconstruction level of the pixel at vector coordinate (q). Then the probability of occurrence of the possible states of the image vector can be written in terms of the joint probability distribution as

$$P(\mathbf{f}) = p\{f(1) = r_{j_1}(1)\}p\{f(2) = r_{j_2}(2)\}...p\{f(Q) = r_{j_Q}(Q)\} \tag{5.4-4}$$

where $0 \le j_q \le j_Q = J - 1$. Normally, the reconstruction levels are set identically for each vector component and the joint probability distribution reduces to

$$P(\mathbf{f}) = p\{f(1) = r_{j_1}\}p\{f(2) = r_{j_2}\}...p\{f(Q) = r_{j_Q}\} \tag{5.4-5}$$

Probability distributions of image values can be estimated by *histogram measurements*. For example, the first-order probability distribution

$$P[f(q)] = P_R[f(q) = r_j] \qquad (5.4-6)$$

of the amplitude value at vector coordinate q can be estimated by examining a large collection of images representative of a given image class (e.g., chest x-rays, aerial scenes of crops). The *first-order histogram* estimate of the probability distribution is the frequency ratio

$$H_E(j; q) = \frac{N_p(j)}{N_p} \qquad (5.4-7)$$

where N_p represents the total number of images examined and $N_p(j)$ denotes the number for which $f(q) = r_j$ for $j = 0, 1,..., J - 1$. If the image source is statistically stationary, the first-order probability distribution of Eq. 5.4-6 will be the same for all vector components q. Furthermore, if the image source is *ergodic*, ensemble averages (measurements over a collection of pictures) can be replaced by spatial averages. Under the ergodic assumption, the first-order probability distribution can be estimated by measurement of the spatial histogram

$$H_S(j) = \frac{N_S(j)}{Q} \qquad (5.4-8)$$

where $N_S(j)$ denotes the number of pixels in an image for which $f(q) = r_j$ for $1 \leq q \leq Q$ and $0 \leq j \leq J - 1$. For example, for an image with 256 gray levels, $H_S(j)$ denotes the number of pixels possessing gray level j for $0 \leq j \leq 255$.

Figure 5.4-1 shows first-order histograms of the red, green, and blue components of a color image. Most natural images possess many more dark pixels than bright pixels, and their histograms tend to fall off exponentially at higher luminance levels.

Estimates of the second-order probability distribution for ergodic image sources can be obtained by measurement of the *second-order spatial histogram*, which is a measure of the joint occurrence of pairs of pixels separated by a specified distance. With reference to Figure 5.4-2, let $F(n_1, n_2)$ and $F(n_3, n_4)$ denote a pair of pixels separated by r radial units at an angle θ with respect to the horizontal axis. As a consequence of the rectilinear grid, the separation parameters may only assume certain discrete values. The second-order spatial histogram is then the frequency ratio

$$H_S(j_1, j_2; r, \theta) = \frac{N_S(j_1, j_2)}{Q_T} \qquad (5.4-9)$$

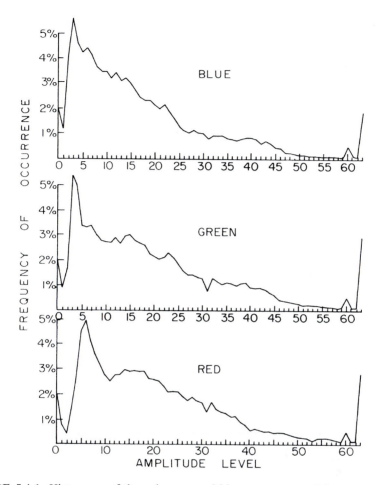

FIGURE 5.4-1. Histograms of the red, green and blue components of the `smpte_girl _linear` color image.

where $N_S(j_1, j_2)$ denotes the number of pixel pairs for which $F(n_1, n_2) = r_{j_1}$ and $F(n_3, n_4) = r_{j_2}$. The factor Q_T in the denominator of Eq. 5.4-9 represents the total number of pixels lying in an image region for which the separation is (r, θ). Because of boundary effects, $Q_T < Q$.

Second-order spatial histograms of a monochrome image are presented in Figure 5.4-3 as a function of pixel separation distance and angle. As the separation increases, the pairs of pixels become less correlated and the histogram energy tends to spread more uniformly about the plane.

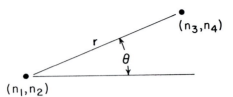

FIGURE 5.4-2. Geometric relationships of pixel pairs.

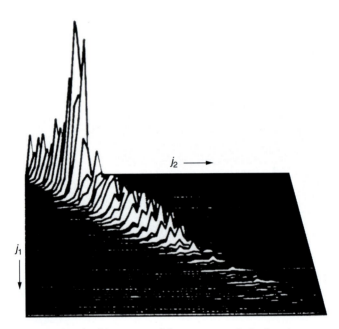

FIGURE 5.4-3. Second-order histogram of the `smpte_girl_luminance` monochrome image; $r = 1$ and $\theta = 0$.

5.5. LINEAR OPERATOR STATISTICAL REPRESENTATION

If an input image array is considered to be a sample of a random process with known first and second-order moments, the first- and second-order moments of the output image array can be determined for a given linear transformation. First, the mean of the output image array is

$$E\{P(m_1, m_2)\} = E\left\{ \sum_{n_1=1}^{N_1} \sum_{n_2=1}^{N_2} F(n_1, n_2) O(n_1, n_2 ; m_1, m_2) \right\} \qquad (5.5\text{-}1a)$$

Because the expectation operator is linear,

$$E\{P(m_1, m_2)\} = \sum_{n_1=1}^{N_1} \sum_{n_2=1}^{N_2} E\{F(n_1, n_2)\}O(n_1, n_2; m_1, m_2) \tag{5.5-1b}$$

The correlation function of the output image array is

$$R_P(m_1, m_2; m_3, m_4) = E\{P(m_1, m_2)P^*(m_3, m_4)\} \tag{5.5-2a}$$

or in expanded form

$$R_P(m_1, m_2; m_3, m_4) = E\left\{\left[\sum_{n_1=1}^{N_1} \sum_{n_2=1}^{N_2} F(n_1, n_2)O(n_1, n_2; m_1, m_2)\right] \times \right.$$

$$\left. \left[\sum_{n_3=1}^{N_1} \sum_{n_4=1}^{N_2} F^*(n_3, n_4)O^*(n_3, n_3; m_3, m_4)\right]\right\} \tag{5.5-2b}$$

After multiplication of the series and performance of the expectation operation, one obtains

$$R_P(m_1, m_2; m_3, m_4) = \sum_{n_1=1}^{N_1} \sum_{n_2=1}^{N_2} \sum_{n_3=1}^{N_1} \sum_{n_4=1}^{N_2} R_F(n_1, n_2, n_3, n_4)O(n_1, n_2; m_1, m_2)$$

$$\times O^*(n_3, n_3; m_3, m_4) \tag{5.5-3}$$

where $R_F(n_1, n_2; n_3, n_4)$ represents the correlation function of the input image array. In a similar manner, the covariance function of the output image is found to be

$$K_P(m_1, m_2; m_3, m_4) = \sum_{n_1=1}^{N_1} \sum_{n_2=1}^{N_2} \sum_{n_3=1}^{N_1} \sum_{n_4=1}^{N_2} K_F(n_1, n_2, n_3, n_4)O(n_1, n_2; m_1, m_2)$$

$$\times O^*(n_3, n_3; m_3, m_4) \tag{5.5-4}$$

If the input and output image arrays are expressed in vector form, the formulation of the moments of the transformed image becomes much more compact. The mean of the output vector $\mathbf{p}$ is

$$\boldsymbol{\eta}_\mathbf{p} = E\{\mathbf{p}\} = E\{\mathbf{Tf}\} = \mathbf{T}E\{\mathbf{f}\} = \mathbf{T}\boldsymbol{\eta}_\mathbf{f} \qquad (5.5\text{-}5)$$

and the correlation matrix of $\mathbf{p}$ is

$$\mathbf{R}_\mathbf{p} = E\{\mathbf{pp}^{*T}\} = E\{\mathbf{Tff}^{*T}\mathbf{T}^{*T}\} = \mathbf{TR}_\mathbf{f}\mathbf{T}^{*T} \qquad (5.5\text{-}6)$$

Finally, the covariance matrix of $\mathbf{p}$ is

$$\mathbf{K}_\mathbf{p} = \mathbf{TK}_\mathbf{f}\mathbf{T}^{*T} \qquad (5.5\text{-}7)$$

Applications of this theory to superposition and unitary transform operators are given in following chapters.

A special case of the general linear transformation $\mathbf{p} = \mathbf{Tf}$, of fundamental importance, occurs when the covariance matrix of Eq. 5.5-7 assumes the form

$$\mathbf{K}_\mathbf{p} = \mathbf{TK}_\mathbf{f}\mathbf{T}^{*T} = \boldsymbol{\Lambda} \qquad (5.5\text{-}8)$$

where $\boldsymbol{\Lambda}$ is a diagonal matrix. In this case, the elements of $\mathbf{p}$ are uncorrelated. From Appendix A1.2, it is found that the transformation $\mathbf{T}$, which produces the diagonal matrix $\boldsymbol{\Lambda}$, has rows that are eigenvectors of $\mathbf{K}_\mathbf{f}$. The diagonal elements of $\boldsymbol{\Lambda}$ are the corresponding eigenvalues of $\mathbf{K}_\mathbf{f}$. This operation is called both a *matrix diagonalization* and a *principal components transformation*.

REFERENCES

1. W. K. Pratt, "Vector Formulation of Two Dimensional Signal Processing Operations," *Computer Graphics and Image Processing*, **4**, 1, March 1975, 1–24.

2. J. O. Eklundh, "A Fast Computer Method for Matrix Transposing," *IEEE Trans. Computers*, **C-21**, 7, July 1972, 801–803.

3. R. E. Twogood and M. P. Ekstrom, "An Extension of Eklundh's Matrix Transposition Algorithm and Its Applications in Digital Image Processing," *IEEE Trans. Computers*, **C-25**, 9, September 1976, 950–952.

4. A. Papoulis, *Probability, Random Variables, and Stochastic Processes*, 3rd ed., McGraw-Hill, New York, 1991.

5. U. Grenander and G. Szego, *Toeplitz Forms and Their Applications*, University of California Press, Berkeley, CA, 1958.

6. L. D. Davisson, private communication.

7. M. N. Huhns, "Optimum Restoration of Quantized Correlated Signals," USCIPI Report 600, University of Southern California, Image Processing Institute, Los Angeles August 1975.

6

IMAGE QUANTIZATION

Any analog quantity that is to be processed by a digital computer or digital system must be converted to an integer number proportional to its amplitude. The conversion process between analog samples and discrete-valued samples is called *quantization*. The following section includes an analytic treatment of the quantization process, which is applicable not only for images but for a wide class of signals encountered in image processing systems. Section 6.2 considers the processing of quantized variables. The last section discusses the subjective effects of quantizing monochrome and color images.

6.1. SCALAR QUANTIZATION

Figure 6.1-1 illustrates a typical example of the quantization of a scalar signal. In the quantization process, the amplitude of an analog signal sample is compared to a set of decision levels. If the sample amplitude falls between two decision levels, it is quantized to a fixed reconstruction level lying in the quantization band. In a digital system, each quantized sample is assigned a binary code. An equal-length binary code is indicated in the example.

For the development of quantitative scalar signal quantization techniques, let f and $\hat{f}$ represent the amplitude of a real, scalar signal sample and its quantized value, respectively. It is assumed that f is a sample of a random process with known probability density $p(f)$. Furthermore, it is assumed that f is constrained to lie in the range

$$a_L \leq f \leq a_U \tag{6.1-1}$$

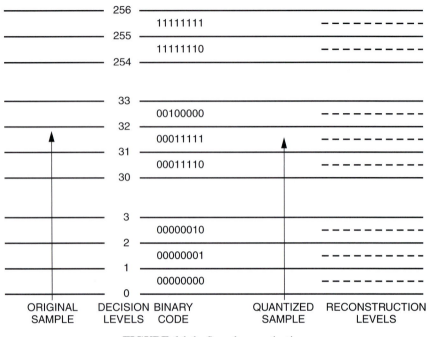

FIGURE 6.1-1. Sample quantization.

where a_U and a_L represent upper and lower limits.

Quantization entails specification of a set of decision levels d_j and a set of reconstruction levels r_j such that if

$$d_j \le f < d_{j+1} \qquad (6.1\text{-}2)$$

the sample is quantized to a reconstruction value r_j. Figure 6.1-2*a* illustrates the placement of decision and reconstruction levels along a line for J quantization levels. The staircase representation of Figure 6.1-2*b* is another common form of description.

Decision and reconstruction levels are chosen to minimize some desired quantization error measure between f and $\hat{f}$. The quantization error measure usually employed is the mean-square error because this measure is tractable, and it usually correlates reasonably well with subjective criteria. For J quantization levels, the mean-square quantization error is

$$\mathcal{E} = E\{(f-\hat{f})^2\} = \int_{a_L}^{a_U} (f-\hat{f})^2 p(f) \ df = \sum_{j=0}^{J-1} (f-r_j)^2 p(f) \ df \qquad (6.1\text{-}3)$$

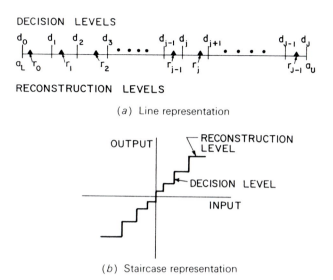

(a) Line representation

(b) Staircase representation

FIGURE 6.1-2. Quantization decision and reconstruction levels.

For a large number of quantization levels J, the probability density may be represented as a constant value $p(r_j)$ over each quantization band. Hence

$$\mathcal{E} = \sum_{j=0}^{J-1} p(r_j) \int_{d_j}^{d_{j+1}} (f - r_j)^2 \, df \tag{6.1-4}$$

which evaluates to

$$\mathcal{E} = \frac{1}{3} \sum_{j=0}^{J-1} p(r_j)[(d_{j+1} - r_j)^3 - (d_j - r_j)^3] \tag{6.1-5}$$

The optimum placing of the reconstruction level r_j within the range d_{j-1} to d_j can be determined by minimization of $\mathcal{E}$ with respect to r_j. Setting

$$\frac{d\mathcal{E}}{dr_j} = 0 \tag{6.1-6}$$

yields

$$r_j = \frac{d_{j+1} + d_j}{2} \tag{6.1-7}$$

Therefore, the optimum placement of reconstruction levels is at the midpoint between each pair of decision levels. Substitution for this choice of reconstruction levels into the expression for the quantization error yields

$$\mathcal{E} = \frac{1}{12} \sum_{j=0}^{J-1} p(r_j)(d_{j+1} - d_j)^3 \tag{6.1-8}$$

The optimum choice for decision levels may be found by minimization of $\mathcal{E}$ in Eq. 6.1-8 by the method of Lagrange multipliers. Following this procedure, Panter and Dite (1) found that the decision levels may be computed to a good approximation from the integral equation

$$d_j = \frac{(a_U - a_L)\int_{a_L}^{a_j} [p(f)]^{-1/3}\, df}{\int_{a_L}^{a_U} [p(f)]^{-1/3}\, df} \tag{6.1-9a}$$

where

$$a_j = \frac{j(a_U - a_L)}{J} + a_L \tag{6.1-9b}$$

for $j = 0, 1,..., J$. If the probability density of the sample is uniform, the decision levels will be uniformly spaced. For nonuniform probability densities, the spacing of decision levels is narrow in large-amplitude regions of the probability density function and widens in low-amplitude portions of the density. Equation 6.1-9 does not reduce to closed form for most probability density functions commonly encountered in image processing systems models, and hence the decision levels must be obtained by numerical integration.

If the number of quantization levels is not large, the approximation of Eq. 6.1-4 becomes inaccurate, and exact solutions must be explored. From Eq. 6.1-3, setting the partial derivatives of the error expression with respect to the decision and reconstruction levels equal to zero yields

$$\frac{\partial \mathcal{E}}{\partial d_j} = (d_j - r_j)^2 p(d_j) - (d_j - r_{j-1})^2 p(d_j) = 0 \tag{6.1-10a}$$

$$\frac{\partial \mathcal{E}}{\partial r_j} = 2\int_{d_j}^{d_{j+1}} (f - r_j) p(f)\, df = 0 \tag{6.1-10b}$$

Upon simplification, the set of equations

$$r_j = 2d_j - r_{j-1} \tag{6.1-11a}$$

$$r_j = \frac{\int_{d_j}^{d_{j+1}} f p(f)\, df}{\int_{d_j}^{d_{j+1}} p(f)\, df} \tag{6.1-11b}$$

is obtained. Recursive solution of these equations for a given probability distribution $p(f)$ provides optimum values for the decision and reconstruction levels. Max (2) has developed a solution for optimum decision and reconstruction levels for a Gaussian density and has computed tables of optimum levels as a function of the number of quantization steps. Table 6.1-1 lists placements of decision and quantization levels for uniform, Gaussian, Laplacian, and Rayleigh densities for the *Max quantizer.*

If the decision and reconstruction levels are selected to satisfy Eq. 6.1-11, it can easily be shown that the mean-square quantization error becomes

$$\mathcal{E}_{\min} = \sum_{j=0}^{J-1} \left[\int_{d_j}^{d_{j+1}} f^2 p(f)\, df - r_j^2 \int_{d_j}^{d_{j+1}} p(f)\, df \right] \tag{6.1-12}$$

In the special case of a uniform probability density, the minimum mean-square quantization error becomes

$$\mathcal{E}_{\min} = \frac{1}{12 J^2} \tag{6.1-13}$$

Quantization errors for most other densities must be determined by computation.

It is possible to perform nonlinear quantization by a companding operation, as shown in Figure 6.1-3, in which the sample is transformed nonlinearly, linear quantization is performed, and the inverse nonlinear transformation is taken (3). In the companding system of quantization, the probability density of the transformed samples is forced to be uniform. Thus, from Figure 6.1-3, the transformed sample value is

$$g = T\{f\} \tag{6.1-14}$$

where the nonlinear transformation $T\{\cdot\}$ is chosen such that the probability density of g is uniform. Thus,

FIGURE 6.1-3. Companding quantizer.

TABLE 6.1-1. Placement of Decision and Reconstruction Levels for Max Quantizer

Bits	Uniform		Gaussian		Laplacian		Rayleigh	
	d_i	r_i	d_i	r_i	d_i	r_i	d_i	r_i
1	−1.0000	−0.5000	−∞	−0.7979	−∞	−0.7071	0.0000	1.2657
	0.0000	0.5000	0.0000	0.7979	0.0000	0.7071	2.0985	2.9313
	1.0000		∞		−∞		∞	
2	−1.0000	−0.7500	−∞	−1.5104	∞	−1.8340	0.0000	0.8079
	−0.5000	−0.2500	−0.9816	−0.4528	−1.1269	−0.4198	1.2545	1.7010
	−0.0000	0.2500	0.0000	0.4528	0.0000	0.4198	2.1667	2.6325
	0.5000	0.7500	0.9816	1.5104	1.1269	1.8340	3.2465	3.8604
	1.0000		∞		∞		∞	
3	−1.0000	−0.8750	−∞	−2.1519	−∞	−3.0867	0.0000	0.5016
	−0.7500	−0.6250	−1.7479	−1.3439	−2.3796	−1.6725	0.7619	1.0222
	−0.5000	−0.3750	−1.0500	−0.7560	−1.2527	−0.8330	1.2594	1.4966
	−0.2500	−0.1250	−0.5005	−0.2451	−0.5332	−0.2334	1.7327	1.9688
	0.0000	0.1250	0.0000	0.2451	0.0000	0.2334	2.2182	2.4675
	0.2500	0.3750	0.5005	0.7560	0.5332	0.8330	2.7476	3.0277
	0.5000	0.6250	1.0500	1.3439	1.2527	1.6725	3.3707	3.7137
	0.7500	0.8750	1.7479	2.1519	2.3796	3.0867	4.2124	4.7111
	1.0000		∞		∞		∞	
4	−1.0000	−0.9375	−∞	−2.7326	−∞	−4.4311	0.0000	0.3057
	−0.8750	−0.8125	−2.4008	−2.0690	−3.7240	−3.0169	0.4606	0.6156
	−0.7500	−0.6875	−1.8435	−1.6180	−2.5971	−2.1773	0.7509	0.8863
	−0.6250	−0.5625	−1.4371	−1.2562	−1.8776	−1.5778	1.0130	1.1397
	−0.5000	−0.4375	−1.0993	−0.9423	−1.3444	−1.1110	1.2624	1.3850
	−0.3750	−0.3125	−0.7995	−0.6568	−0.9198	−0.7287	1.5064	1.6277
	−0.2500	−0.1875	−0.5224	−0.3880	−0.5667	−0.4048	1.7499	1.8721
	−0.1250	−0.0625	−0.2582	−0.1284	−0.2664	−0.1240	1.9970	2.1220
	0.0000	0.0625	0.0000	0.1284	0.0000	0.1240	2.2517	2.3814
	0.1250	0.1875	0.2582	0.3880	0.2644	0.4048	2.5182	2.6550
	0.2500	0.3125	0.5224	0.6568	0.5667	0.7287	2.8021	2.9492
	0.3750	0.4375	0.7995	0.9423	0.9198	1.1110	3.1110	3.2729
	0.5000	0.5625	1.0993	1.2562	1.3444	1.5778	3.4566	3.6403
	0.6250	0.6875	1.4371	1.6180	1.8776	2.1773	3.8588	4.0772
	0.7500	0.8125	1.8435	2.0690	2.5971	3.0169	4.3579	4.6385
	0.8750	0.9375	2.4008	2.7326	3.7240	4.4311	5.0649	5.4913
	1.0000		∞		∞		∞	

$$p(g) = 1 \tag{6.1-15}$$

for $-\frac{1}{2} \le g \le \frac{1}{2}$. If f is a zero mean random variable, the proper transformation function is (4)

$$T\{f\} = \int_{-\infty}^{f} p(z) \, dz - \frac{1}{2} \tag{6.1-16}$$

That is, the nonlinear transformation function is equivalent to the cumulative probability distribution of f. Table 6.1-2 contains the companding transformations and inverses for the Gaussian, Rayleigh, and Laplacian probability densities. It should be noted that nonlinear quantization by the companding technique is an approximation to optimum quantization, as specified by the Max solution. The accuracy of the approximation improves as the number of quantization levels increases.

6.2. PROCESSING QUANTIZED VARIABLES

Numbers within a digital computer that represent image variables, such as luminance or tristimulus values, normally are input as the integer codes corresponding to the quantization reconstruction levels of the variables, as illustrated in Figure 6.1-1. If the quantization is linear, the jth integer value is given by

$$j = \left[(J-1) \frac{f - a_L}{a_U - a_L} \right]_N \tag{6.2-1}$$

where J is the maximum integer value, f is the unquantized pixel value over a lower-to-upper range of a_L to a_U, and $[\cdot]_N$ denotes the nearest integer value of the argument. The corresponding reconstruction value is

$$r_j = \frac{a_U - a_L}{J} j + \frac{a_U - a_L}{2J} + a_L \tag{6.2-2}$$

Hence, r_j is linearly proportional to j. If the computer processing operation is itself linear, the integer code j can be numerically processed rather than the real number r_j. However, if nonlinear processing is to be performed, for example, taking the logarithm of a pixel, it is necessary to process r_j as a real variable rather than the integer j because the operation is scale dependent. If the quantization is nonlinear, all processing must be performed in the real variable domain.

In a digital computer, there are two major forms of numeric representation: real and integer. Real numbers are stored in floating-point form, and typically have a large dynamic range with fine precision. Integer numbers can be strictly positive or bipolar (negative or positive). The two's complement number system is commonly

TABLE 6.1.-2. Companding Quantization Transformations

	Probability Density	Forward Transformation	Inverse Transformation
Gaussian	$p(f) = (2\pi\sigma^2)^{-1/2}\exp\left\{-\dfrac{f^2}{2\sigma^2}\right\}$	$g = \dfrac{1}{2}\,\mathrm{erf}\left\{\dfrac{f}{\sqrt{2}\sigma}\right\}$	$\hat{f} = \sqrt{2}\,\sigma\,\mathrm{erf}^{-1}\{2\hat{g}\}$
Rayleigh	$p(f) = \dfrac{f}{\sigma^2}\exp\left\{-\dfrac{f^2}{2\sigma^2}\right\}$	$g = \dfrac{1}{2} - \exp\left\{-\dfrac{f^2}{2\sigma^2}\right\}$	$\hat{f} = \left[\sqrt{2}\sigma^2\ln\left\{1/\left(\dfrac{1}{2}-\hat{g}\right)\right\}\right]^{1/2}$
Laplacian	$p(f) = \dfrac{\alpha}{2}\exp\{-\alpha|f|\}$	$g = \dfrac{1}{2}[1 - \exp\{-\alpha f\}] \quad f \geq 0$	$\hat{f} = -\dfrac{1}{\alpha}\ln\{1-2\hat{g}\} \quad \hat{g} \geq 0$
		$g = -\dfrac{1}{2}[1 - \exp\{\alpha f\}] \quad f < 0$	$\hat{f} = \dfrac{1}{\alpha}\ln\{1+2\hat{g}\} \quad \hat{g} < 0$

where $\mathrm{erf}\{x\} = \dfrac{2}{\sqrt{\pi}}\displaystyle\int_0^x \exp\{-y^2\}\,dy$ and $\alpha = \dfrac{\sqrt{2}}{\sigma}$

used in computers and digital processing hardware for representing bipolar integers. The general format is as follows:

$$S.M_1,M_2,...,M_{B-1}$$

where S is a sign bit (0 for positive, 1 for negative), followed, conceptually, by a binary point, M_b denotes a magnitude bit, and B is the number of bits in the computer word. Table 6.2-1 lists the two's complement correspondence between integer, fractional, and decimal numbers for a 4-bit word. In this representation, all pixels are scaled in amplitude between -1.0 and $1.0 - 2^{-(B-1)}$. One of the advantages of

TABLE 6.2-1.   Two's Complement Code for 4-Bit Code Word

Code	Fractional Value	Decimal Value
0.111	$+\dfrac{7}{8}$	+0.875
0.110	$+\dfrac{6}{8}$	+0.750
0.101	$+\dfrac{5}{8}$	+0.625
0.100	$+\dfrac{4}{8}$	+0.500
0.011	$+\dfrac{3}{8}$	+0.375
0.010	$+\dfrac{2}{8}$	+0.250
0.001	$+\dfrac{1}{8}$	+0.125
0.000	0	0.000
1.111	$-\dfrac{1}{8}$	−0.125
1.110	$-\dfrac{2}{8}$	−0.250
1.101	$-\dfrac{3}{8}$	−0.375
1.100	$-\dfrac{4}{8}$	−0.500
1.011	$-\dfrac{5}{8}$	−0.625
1.010	$-\dfrac{6}{8}$	−0.750
1.001	$-\dfrac{7}{8}$	−0.875
1.000	$-\dfrac{8}{8}$	−1.000

this representation is that pixel scaling is independent of precision in the sense that a pixel $F(j, k)$ is bounded over the range

$$-1.0 \leq F(j, k) < 1.0$$

regardless of the number of bits in a word.

6.3. MONOCHROME AND COLOR IMAGE QUANTIZATION

This section considers the subjective and quantitative effects of the quantization of monochrome and color images.

6.3.1. Monochrome Image Quantization

Monochrome images are typically input to a digital image processor as a sequence of uniform-length binary code words. In the literature, the binary code is often called a *pulse code modulation* (PCM) code. Because uniform-length code words are used for each image sample, the number of amplitude quantization levels is determined by the relationship

$$L = 2^B \tag{6.3-1}$$

where B represents the number of code bits allocated to each sample.

A bit rate compression can be achieved for PCM coding by the simple expedient of restricting the number of bits assigned to each sample. If image quality is to be judged by an analytic measure, B is simply taken as the smallest value that satisfies the minimal acceptable image quality measure. For a subjective assessment, B is lowered until quantization effects become unacceptable. The eye is only capable of judging the absolute brightness of about 10 to 15 shades of gray, but it is much more sensitive to the difference in the brightness of adjacent gray shades. For a reduced number of quantization levels, the first noticeable artifact is a *gray scale contouring* caused by a jump in the reconstructed image brightness between quantization levels in a region where the original image is slowly changing in brightness. The minimal number of quantization bits required for basic PCM coding to prevent gray scale contouring is dependent on a variety of factors, including the linearity of the image display and noise effects before and after the image digitizer.

Assuming that an image sensor produces an output pixel sample proportional to the image intensity, a question of concern then is: Should the image intensity itself, or some function of the image intensity, be quantized? Furthermore, should the quantization scale be linear or nonlinear? Linearity or nonlinearity of the quantization scale can

 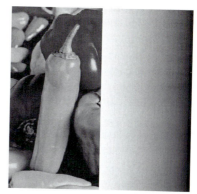

(*a*) 8 bit, 256 levels   (*b*) 7 bit, 128 levels

(*c*) 6 bit, 64 levels   (*d*) 5 bit, 32 levels

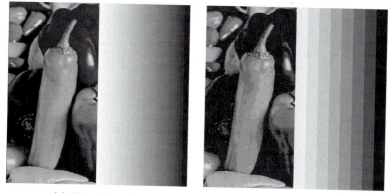

(*e*) 4 bit, 16 levels   (*f*) 3 bit, 8 levels

FIGURE 6.3-1. Uniform quantization of the `peppers_ramp_luminance` monochrome image.

be viewed as a matter of implementation. A given nonlinear quantization scale can be realized by the companding operation of Figure 6.1-3, in which a nonlinear amplification weighting of the continuous signal to be quantized is performed, followed by linear quantization, followed by an inverse weighting of the quantized amplitude. Thus, consideration is limited here to linear quantization of companded pixel samples.

There have been many experimental studies to determine the number and placement of quantization levels required to minimize the effect of gray scale contouring (5–8). Goodall (5) performed some of the earliest experiments on digital television and concluded that 6 bits of intensity quantization (64 levels) were required for good quality and that 5 bits (32 levels) would suffice for a moderate amount of contouring. Other investigators have reached similar conclusions. In most studies, however, there has been some question as to the linearity and calibration of the imaging system. As noted in Section 3.5.3, most television cameras and monitors exhibit a nonlinear response to light intensity. Also, the photographic film that is often used to record the experimental results is highly nonlinear. Finally, any camera or monitor noise tends to diminish the effects of contouring.

Figure 6.3-1 contains photographs of an image linearly quantized with a variable number of quantization levels. The source image is a split image in which the left side is a luminance image and the right side is a computer-generated linear ramp. In Figure 6.3-1, the luminance signal of the image has been uniformly quantized with from 8 to 256 levels (3 to 8 bits). Gray scale contouring in these pictures is apparent in the ramp part of the split image for 6 or fewer bits. The contouring of the luminance image part of the split image becomes noticeable for 5 bits.

As discussed in Section 2-4, it has been postulated that the eye responds logarithmically or to a power law of incident light amplitude. There have been several efforts to quantitatively model this nonlinear response by a *lightness function* Λ, which is related to incident luminance. Priest et al. (9) have proposed a square-root nonlinearity

$$\Lambda = (100.0Y)^{1/2} \tag{6.3-2}$$

where $0.0 \leq Y \leq 1.0$ and $0.0 \leq \Lambda \leq 10.0$. Ladd and Pinney (10) have suggested a cube-root scale

$$\Lambda = 2.468(100.0Y)^{1/3} - 1.636 \tag{6.3-3}$$

A logarithm scale

$$\Lambda = 5.0[\log_{10}\{100.0Y\}] \tag{6.3-4}$$

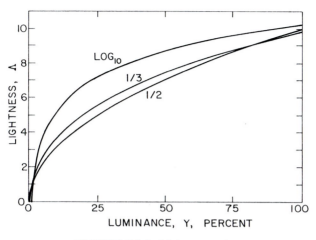

FIGURE 6.3-2. Lightness scales.

where $0.01 \leq Y \leq 1.0$ has also been proposed by Foss et al. (11). Figure 6.3-2 compares these three scaling functions.

In an effort to reduce the grey scale contouring of linear quantization, it is reasonable to apply a lightness scaling function prior to quantization, and then to apply its inverse to the reconstructed value in correspondence to the companding quantizer of Figure 6.1-3. Figure 6.3-3 presents a comparison of linear, square-root, cube-root, and logarithmic quantization for a 4-bit quantizer. Among the lightness scale quantizers, the gray scale contouring appears least for the square-root scaling. The lightness quantizers exhibit less contouring than the linear quantizer in dark areas but worse contouring for bright regions.

6.3.2. Color Image Quantization

A color image may be represented by its red, green, and blue source tristimulus values or any linear or nonlinear invertible function of the source tristimulus values. If the red, green, and blue tristimulus values are to be quantized individually, the selection of the number and placement of quantization levels follows the same general considerations as for a monochrome image. The eye exhibits a nonlinear response to spectral lights as well as white light, and therefore, it is subjectively preferable to compand the tristimulus values before quantization. It is known, however, that the eye is most sensitive to brightness changes in the blue region of the spectrum, moderately sensitive to brightness changes in the green spectral region, and least sensitive to red changes. Thus, it is possible to assign quantization levels on this basis more efficiently than simply using an equal number for each tristimulus value.

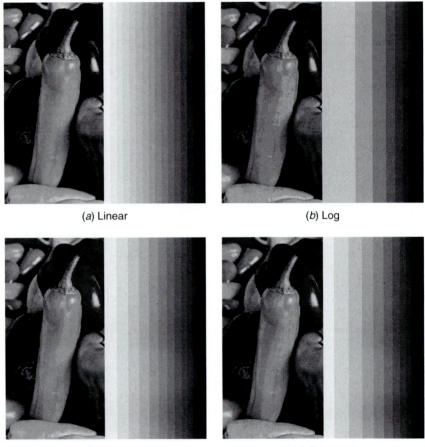

(a) Linear   (b) Log

(c) Square root   (d) Cube root

FIGURE 6.3-3. Comparison of lightness scale quantization of the `peppers_ramp` `_luminance` image for 4 bit quantization.

Figure 6.3-4 is a general block diagram for a color image quantization system. A source image described by source tristimulus values R, G, B is converted to three components $x(1)$, $x(2)$, $x(3)$, which are then quantized. Next, the quantized components $\hat{x}(1)$, $\hat{x}(2)$, $\hat{x}(3)$ are converted back to the original color coordinate system, producing the quantized tristimulus values $\hat{R}$, $\hat{G}$, $\hat{B}$. The quantizer in Figure 6.3-4 effectively partitions the color space of the color coordinates $x(1)$, $x(2)$, $x(3)$ into quantization cells and assigns a single color value to all colors within a cell. To be most efficient, the three color components $x(1)$, $x(2)$, $x(3)$ should be quantized jointly. However, implementation considerations often dictate separate quantization of the color components. In such a system, $x(1)$, $x(2)$, $x(3)$ are individually quantized over

FIGURE 6.3-4 Color image quantization model.

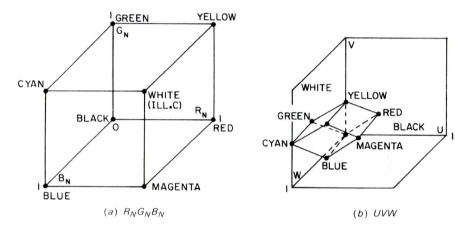

FIGURE 6.3-5. Loci of reproducible colors for $R_N G_N B_N$ and UVW coordinate systems.

their maximum ranges. In effect, the physical color solid is enclosed in a rectangular solid, which is then divided into rectangular quantization cells.

If the source tristimulus values are converted to some other coordinate system for quantization, some immediate problems arise. As an example, consider the quantization of the UVW tristimulus values. Figure 6.3-5 shows the locus of reproducible colors for the RGB source tristimulus values plotted as a cube and the transformation of this color cube into the UVW coordinate system. It is seen that the RGB cube becomes a parallelepiped. If the UVW tristimulus values are to be quantized individually over their maximum and minimum limits, many of the quantization cells represent nonreproducible colors and hence are wasted. It is only worthwhile to quantize colors within the parallelepiped, but this generally is a difficult operation to implement efficiently.

In the present analysis, it is assumed that each color component is linearly quantized over its maximum range into $2^{B(i)}$ levels, where $B(i)$ represents the number of bits assigned to the component $x(i)$. The total number of bits allotted to the coding is fixed at

$$B_T = B(1) + B(2) + B(3) \qquad (6.3\text{-}5)$$

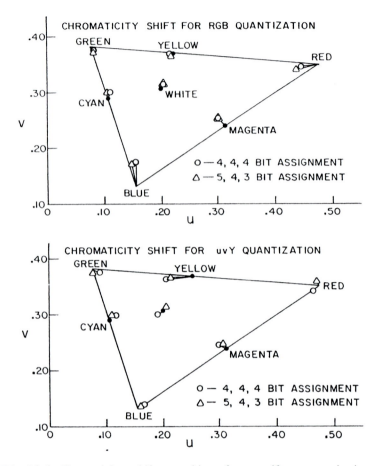

FIGURE 6.3-6. Chromaticity shifts resulting from uniform quantization of the smpte_girl_linear color image.

Let $a_U(i)$ represent the upper bound of $x(i)$ and $a_L(i)$ the lower bound. Then each quantization cell has dimension

$$q(i) = \frac{a_U(i) - a_L(i)}{2^{B(i)}} \tag{6.3-6}$$

Any color with color component $x(i)$ within the quantization cell will be quantized to the color component value $\hat{x}(i)$. The maximum quantization error along each color coordinate axis is then

$$\varepsilon(i) = |x(i) - \hat{x}(i)| = \frac{a_U(i) - a_L(i)}{2^{B(i)+1}}$$ (6.3-7)

Thus, the coordinates of the quantized color become

$$\hat{x}(i) = x(i) \pm \varepsilon(i)$$ (6.3-8)

subject to the conditions $a_L(i) \leq \hat{x}(i) \leq a_U(i)$. It should be observed that the values of $\hat{x}(i)$ will always lie within the smallest cube enclosing the color solid for the given color coordinate system. Figure 6.3-6 illustrates chromaticity shifts of various colors for quantization in the $R_N G_N B_N$ and Yuv coordinate systems (12).

Jain and Pratt (12) have investigated the optimal assignment of quantization decision levels for color images in order to minimize the geodesic color distance between an original color and its reconstructed representation. Interestingly enough, it was found that quantization of the $R_N G_N B_N$ color coordinates provided better results than for other common color coordinate systems. The primary reason was that all quantization levels were occupied in the $R_N G_N B_N$ system, but many levels were unoccupied with the other systems. This consideration seemed to override the metric nonuniformity of the $R_N G_N B_N$ color space.

REFERENCES

1. P. F. Panter and W. Dite, "Quantization Distortion in Pulse Code Modulation with Non-uniform Spacing of Levels," *Proc. IRE*, **39**, 1, January 1951, 44–48.

2. J. Max, "Quantizing for Minimum Distortion," *IRE Trans. Information Theory*, **IT-6**, 1, March 1960, 7–12.

3. V. R. Algazi, "Useful Approximations to Optimum Quantization," *IEEE Trans. Communication Technology*, **COM-14**, 3, June 1966, 297–301.

4. R. M. Gray, "Vector Quantization," *IEEE ASSP Magazine*, April 1984, 4–29.

5. W. M. Goodall, "Television by Pulse Code Modulation," *Bell System Technical J.*, January 1951.

6. R. L. Cabrey, "Video Transmission over Telephone Cable Pairs by Pulse Code Modulation," *Proc. IRE*, **48**, 9, September 1960, 1546–1551.

7. L. H. Harper, "PCM Picture Transmission," *IEEE Spectrum*, **3**, 6, June 1966, 146.

8. F. W. Scoville and T. S. Huang, "The Subjective Effect of Spatial and Brightness Quantization in PCM Picture Transmission," *NEREM Record*, 1965, 234–235.

9. I. G. Priest, K. S. Gibson, and H. J. McNicholas, "An Examination of the Munsell Color System, I. Spectral and Total Reflection and the Munsell Scale of Value," Technical Paper 167, National Bureau of Standards, Washington, DC, 1920.

10. J. H. Ladd and J. E. Pinney, "Empherical Relationships with the Munsell Value Scale," *Proc. IRE* (Correspondence), **43**, 9, 1955, 1137.

11. C. E. Foss, D. Nickerson, and W. C. Granville, "Analysis of the Oswald Color System," *J. Optical Society of America*, **34**, 1, July 1944, 361–381.

12. A. K. Jain and W. K. Pratt, "Color Image Quantization," IEEE Publication 72 CH0 601-5-NTC, *National Telecommunications Conference 1972 Record*, Houston, TX, December 1972.

PART 3

DISCRETE TWO-DIMENSIONAL LINEAR PROCESSING

Part 3 of the book is concerned with a unified analysis of discrete two-dimensional linear processing operations. Several forms of discrete two-dimensional superposition and convolution operators are developed and related to one another. Two-dimensional transforms, such as the Fourier, Hartley, cosine, and Karhunen–Loeve transforms, are introduced. Consideration is given to the utilization of two-dimensional transforms as an alternative means of achieving convolutional processing more efficiently.

7

SUPERPOSITION AND CONVOLUTION

In Chapter 1, superposition and convolution operations were derived for continuous two-dimensional image fields. This chapter provides a derivation of these operations for discrete two-dimensional images. Three types of superposition and convolution operators are defined: finite area, sampled image, and circulant area. The finite-area operator is a linear filtering process performed on a discrete image data array. The sampled image operator is a discrete model of a continuous two-dimensional image filtering process. The circulant area operator provides a basis for a computationally efficient means of performing either finite-area or sampled image superposition and convolution.

7.1. FINITE-AREA SUPERPOSITION AND CONVOLUTION

Mathematical expressions for finite-area superposition and convolution are developed below for both series and vector-space formulations.

7.1.1. Finite-Area Superposition and Convolution: Series Formulation

Let $F(n_1, n_2)$ denote an image array for $n_1, n_2 = 1, 2,..., N$. For notational simplicity, all arrays in this chapter are assumed square. In correspondence with Eq. 1.2-6, the image array can be represented at some point (m_1, m_2) as a sum of amplitude weighted Dirac delta functions by the discrete sifting summation

$$F(m_1, m_2) = \sum_{n_1} \sum_{n_2} F(n_1, n_2)\delta(m_1 - n_1 + 1, m_2 - n_2 + 1) \qquad (7.1-1)$$

The term

$$
\delta(m_1 - n_1 + 1, m_2 - n_2 + 1) = \begin{cases} 1 & \text{if } m_1 = n_1 \text{ and } m_2 = n_2 & \text{(7.1-2a)} \\ \\ 0 & \text{otherwise} & \text{(7.1-2b)} \end{cases}
$$

is a discrete delta function. Now consider a spatial linear operator $O\{\cdot\}$ that produces an output image array

$$
Q(m_1, m_2) = O\{F(m_1, m_2)\} \tag{7.1-3}
$$

by a linear spatial combination of pixels within a neighborhood of (m_1, m_2). From the sifting summation of Eq. 7.1-1,

$$
Q(m_1, m_2) = O\left\{\sum_{n_1} \sum_{n_2} F(n_1, n_2)\delta(m_1 - n_1 + 1, m_2 - n_2 + 1)\right\} \tag{7.1-4a}
$$

or

$$
Q(m_1, m_2) = \sum_{n_1} \sum_{n_2} F(n_1, n_2)O\{\delta(m_1 - n_1 + 1, m_2 - n_2 + 1)\} \tag{7.1-4b}
$$

recognizing that $O\{\cdot\}$ is a linear operator and that $F(n_1, n_2)$ in the summation of Eq. 7.1-4a is a constant in the sense that it does not depend on (m_1, m_2). The term $O\{\delta(t_1, t_2)\}$ for $t_i = m_i - n_i + 1$ is the response at output coordinate (m_1, m_2) to a unit amplitude input at coordinate (n_1, n_2). It is called the *impulse response function array* of the linear operator and is written as

$$
\delta(m_1 - n_1 + 1, m_2 - n_2 + 1; m_1, m_2) = O\{\delta(t_1, t_2)\} \quad \text{for } 1 \le t_1, t_2 \le L \tag{7.1-5}
$$

and is zero otherwise. For notational simplicity, the impulse response array is considered to be square.

In Eq. 7.1-5 it is assumed that the impulse response array is of limited spatial extent. This means that an output image pixel is influenced by input image pixels only within some finite area $L \times L$ neighborhood of the corresponding output image pixel. The output coordinates (m_1, m_2) in Eq. 7.1-5 following the semicolon indicate that in the general case, called *finite area superposition*, the impulse response array can change form for each point (m_1, m_2) in the processed array $Q(m_1, m_2)$. Following this nomenclature, the finite area superposition operation is defined as

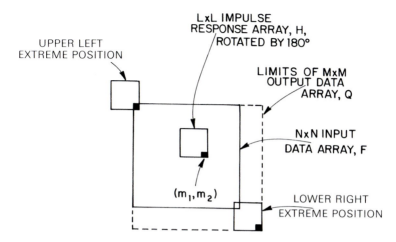

FIGURE 7.1-1. Relationships between input data, output data, and impulse response arrays for finite-area superposition; upper left corner justified array definition.

$$Q(m_1, m_2) = \sum_{n_1} \sum_{n_2} F(n_1, n_2)H(m_1 - n_1 + 1, m_2 - n_2 + 1; \; m_1, m_2) \qquad (7.1\text{-}6)$$

The limits of the summation are

$$\text{MAX}\{1, m_i - L + 1\} \le n_i \le \text{MIN}\{N, m_i\} \qquad (7.1\text{-}7)$$

where $\text{MAX}\{a, b\}$ and $\text{MIN}\{a, b\}$ denote the maximum and minimum of the arguments, respectively. Examination of the indices of the impulse response array at its extreme positions indicates that $M = N + L - 1$, and hence the processed output array Q is of larger dimension than the input array F. Figure 7.1-1 illustrates the geometry of finite-area superposition. If the impulse response array H is spatially invariant, the superposition operation reduces to the convolution operation.

$$Q(m_1, m_2) = \sum_{n_1} \sum_{n_2} F(n_1, n_2)H(m_1 - n_1 + 1, m_2 - n_2 + 1) \qquad (7.1\text{-}8)$$

Figure 7.1-2 presents a graphical example of convolution with a 3×3 impulse response array.

Equation 7.1-6 expresses the finite-area superposition operation in *left-justified form* in which the input and output arrays are aligned at their upper left corners. It is often notationally convenient to utilize a definition in which the output array is centered with respect to the input array. This definition of *centered superposition* is given by

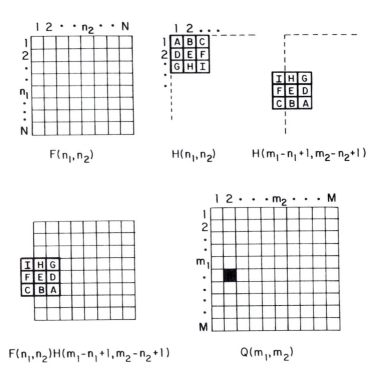

FIGURE 7.1-2. Graphical example of finite-area convolution with a 3×3 impulse response array; upper left corner justified array definition.

$$Q_c(j_1, j_2) = \sum_{n_1} \sum_{n_2} F(n_1, n_2) H(j_1 - n_1 + L_c, j_2 - n_2 + L_c; j_1, j_2) \qquad (7.1\text{-}9)$$

where $-(L-3)/2 \le j_i \le N + (L-1)/2$ and $L_c = (L+1)/2$. The limits of the summation are

$$\text{MAX}\{1, j_i - (L-1)/2\} \le n_i \le \text{MIN}\{N, j_i + (L-1)/2\} \qquad (7.1\text{-}10)$$

Figure 7.1-3 shows the spatial relationships between the arrays F, H, and Q_c for centered superposition with a 5×5 impulse response array.

In digital computers and digital image processors, it is often convenient to restrict the input and output arrays to be of the same dimension. For such systems, Eq. 7.1-9 needs only to be evaluated over the range $1 \le j_i \le N$. When the impulse response

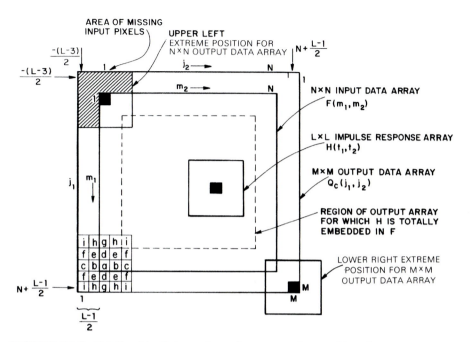

FIGURE 7.1-3. Relationships between input data, output data, and impulse response arrays for finite-area superposition; centered array definition.

array is located on the border of the input array, the product computation of Eq. 7.1-9 does not involve all of the elements of the impulse response array. This situation is illustrated in Figure 7.1-3, where the impulse response array is in the upper left corner of the input array. The input array pixels "missing" from the computation are shown crosshatched in Figure 7.1-3. Several methods have been proposed to deal with this border effect. One method is to perform the computation of all of the impulse response elements as if the missing pixels are of some constant value. If the constant value is zero, the result is called *centered, zero padded superposition*. A variant of this method is to regard the missing pixels to be mirror images of the input array pixels, as indicated in the lower left corner of Figure 7.1-3. In this case the *centered, reflected boundary superposition* definition becomes

$$Q_c(j_1, j_2) = \sum_{n_1} \sum_{n_2} F(n_1', n_2') H(j_1 - n_1 + L_c, j_2 - n_2 + L_c; j_1, j_2) \quad (7.1\text{-}11)$$

where the summation limits are

$$j_i - (L-1)/2 \le n_i \le j_i + (L-1)/2 \quad (7.1\text{-}12)$$

and

$$
n'_i = \begin{cases} 2 - n_i & \text{for } n_i \leq 0 & (7.1\text{-}13a) \\[2mm] n_i & \text{for } 1 \leq n_i \leq N & (7.1\text{-}13b) \\[2mm] 2N - n_i & \text{for } n_i > N & (7.1\text{-}13c) \end{cases}
$$

In many implementations, the superposition computation is limited to the range $(L+1)/2 \leq j_i \leq N - (L-1)/2$, and the border elements of the $N \times N$ array Q_c are set to zero. In effect, the superposition operation is computed only when the impulse response array is fully embedded within the confines of the input array. This region is described by the dashed lines in Figure 7.1-3. This form of superposition is called *centered, zero boundary superposition.*

If the impulse response array H is spatially invariant, the centered definition for convolution becomes

$$
Q_c(j_1, j_2) = \sum_{n_1} \sum_{n_2} F(n_1, n_2) H(j_1 - n_1 + L_c, j_2 - n_2 + L_c) \qquad (7.1\text{-}14)
$$

The 3×3 impulse response array, which is called a *small generating kernel* (SGK), is fundamental to many image processing algorithms (1). When the SGK is totally embedded within the input data array, the general term of the centered convolution operation can be expressed explicitly as

$$
\begin{aligned}
Q_c(j_1, j_2) = \; & H(3,3)F(j_1 - 1, j_2 - 1) + H(3,2)F(j_1 - 1, j_2) + H(3,1)F(j_1 - 1, j_2 + 1) \\[2mm]
& + H(2,3)F(j_1, j_2 - 1) + H(2,2)F(j_1, j_2) + H(2,1)F(j_1, j_2 + 1) \\[2mm]
& + H(1,3)F(j_1 + 1, j_2 - 1) + H(1,2)F(j_1 + 1, j_2) + H(1,1)F(j_1 + 1, j_2 + 1)
\end{aligned}
$$

$$(7.1\text{-}15)$$

for $2 \leq j_i \leq N - 1$. In Chapter 9 it will be shown that convolution with arbitrary-size impulse response arrays can be achieved by sequential convolutions with SGKs.

The four different forms of superposition and convolution are each useful in various image processing applications. The upper left corner–justified definition is appropriate for computing the correlation function between two images. The centered, zero padded and centered, reflected boundary definitions are generally employed for image enhancement filtering. Finally, the centered, zero boundary definition is used for the computation of spatial derivatives in edge detection. In this application, the derivatives are not meaningful in the border region.

```
0.040 0.080 0.120 0.160 0.200 0.200 0.200        0.000 0.000 0.000 0.000 0.000 0.000 0.000
0.080 0.160 0.240 0.320 0.400 0.400 0.400        0.000 0.000 0.000 0.000 0.000 0.000 0.000
0.120 0.240 0.360 0.480 0.600 0.600 0.600        0.000 0.000 1.000 1.000 1.000 1.000 1.000
0.160 0.320 0.480 0.640 0.800 0.800 0.800        0.000 0.000 1.000 1.000 1.000 1.000 1.000
0.200 0.400 0.600 0.800 1.000 1.000 1.000        0.000 0.000 1.000 1.000 1.000 1.000 1.000
0.200 0.400 0.600 0.800 1.000 1.000 1.000        0.000 0.000 1.000 1.000 1.000 1.000 1.000
0.200 0.400 0.600 0.800 1.000 1.000 1.000        0.000 0.000 1.000 1.000 1.000 1.000 1.000
```

(*a*) Upper left corner justified   (*b*) Centered, zero boundary

```
0.360 0.480 0.600 0.600 0.600 0.600 0.600        1.000 1.000 1.000 1.000 1.000 1.000 1.000
0.480 0.640 0.800 0.800 0.800 0.800 0.800        1.000 1.000 1.000 1.000 1.000 1.000 1.000
0.600 0.800 1.000 1.000 1.000 1.000 1.000        1.000 1.000 1.000 1.000 1.000 1.000 1.000
0.600 0.800 1.000 1.000 1.000 1.000 1.000        1.000 1.000 1.000 1.000 1.000 1.000 1.000
0.600 0.800 1.000 1.000 1.000 1.000 1.000        1.000 1.000 1.000 1.000 1.000 1.000 1.000
0.600 0.800 1.000 1.000 1.000 1.000 1.000        1.000 1.000 1.000 1.000 1.000 1.000 1.000
0.600 0.800 1.000 1.000 1.000 1.000 1.000        1.000 1.000 1.000 1.000 1.000 1.000 1.000
```

(*c*) Centered, zero padded   (*d*) Centered, reflected

FIGURE 7.1-4 Finite-area convolution boundary conditions, upper left corner of convolved image.

Figure 7.1-4 shows computer printouts of pixels in the upper left corner of a convolved image for the four types of convolution boundary conditions. In this example, the source image is constant of maximum value 1.0. The convolution impulse response array is a 5×5 uniform array.

7.1.2. Finite-Area Superposition and Convolution: Vector-Space Formulation

If the arrays F and Q of Eq. 7.1-6 are represented in vector form by the $N^2 \times 1$ vector $\mathbf{f}$ and the $M^2 \times 1$ vector $\mathbf{q}$, respectively, the finite-area superposition operation can be written as (2)

$$\mathbf{q} = \mathbf{Df} \qquad (7.1-16)$$

where $\mathbf{D}$ is a $M^2 \times N^2$ matrix containing the elements of the impulse response. It is convenient to partition the superposition operator matrix $\mathbf{D}$ into submatrices of dimension $M \times N$. Observing the summation limits of Eq. 7.1-7, it is seen that

$$\mathbf{D} = \begin{bmatrix} \mathbf{D}_{1,1} & \mathbf{0} & \cdots & \mathbf{0} \\ \mathbf{D}_{2,1} & \mathbf{D}_{2,2} & & \vdots \\ \vdots & \vdots & & \mathbf{0} \\ \mathbf{D}_{L,1} & \mathbf{D}_{L,2} & \mathbf{D}_{M-L+1,N} & \\ \mathbf{0} & \mathbf{D}_{L+1,1} & & \vdots \\ \vdots & \vdots & & \vdots \\ \mathbf{0} & \cdots & \mathbf{0} & \mathbf{D}_{M,N} \end{bmatrix} \qquad (7.1-17)$$

$$H = \begin{bmatrix} 11 & 12 & 13 \\ 21 & 22 & 23 \\ 31 & 32 & 33 \end{bmatrix}$$

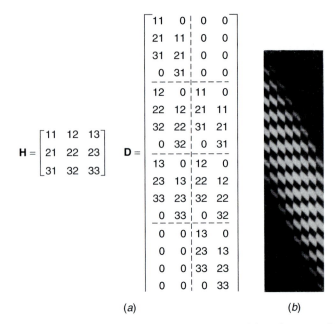

(a)   (b)

FIGURE 7.1-5 Finite-area convolution operators: (a) general impulse array, $M = 4$, $N = 2$, $L = 3$; (b) Gaussian-shaped impulse array, $M = 16$, $N = 8$, $L = 9$.

The general nonzero term of **D** is then given by

$$D_{m_2,n_2}(m_1, n_1) = H(m_1 - n_1 + 1, m_2 - n_2 + 1;\; m_1, m_2) \qquad (7.1\text{-}18)$$

Thus, it is observed that **D** is highly structured and quite sparse, with the center band of submatrices containing stripes of zero-valued elements.

If the impulse response is position invariant, the structure of **D** does not depend explicitly on the output array coordinate (m_1, m_2). Also,

$$\mathbf{D}_{m_2,n_2} = \mathbf{D}_{m_2 + 1, n_2 + 1} \qquad (7.1\text{-}19)$$

As a result, the columns of **D** are shifted versions of the first column. Under these conditions, the finite-area superposition operator is known as the *finite-area convolution operator*. Figure 7.1-5a contains a notational example of the finite-area convolution operator for a 2×2 $(N = 2)$ input data array, a 4×4 $(M = 4)$ output data array, and a 3×3 $(L = 3)$ impulse response array. The integer pairs (i, j) at each element of **D** represent the element (i, j) of $H(i, j)$. The basic structure of **D** can be seen more clearly in the larger matrix depicted in Figure 7.1-5b. In this example, $M = 16$,

$N = 8$, $L = 9$, and the impulse response has a symmetrical Gaussian shape. Note that **D** is a 256×64 matrix in this example.

Following the same technique as that leading to Eq. 5.4-7, the matrix form of the superposition operator may be written as

$$Q = \sum_{m=1}^{M} \sum_{n=1}^{N} D_{m,n} F v_n u_m^T \qquad (7.1\text{-}20)$$

If the impulse response is spatially invariant and is of separable form such that

$$H = h_C h_R^T \qquad (7.1\text{-}21)$$

where h_R and h_C are column vectors representing row and column impulse responses, respectively, then

$$D = D_C \otimes D_R \qquad (7.1\text{-}22)$$

The matrices D_R and D_C are $M \times N$ matrices of the form

$$D_R = \begin{bmatrix} h_R(1) & 0 & \cdots & 0 \\ h_R(2) & h_R(1) & & \vdots \\ h_R(3) & h_R(2) & \cdots & 0 \\ \vdots & & & h_R(1) \\ h_R(L) & & & \vdots \\ 0 & & & \vdots \\ \vdots & & & \vdots \\ 0 & \cdots & 0 & h_R(L) \end{bmatrix} \qquad (7.1\text{-}23)$$

The two-dimensional convolution operation may then be computed by sequential row and column one-dimensional convolutions. Thus

$$Q = D_C F D_R^T \qquad (7.1\text{-}24)$$

In vector form, the general finite-area superposition or convolution operator requires $N^2 L^2$ operations if the zero-valued multiplications of **D** are avoided. The separable operator of Eq. 7.1-24 can be computed with only $NL(M + N)$ operations.

7.2. SAMPLED IMAGE SUPERPOSITION AND CONVOLUTION

Many applications in image processing require a discretization of the superposition integral relating the input and output continuous fields of a linear system. For example, image blurring by an optical system, sampling with a finite-area aperture or imaging through atmospheric turbulence, may be modeled by the superposition integral equation

$$\tilde{G}(x, y) = \int_{-\infty}^{\infty} \int_{-\infty}^{\infty} \tilde{F}(\alpha, \beta) \tilde{J}(x, y; \alpha, \beta) \, d\alpha \, d\beta \qquad (7.2\text{-}1a)$$

where $\tilde{F}(x,y)$ and $\tilde{G}(x, y)$ denote the input and output fields of a linear system, respectively, and the kernel $\tilde{J}(x, y; \alpha, \beta)$ represents the impulse response of the linear system model. In this chapter, a tilde over a variable indicates that the spatial indices of the variable are bipolar; that is, they range from negative to positive spatial limits. In this formulation, the impulse response may change form as a function of its four indices: the input and output coordinates. If the linear system is space invariant, the output image field may be described by the convolution integral

$$\tilde{G}(x, y) = \int_{-\infty}^{\infty} \int_{-\infty}^{\infty} \tilde{F}(\alpha, \beta) \tilde{J}(x - \alpha, y - \beta) \, d\alpha \, d\beta \qquad (7.2\text{-}1b)$$

For discrete processing, physical image sampling will be performed on the output image field. Numerical representation of the integral must also be performed in order to relate the physical samples of the output field to points on the input field.

Numerical representation of a superposition or convolution integral is an important topic because improper representations may lead to gross modeling errors or numerical instability in an image processing application. Also, selection of a numerical representation algorithm usually has a significant impact on digital processing computational requirements.

As a first step in the discretization of the superposition integral, the output image field is physically sampled by a $(2J + 1) \times (2J + 1)$ array of Dirac pulses at a resolution ΔS to obtain an array whose general term is

$$\tilde{G}(j_1 \Delta S, j_2 \Delta S) = \tilde{G}(x, y) \delta(x - j_1 \Delta S, y - j_2 \Delta S) \qquad (7.2\text{-}2)$$

where $-J \le j_i \le J$. Equal horizontal and vertical spacing of sample pulses is assumed for notational simplicity. The effect of finite area sample pulses can easily be incorporated by replacing the impulse response with $\tilde{J}(x, y; \alpha, \beta) \circledast P(-x,-y)$, where $P(-x,-y)$ represents the pulse shape of the sampling pulse. The delta function may be brought under the integral sign of the superposition integral of Eq. 7.2-1a to give

$$\tilde{G}(j_1 \Delta S, j_2 \Delta S) = \int_{-\infty}^{\infty} \int_{-\infty}^{\infty} \tilde{F}(\alpha, \beta) \tilde{J}(j_1 \Delta S, j_2 \Delta S; \alpha, \beta) \, d\alpha \, d\beta \qquad (7.2\text{-}3)$$

It should be noted that the physical sampling is performed on the observed image spatial variables (x, y); physical sampling does not affect the dummy variables of integration (α, β).

Next, the impulse response must be truncated to some spatial bounds. Thus, let

$$\tilde{J}(x, y; \alpha, \beta) = 0 \tag{7.2-4}$$

for $|x| > T$ and $|y| > T$. Then,

$$\tilde{G}(j_1 \, \Delta S, j_1 \, \Delta S) = \int_{j_1 \Delta S - T}^{j_1 \Delta S + T} \int_{j_2 \Delta S - T}^{j_2 \Delta S + T} \tilde{F}(\alpha, \beta) \, \tilde{J}(j_1 \, \Delta S, j_2 \, \Delta S; \alpha, \beta) \, d\alpha \, d\beta \tag{7.2-5}$$

Truncation of the impulse response is equivalent to multiplying the impulse response by a window function $V(x, y)$, which is unity for $|x| < T$ and $|y| < T$ and zero elsewhere. By the Fourier convolution theorem, the Fourier spectrum of $G(x, y)$ is equivalently convolved with the Fourier transform of $V(x, y)$, which is a two-dimensional sinc function. This distortion of the Fourier spectrum of $G(x, y)$ results in the introduction of high-spatial-frequency artifacts (a Gibbs phenomenon) at spatial frequency multiples of $2\pi/T$. Truncation distortion can be reduced by using a shaped window, such as the Bartlett, Blackman, Hamming, or Hanning windows (3), which smooth the sharp cutoff effects of a rectangular window. This step is especially important for image restoration modeling because ill-conditioning of the superposition operator may lead to severe amplification of the truncation artifacts.

In the next step of the discrete representation, the continuous ideal image array $\tilde{F}(\alpha, \beta)$ is represented by mesh points on a rectangular grid of resolution ΔI and dimension $(2K + 1) \times (2K + 1)$. This is not a physical sampling process, but merely an abstract numerical representation whose general term is described by

$$\tilde{F}(k_1 \, \Delta I, k_2 \, \Delta I) = \tilde{F}(\alpha, \beta)\delta(\alpha - k_1 \, \Delta I, \alpha - k_2 \, \Delta I) \tag{7.2-6}$$

where $K_{iL} \le k_i \le K_{iU}$, with K_{iU} and K_{iL} denoting the upper and lower index limits.

If the ultimate objective is to estimate the continuous ideal image field by processing the physical observation samples, the mesh spacing ΔI should be fine enough to satisfy the Nyquist criterion for the ideal image. That is, if the spectrum of the ideal image is bandlimited and the limits are known, the mesh spacing should be set at the corresponding Nyquist spacing. Ideally, this will permit perfect interpolation of the estimated points $\tilde{F}(k_1 \, \Delta I, k_2 \, \Delta I)$ to reconstruct $\tilde{F}(x, y)$.

The continuous integration of Eq. 7.2-5 can now be approximated by a discrete summation by employing a quadrature integration formula (4). The physical image samples may then be expressed as

$$\tilde{G}(j_1 \, \Delta S, j_2 \, \Delta S) = \sum_{k_1 = K_{1L}}^{K_{1U}} \sum_{k_2 = K_{2L}}^{K_{2U}} \tilde{F}(k_1 \, \Delta I, k_2 \, \Delta I)\tilde{W}(k_1, k_2)\tilde{J}(j_1 \, \Delta S, j_2 \, \Delta S; k_1 \, \Delta I, k_2 \, \Delta I)$$

$$\tag{7.2-7}$$

where $\tilde{W}(k_1, k_2)$ is a weighting coefficient for the particular quadrature formula employed. Usually, a rectangular quadrature formula is used, and the weighting coefficients are unity. In any case, it is notationally convenient to lump the weighting coefficient and the impulse response function together so that

$$\tilde{H}(j_1 \, \Delta S, j_2 \, \Delta S; k_1 \, \Delta I, k_2 \, \Delta I) = \tilde{W}(k_1, k_2) \tilde{J}(j_1 \, \Delta S, j_2 \, \Delta S; k_1 \Delta I, k_2 \, \Delta I) \qquad (7.2\text{-}8)$$

Then,

$$\tilde{G}(j_1 \, \Delta S, j_2 \, \Delta S) = \sum_{k_1 = K_{1L}}^{K_{1U}} \sum_{k_2 = K_{2L}}^{K_{2U}} \tilde{F}(k_1 \, \Delta I, k_2 \, \Delta I) \tilde{H}(j_1 \, \Delta S, \ j_2 \, \Delta S; k_1 \, \Delta I, k_2 \, \Delta I) \qquad (7.2\text{-}9)$$

Again, it should be noted that $\tilde{H}$ is not spatially discretized; the function is simply evaluated at its appropriate spatial argument. The limits of summation of Eq. 7.2-9 are

$$K_{iL} = \left[j_i \frac{\Delta S}{\Delta I} - \frac{T}{\Delta I} \right]_N \qquad K_{iU} = \left[j_i \frac{\Delta S}{\Delta I} + \frac{T}{\Delta I} \right]_N \qquad (7.2\text{-}10)$$

where $[\cdot]_N$ denotes the nearest integer value of the argument.

Figure 7.2-1 provides an example relating actual physical sample values $\tilde{G}(j_1 \, \Delta S, j_2 \, \Delta S)$ to mesh points $\tilde{F}(k_1 \, \Delta I, k_2 \, \Delta I)$ on the ideal image field. In this example, the mesh spacing is twice as large as the physical sample spacing. In the figure,

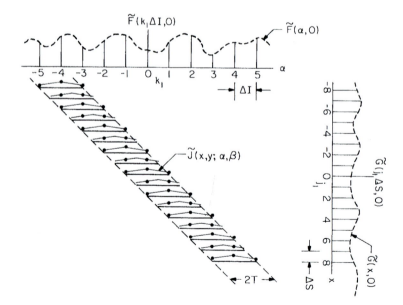

FIGURE 7.2-1. Relationship of physical image samples to mesh points on an ideal image field for numerical representation of a superposition integral.

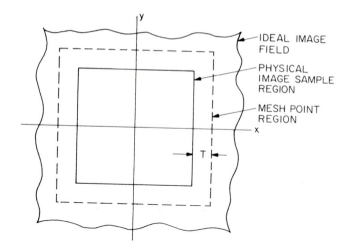

FIGURE 7.2-2. Relationship between regions of physical samples and mesh points for numerical representation of a superposition integral.

the values of the impulse response function that are utilized in the summation of Eq. 7.2-9 are represented as dots.

An important observation should be made about the discrete model of Eq. 7.2-9 for a sampled superposition integral; the physical area of the ideal image field $\tilde{F}(x, y)$ containing mesh points contributing to physical image samples is larger than the sample image $\tilde{G}(j_1 \, \Delta S, j_2 \, \Delta S)$ regardless of the relative number of physical samples and mesh points. The dimensions of the two image fields, as shown in Figure 7.2-2, are related by

$$ J \, \Delta S + T = K \, \Delta I \tag{7.2-11} $$

to within an accuracy of one sample spacing.

At this point in the discussion, a discrete and finite model for the sampled superposition integral has been obtained in which the physical samples $\tilde{G}(j_1 \, \Delta S, j_2 \, \Delta S)$ are related to points on an ideal image field $\tilde{F}(k_1 \, \Delta I, k_2 \, \Delta I)$ by a discrete mathematical superposition operation. This discrete superposition is an approximation to continuous superposition because of the truncation of the impulse response function $\tilde{J}(x, y; \alpha, \beta)$ and quadrature integration. The truncation approximation can, of course, be made arbitrarily small by extending the bounds of definition of the impulse response, but at the expense of large dimensionality. Also, the quadrature integration approximation can be improved by use of complicated formulas of quadrature, but again the price paid is computational complexity. It should be noted, however, that discrete superposition is a perfect approximation to continuous superposition if the spatial functions of Eq. 7.2-1 are all bandlimited and the physical sampling and numerical representation periods are selected to be the corresponding Nyquist period (5).

It is often convenient to reformulate Eq. 7.2-9 into vector-space form. Toward this end, the arrays $\tilde{G}$ and $\tilde{F}$ are reindexed to $M \times M$ and $N \times N$ arrays, respectively, such that all indices are positive. Let

$$F(n_1 \Delta I, n_2 \Delta I) = \tilde{F}(k_1 \Delta I, k_2 \Delta I) \tag{7.2-12a}$$

where $n_i = k_i + K + 1$ and let

$$G(m_1 \Delta S, m_2 \Delta S) = G(j_1 \Delta S, j_2 \Delta S) \tag{7.2-12b}$$

where $m_i = j_i + J + 1$. Also, let the impulse response be redefined such that

$$H(m_1 \Delta S, m_2 \Delta S; n_1 \Delta I, n_2 \Delta I) = \tilde{H}(j_1 \Delta S, j_2 \Delta S; k_1 \Delta I, k_2 \Delta I) \tag{7.2-12c}$$

Figure 7.2-3 illustrates the geometrical relationship between these functions.

The discrete superposition relationship of Eq. 7.2-9 for the shifted arrays becomes

$$G(m_1 \Delta S, m_2 \Delta S) = \sum_{n_1 = N_{1L}}^{N_{1U}} \sum_{n_2 = N_{2L}}^{N_{2U}} F(n_1 \Delta I, n_2 \Delta I) H(m_1 \Delta S, m_2 \Delta S; n_1 \Delta I, n_2 \Delta I) \tag{7.2-13}$$

for $(1 \leq m_i \leq M)$ where

$$N_{iL} = \left[m_i \frac{\Delta S}{\Delta I} \right]_N \qquad N_{iU} = \left[m_i \frac{\Delta S}{\Delta I} + \frac{2T}{\Delta I} \right]_N$$

Following the techniques outlined in Chapter 5, the vectors $\mathbf{g}$ and $\mathbf{f}$ may be formed by column scanning the matrices $\mathbf{G}$ and $\mathbf{F}$ to obtain

$$\mathbf{g} = \mathbf{B}\mathbf{f} \tag{7.2-14}$$

where $\mathbf{B}$ is a $M^2 \times N^2$ matrix of the form

$$\mathbf{B} = \begin{bmatrix} \mathbf{B}_{1,1} & \mathbf{B}_{1,2} & \cdots & \mathbf{B}_{(1,L)} & \mathbf{0} & \cdots & \mathbf{0} \\ \mathbf{0} & \mathbf{B}_{2,2} & & \vdots & & & \vdots \\ \vdots & & & & & & \mathbf{0} \\ \mathbf{0} & \cdots & \mathbf{0} & \mathbf{B}_{M,N-L+1} & \cdots & \mathbf{B}_{M,N} \end{bmatrix} \tag{7.2-15}$$

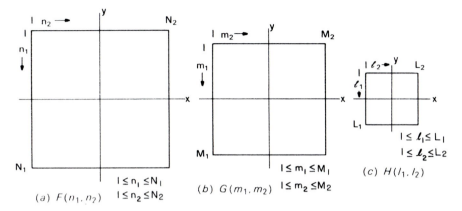

FIGURE 7.2-3. Sampled image arrays.

The general term of **B** is defined as

$$B_{m_2, n_2}(m_1, n_1) = H(m_1 \, \Delta S, m_2 \, \Delta S; n_1 \, \Delta I, n_2 \, \Delta I) \tag{7.2-16}$$

for $1 \le m_i \le M$ and $m_i \le n_i \le m_i + L - 1$ where $L = [2T/\Delta I]_N$ represents the nearest odd integer dimension of the impulse response in resolution units ΔI. For descriptional simplicity, **B** is called the *blur matrix* of the superposition integral.

If the impulse response function is translation invariant such that

$$H(m_1 \, \Delta S, m_2 \, \Delta S; n_1 \, \Delta I, n_2 \, \Delta I) = H(m_1 \, \Delta S - n_1 \, \Delta I, m_2 \, \Delta S - n_2 \, \Delta I) \tag{7.2-17}$$

then the discrete superposition operation of Eq. 7.2-13 becomes a discrete convolution operation of the form

$$G(m_1 \, \Delta S, m_2 \, \Delta S) = \sum_{n_1 = N_{1L}}^{N_{1U}} \sum_{n_2 = N_{2L}}^{N_{2U}} F(n_1 \, \Delta I, n_2 \, \Delta I) H(m_1 \, \Delta S - n_1 \, \Delta I, m_2 \, \Delta S - n_2 \, \Delta I)$$

$$\tag{7.2-18}$$

If the physical sample and quadrature mesh spacings are equal, the general term of the blur matrix assumes the form

$$B_{m_2, n_2}(m_1, n_1) = H(m_1 - n_1 + L, m_2 - n_2 + L) \tag{7.2-19}$$

$$\mathbf{H} = \begin{bmatrix} 11 & 12 & 13 \\ 21 & 22 & 23 \\ 31 & 32 & 33 \end{bmatrix}$$

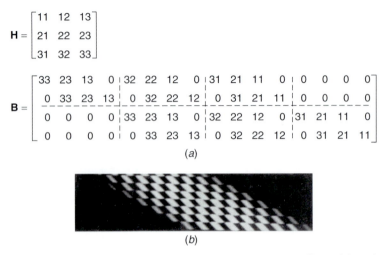

$$\mathbf{B} = \begin{bmatrix} 33 & 23 & 13 & 0 & 32 & 22 & 12 & 0 & 31 & 21 & 11 & 0 & 0 & 0 & 0 & 0 \\ 0 & 33 & 23 & 13 & 0 & 32 & 22 & 12 & 0 & 31 & 21 & 11 & 0 & 0 & 0 & 0 \\ 0 & 0 & 0 & 0 & 33 & 23 & 13 & 0 & 32 & 22 & 12 & 0 & 31 & 21 & 11 & 0 \\ 0 & 0 & 0 & 0 & 0 & 33 & 23 & 13 & 0 & 32 & 22 & 12 & 0 & 31 & 21 & 11 \end{bmatrix}$$

(a)

(b)

FIGURE 7.2-4. Sampled infinite area convolution operators: (a) General impulse array, $M = 2$, $N = 4$, $L = 3$; (b) Gaussian-shaped impulse array, $M = 8$, $N = 16$, $L = 9$.

In Eq. 7.2-19, the mesh spacing variable ΔI is understood. In addition,

$$\mathbf{B}_{m_2, n_2} = \mathbf{B}_{m_2 + 1, n_2 + 1} \tag{7.2-20}$$

Consequently, the rows of **B** are shifted versions of the first row. The operator **B** then becomes a sampled infinite area convolution operator, and the series form representation of Eq. 7.2-19 reduces to

$$G(m_1 \, \Delta S, m_2 \, \Delta S) = \sum_{n_1 = m_1}^{m_1 + L - 1} \sum_{n_2 = m_2}^{m_2 + L - 1} F(n_1, n_2) H(m_1 - n_1 + L, m_2 - n_2 + L) \tag{7.2-21}$$

where the sampling spacing is understood.

Figure 7.2-4a is a notational example of the sampled image convolution operator for a 4×4 ($N = 4$) data array, a 2×2 ($M = 2$) filtered data array, and a 3×3 ($L = 3$) impulse response array. An extension to larger dimension is shown in Figure 7.2-4b for $M = 8$, $N = 16$, $L = 9$ and a Gaussian-shaped impulse response.

When the impulse response is spatially invariant and orthogonally separable,

$$\mathbf{B} = \mathbf{B}_C \otimes \mathbf{B}_R \tag{7.2-22}$$

where $\mathbf{B}_R$ and $\mathbf{B}_C$ are $M \times N$ matrices of the form

$$\mathbf{B}_R = \begin{bmatrix} h_R(L) & h_R(L-1) & \cdots & h_R(1) & 0 & \cdots & 0 \\ 0 & h_R(L) & & & & & \vdots \\ \vdots & & & & & & 0 \\ 0 & \cdots & 0 & h_R(L) & & \cdots & h_R(1) \end{bmatrix} \qquad (7.2\text{-}23)$$

The two-dimensional convolution operation then reduces to sequential row and column convolutions of the matrix form of the image array. Thus

$$\mathbf{G} = \mathbf{B}_C \mathbf{F} \mathbf{B}_R^T \qquad (7.2\text{-}24)$$

The superposition or convolution operator expressed in vector form requires $M^2 L^2$ operations if the zero multiplications of $\mathbf{B}$ are avoided. A separable convolution operator can be computed in matrix form with only $ML(M+N)$ operations.

7.3. CIRCULANT SUPERPOSITION AND CONVOLUTION

In circulant superposition (2), the input data, the processed output, and the impulse response arrays are all assumed spatially periodic over a common period. To unify the presentation, these arrays will be defined in terms of the spatially limited arrays considered previously. First, let the $N \times N$ data array $F(n_1, n_2)$ be embedded in the upper left corner of a $J \times J$ array $(J > N)$ of zeros, giving

$$F_E(n_1, n_2) = \begin{cases} F(n_1, n_2) & \text{for } 1 \le n_i \le N & (7.3\text{-}1a) \\ \\ 0 & \text{for } N+1 \le n_i \le J & (7.3\text{-}1b) \end{cases}$$

In a similar manner, an extended impulse response array is created by embedding the spatially limited impulse array in a $J \times J$ matrix of zeros. Thus, let

$$H_E(l_1, l_2; m_1, m_2) = \begin{cases} H(l_1, l_2; m_1, m_2) & \text{for } 1 \le l_i \le L & (7.3\text{-}2a) \\ \\ 0 & \text{for } L+1 \le l_i \le J & (7.3\text{-}2b) \end{cases}$$

Periodic arrays $F_E(n_1, n_2)$ and $H_E(l_1, l_2; m_1, m_2)$ are now formed by replicating the extended arrays over the spatial period J. Then, the circulant superposition of these functions is defined as

$$K_E(m_1, m_2) = \sum_{n_1 = 1}^{J} \sum_{n_2 = 1}^{J} F_E(n_1, n_2) H_E(m_1 - n_1 + 1, m_2 - n_2 + 1; m_1, m_2) \quad (7.3\text{-}3)$$

Similarity of this equation with Eq. 7.1-6 describing finite-area superposition is evident. In fact, if J is chosen such that $J = N + L - 1$, the terms $F_E(n_1, n_2) = F(n_1, n_2)$ for $1 \le n_i \le N$. The similarity of the circulant superposition operation and the sampled image superposition operation should also be noted. These relations become clearer in the vector-space representation of the circulant superposition operation.

Let the arrays F_E and K_E be expressed in vector form as the $J^2 \times 1$ vectors $\mathbf{f}_E$ and $\mathbf{k}_E$, respectively. Then, the circulant superposition operator can be written as

$$\mathbf{k}_E = \mathbf{C} \mathbf{f}_E \quad (7.3\text{-}4)$$

where $\mathbf{C}$ is a $J^2 \times J^2$ matrix containing elements of the array H_E. The circulant superposition operator can then be conveniently expressed in terms of $J \times J$ submatrices $\mathbf{C}_{mn}$ as given by

$$
\mathbf{C} =
\begin{bmatrix}
\mathbf{C}_{1,1} & \mathbf{0} & \mathbf{0} & \cdots & \mathbf{0} & \mathbf{C}_{1,J-L+2} & \cdots & \mathbf{C}_{1,J} \\
\mathbf{C}_{2,1} & \mathbf{C}_{2,2} & \mathbf{0} & \cdots & \mathbf{0} & \mathbf{0} & \vdots & \vdots \\
\vdots & & \cdot & & & & \mathbf{0} & \mathbf{C}_{L-1,J} \\
\mathbf{C}_{2,1} & \mathbf{C}_{L,2} & \mathbf{0} & & & \cdots & & \mathbf{0} \\
\mathbf{0} & \mathbf{C}_{L+1,2} & & & & & & \vdots \\
\mathbf{0} & & & & & & & \vdots \\
\vdots & & & & & & & \vdots \\
\mathbf{0} & & \cdots & \mathbf{0} & \mathbf{C}_{J,J-L+1} & \mathbf{C}_{J,J-L+2} & \cdots & \mathbf{C}_{J,J}
\end{bmatrix}
\quad (7.3\text{-}5)
$$

where

$$\mathbf{C}_{m_2, n_2}(m_1, n_1) = H_E(k_1, k_2; m_1, m_2) \quad (7.3\text{-}6)$$

$$H = \begin{bmatrix} 11 & 12 & 13 \\ 21 & 22 & 23 \\ 31 & 32 & 33 \end{bmatrix}$$

$$C = \begin{bmatrix}
11 & 0 & 31 & 21 & 0 & 0 & 0 & 0 & 13 & 0 & 33 & 23 & 12 & 0 & 32 & 22 \\
21 & 11 & 0 & 31 & 0 & 0 & 0 & 0 & 23 & 13 & 0 & 33 & 22 & 12 & 0 & 32 \\
31 & 21 & 11 & 0 & 0 & 0 & 0 & 0 & 33 & 23 & 13 & 0 & 32 & 22 & 12 & 0 \\
0 & 31 & 21 & 11 & 0 & 0 & 0 & 0 & 0 & 33 & 23 & 13 & 0 & 32 & 22 & 12 \\
12 & 0 & 32 & 22 & 11 & 0 & 31 & 21 & 0 & 0 & 0 & 0 & 13 & 0 & 33 & 23 \\
22 & 12 & 0 & 32 & 21 & 11 & 0 & 31 & 0 & 0 & 0 & 0 & 23 & 13 & 0 & 33 \\
32 & 22 & 12 & 0 & 31 & 21 & 11 & 0 & 0 & 0 & 0 & 0 & 33 & 23 & 13 & 0 \\
0 & 32 & 22 & 12 & 0 & 31 & 21 & 11 & 0 & 0 & 0 & 0 & 0 & 33 & 23 & 13 \\
13 & 0 & 33 & 23 & 12 & 0 & 32 & 22 & 11 & 0 & 31 & 21 & 0 & 0 & 0 & 0 \\
23 & 13 & 0 & 33 & 22 & 12 & 0 & 32 & 21 & 11 & 0 & 31 & 0 & 0 & 0 & 0 \\
33 & 23 & 13 & 0 & 32 & 22 & 12 & 0 & 31 & 21 & 11 & 0 & 0 & 0 & 0 & 0 \\
0 & 33 & 23 & 13 & 0 & 32 & 22 & 12 & 0 & 31 & 21 & 11 & 0 & 0 & 0 & 0 \\
0 & 0 & 0 & 0 & 13 & 0 & 33 & 23 & 12 & 0 & 32 & 22 & 11 & 0 & 31 & 21 \\
0 & 0 & 0 & 0 & 23 & 13 & 0 & 33 & 22 & 12 & 0 & 32 & 21 & 11 & 0 & 31 \\
0 & 0 & 0 & 0 & 33 & 23 & 13 & 0 & 32 & 22 & 12 & 0 & 31 & 21 & 11 & 0 \\
0 & 0 & 0 & 0 & 0 & 33 & 23 & 13 & 0 & 32 & 22 & 12 & 0 & 31 & 21 & 11
\end{bmatrix}$$

(a)

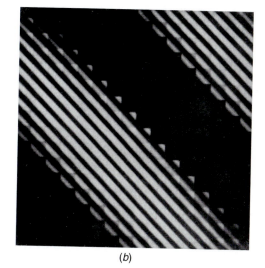

(b)

FIGURE 7.3-1. Circulant convolution operators: (a) General impulse array, $J = 4$, $L = 3$; (b) Gaussian-shaped impulse array, $J = 16$, $L = 9$.

for $1 \leq n_i \leq J$ and $1 \leq m_i \leq J$ with $k_i = (m_i - n_i + 1)$ modulo J and $H_E(0, 0) = 0$. It should be noted that each row and column of $\mathbf{C}$ contains L nonzero submatrices. If the impulse response array is spatially invariant, then

$$\mathbf{C}_{m_2, n_2} = \mathbf{C}_{m_2 + 1, n_2 + 1} \tag{7.3-7}$$

and the submatrices of the rows (columns) can be obtained by a circular shift of the first row (column). Figure 7.3-1a illustrates the circulant convolution operator for 16×16 ($J = 4$) data and filtered data arrays and for a 3×3 ($L = 3$) impulse response array. In Figure 7.3-1b, the operator is shown for $J = 16$ and $L = 9$ with a Gaussian-shaped impulse response.

Finally, when the impulse response is spatially invariant and orthogonally separable,

$$\mathbf{C} = \mathbf{C}_C \otimes \mathbf{C}_R \tag{7.3-8}$$

where $\mathbf{C}_R$ and $\mathbf{C}_C$ are $J \times J$ matrices of the form

$$\mathbf{C}_R = \begin{bmatrix} h_R(1) & 0 & \cdots & 0 & h_R(L) & \cdots & h_R(3) & h_R(2) \\ h_R(2) & h_R(1) & \cdots & 0 & 0 & & \vdots & h_R(3) \\ \vdots & \vdots & & & & & & \vdots \\ h_R(L-1) & & & & \cdots & 0 & h_R(L) \\ h_R(L) & h_R(L-1) & & & & & 0 \\ 0 & h_R(L) & & & & & \vdots \\ \vdots & & & & & & 0 \\ 0 & & \cdots & 0 & h_R(L) & \cdots & \cdots & h_R(2) & h_R(1) \end{bmatrix} \tag{7.3-9}$$

Two-dimensional circulant convolution may then be computed as

$$\mathbf{K}_E = \mathbf{C}_C \mathbf{F}_E \mathbf{C}_R^T \tag{7.3-10}$$

7.4. SUPERPOSITION AND CONVOLUTION OPERATOR RELATIONSHIPS

The elements of the finite-area superposition operator $\mathbf{D}$ and the elements of the sampled image superposition operator $\mathbf{B}$ can be extracted from circulant superposition operator $\mathbf{C}$ by use of selection matrices defined as (2)

$$\mathbf{S1}_J^{(K)} = \begin{bmatrix} \mathbf{I}_K \vert \ \mathbf{0} \end{bmatrix} \tag{7.4-1a}$$

$$\mathbf{S2}_J^{(K)} = \begin{bmatrix} \mathbf{0}_A \vert \mathbf{I}_K \vert \ \mathbf{0} \end{bmatrix} \tag{7.4-1b}$$

where $\mathbf{S1}_J^{(K)}$ and $\mathbf{S2}_J^{(K)}$ are $K \times J$ matrices, $\mathbf{I}_K$ is a $K \times K$ identity matrix, and $\mathbf{0}_A$ is a $K \times L - 1$ matrix. For future reference, it should be noted that the generalized inverses of $\mathbf{S1}$ and $\mathbf{S2}$ and their transposes are

$$[\mathbf{S1}_J^{(K)}]^- = [\mathbf{S1}_J^{(K)}]^T \tag{7.4-2a}$$

$$[[\mathbf{S1}_J^{(K)}]^T]^- = \mathbf{S1}_J^K \tag{7.4-2b}$$

$$[\mathbf{S2}_J^{(K)}]^- = [\mathbf{S2}_J^{(K)}]^T \tag{7.4-2c}$$

$$[[\mathbf{S2}_J^{(K)}]^T]^- = \mathbf{S2}_J^K \tag{7.4-2d}$$

Examination of the structure of the various superposition operators indicates that

$$\mathbf{D} = [\mathbf{S1}_J^{(M)} \otimes \mathbf{S1}_J^{(M)}]\mathbf{C}[\mathbf{S1}_J^{(N)} \otimes \mathbf{S1}_J^{(N)}]^T \tag{7.4-3a}$$

$$\mathbf{B} = [\mathbf{S2}_J^{(M)} \otimes \mathbf{S2}_J^{(M)}]\mathbf{C}[\mathbf{S1}_J^{(N)} \otimes \mathbf{S1}_J^{(N)}]^T \tag{7.4-3b}$$

That is, the matrix $\mathbf{D}$ is obtained by extracting the first M rows and N columns of submatrices $\mathbf{C}_{mn}$ of $\mathbf{C}$. The first M rows and N columns of each submatrix are also extracted. A similar explanation holds for the extraction of $\mathbf{B}$ from $\mathbf{C}$. In Figure 7.3-1, the elements of $\mathbf{C}$ to be extracted to form $\mathbf{D}$ and $\mathbf{B}$ are indicated by boxes.

From the definition of the extended input data array of Eq. 7.3-1, it is obvious that the spatially limited input data vector $\mathbf{f}$ can be obtained from the extended data vector $\mathbf{f}_E$ by the selection operation

$$\mathbf{f} = [\mathbf{S1}_J^{(N)} \otimes \mathbf{S1}_J^{(N)}]\mathbf{f}_E \tag{7.4-4a}$$

and furthermore,

$$\mathbf{f}_E = [\mathbf{S1}_J^{(N)} \otimes \mathbf{S1}_J^{(N)}]^T \mathbf{f} \tag{7.4-4b}$$

It can also be shown that the output vector for finite-area superposition can be obtained from the output vector for circulant superposition by the selection operation

$$\mathbf{q} = [\mathbf{S1}_J^{(M)} \otimes \mathbf{S1}_J^{(M)}]\mathbf{k}_E \qquad (7.4\text{-}5a)$$

The inverse relationship also exists in the form

$$\mathbf{k}_E = [\mathbf{S1}_J^{(M)} \otimes \mathbf{S1}_J^{(M)}]^T \mathbf{q} \qquad (7.4\text{-}5b)$$

For sampled image superposition

$$\mathbf{g} = [\mathbf{S2}_J^{(M)} \otimes \mathbf{S2}_J^{(M)}]\mathbf{k}_E \qquad (7.4\text{-}6)$$

but it is not possible to obtain $\mathbf{k}_E$ from $\mathbf{g}$ because of the underdeterminacy of the sampled image superposition operator. Expressing both $\mathbf{q}$ and $\mathbf{k}_E$ of Eq. 7.4-5a in matrix form leads to

$$\mathbf{Q} = \sum_{m=1}^{M} \sum_{n=1}^{J} \mathbf{M}_m^T[\mathbf{S1}_J^{(M)} \otimes \mathbf{S1}_J^{(M)}]\mathbf{N}_n\mathbf{K}_E\mathbf{v}_n\mathbf{u}_m^T \qquad (7.4\text{-}7)$$

As a result of the separability of the selection operator, Eq. 7.4-7 reduces to

$$\mathbf{Q} = [\mathbf{S1}_J^{(M)}]\mathbf{K}_E[\mathbf{S1}_J^{(M)}]^T \qquad (7.4\text{-}8)$$

Similarly, for Eq. 7.4-6 describing sampled infinite-area superposition,

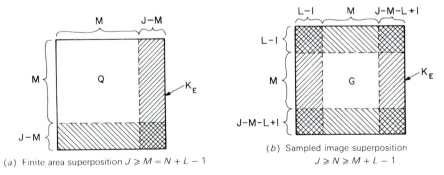

(a) Finite area superposition $J \geqslant M = N + L - 1$

(b) Sampled image superposition $J \geqslant N \geqslant M + L - 1$

FIGURE 7.4-1. Location of elements of processed data Q and G from K_E.

$$\mathbf{G} = [\mathbf{S2}_J^{(M)}]\mathbf{K}_E[\mathbf{S2}_J^{(M)}]^T \tag{7.4-9}$$

Figure 7.4-1 illustrates the locations of the elements of $\mathbf{G}$ and $\mathbf{Q}$ extracted from $\mathbf{K}_E$ for finite-area and sampled infinite-area superposition.

In summary, it has been shown that the output data vectors for either finite-area or sampled image superposition can be obtained by a simple selection operation on the output data vector of circulant superposition. Computational advantages that can be realized from this result are considered in Chapter 9.

REFERENCES

1. J. F. Abramatic and O. D. Faugeras, "Design of Two-Dimensional FIR Filters from Small Generating Kernels," *Proc. IEEE Conference on Pattern Recognition and Image Processing*, Chicago, May 1978.

2. W. K. Pratt, "Vector Formulation of Two Dimensional Signal Processing Operations," *Computer Graphics and Image Processing*, **4**, 1, March 1975, 1–24.

3. A. V. Oppenheim and R. W. Schaefer (Contributor), *Digital Signal Processing*, Prentice Hall, Englewood Cliffs, NJ, 1975.

4. T. R. McCalla, *Introduction to Numerical Methods and FORTRAN Programming*, Wiley, New York, 1967.

5. A. Papoulis, *Systems and Transforms with Applications in Optics*, 2nd ed., McGraw-Hill, New York, 1981.

8

UNITARY TRANSFORMS

Two-dimensional unitary transforms have found two major applications in image processing. Transforms have been utilized to extract features from images. For example, with the Fourier transform, the average value or *dc term* is proportional to the average image amplitude, and the high-frequency terms (*ac term*) give an indication of the amplitude and orientation of edges within an image. Dimensionality reduction in computation is a second image processing application. Stated simply, those transform coefficients that are small may be excluded from processing operations, such as filtering, without much loss in processing accuracy. Another application in the field of image coding is transform image coding, in which a bandwidth reduction is achieved by discarding or grossly quantizing low-magnitude transform coefficients. In this chapter we consider the properties of unitary transforms commonly used in image processing.

8.1. GENERAL UNITARY TRANSFORMS

A *unitary transform* is a specific type of linear transformation in which the basic linear operation of Eq. 5.4-1 is exactly invertible and the operator kernel satisfies certain orthogonality conditions (1,2). The forward unitary transform of the $N_1 \times N_2$ image array $F(n_1, n_2)$ results in a $N_1 \times N_2$ transformed image array as defined by

$$\mathcal{F}(m_1, m_2) = \sum_{n_1=1}^{N_1} \sum_{n_2=1}^{N_2} F(n_1, n_2)A(n_1, n_2; m_1, m_2) \qquad (8.1\text{-}1)$$

where $A(n_1, n_2; m_1, m_2)$ represents the forward transform kernel. A reverse or inverse transformation provides a mapping from the transform domain to the image space as given by

$$F(n_1, n_2) = \sum_{m_1=1}^{N_1} \sum_{m_2=1}^{N_2} \mathcal{F}(m_1, m_2)B(n_1, n_2; m_1, m_2) \qquad (8.1\text{-}2)$$

where $B(n_1, n_2; m_1, m_2)$ denotes the inverse transform kernel. The transformation is unitary if the following orthonormality conditions are met:

$$\sum_{m_1} \sum_{m_2} A(n_1, n_2; m_1, m_2)A^*(j_1, j_2; m_1, m_2) = \delta(n_1 - j_1, n_2 - j_2) \qquad (8.1\text{-}3a)$$

$$\sum_{m_1} \sum_{m_2} B(n_1, n_2; m_1, m_2)B^*(j_1, j_2; m_1, m_2) = \delta(n_1 - j_1, n_2 - j_2) \qquad (8.1\text{-}3b)$$

$$\sum_{n_1} \sum_{n_2} A(n_1, n_2; m_1, m_2)A^*(n_1, n_2; k_1, k_2) = \delta(m_1 - k_1, m_2 - k_2) \qquad (8.1\text{-}3c)$$

$$\sum_{n_1} \sum_{n_2} B(n_1, n_2; m_1, m_2)B^*(n_1, n_2; k_1, k_2) = \delta(m_1 - k_1, m_2 - k_2) \qquad (8.1\text{-}3c)$$

The transformation is said to be separable if its kernels can be written in the form

$$A(n_1, n_2; m_1, m_2) = A_C(n_1, m_1)A_R(n_2, m_2) \qquad (8.1\text{-}4a)$$

$$B(n_1, n_2; m_1, m_2) = B_C(n_1, m_1)B_R(n_2, m_2) \qquad (8.1\text{-}4b)$$

where the kernel subscripts indicate row and column one-dimensional transform operations. A separable two-dimensional unitary transform can be computed in two steps. First, a one-dimensional transform is taken along each column of the image, yielding

$$P(m_1, n_2) = \sum_{n_1=1}^{N_1} F(n_1, n_2)A_C(n_1, m_1) \qquad (8.1\text{-}5)$$

Next, a second one-dimensional unitary transform is taken along each row of $P(m_1, n_2)$, giving

$$\mathcal{F}(m_1, m_2) = \sum_{n_2=1}^{N_2} P(m_1, n_2)A_R(n_2, m_2) \qquad (8.1\text{-}6)$$

Unitary transforms can conveniently be expressed in vector-space form (3). Let **F** and **f** denote the matrix and vector representations of an image array, and let $\mathcal{F}$ and f be the matrix and vector forms of the transformed image. Then, the two-dimensional unitary transform written in vector form is given by

$$f = \mathbf{A}\mathbf{f} \qquad (8.1\text{-}7)$$

where **A** is the forward transformation matrix. The reverse transform is

$$\mathbf{f} = \mathbf{B}f \qquad (8.1\text{-}8)$$

where **B** represents the inverse transformation matrix. It is obvious then that

$$\mathbf{B} = \mathbf{A}^{-1} \qquad (8.1\text{-}9)$$

For a unitary transformation, the matrix inverse is given by

$$\mathbf{A}^{-1} = \mathbf{A}^{*T} \qquad (8.1\text{-}10)$$

and **A** is said to be a *unitary matrix*. A real unitary matrix is called an *orthogonal matrix*. For such a matrix,

$$\mathbf{A}^{-1} = \mathbf{A}^{T} \qquad (8.1\text{-}11)$$

If the transform kernels are separable such that

$$\mathbf{A} = \mathbf{A}_C \otimes \mathbf{A}_R \qquad (8.1\text{-}12)$$

where $\mathbf{A}_R$ and $\mathbf{A}_C$ are row and column unitary transform matrices, then the transformed image matrix can be obtained from the image matrix by

$$\mathcal{F} = \mathbf{A}_C \mathbf{F} \mathbf{A}_R^{T} \qquad (8.1\text{-}13a)$$

The inverse transformation is given by

$$\mathbf{F} = \mathbf{B}_C \mathcal{F} \mathbf{B}_R^{T} \qquad (8.1\text{-}13b)$$

where $\mathbf{B}_C = \mathbf{A}_C^{-1}$ and $\mathbf{B}_R = \mathbf{A}_R^{-1}$.

Separable unitary transforms can also be expressed in a hybrid series–vector space form as a sum of vector outer products. Let $\mathbf{a}_C(n_1)$ and $\mathbf{a}_R(n_2)$ represent rows n_1 and n_2 of the unitary matrices $\mathbf{A}_R$ and $\mathbf{A}_R$, respectively. Then, it is easily verified that

$$\mathcal{F} = \sum_{n_1=1}^{N_1} \sum_{n_2=1}^{N_2} F(n_1, n_2)\mathbf{a}_C(n_1)\mathbf{a}_R^T(n_2) \tag{8.1-14a}$$

Similarly,

$$\mathbf{F} = \sum_{m_1=1}^{N_1} \sum_{m_2=1}^{N_2} \mathcal{F}(m_1, m_2)\mathbf{b}_C(m_1)\mathbf{b}_R^T(m_2) \tag{8.1-14b}$$

where $\mathbf{b}_C(m_1)$ and $\mathbf{b}_R(m_2)$ denote rows m_1 and m_2 of the unitary matrices $\mathbf{B}_C$ and $\mathbf{B}_R$, respectively. The vector outer products of Eq. 8.1-14 form a series of matrices, called *basis matrices*, that provide matrix decompositions of the image matrix $\mathbf{F}$ or its unitary transformation $\mathcal{F}$.

There are several ways in which a unitary transformation may be viewed. An image transformation can be interpreted as a decomposition of the image data into a generalized two-dimensional spectrum (4). Each spectral component in the transform domain corresponds to the amount of energy of the spectral function within the original image. In this context, the concept of frequency may now be generalized to include transformations by functions other than sine and cosine waveforms. This type of generalized spectral analysis is useful in the investigation of specific decompositions that are best suited for particular classes of images. Another way to visualize an image transformation is to consider the transformation as a multidimensional rotation of coordinates. One of the major properties of a unitary transformation is that measure is preserved. For example, the mean-square difference between two images is equal to the mean-square difference between the unitary transforms of the images. A third approach to the visualization of image transformation is to consider Eq. 8.1-2 as a means of synthesizing an image with a set of two-dimensional mathematical functions $B(n_1, n_2; m_1, m_2)$ for a fixed transform domain coordinate (m_1, m_2). In this interpretation, the kernel $B(n_1, n_2; m_1, m_2)$ is called a *two-dimensional basis function* and the transform coefficient $\mathcal{F}(m_1, m_2)$ is the amplitude of the basis function required in the synthesis of the image.

In the remainder of this chapter, to simplify the analysis of two-dimensional unitary transforms, all image arrays are considered square of dimension N. Furthermore, when expressing transformation operations in series form, as in Eqs. 8.1-1 and 8.1-2, the indices are renumbered and renamed. Thus the input image array is denoted by $F(j, k)$ for $j, k = 0, 1, 2,..., N - 1$, and the transformed image array is represented by $\mathcal{F}(u, v)$ for $u, v = 0, 1, 2,..., N - 1$. With these definitions, the forward unitary transform becomes

$$\mathcal{F}(u, v) = \sum_{j=0}^{N-1} \sum_{k=0}^{N-1} F(j, k) A(j, k; u, v) \tag{8.1-15a}$$

and the inverse transform is

$$F(j, k) = \sum_{u=0}^{N-1} \sum_{v=0}^{N-1} \mathcal{F}(u, v) B(j, k; u, v) \tag{8.1-15b}$$

8.2. FOURIER TRANSFORM

The discrete two-dimensional *Fourier transform* of an image array is defined in series form as (5–10)

$$\mathcal{F}(u, v) = \frac{1}{N} \sum_{j=0}^{N-1} \sum_{k=0}^{N-1} F(j, k) \exp\left\{ \frac{-2\pi i}{N}(uj + vk) \right\} \tag{8.2-1a}$$

where $i = \sqrt{-1}$, and the discrete inverse transform is given by

$$F(j, k) = \frac{1}{N} \sum_{u=0}^{N-1} \sum_{v=0}^{N-1} \mathcal{F}(u, v) \exp\left\{ \frac{2\pi i}{N}(uj + vk) \right\} \tag{8.2-1b}$$

The indices (u, v) are called the *spatial frequencies* of the transformation in analogy with the continuous Fourier transform. It should be noted that Eq. 8.2-1 is not universally accepted by all authors; some prefer to place all scaling constants in the inverse transform equation, while still others employ a reversal in the sign of the kernels.

Because the transform kernels are separable and symmetric, the two dimensional transforms can be computed as sequential row and column one-dimensional transforms. The basis functions of the transform are complex exponentials that may be decomposed into sine and cosine components. The resulting Fourier transform pairs then become

$$A(j, k; u, v) = \exp\left\{ \frac{-2\pi i}{N}(uj + vk) \right\} = \cos\left\{ \frac{2\pi}{N}(uj + vk) \right\} - i\sin\left\{ \frac{2\pi}{N}(uj + vk) \right\} \tag{8.2-2a}$$

$$B(j, k; u, v) = \exp\left\{ \frac{2\pi i}{N}(uj + vk) \right\} = \cos\left\{ \frac{2\pi}{N}(uj + vk) \right\} + i\sin\left\{ \frac{2\pi}{N}(uj + vk) \right\} \tag{8.2-2b}$$

Figure 8.2-1 shows plots of the sine and cosine components of the one-dimensional Fourier basis functions for $N = 16$. It should be observed that the basis functions are a rough approximation to continuous sinusoids only for low frequencies; in fact, the

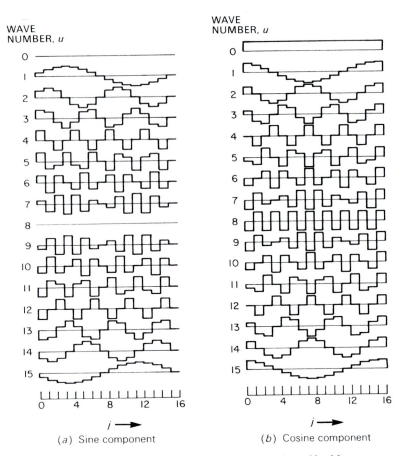

FIGURE 8.2-1 Fourier transform basis functions, $N = 16$.

highest-frequency basis function is a square wave. Also, there are obvious redundancies between the sine and cosine components.

The Fourier transform plane possesses many interesting structural properties. The spectral component at the origin of the Fourier domain

$$\mathcal{F}(0, 0) = \frac{1}{N} \sum_{j=0}^{N-1} \sum_{k=0}^{N-1} F(j, k) \tag{8.2-3}$$

is equal to N times the spatial average of the image plane. Making the substitutions $u = u + mN$, $v = v + nN$ in Eq. 8.2-1, where m and n are constants, results in

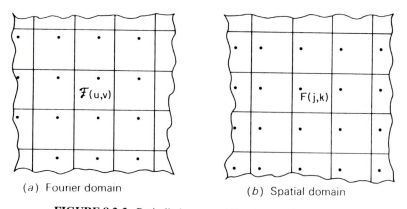

(a) Fourier domain   (b) Spatial domain

FIGURE 8.2-2. Periodic image and Fourier transform arrays.

$$\mathcal{F}(u + mN, v + nN) = \frac{1}{N} \sum_{j=0}^{N-1} \sum_{k=0}^{N-1} F(j, k) \exp\left\{\frac{-2\pi i}{N}(uj + vk)\right\} \exp\{-2\pi i(mj + nk)\}$$

(8.2-4)

For all integer values of m and n, the second exponential term of Eq. 8.2-5 assumes a value of unity, and the transform domain is found to be periodic. Thus, as shown in Figure 8.2-2a,

$$\mathcal{F}(u + mN, v + nN) = \mathcal{F}(u, v)$$

(8.2-5)

for $m, n = 0, \pm 1, \pm 2, \ldots$.

The two-dimensional Fourier transform of an image is essentially a Fourier series representation of a two-dimensional field. For the Fourier series representation to be valid, the field must be periodic. Thus, as shown in Figure 8.2-2b, the original image must be considered to be periodic horizontally and vertically. The right side of the image therefore abuts the left side, and the top and bottom of the image are adjacent. Spatial frequencies along the coordinate axes of the transform plane arise from these transitions.

If the image array represents a luminance field, $F(j, k)$ will be a real positive function. However, its Fourier transform will, in general, be complex. Because the transform domain contains $2N^2$ components, the real and imaginary, or phase and magnitude components, of each coefficient, it might be thought that the Fourier transformation causes an increase in dimensionality. This, however, is not the case because $\mathcal{F}(u, v)$ exhibits a property of conjugate symmetry. From Eq. 8.2-4, with m and n set to integer values, conjugation yields

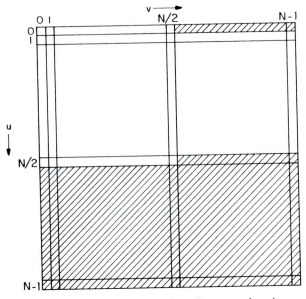

FIGURE 8.2-3. Fourier transform frequency domain.

$$\mathcal{F}^*(u + mN, v + nN) = \frac{1}{N} \sum_{j=0}^{N-1} \sum_{k=0}^{N-1} F(j, k) \exp\left\{\frac{-2\pi i}{N}(uj + vk)\right\} \qquad (8.2\text{-}6)$$

By the substitution $u = -u$ and $v = -v$ it can be shown that

$$\mathcal{F}(u, v) = \mathcal{F}^*(-u + mN, -v + nN) \qquad (8.2\text{-}7)$$

for $n = 0, \pm 1, \pm 2, \ldots$. As a result of the conjugate symmetry property, almost one-half of the transform domain samples are redundant; that is, they can be generated from other transform samples. Figure 8.2-3 shows the transform plane with a set of redundant components crosshatched. It is possible, of course, to choose the left half-plane samples rather than the upper plane samples as the nonredundant set.

Figure 8.2-4 shows a monochrome test image and various versions of its Fourier transform, as computed by Eq. 8.2-1a, where the test image has been scaled over unit range $0.0 \leq F(j, k) \leq 1.0$. Because the dynamic range of transform components is much larger than the exposure range of photographic film, it is necessary to compress the coefficient values to produce a useful display. Amplitude compression to a unit range display array $\mathcal{D}(u, v)$ can be obtained by clipping large-magnitude values according to the relation

(a) Original   (b) Clipped magnitude, nonordered

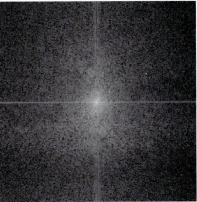

(c) Log magnitude, nonordered   (d) Log magnitude, ordered

FIGURE 8.2-4.   Fourier transform of the `smpte_girl_luma` image.

$$\mathcal{D}(u, v) = \begin{cases} 1.0 & \text{if } |\mathcal{F}(u, v)| \ge c|\mathcal{F}_{max}| \quad (8.2\text{-}8a) \\ \dfrac{|\mathcal{F}(u, v)|}{c|\mathcal{F}_{max}|} & \text{if } |\mathcal{F}(u, v)| < c|\mathcal{F}_{max}| \quad (8.2\text{-}8b) \end{cases}$$

where $0.0 < c \le 1.0$ is the clipping factor and $|\mathcal{F}_{max}|$ is the maximum coefficient magnitude. Another form of amplitude compression is to take the logarithm of each component as given by

$$\mathcal{D}(u, v) = \frac{\log\{a + b|\mathcal{F}(u, v)|\}}{\log\{a + b|\mathcal{F}_{max}|\}} \qquad (8.2\text{-}9)$$

where a and b are scaling constants. Figure 8.2-4b is a clipped magnitude display of the magnitude of the Fourier transform coefficients. Figure 8.2-4c is a logarithmic display for $a = 1.0$ and $b = 100.0$.

In mathematical operations with continuous signals, the origin of the transform domain is usually at its geometric center. Similarly, the Fraunhofer diffraction pattern of a photographic transparency of transmittance $F(x, y)$ produced by a coherent optical system has its zero-frequency term at the center of its display. A computer-generated two-dimensional discrete Fourier transform with its origin at its center can be produced by a simple reordering of its transform coefficients. Alternatively, the quadrants of the Fourier transform, as computed by Eq. 8.2-la, can be reordered automatically by multiplying the image function by the factor $(-1)^{j+k}$ prior to the Fourier transformation. The proof of this assertion follows from Eq. 8.2-4 with the substitution $m = n = \frac{1}{2}$. Then, by the identity

$$\exp\{i\pi(j + k)\} = (-1)^{j+k} \tag{8.2-10}$$

Eq. 8.2-5 can be expressed as

$$\mathcal{F}(u + N/2, v + N/2) = \frac{1}{N} \sum_{j=0}^{N-1} \sum_{k=0}^{N-1} F(j, k)(-1)^{j+k} \exp\left\{\frac{-2\pi i}{N}(uj + vk)\right\} \tag{8.2-11}$$

Figure 8.2-4d contains a log magnitude display of the reordered Fourier components. The conjugate symmetry in the Fourier domain is readily apparent from the photograph.

The Fourier transform written in series form in Eq. 8.2-1 may be redefined in vector-space form as

$$f = \mathbf{A}f \tag{8.2-12a}$$

$$\mathbf{f} = \mathbf{A}^{*T} f \tag{8.2-12b}$$

where $\mathbf{f}$ and f are vectors obtained by column scanning the matrices $\mathbf{F}$ and $\mathcal{F}$, respectively. The transformation matrix $\mathbf{A}$ can be written in direct product form as

$$\mathbf{A} = \mathbf{A}_C \otimes \mathbf{A}_R \tag{8.2-13}$$

where

$$
\mathbf{A}_R = \mathbf{A}_C =
\begin{bmatrix}
\mathcal{W}^0 & \mathcal{W}^0 & \mathcal{W}^0 & \cdots & \mathcal{W}^0 \\
\mathcal{W}^0 & \mathcal{W}^1 & \mathcal{W}^2 & \cdots & \mathcal{W}^{N-1} \\
\mathcal{W}^0 & \mathcal{W}^2 & \mathcal{W}^4 & \cdots & \mathcal{W}^{2(N-1)} \\
\vdots & & & & \vdots \\
\mathcal{W}^0 & \cdot & \cdot & \cdots & \mathcal{W}^{(N-1)^2}
\end{bmatrix}
\tag{8.2-14}
$$

with $\mathcal{W} = \exp\{-2\pi i/N\}$. As a result of the direct product decomposition of $\mathbf{A}$, the image matrix and transformed image matrix are related by

$$
\mathcal{F} = \mathbf{A}_C \mathbf{F} \mathbf{A}_R
\tag{8.2-15a}
$$

$$
\mathbf{F} = \mathbf{A}_C^* \mathcal{F} \mathbf{A}_R^*
\tag{8.2-15b}
$$

The properties of the Fourier transform previously proved in series form obviously hold in the matrix formulation.

One of the major contributions to the field of image processing was the discovery (5) of an efficient computational algorithm for the discrete Fourier transform (DFT). Brute-force computation of the discrete Fourier transform of a one-dimensional sequence of N values requires on the order of N^2 complex multiply and add operations. A fast Fourier transform (FFT) requires on the order of $N \log N$ operations. For large images the computational savings are substantial. The original FFT algorithms were limited to images whose dimensions are a power of 2 (e.g., $N = 2^9 = 512$). Modern algorithms exist for less restrictive image dimensions.

Although the Fourier transform possesses many desirable analytic properties, it has a major drawback: Complex, rather than real number computations are necessary. Also, for image coding it does not provide as efficient image energy compaction as other transforms.

8.3. COSINE, SINE, AND HARTLEY TRANSFORMS

The cosine, sine, and Hartley transforms are unitary transforms that utilize sinusoidal basis functions, as does the Fourier transform. The cosine and sine transforms are not simply the cosine and sine parts of the Fourier transform. In fact, the cosine and sine parts of the Fourier transform, individually, are not orthogonal functions. The Hartley transform jointly utilizes sine and cosine basis functions, but its coefficients are real numbers, as contrasted with the Fourier transform whose coefficients are, in general, complex numbers.

8.3.1. Cosine Transform

The *cosine transform*, discovered by Ahmed et al. (12), has found wide application in transform image coding. In fact, it is the foundation of the JPEG standard (13) for still image coding and the MPEG standard for the coding of moving images (14). The forward cosine transform is defined as (12)

$$\mathcal{F}(u, v) = \frac{2}{N} C(u)C(v) \sum_{j=0}^{N-1} \sum_{k=0}^{N-1} F(j, k) \cos\left\{\frac{\pi}{N}[u(j + \tfrac{1}{2})]\right\} \cos\left\{\frac{\pi}{N}[v(k + \tfrac{1}{2})]\right\}$$

$$(8.3\text{-}1a)$$

$$F(j, k) = \frac{2}{N} \sum_{j=0}^{N-1} \sum_{k=0}^{N-1} C(u)C(v) \mathcal{F}(u, v) \cos\left\{\frac{\pi}{N}[u(j + \tfrac{1}{2})]\right\} \cos\left\{\frac{\pi}{N}[v(k + \tfrac{1}{2})]\right\}$$

$$(8.3\text{-}1b)$$

where $C(0) = (2)^{-1/2}$ and $C(w) = 1$ for $w = 1, 2,..., N - 1$. It has been observed that the basis functions of the cosine transform are actually a class of discrete Chebyshev polynomials (12).

Figure 8.3-1 is a plot of the cosine transform basis functions for $N = 16$. A photograph of the cosine transform of the test image of Figure 8.2-4a is shown in Figure 8.3-2a. The origin is placed in the upper left corner of the picture, consistent with matrix notation. It should be observed that as with the Fourier transform, the image energy tends to concentrate toward the lower spatial frequencies.

The cosine transform of a $N \times N$ image can be computed by reflecting the image about its edges to obtain a $2N \times 2N$ array, taking the FFT of the array and then extracting the real parts of the Fourier transform (15). Algorithms also exist for the direct computation of each row or column of Eq. 8.3-1 with on the order of $N \log N$ real arithmetic operations (12,16).

8.3.2. Sine Transform

The *sine transform*, introduced by Jain (17), as a fast algorithmic substitute for the Karhunen–Loeve transform of a Markov process is defined in one-dimensional form by the basis functions

$$A(u, j) = \sqrt{\frac{2}{N + 1}} \sin\left\{\frac{(j + 1)(u + 1)\pi}{N + 1}\right\}$$

$$(8.3\text{-}2)$$

for $u, j = 0, 1, 2,..., N - 1$. Consider the tridiagonal matrix

WAVE
NUMBER, u

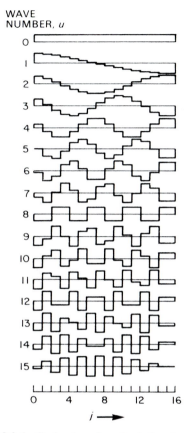

FIGURE 8.3-1. Cosine transform basis functions, $N = 16$.

$$\mathbf{T} = \begin{bmatrix} 1 & -\alpha & 0 & \cdots & & & 0 \\ -\alpha & 1 & -\alpha & & & & \\ \cdot & \cdot & \cdot & & & & \\ \cdot & \cdot & \cdot & & & & \cdot \\ \cdot & & & & & & \cdot \\ \cdot & & & & & & \\ \cdot & & & & -\alpha & 1 & -\alpha \\ 0 & & & \cdots & 0 & -\alpha & 1 \end{bmatrix} \qquad (8.3\text{-}3)$$

where $\alpha = \rho/(1 + \rho^2)$ and $0.0 \leq \rho \leq 1.0$ is the adjacent element correlation of a Markov process covariance matrix. It can be shown (18) that the basis functions of

(*a*) Cosine

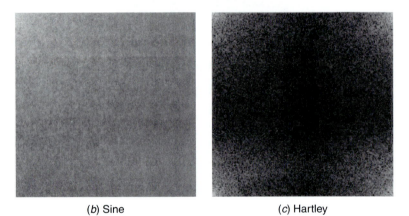

(*b*) Sine   (*c*) Hartley

FIGURE 8.3-2. Cosine, sine, and Hartley transforms of the `smpte_girl_luma` image, log magnitude displays

Eq. 8.3-2, inserted as the elements of a unitary matrix $\mathbf{A}$, diagonalize the matrix $\mathbf{T}$ in the sense that

$$\mathbf{A}\mathbf{T}\mathbf{A}^T = \mathbf{D} \tag{8.3-4}$$

Matrix $\mathbf{D}$ is a diagonal matrix composed of the terms

$$D(k, k) = \frac{1 - \rho^2}{1 - 2\rho \cos\{k\pi/(N + 1)\} + \rho^2} \tag{8.3-5}$$

for $k = 1, 2,..., N$. Jain (17) has shown that the cosine and sine transforms are interrelated in that they diagonalize a family of tridiagonal matrices.

WAVE
NUMBER, u

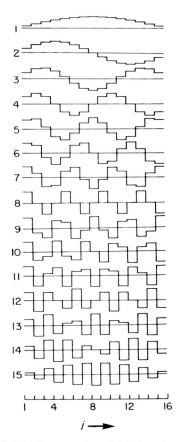

FIGURE 8.3-3. Sine transform basis functions, $N = 15$.

The two-dimensional sine transform is defined as

$$\mathcal{F}(u, v) = \frac{2}{N+1} \sum_{j=0}^{N-1} \sum_{k=0}^{N-1} F(j, k) \sin\left\{\frac{(j+1)(u+1)\pi}{N+1}\right\} \sin\left\{\frac{(k+1)(v+1)\pi}{N+1}\right\} \quad (8.3\text{-}6)$$

Its inverse is of identical form.

Sine transform basis functions are plotted in Figure 8.3-3 for $N = 15$. Figure 8.3-2b is a photograph of the sine transform of the test image. The sine transform can also be computed directly from Eq. 8.3-10, or efficiently with a Fourier transform algorithm (17).

8.3.3. Hartley Transform

Bracewell (19,20) has proposed a discrete real-valued unitary transform, called the *Hartley transform*, as a substitute for the Fourier transform in many filtering applications. The name derives from the continuous integral version introduced by Hartley in 1942 (21). The discrete two-dimensional Hartley transform is defined by the transform pair

$$\mathcal{F}(u, v) = \frac{1}{N} \sum_{j=0}^{N-1} \sum_{k=0}^{N-1} F(j, k) \, \text{cas}\left\{ \frac{2\pi}{N}(uj + vk) \right\} \tag{8.3-7a}$$

$$F(j, k) = \frac{1}{N} \sum_{u=0}^{N-1} \sum_{v=0}^{N-1} \mathcal{F}(u, v) \, \text{cas}\left\{ \frac{2\pi}{N}(uj + vk) \right\} \tag{8.3-7b}$$

where $\text{cas}\,\theta \equiv \cos\theta + \sin\theta$. The structural similarity between the Fourier and Hartley transforms becomes evident when comparing Eq. 8.3-7 and Eq. 8.2-2.

It can be readily shown (17) that the $\text{cas}\,\theta$ function is an orthogonal function. Also, the Hartley transform possesses equivalent but not mathematically identical structural properties of the discrete Fourier transform (20). Figure 8.3-2c is a photograph of the Hartley transform of the test image.

The Hartley transform can be computed efficiently by a FFT-like algorithm (20). The choice between the Fourier and Hartley transforms for a given application is usually based on computational efficiency. In some computing structures, the Hartley transform may be more efficiently computed, while in other computing environments, the Fourier transform may be computationally superior.

8.4. HADAMARD, HAAR, AND DAUBECHIES TRANSFORMS

The Hadamard, Haar, and Daubechies transforms are related members of a family of nonsinusoidal transforms.

8.4.1. Hadamard Transform

The *Hadamard transform* (22,23) is based on the *Hadamard matrix* (24), which is a square array of plus and minus 1s whose rows and columns are orthogonal. A normalized $N \times N$ Hadamard matrix satisfies the relation

$$\mathbf{H}\mathbf{H}^T = \mathbf{I} \tag{8.4-1}$$

The smallest orthonormal Hadamard matrix is the 2×2 Hadamard matrix given by

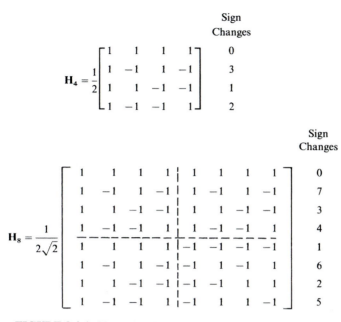

FIGURE 8.4-1. Nonordered Hadamard matrices of size 4 and 8.

$$\mathbf{H}_2 = \frac{1}{\sqrt{2}} \begin{bmatrix} 1 & 1 \\ 1 & -1 \end{bmatrix} \tag{8.4-2}$$

It is known that if a Hadamard matrix of size N exists ($N > 2$), then $N = 0$ modulo 4 (22). The existence of a Hadamard matrix for every value of N satisfying this requirement has not been shown, but constructions are available for nearly all permissible values of N up to 200. The simplest construction is for a Hadamard matrix of size $N = 2n$, where n is an integer. In this case, if $\mathbf{H}_N$ is a Hadamard matrix of size N, the matrix

$$\mathbf{H}_{2N} = \frac{1}{\sqrt{2}} \begin{bmatrix} \mathbf{H}_N & \mathbf{H}_N \\ \mathbf{H}_N & -\mathbf{H}_N \end{bmatrix} \tag{8.4-3}$$

is a Hadamard matrix of size $2N$. Figure 8.4-1 shows Hadamard matrices of size 4 and 8 obtained by the construction of Eq. 8.4-3.

Harmuth (25) has suggested a frequency interpretation for the Hadamard matrix generated from the core matrix of Eq. 8.4-3; the number of sign changes along each row of the Hadamard matrix divided by 2 is called the *sequency* of the row. It is possible to construct a Hadamard matrix of order $N = 2^n$ whose number of sign changes per row increases from 0 to $N - 1$. This attribute is called the *sequency property* of the unitary matrix.

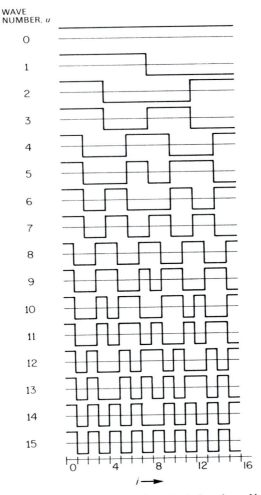

WAVE
NUMBER, u

0

1

2

3

4

5

6

7

8

9

10

11

12

13

14

15

$i \longrightarrow$

FIGURE 8.4-2. Hadamard transform basis functions, $N = 16$.

The rows of the Hadamard matrix of Eq. 8.4-3 can be considered to be samples of rectangular waves with a subperiod of $1/N$ units. These continuous functions are called *Walsh functions* (26). In this context, the Hadamard matrix merely performs the decomposition of a function by a set of rectangular waveforms rather than the sine–cosine waveforms with the Fourier transform. A series formulation exists for the Hadamard transform (23).

Hadamard transform basis functions for the ordered transform with $N = 16$ are shown in Figure 8.4-2. The ordered Hadamard transform of the test image in shown in Figure 8.4-3*a*.

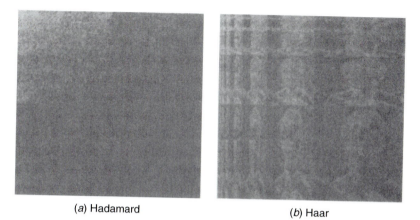

(a) Hadamard   (b) Haar

FIGURE 8.4-3. Hadamard and Haar transforms of the `smpte_girl_luma` image, log magnitude displays.

8.4.2. Haar Transform

The *Haar transform* (1,26,27) is derived from the *Haar matrix*. The following are 4×4 and 8×8 orthonormal Haar matrices:

$$\mathbf{H}_4 = \frac{1}{2} \begin{bmatrix} 1 & 1 & 1 & 1 \\ 1 & 1 & -1 & -1 \\ \sqrt{2} & -\sqrt{2} & 0 & 0 \\ 0 & 0 & \sqrt{2} & -\sqrt{2} \end{bmatrix} \tag{8.4-4}$$

$$\mathbf{H}_8 = \frac{1}{\sqrt{8}} \begin{bmatrix} 1 & 1 & 1 & 1 & 1 & 1 & 1 & 1 \\ 1 & 1 & 1 & 1 & -1 & -1 & -1 & -1 \\ \sqrt{2} & \sqrt{2} & -\sqrt{2} & -\sqrt{2} & 0 & 0 & 0 & 0 \\ 0 & 0 & 0 & 0 & \sqrt{2} & \sqrt{2} & -\sqrt{2} & -\sqrt{2} \\ 2 & -2 & 0 & 0 & 0 & 0 & 0 & 0 \\ 0 & 0 & 2 & -2 & 0 & 0 & 0 & 0 \\ 0 & 0 & 0 & 0 & 2 & -2 & 0 & 0 \\ 0 & 0 & 0 & 0 & 0 & 0 & 2 & -2 \end{bmatrix} \tag{8.4-5}$$

Extensions to higher-order Haar matrices follow the structure indicated by Eqs. 8.4-4 and 8.4-5. Figure 8.4-4 is a plot of the Haar basis functions for $N = 16$.

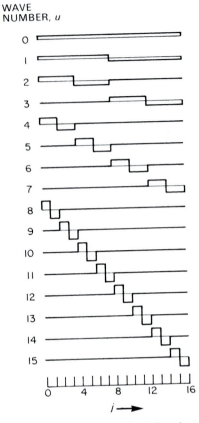

FIGURE 8.4-4. Haar transform basis functions, $N = 16$.

The Haar transform can be computed recursively (29) using the following $N \times N$ recursion matrix

$$\mathbf{R}_N = \begin{bmatrix} \mathbf{V}_N \\ \mathbf{W}_N \end{bmatrix} \tag{8.4-6}$$

where $\mathbf{V}_N$ is a $N/2 \times N$ scaling matrix and $\mathbf{W}_N$ is a $N/2 \times N$ *wavelet matrix* defined as

$$\mathbf{V}_N = \frac{1}{\sqrt{2}} \begin{bmatrix} 1 & 1 & 0 & 0 & 0 & 0 & \cdots & 0 & 0 & 0 & 0 \\ 0 & 0 & 1 & 1 & 0 & 0 & \cdots & 0 & 0 & 0 & 0 \\ & & & & & & & & & & \\ 0 & 0 & 0 & 0 & 0 & 0 & \cdots & 1 & 1 & 0 & 0 \\ 0 & 0 & 0 & 0 & 0 & 0 & \cdots & 0 & 0 & 1 & 1 \end{bmatrix} \tag{8.4-7a}$$

$$\mathbf{W}_N = \frac{1}{\sqrt{2}} \begin{bmatrix} 1 & -1 & 0 & 0 & 0 & 0 & \cdots & 0 & 0 & 0 & 0 \\ 0 & 0 & 1 & -1 & 0 & 0 & \cdots & 0 & 0 & 0 & 0 \\ & & \vdots & & & & & & \vdots & & \\ 0 & 0 & 0 & 0 & 0 & 0 & \cdots & 1 & -1 & 0 & 0 \\ 0 & 0 & 0 & 0 & 0 & 0 & \cdots & 0 & 0 & 1 & -1 \end{bmatrix} \tag{8.4-7b}$$

The elements of the rows of $\mathbf{V}_N$ are called *first-level scaling signals*, and the elements of the rows of W_N are called *first-level Haar wavelets* (29).

The first-level Haar transform of a $N \times 1$ vector $\mathbf{f}$ is

$$\mathbf{f}_1 = \mathbf{R}_N \mathbf{f} = [\mathbf{a}_1 | \mathbf{d}_1]^T \tag{8.4-8}$$

where

$$\mathbf{a}_1 = \mathbf{V}_N \mathbf{f} \tag{8.4-9a}$$

$$\mathbf{d}_1 = W_N \mathbf{f} \tag{8.4-9b}$$

The vector $\mathbf{a}_1$ represents the running average or *trend* of the elements of $\mathbf{f}$, and the vector $\mathbf{d}_1$ represents the running fluctuation of the elements of $\mathbf{f}$. The next step in the recursion process is to compute the second-level Haar transform from the trend part of the first-level transform and concatenate it with the first-level fluctuation vector. This results in

$$\mathbf{f}_2 = [\mathbf{a}_2 | \mathbf{d}_2 | \mathbf{d}_1]^T \tag{8.4-10}$$

where

$$\mathbf{a}_2 = \mathbf{V}_{N/2} \mathbf{a}_1 \tag{8.4-11a}$$

$$\mathbf{d}_2 = \mathbf{W}_{N/2} \mathbf{a}_1 \tag{8.4-11b}$$

are $N/4 \times 1$ vectors. The process continues until the full transform

$$f \equiv \mathbf{f}_n = [\mathbf{a}_n | \mathbf{d}_n | \mathbf{d}_{n-1} | \cdots | \mathbf{d}_1]^T \tag{8.4-12}$$

is obtained where $N = 2^n$. It should be noted that the intermediate levels are unitary transforms.

The Haar transform can be likened to a sampling process in which rows of the transform matrix sample an input data sequence with finer and finer resolution increasing of powers of 2. In image processing applications, the Haar transform provides a transform domain in which a type of differential energy is concentrated in localized regions.

8.4.3. Daubechies Transforms

Daubechies (30) has discovered a class of *wavelet transforms* that utilize running averages and running differences of the elements of a vector, as with the Haar transform. The difference between the Haar and Daubechies transforms is that the averages and differences are grouped in four or more elements.

The *Daubechies transform* of support four, called *Daub4*, can be defined in a manner similar to the Haar recursive generation process. The first-level scaling and wavelet matrices are defined as

$$
V_N = \begin{bmatrix}
\alpha_1 & \alpha_2 & \alpha_3 & \alpha_4 & 0 & 0 & \cdots & 0 & 0 & 0 & 0 \\
0 & 0 & \alpha_1 & \alpha_2 & \alpha_3 & \alpha_4 & \cdots & 0 & 0 & 0 & 0 \\
\vdots & \vdots & \vdots & \vdots & \vdots & \vdots & & \vdots & \vdots & \vdots & \vdots \\
0 & 0 & 0 & 0 & 0 & 0 & \cdots & \alpha_1 & \alpha_2 & \alpha_3 & \alpha_4 \\
\alpha_3 & \alpha_4 & 0 & 0 & 0 & 0 & \cdots & 0 & 0 & \alpha_1 & \alpha_2
\end{bmatrix}
\tag{8.4-13a}
$$

$$
W_N = \begin{bmatrix}
\beta_1 & \beta_2 & \beta_3 & \beta_4 & 0 & 0 & \cdots & 0 & 0 & 0 & 0 \\
0 & 0 & \beta_1 & \beta_2 & \beta_3 & \beta_4 & \cdots & 0 & 0 & 0 & 0 \\
\vdots & \vdots & \vdots & \vdots & \vdots & \vdots & & \vdots & \vdots & \vdots & \vdots \\
0 & 0 & 0 & 0 & 0 & 0 & \cdots & \beta_1 & \beta_2 & \beta_3 & \beta_4 \\
\beta_3 & \beta_4 & 0 & 0 & 0 & 0 & \cdots & 0 & 0 & \beta_1 & \beta_2
\end{bmatrix}
\tag{8.4-13b}
$$

where

$$
\alpha_1 = -\beta_4 = \frac{1 + \sqrt{3}}{4\sqrt{2}}
\tag{8.4-14a}
$$

$$
\alpha_2 = \beta_3 = \frac{3 + \sqrt{3}}{4\sqrt{2}}
\tag{8.4-14b}
$$

$$
\alpha_3 = -\beta_2 = \frac{3 - \sqrt{3}}{4\sqrt{2}}
\tag{8.4-14c}
$$

$$
\alpha_4 = \beta_1 = \frac{1 - \sqrt{3}}{4\sqrt{2}}
\tag{8.4-14d}
$$

In Eqs. 8.4-13*a* and 8.4-13*b*, the row-to-row shift is by two elements, and the last two scale factors wrap around on the last rows. Following the recursion process of the Haar transform results in the Daub4 transform final stage:

$$f \equiv \mathbf{f}_n = [\mathbf{a}_n | \mathbf{d}_n | \mathbf{d}_{n-1} | \cdots | \mathbf{d}_1]^T \tag{8.4-15}$$

Daubechies has extended the wavelet transform concept for higher degrees of support, 6, 8, 10,..., by straightforward extension of Eq. 8.4-13 (29). Daubechies also has also constructed another family of wavelets, called *coiflets*, after a suggestion of Coifman (29).

8.5. KARHUNEN–LOEVE TRANSFORM

Techniques for transforming continuous signals into a set of uncorrelated representational coefficients were originally developed by Karhunen (31) and Loeve (32). Hotelling (33) has been credited (34) with the conversion procedure that transforms discrete signals into a sequence of uncorrelated coefficients. However, most of the literature in the field refers to both discrete and continuous transformations as either a *Karhunen–Loeve transform* or an *eigenvector transform*.

The Karhunen–Loeve transformation is a transformation of the general form

$$\mathcal{F}(u, v) = \sum_{j=0}^{N-1} \sum_{k=0}^{N-1} F(j, k) A(j, k; u, v) \tag{8.5-1}$$

for which the kernel $A(j, k; u, v)$ satisfies the equation

$$\lambda(u, v) A(j, k; u, v) = \sum_{j'=0}^{N-1} \sum_{k'=0}^{N-1} K_F(j, k; j', k') A(j', k'; u, v) \tag{8.5-2}$$

where $K_F(j, k; j', k')$ denotes the covariance function of the image array and $\lambda(u, v)$ is a constant for fixed (u, v). The set of functions defined by the kernel are the eigenfunctions of the covariance function, and $\lambda(u, v)$ represents the eigenvalues of the covariance function. It is usually not possible to express the kernel in explicit form. If the covariance function is separable such that

$$K_F(j, k; j', k') = K_C(j, j') K_R(k, k') \tag{8.5-3}$$

then the Karhunen-Loeve kernel is also separable and

$$A(j, k; u, v) = A_C(u, j) A_R(v, k) \tag{8.5-4}$$

The row and column kernels satisfy the equations

$$\lambda_R(u)A_R(v, k) = \sum_{k' = 0}^{N-1} K_R(k, k')A_R(v, k') \tag{8.5-5a}$$

$$\lambda_C(v)A_C(u, j) = \sum_{j' = 0}^{N-1} K_C(j, j')A_C(u, j') \tag{8.5-5b}$$

In the special case in which the covariance matrix is of separable first-order Markov process form, the eigenfunctions can be written in explicit form. For a one-dimensional Markov process with correlation factor ρ, the eigenfunctions and eigenvalues are given by (35)

$$A(u, j) = \left[\frac{2}{N + \lambda^2(u)} \right]^{1/2} \sin\left\{ w(u)\left(j - \frac{N-1}{2}\right) + \frac{(u + 1)\pi}{2} \right\} \tag{8.5-6}$$

and

$$\lambda(u) = \frac{1 - \rho^2}{1 - 2\rho \cos\{w(u)\} + \rho^2} \qquad \text{for } 0 \le j, u \le N - 1 \tag{8.5-7}$$

where $w(u)$ denotes the root of the transcendental equation

$$\tan\{Nw\} = \frac{(1 - \rho^2) \sin w}{\cos w - 2\rho + \rho^2 \cos w} \tag{8.5-8}$$

The eigenvectors can also be generated by the recursion formula (36)

$$A(u, 0) = \frac{\lambda(u)}{1 - \rho^2}[A(u, 0) - \rho A(u, 1)] \tag{8.5-9a}$$

$$A(u, j) = \frac{\lambda(u)}{1 - \rho^2}[-\rho A(u, j - 1) + (1 + \rho^2)A(u, j) - \rho A(u, j + 1)] \quad \text{for } 0 < j < N - 1 \tag{8.5-9b}$$

$$A(u, N - 1) = \frac{\lambda(u)}{1 - \rho^2}[-\rho A(u, N - 2) + \rho A(u, N - 1)] \tag{8.5-9c}$$

by initially setting $A(u, 0) = 1$ and subsequently normalizing the eigenvectors.

If the image array and transformed image array are expressed in vector form, the Karhunen–Loeve transform pairs are

$$f = \mathbf{Af} \tag{8.5-10}$$

$$\mathbf{f} = \mathbf{A}^T f \tag{8.5-11}$$

The transformation matrix $\mathbf{A}$ satisfies the relation

$$\mathbf{AK_f} = \mathbf{\Lambda A} \tag{8.5-12}$$

where $\mathbf{K_f}$ is the covariance matrix of $\mathbf{f}$, $\mathbf{A}$ is a matrix whose rows are eigenvectors of $\mathbf{K_f}$, and $\mathbf{\Lambda}$ is a diagonal matrix of the form

$$\mathbf{\Lambda} = \begin{bmatrix} \lambda(1) & 0 & \cdots & 0 \\ 0 & \lambda(2) & & \vdots \\ \vdots & & \cdots & 0 \\ 0 & \cdots & 0 & \lambda(N^2) \end{bmatrix} \tag{8.5-13}$$

If $\mathbf{K_f}$ is of separable form, then

$$\mathbf{A} = \mathbf{A}_C \otimes \mathbf{A}_R \tag{8.5-14}$$

where $\mathbf{A}_R$ and $\mathbf{A}_C$ satisfy the relations

$$\mathbf{A}_R \mathbf{K}_R = \mathbf{\Lambda}_R \mathbf{A}_R \tag{8.5-15a}$$

$$\mathbf{A}_C \mathbf{K}_C = \mathbf{\Lambda}_C \mathbf{A}_C \tag{8.5-15b}$$

and $\lambda(w) = \lambda_R(v)\lambda_C(u)$ for $u, v = 1, 2, \dots, N$.

Figure 8.5-1 is a plot of the Karhunen–Loeve basis functions for a one-dimensional Markov process with adjacent element correlation $\rho = 0.9$.

VECTOR
NUMBER, μ

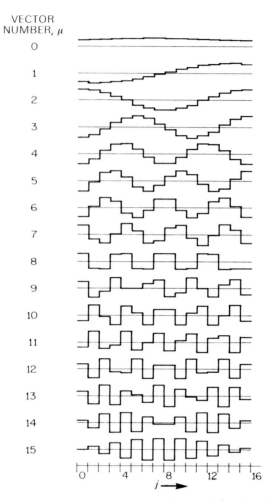

FIGURE 8.5-1. Karhunen–Loeve transform basis functions, $N = 16$.

REFERENCES

1. H. C. Andrews, *Computer Techniques in Image Processing*, Academic Press, New York, 1970.

2. H. C. Andrews, "Two Dimensional Transforms," in *Topics in Applied Physics: Picture Processing and Digital Filtering*, Vol. 6, T. S. Huang, Ed., Springer-Verlag, New York, 1975.

3. R. Bellman, *Introduction to Matrix Analysis*, 2nd ed., Society for Industrial and Applied Mathematics, Philadelphia, 1997.

4. H. C. Andrews and K. Caspari, "A Generalized Technique for Spectral Analysis," *IEEE Trans. Computers*, **C-19**, 1, January 1970, 16–25.

5. J. W. Cooley and J. W. Tukey, "An Algorithm for the Machine Calculation of Complex Fourier Series," *Mathematics of Computation* **19**, 90, April 1965, 297–301.

6. *IEEE Trans. Audio and Electroacoustics*, Special Issue on Fast Fourier Transforms, **AU-15**, 2, June 1967.

7. W. T. Cochran et al., "What Is the Fast Fourier Transform?" *Proc. IEEE*, **55**, 10, 1967, 1664–1674.

8. *IEEE Trans. Audio and Electroacoustics*, Special Issue on Fast Fourier Transforms, **AU-17**, 2, June 1969.

9. J. W. Cooley, P. A. Lewis, and P. D. Welch, "Historical Notes on the Fast Fourier Transform," *Proc. IEEE*, **55**, 10, October 1967, 1675–1677.

10. B. O. Brigham and R. B. Morrow, "The Fast Fourier Transform," *IEEE Spectrum*, **4**, 12, December 1967, 63–70.

11. C. S. Burrus and T. W. Parks, *DFT/FFT and Convolution Algorithms*, Wiley-Interscience, New York, 1985.

12. N. Ahmed, T. Natarajan, and K. R. Rao, "On Image Processing and a Discrete Cosine Transform," *IEEE Trans. Computers*, **C-23**, 1, January 1974, 90–93.

13. W. B. Pennebaker and J. L. Mitchell, *JPEG Still Image Data Compression Standard*, Van Nostrand Reinhold, New York, 1993.

14. K. R. Rao and J. J. Hwang, *Techniques and Standards for Image, Video, and Audio Coding*, Prentice Hall, Upper Saddle River, NJ, 1996.

15. R. W. Means, H. J. Whitehouse, and J. M. Speiser, "Television Encoding Using a Hybrid Discrete Cosine Transform and a Differential Pulse Code Modulator in Real Time," *Proc. National Telecommunications Conference*, San Diego, CA, December 1974, 61–66.

16. W. H. Chen, C. Smith, and S. C. Fralick, "Fast Computational Algorithm for the Discrete Cosine Transform," *IEEE Trans. Communications.*, **COM-25**, 9, September 1977, 1004–1009.

17. A. K. Jain, "A Fast Karhunen–Loeve Transform for Finite Discrete Images," *Proc. National Electronics Conference*, Chicago, October 1974, 323–328.

18. A. K. Jain and E. Angel, "Image Restoration, Modeling, and Reduction of Dimensionality," *IEEE Trans. Computers*, **C-23**, 5, May 1974, 470–476.

19. R. M. Bracewell, "The Discrete Hartley Transform," *J. Optical Society of America*, **73**, 12, December 1983, 1832–1835.

20. R. M. Bracewell, *The Hartley Transform*, Oxford University Press, Oxford, 1986.

21. R. V. L. Hartley, "A More Symmetrical Fourier Analysis Applied to Transmission Problems," *Proc. IRE*, **30**, 1942, 144–150.

22. J. E. Whelchel, Jr. and D. F. Guinn, "The Fast Fourier–Hadamard Transform and Its Use in Signal Representation and Classification," *EASCON 1968 Convention Record*, 1968, 561–573.

23. W. K. Pratt, H. C. Andrews, and J. Kane, "Hadamard Transform Image Coding," *Proc. IEEE*, **57**, 1, January 1969, 58–68.

24. J. Hadamard, "Resolution d'une question relative aux determinants," *Bull. Sciences Mathematiques,* Ser. 2, 17, Part I, 1893, 240–246.

25. H. F. Harmuth, *Transmission of Information by Orthogonal Functions*, Springer-Verlag, New York, 1969.

26. J. L. Walsh, "A Closed Set of Orthogonal Functions," *American J. Mathematics*, **45**, 1923, 5–24.

27. A. Haar, "Zur Theorie der Orthogonalen-Funktionen," *Mathematische Annalen*, **5**, 1955, 17–31.

28. K. R. Rao, M. A. Narasimhan, and K. Revuluri, "Image Data Processing by Hadamard–Haar Transforms," *IEEE Trans. Computers*, **C-23**, 9, September 1975, 888–896.

29. J. S. Walker, *A Primer on Wavelets and Their Scientific Applications*, Chapman & Hall/CRC, Press, Boca Raton, FL, 1999.

30. I. Daubechies, *Ten Lectures on Wavelets*, SIAM, Philadelphia, 1992.

31. H. Karhunen, 1947, English translation by I. Selin, "On Linear Methods in Probability Theory," Doc. T-131, Rand Corporation, Santa Monica, CA, August 11, 1960.

32. M. Loeve, *Fonctions aldatories de seconde ordre*, Hermann, Paris, 1948.

33. H. Hotelling, "Analysis of a Complex of Statistical Variables into Principal Components," *J. Educational Psychology*, **24**, 1933, 417–441, 498–520.

34. P. A. Wintz, "Transform Picture Coding," *Proc. IEEE*, **60**, 7, July 1972, 809–820.

35. W. D. Ray and R. M. Driver, "Further Decomposition of the Karhunen–Loeve Series Representation of a Stationary Random Process," *IEEE Trans. Information Theory*, **IT-16**, 6, November 1970, 663–668.

36. W. K. Pratt, "Generalized Wiener Filtering Computation Techniques," *IEEE Trans. Computers*, **C-21**, 7, July 1972, 636–641.

9

LINEAR PROCESSING TECHNIQUES

Most discrete image processing computational algorithms are linear in nature; an output image array is produced by a weighted linear combination of elements of an input array. The popularity of linear operations stems from the relative simplicity of spatial linear processing as opposed to spatial nonlinear processing. However, for image processing operations, conventional linear processing is often computationally infeasible without efficient computational algorithms because of the large image arrays. This chapter considers indirect computational techniques that permit more efficient linear processing than by conventional methods.

9.1. TRANSFORM DOMAIN PROCESSING

Two-dimensional linear transformations have been defined in Section 5.4 in series form as

$$P(m_1, m_2) = \sum_{n_1 = 1}^{N_1} \sum_{n_2 = 1}^{N_2} F(n_1, n_2) T(n_1, n_2 \,; m_1, m_2) \qquad (9.1\text{-}1)$$

and defined in vector form as

$$\mathbf{p} = \mathbf{Tf} \qquad (9.1\text{-}2)$$

It will now be demonstrated that such linear transformations can often be computed more efficiently by an indirect computational procedure utilizing two-dimensional unitary transforms than by the direct computation indicated by Eq. 9.1-1 or 9.1-2.

213

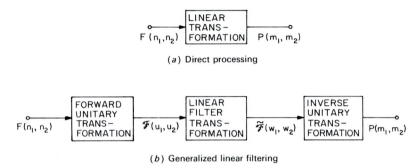

FIGURE 9.1-1. Direct processing and generalized linear filtering; series formulation.

Figure 9.1-1 is a block diagram of the indirect computation technique called *generalized linear filtering* (1). In the process, the input array $F(n_1, n_2)$ undergoes a two-dimensional unitary transformation, resulting in an array of transform coefficients $\mathcal{F}(u_1, u_2)$. Next, a linear combination of these coefficients is taken according to the general relation

$$\tilde{\mathcal{F}}(w_1, w_2) = \sum_{u_1 = 1}^{M_1} \sum_{u_2 = 1}^{M_2} \mathcal{F}(u_1, u_2) T(u_1, u_2; w_1, w_2) \tag{9.1-3}$$

where $T(u_1, u_2; w_1, w_2)$ represents the linear filtering transformation function. Finally, an inverse unitary transformation is performed to reconstruct the processed array $P(m_1, m_2)$. If this computational procedure is to be more efficient than direct computation by Eq. 9.1-1, it is necessary that fast computational algorithms exist for the unitary transformation, and also the kernel $T(u_1, u_2; w_1, w_2)$ must be reasonably sparse; that is, it must contain many zero elements.

The generalized linear filtering process can also be defined in terms of vector-space computations as shown in Figure 9.1-2. For notational simplicity, let $N_1 = N_2 = N$ and $M_1 = M_2 = M$. Then the generalized linear filtering process can be described by the equations

$$f = [A_{N^2}]f \tag{9.1-4a}$$

$$\tilde{f} = Tf \tag{9.1-4b}$$

$$p = [A_{M^2}]^{-1}\tilde{f} \tag{9.1-4c}$$

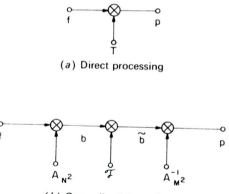

(a) Direct processing

(b) Generalized linear filtering

FIGURE 9.1-2. Direct processing and generalized linear filtering; vector formulation.

where $\mathbf{A}_{N^2}$ is a $N^2 \times N^2$ unitary transform matrix, $\mathcal{T}$ is a $M^2 \times N^2$ linear filtering transform operation, and $\mathbf{A}_{M^2}$ is a $M^2 \times M^2$ unitary transform matrix. From Eq. 9.1-4, the input and output vectors are related by

$$\mathbf{p} = [\mathbf{A}_{M^2}]^{-1} \mathcal{T}[\mathbf{A}_{N^2}]\mathbf{f} \qquad (9.1-5)$$

Therefore, equating Eqs. 9.1-2 and 9.1-5 yields the relations between $\mathcal{T}$ and $\mathbf{T}$ given by

$$\mathbf{T} = [\mathbf{A}_{M^2}]^{-1} \mathcal{T}[\mathbf{A}_{N^2}] \qquad (9.1-6a)$$

$$\mathcal{T} = [\mathbf{A}_{M^2}]\mathbf{T}[\mathbf{A}_{N^2}]^{-1} \qquad (9.1-6b)$$

If direct processing is employed, computation by Eq. 9.1-2 requires $k_P(M^2 N^2)$ operations, where $0 \le k_P \le 1$ is a measure of the sparseness of $\mathbf{T}$. With the generalized linear filtering technique, the number of operations required for a given operator are:

Forward transform: N^4 by direct transformation

            $2N^2 \log_2 N$ by fast transformation

Filter multiplication: $k_T M^2 N^2$

Inverse transform: M^4 by direct transformation

            $2M^2 \log_2 M$ by fast transformation

where $0 \leq k_T \leq 1$ is a measure of the sparseness of $\mathcal{T}$. If $k_T = 1$ and direct unitary transform computation is performed, it is obvious that the generalized linear filtering concept is not as efficient as direct computation. However, if fast transform algorithms, similar in structure to the fast Fourier transform, are employed, generalized linear filtering will be more efficient than direct processing if the sparseness index satisfies the inequality

$$k_T < k_P - \frac{2}{M^2}\log_2 N - \frac{2}{N^2}\log_2 M \qquad (9.1\text{-}7)$$

In many applications, $\mathcal{T}$ will be sufficiently sparse such that the inequality will be satisfied. In fact, unitary transformation tends to decorrelate the elements of $\mathbf{T}$ causing $\mathcal{T}$ to be sparse. Also, it is often possible to render the filter matrix sparse by setting small-magnitude elements to zero without seriously affecting computational accuracy (1).

   In subsequent sections, the structure of superposition and convolution operators is analyzed to determine the feasibility of generalized linear filtering in these applications.

9.2. TRANSFORM DOMAIN SUPERPOSITION

The superposition operations discussed in Chapter 7 can often be performed more efficiently by transform domain processing rather than by direct processing. Figure 9.2-1a and b illustrate block diagrams of the computational steps involved in direct finite area or sampled image superposition. In Figure 9.2-1d and e, an alternative form of processing is illustrated in which a unitary transformation operation is performed on the data vector $\mathbf{f}$ before multiplication by a finite area filter matrix $\mathcal{D}$ or sampled image filter matrix $\mathcal{B}$. An inverse transform reconstructs the output vector. From Figure 9.2-1, for finite-area superposition, because

$$\mathbf{q} = \mathbf{Df} \qquad (9.2\text{-}1a)$$

and

$$\mathbf{q} = [\mathbf{A}_{M^2}]^{-1}\mathcal{D}[\mathbf{A}_{N^2}]\mathbf{f} \qquad (9.2\text{-}1b)$$

then clearly the finite-area filter matrix may be expressed as

$$\mathcal{D} = [\mathbf{A}_{M^2}]\mathbf{D}[\mathbf{A}_{N^2}]^{-1} \qquad (9.2\text{-}2a)$$

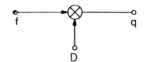

(a) Finite area superposition

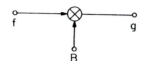

(b) Sampled image superposition

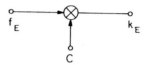

(c) Circulant superposition

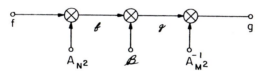

(d) Transform domain finite area superposition

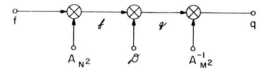

(e) Transform domain sampled image superposition

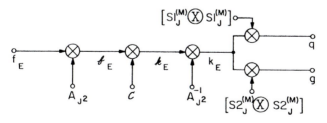

(f) Transform domain circulant superposition

FIGURE 9.2-1. Data and transform domain superposition.

Similarly,

$$\mathcal{B} = [\mathbf{A}_{M^2}]\mathbf{B}[\mathbf{A}_{N^2}]^{-1} \tag{9.2-2b}$$

If direct finite-area superposition is performed, the required number of computational operations is approximately N^2L^2, where L is the dimension of the impulse response matrix. In this case, the sparseness index of $\mathbf{D}$ is

$$k_D = \left(\frac{L}{N}\right)^2 \tag{9.2-3a}$$

Direct sampled image superposition requires on the order of M^2L^2 operations, and the corresponding sparseness index of $\mathbf{B}$ is

$$k_B = \left(\frac{L}{M}\right)^2 \tag{9.2-3b}$$

Figure 9.2-1f is a block diagram of a system for performing circulant superposition by transform domain processing. In this case, the input vector $\mathbf{k}_E$ is the *extended data vector*, obtained by embedding the input image array $F(n_1, n_2)$ in the left corner of a $J \times J$ array of zeros and then column scanning the resultant matrix. Following the same reasoning as above, it is seen that

$$\mathbf{k}_E = \mathbf{C}\mathbf{f}_E = [\mathbf{A}_{J^2}]^{-1}\mathcal{C}[\mathbf{A}_{J^2}]\mathbf{f} \tag{9.2-4a}$$

and hence,

$$\mathcal{C} = [\mathbf{A}_{J^2}]\mathbf{C}[\mathbf{A}_{J^2}]^{-1} \tag{9.2-4b}$$

As noted in Chapter 7, the equivalent output vector for either finite-area or sampled image superposition can be obtained by an element selection operation of $\mathbf{k}_E$. For finite-area superposition,

$$\mathbf{q} = [\mathbf{S1}_J^{(M)} \otimes \mathbf{S1}_J^{(M)}]\mathbf{k}_E \tag{9.2-5a}$$

and for sampled image superposition

$$\mathbf{g} = [\mathbf{S2}_J^{(M)} \otimes \mathbf{S2}_J^{(M)}]\mathbf{k}_E \tag{9.2-5b}$$

Also, the matrix form of the output for finite-area superposition is related to the extended image matrix $\mathbf{K}_E$ by

$$\mathbf{Q} = [\mathbf{S1}_J^{(M)}]\mathbf{K}_E[\mathbf{S1}_J^{(M)}]^T \qquad (9.2\text{-}6a)$$

For sampled image superposition,

$$\mathbf{G} = [\mathbf{S2}_J^{(M)}]\mathbf{K}_E[\mathbf{S2}_J^{(M)}]^T \qquad (9.2\text{-}6b)$$

The number of computational operations required to obtain $\mathbf{k}_E$ by transform domain processing is given by the previous analysis for $M = N = J$.

Direct transformation $\qquad\qquad 3J^4$

Fast transformation: $\qquad\qquad J^2 + 4J^2 \log_2 J$

If $\mathcal{C}$ is sparse, many of the J^2 filter multiplication operations can be avoided.

From the discussion above, it can be seen that the secret to computationally efficient superposition is to select a transformation that possesses a fast computational algorithm that results in a relatively sparse transform domain superposition filter matrix. As an example, consider finite-area convolution performed by Fourier domain processing (2,3). Referring to Figure 9.2-1, let

$$\mathbf{A}_{K^2} = \mathbf{A}_K \otimes \mathbf{A}_K \qquad (9.2\text{-}7)$$

where

$$\mathbf{A}_K = \left[\frac{1}{\sqrt{K}} W^{(x-1)(y-1)} \right] \qquad \text{with } W \equiv \exp\left\{ \frac{-2\pi i}{K} \right\}$$

for $x, y = 1, 2,..., K$. Also, let $\mathbf{h}_E^{(K)}$ denote the $K^2 \times 1$ vector representation of the extended spatially invariant impulse response array of Eq. 7.3-2 for $J = K$. The Fourier transform of $\mathbf{h}_E^{(K)}$ is denoted as

$$\hbar_E^{(K)} = [\mathbf{A}_{K^2}]\mathbf{h}_E^{(K)} \qquad (9.2\text{-}8)$$

These transform components are then inserted as the diagonal elements of a $K^2 \times K^2$ matrix

$$\mathcal{H}^{(K)} = \text{diag}[\hbar_E^{(K)}(1), ..., \hbar_E^{(K)}(K^2)] \qquad (9.2\text{-}9)$$

Then, it can be shown, after considerable manipulation, that the Fourier transform domain superposition matrices for finite area and sampled image convolution can be written as (4)

$$\mathcal{D} = \mathcal{H}^{(M)}[\mathbf{P}_D \otimes \mathbf{P}_D] \tag{9.2-10}$$

for $N = M - L + 1$ and

$$\mathcal{B} = [\mathbf{P}_B \otimes \mathbf{P}_B] \mathcal{H}^{(N)} \tag{9.2-11}$$

where $N = M + L + 1$ and

$$P_D(u, v) = \frac{1}{\sqrt{M}} \frac{1 - W_M^{-(u-1)(L-1)}}{1 - W_M^{-(u-1)} - W_N^{-(v-1)}} \tag{9.2-12a}$$

$$P_B(u, v) = \frac{1}{\sqrt{N}} \frac{1 - W_N^{-(v-1)(L-1)}}{1 - W_M^{-(u-1)} - W_N^{-(v-1)}} \tag{9.2-12b}$$

Thus the transform domain convolution operators each consist of a scalar weighting matrix $\mathcal{H}^{(K)}$ and an interpolation matrix $(\mathbf{P} \otimes \mathbf{P})$ that performs the dimensionality conversion between the N^2-element input vector and the M^2-element output vector. Generally, the interpolation matrix is relatively sparse, and therefore, transform domain superposition is quite efficient.

Now, consider circulant area convolution in the transform domain. Following the previous analysis it is found (4) that the circulant area convolution filter matrix reduces to a scalar operator

$$\mathcal{C} = J\mathcal{H}^{(J)} \tag{9.2-13}$$

Thus, as indicated in Eqs. 9.2-10 to 9.2-13, the Fourier domain convolution filter matrices can be expressed in a compact closed form for analysis or operational storage. No closed-form expressions have been found for other unitary transforms.

Fourier domain convolution is computationally efficient because the convolution operator $\mathbf{C}$ is a circulant matrix, and the corresponding filter matrix $\mathcal{C}$ is of diagonal form. Actually, as can be seen from Eq. 9.1-6, the Fourier transform basis vectors are eigenvectors of $\mathbf{C}$ (5). This result does not hold true for superposition in general, nor for convolution using other unitary transforms. However, in many instances, the filter matrices $\mathcal{D}$, $\mathcal{B}$, and $\mathcal{C}$ are relatively sparse, and computational savings can often be achieved by transform domain processing.

Signal   Fourier   Hadamard

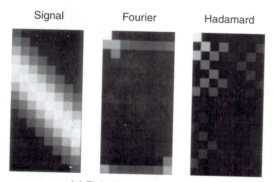

(*a*) Finite length convolution

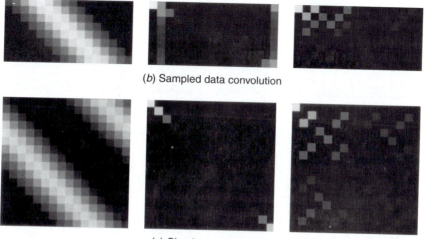

(*b*) Sampled data convolution

(*c*) Circulant convolution

FIGURE 9.2-2. One-dimensional Fourier and Hadamard domain convolution matrices.

Figure 9.2-2 shows the Fourier and Hadamard domain filter matrices for the three forms of convolution for a one-dimensional input vector and a Gaussian-shaped impulse response (6). As expected, the transform domain representations are much more sparse than the data domain representations. Also, the Fourier domain circulant convolution filter is seen to be of diagonal form. Figure 9.2-3 illustrates the structure of the three convolution matrices for two-dimensional convolution (4).

9.3. FAST FOURIER TRANSFORM CONVOLUTION

As noted previously, the equivalent output vector for either finite-area or sampled image convolution can be obtained by an element selection operation on the extended output vector $\mathbf{k}_E$ for circulant convolution or its matrix counterpart $\mathbf{K}_E$.

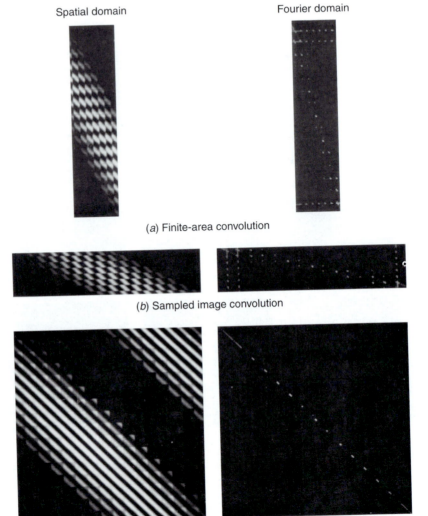

Spatial domain   Fourier domain

(a) Finite-area convolution

(b) Sampled image convolution

(c) Circulant convolution

FIGURE 9.2-3. Two-dimensional Fourier domain convolution matrices.

This result, combined with Eq. 9.2-13, leads to a particularly efficient means of convolution computation indicated by the following steps:

1.   Embed the impulse response matrix in the upper left corner of an all-zero $J \times J$ matrix, $J \geq M$ for finite-area convolution or $J \geq N$ for sampled infinite-area convolution, and take the two-dimensional Fourier transform of the extended impulse response matrix, giving

$$\mathcal{H}_E = \mathbf{A}_J \mathbf{H}_E \mathbf{A}_J \tag{9.3-1}$$

2. Embed the input data array in the upper left corner of an all-zero $J \times J$ matrix, and take the two-dimensional Fourier transform of the extended input data matrix to obtain

$$\mathcal{F}_E = \mathbf{A}_J \mathbf{F}_E \mathbf{A}_J \tag{9.3-2}$$

3. Perform the scalar multiplication

$$\mathcal{K}_E(m, n) = J \mathcal{H}_E(m, n) \mathcal{F}_E(m, n) \tag{9.3-3}$$

where $1 \le m, n \le J$.

4. Take the inverse Fourier transform

$$\mathbf{K}_E = [\mathbf{A}_{J^2}]^{-1} \mathcal{H}_E [\mathbf{A}_{J^2}]^{-1} \tag{9.3-4}$$

5. Extract the desired output matrix

$$\mathbf{Q} = [\mathbf{S1}_J^{(M)}] \mathbf{K}_E [\mathbf{S1}_J^{(M)}]^T \tag{9.3-5a}$$

or

$$\mathbf{G} = [\mathbf{S2}_J^{(M)}] \mathbf{K}_E [\mathbf{S2}_J^{(M)}]^T \tag{9.3-5b}$$

It is important that the size of the extended arrays in steps 1 and 2 be chosen large enough to satisfy the inequalities indicated. If the computational steps are performed with $J = N$, the resulting output array, shown in Figure 9.3-1, will contain erroneous terms in a boundary region of width $L - 1$ elements, on the top and left-hand side of the output field. This is the *wraparound error* associated with incorrect use of the Fourier domain convolution method. In addition, for finite area (D-type) convolution, the bottom and right-hand-side strip of output elements will be missing. If the computation is performed with $J = M$, the output array will be completely filled with the correct terms for D-type convolution. To force $J = M$ for B-type convolution, it is necessary to truncate the bottom and right-hand side of the input array. As a consequence, the top and left-hand-side elements of the output array are erroneous.

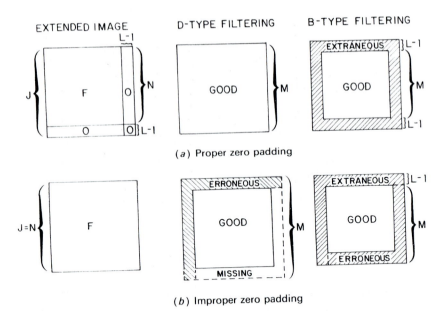

FIGURE 9.3-1. Wraparound error effects.

Figure 9.3-2 illustrates the Fourier transform convolution process with proper zero padding. The example in Figure 9.3-3 shows the effect of no zero padding. In both examples, the image has been filtered using a 11×11 uniform impulse response array. The source image of Figure 9.3-3 is 512×512 pixels. The source image of Figure 9.3-2 is 502×502 pixels. It has been obtained by truncating the bottom 10 rows and right 10 columns of the source image of Figure 9.3-3. Figure 9.3-4 shows computer printouts of the upper left corner of the processed images. Figure 9.3-4a is the result of finite-area convolution. The same output is realized in Figure 9.3-4b for proper zero padding. Figure 9.3-4c shows the wraparound error effect for no zero padding.

In many signal processing applications, the same impulse response operator is used on different data, and hence step 1 of the computational algorithm need not be repeated. The filter matrix $\mathcal{H}_E$ may be either stored functionally or indirectly as a computational algorithm. Using a fast Fourier transform algorithm, the forward and inverse transforms require on the order of $2J^2 \log_2 J$ operations each. The scalar multiplication requires J^2 operations, in general, for a total of $J^2(1 + 4 \log_2 J)$ operations. For an $N \times N$ input array, an $M \times M$ output array, and an $L \times L$ impulse response array, finite-area convolution requires $N^2 L^2$ operations, and sampled image convolution requires $M^2 L^2$ operations. If the dimension of the impulse response L is sufficiently large with respect to the dimension of the input array N, Fourier domain convolution will be more efficient than direct convolution, perhaps by an order of magnitude or more. Figure 9.3-5 is a plot of L versus N for equality

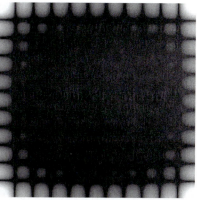

(a) H_E

(b) $\mathcal{H}_E$

(c) F_E

(d) $\mathcal{F}_E$

(e) K_E

(f) $\mathcal{K}_E$

FIGURE 9.3-2. Fourier transform convolution of the `candy_502_luma` image with proper zero padding, clipped magnitude displays of Fourier images.

(a) H_E

(b) $\mathcal{H}_E$

(c) F_E

(d) $\mathcal{F}_E$

(e) k_E

(f) $\mathcal{K}_E$

FIGURE 9.3-3. Fourier transform convolution of the `candy_512_luma` image with improper zero padding, clipped magnitude displays of Fourier images.

```
0.001  0.002  0.003  0.005  0.006  0.007  0.008  0.009  0.010  0.011  0.013  0.013  0.013  0.013  0.013
0.002  0.005  0.007  0.009  0.011  0.014  0.016  0.018  0.021  0.023  0.025  0.025  0.026  0.026  0.026
0.003  0.007  0.010  0.014  0.017  0.020  0.024  0.027  0.031  0.034  0.038  0.038  0.038  0.039  0.039
0.005  0.009  0.014  0.018  0.023  0.027  0.032  0.036  0.041  0.046  0.050  0.051  0.051  0.051  0.051
0.006  0.011  0.017  0.023  0.028  0.034  0.040  0.045  0.051  0.057  0.063  0.063  0.063  0.064  0.064
0.007  0.014  0.020  0.027  0.034  0.041  0.048  0.054  0.061  0.068  0.075  0.076  0.076  0.076  0.076
0.008  0.016  0.024  0.032  0.040  0.048  0.056  0.064  0.072  0.080  0.088  0.088  0.088  0.088  0.088
0.009  0.018  0.027  0.036  0.045  0.054  0.064  0.073  0.082  0.091  0.100  0.100  0.100  0.100  0.101
0.010  0.020  0.031  0.041  0.051  0.061  0.071  0.081  0.092  0.102  0.112  0.112  0.112  0.113  0.113
0.011  0.023  0.034  0.045  0.056  0.068  0.079  0.090  0.102  0.113  0.124  0.124  0.125  0.125  0.125
0.012  0.025  0.037  0.050  0.062  0.074  0.087  0.099  0.112  0.124  0.136  0.137  0.137  0.137  0.137
0.012  0.025  0.037  0.049  0.062  0.074  0.086  0.099  0.111  0.124  0.136  0.136  0.136  0.136  0.136
0.012  0.025  0.037  0.049  0.062  0.074  0.086  0.099  0.111  0.123  0.135  0.135  0.135  0.135  0.135
0.012  0.025  0.037  0.049  0.061  0.074  0.086  0.098  0.110  0.123  0.135  0.135  0.135  0.135  0.134
0.012  0.025  0.037  0.049  0.061  0.074  0.086  0.098  0.110  0.122  0.134  0.134  0.134  0.134  0.134
```

(a) Finite-area convolution

```
0.001  0.002  0.003  0.005  0.006  0.007  0.008  0.009  0.010  0.011  0.013  0.013  0.013  0.013  0.013
0.002  0.005  0.007  0.009  0.011  0.014  0.016  0.018  0.021  0.023  0.025  0.025  0.026  0.026  0.026
0.003  0.007  0.010  0.014  0.017  0.020  0.024  0.027  0.031  0.034  0.038  0.038  0.038  0.039  0.039
0.005  0.009  0.014  0.018  0.023  0.027  0.032  0.036  0.041  0.046  0.050  0.051  0.051  0.051  0.051
0.006  0.011  0.017  0.023  0.028  0.034  0.040  0.045  0.051  0.057  0.063  0.063  0.063  0.064  0.064
0.007  0.014  0.020  0.027  0.034  0.041  0.048  0.054  0.061  0.068  0.075  0.076  0.076  0.076  0.076
0.008  0.016  0.024  0.032  0.040  0.048  0.056  0.064  0.072  0.080  0.088  0.088  0.088  0.088  0.088
0.009  0.018  0.027  0.036  0.045  0.054  0.064  0.073  0.082  0.091  0.100  0.100  0.100  0.100  0.101
0.010  0.020  0.031  0.041  0.051  0.061  0.071  0.081  0.092  0.102  0.112  0.112  0.112  0.113  0.113
0.011  0.023  0.034  0.045  0.056  0.068  0.079  0.090  0.102  0.113  0.124  0.124  0.125  0.125  0.125
0.012  0.025  0.037  0.050  0.062  0.074  0.087  0.099  0.112  0.124  0.136  0.137  0.137  0.137  0.137
0.012  0.025  0.037  0.049  0.062  0.074  0.086  0.099  0.111  0.124  0.136  0.136  0.136  0.136  0.136
0.012  0.025  0.037  0.049  0.062  0.074  0.086  0.099  0.111  0.123  0.135  0.135  0.135  0.135  0.135
0.012  0.025  0.037  0.049  0.061  0.074  0.086  0.098  0.110  0.123  0.135  0.135  0.135  0.135  0.134
0.012  0.025  0.037  0.049  0.061  0.074  0.086  0.098  0.110  0.122  0.134  0.134  0.134  0.134  0.134
```

(b) Fourier transform convolution with proper zero padding

```
0.771  0.700  0.626  0.552  0.479  0.407  0.334  0.260  0.187  0.113  0.040  0.036  0.034  0.033  0.034
0.721  0.655  0.587  0.519  0.452  0.385  0.319  0.252  0.185  0.118  0.050  0.047  0.044  0.044  0.045
0.673  0.612  0.550  0.488  4.426  0.365  0.304  0.243  0.182  0.122  0.061  0.057  0.055  0.055  0.055
0.624  0.569  0.513  0.456  0.399  0.344  0.288  0.234  0.180  0.125  0.071  0.067  0.065  0.065  0.065
0.578  0.528  0.477  0.426  0.374  0.324  0.274  0.225  0.177  0.129  0.081  0.078  0.076  0.075  0.075
0.532  0.488  0.442  0.396  0.350  0.305  0.260  0.217  0.174  0.133  0.091  0.088  0.086  0.085  0.086
0.486  0.448  0.407  0.367  0.326  0.286  0.246  0.208  0.172  0.136  0.101  0.098  0.096  0.096  0.096
0.438  0.405  0.371  0.336  0.301  0.266  0.232  0.200  0.169  0.139  0.110  0108   0.107  0.106  0.106
0.387  0.361  0.333  0.304  0.275  0.246  0.218  0.191  0.166  0.142  0.119  0.118  0.117  0.116  0.116
0.334  0.313  0.292  0.270  0.247  0.225  0.203  0.182  0.163  0.145  0.128  0.127  0.127  0.127  0.127
0.278  0.264  0.249  0.233  0.218  0.202  0.186  0.172  0.159  0.148  0.136  0.137  0.137  0.137  0.137
0.273  0.260  0.246  0.231  0.216  0.200  0.185  0.171  0.158  0.147  0.136  0.136  0.136  0.136  0.136
0.266  0.254  0.241  0.228  0.213  0.198  0.183  0.169  0.157  0.146  0.135  0.135  0.135  0.135  0.135
0.257  0.246  0.234  0.222  0.209  0.195  0.181  0.168  0.156  0.145  0.135  0.135  0.135  0.135  0.134
0.247  0.237  0.227  0.215  0.204  0.192  0.179  0.166  0.155  0.144  0.134  0.134  0.134  0.134  0.134
```

(c) Fourier transform convolution without zero padding

FIGURE 9.3-4. Wraparound error for Fourier transform convolution, upper left corner of processed image.

between direct and Fourier domain finite area convolution. The jaggedness of the plot, in this example, arises from discrete changes in J (64, 128, 256,...) as N increases.

Fourier domain processing is more computationally efficient than direct processing for image convolution if the impulse response is sufficiently large. However, if the image to be processed is large, the relative computational advantage of Fourier domain processing diminishes. Also, there are attendant problems of computational

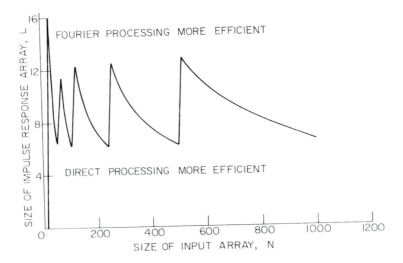

FIGURE 9.3-5. Comparison of direct and Fourier domain processing for finite-area convolution.

accuracy with large Fourier transforms. Both difficulties can be alleviated by a block-mode filtering technique in which a large image is separately processed in adjacent overlapped blocks (2, 7–9).

Figure 9.3-6a illustrates the extraction of a $N_B \times N_B$ pixel block from the upper left corner of a large image array. After convolution with a $L \times L$ impulse response, the resulting $M_B \times M_B$ pixel block is placed in the upper left corner of an output

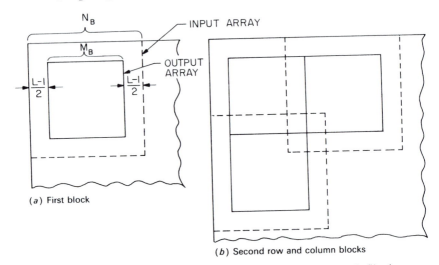

FIGURE 9.3-6. Geometric arrangement of blocks for block-mode filtering.

data array as indicated in Figure 9.3-6a. Next, a second block of $N_B \times N_B$ pixels is extracted from the input array to produce a second block of $M_B \times M_B$ output pixels that will lie adjacent to the first block. As indicated in Figure 9.3-6b, this second input block must be overlapped by $(L-1)$ pixels in order to generate an adjacent output block. The computational process then proceeds until all input blocks are filled along the first row. If a partial input block remains along the row, zero-value elements can be added to complete the block. Next, an input block, overlapped by $(L-1)$ pixels with the first row blocks, is extracted to produce the first block of the second output row. The algorithm continues in this fashion until all output points are computed.

A total of

$$O_F = N^2 + 2N^2 \log_2 N \tag{9.3-6}$$

operations is required for Fourier domain convolution over the full size image array. With block-mode filtering with $N_B \times N_B$ input pixel blocks, the required number of operations is

$$O_B = R^2 (N_B^2 + 2N_B^2 \log_2 N) \tag{9.3-7}$$

where R represents the largest integer value of the ratio $N/(N_B + L - 1)$. Hunt (9) has determined the optimum block size as a function of the original image size and impulse response size.

9.4. FOURIER TRANSFORM FILTERING

The discrete Fourier transform convolution processing algorithm of Section 9.3 is often utilized for computer simulation of continuous Fourier domain filtering. In this section we consider discrete Fourier transform filter design techniques.

9.4.1. Transfer Function Generation

The first step in the discrete Fourier transform filtering process is generation of the discrete domain transfer function. For simplicity, the following discussion is limited to one-dimensional signals. The extension to two dimensions is straightforward.

Consider a one-dimensional continuous signal $f_C(x)$ of wide extent which is bandlimited such that its Fourier transform $f_C(\omega)$ is zero for $|\omega|$ greater than a cut-off frequency ω_0. This signal is to be convolved with a continuous impulse function $h_C(x)$ whose transfer function $h_C(\omega)$ is also bandlimited to ω_0. From Chapter 1 it is known that the convolution can be performed either in the spatial domain by the operation

$$g_C(x) = \int_{-\infty}^{\infty} f_C(\alpha) h_C(x - \alpha) \, d\alpha \qquad (9.4\text{-}1a)$$

or in the continuous Fourier domain by

$$g_C(x) = \frac{1}{2\pi} \int_{-\infty}^{\infty} f_C(\omega) h_C(\omega) \exp\{i\omega x\} \, d\omega \qquad (9.4\text{-}1b)$$

Chapter 7 has presented techniques for the discretization of the convolution integral of Eq. 9.4-1. In this process, the continuous impulse response function $h_C(x)$ must be truncated by spatial multiplication of a window function $y(x)$ to produce the windowed impulse response

$$b_C(x) = h_C(x) y(x) \qquad (9.4\text{-}2)$$

where $y(x) = 0$ for $|x| > T$. The window function is designed to smooth the truncation effect. The resulting convolution integral is then approximated as

$$g_C(x) = \int_{x-T}^{x+T} f_C(\alpha) b_C(x - \alpha) \, d\alpha \qquad (9.4\text{-}3)$$

Next, the output signal $g_C(x)$ is sampled over $2J + 1$ points at a resolution $\Delta = \pi/\omega_0$, and the continuous integration is replaced by a quadrature summation at the same resolution Δ, yielding the discrete representation

$$g_C(j\Delta) = \sum_{k=j-K}^{j+K} f_C(k\Delta) b_C[(j - k)\Delta] \qquad (9.4\text{-}4)$$

where K is the nearest integer value of the ratio T/Δ.

Computation of Eq. 9.4-4 by discrete Fourier transform processing requires formation of the discrete domain transfer function $b_D(u)$. If the continuous domain impulse response function $h_C(x)$ is known analytically, the samples of the windowed impulse response function are inserted as the first $L = 2K + 1$ elements of a J-element sequence and the remaining $J - L$ elements are set to zero. Thus, let

$$b_D(p) = \underbrace{b_C(-K), \ldots, b_C(0), \ldots, b_C(K)}_{L \text{ terms}}, 0, \ldots, 0 \qquad (9.4\text{-}5)$$

where $0 \le p \le P - 1$. The terms of $b_D(p)$ can be extracted from the continuous impulse response function $h_C(x)$ and the window function by the sampling operation

$$b_D(p) = y(x)h_C(x)\delta(x - p\Delta) \tag{9.4-6}$$

The next step in the discrete Fourier transform convolution algorithm is to perform a discrete Fourier transform of $b_D(p)$ over P points to obtain

$$b_D(u) = \frac{1}{\sqrt{P}} \sum_{p=1}^{P-1} b_D(p) \exp\left\{\frac{-2\pi i p u}{P}\right\} \tag{9.4-7}$$

where $0 \leq u \leq P - 1$.

If the continuous domain transfer function $h_C(\omega)$ is known analytically, then $b_D(u)$ can be obtained directly. It can be shown that

$$b_D(u) = \frac{1}{4\sqrt{P}\pi^2} \exp\left\{\frac{-i\pi(L-1)}{P}\right\} h_C\left(\frac{2\pi u}{P\Delta}\right) \tag{9.4-8a}$$

$$b_D(P - u) = b_D^*(u) \tag{9.4-8b}$$

for $u = 0, 1,..., P/2$, where

$$b_C(\omega) = h_C(\omega) \circledast y(\omega) \tag{9.4-8c}$$

and $y(\omega)$ is the continuous domain Fourier transform of the window function $y(x)$. If $h_C(\omega)$ and $y(\omega)$ are known analytically, then, in principle, $h_C(\omega)$ can be obtained by analytically performing the convolution operation of Eq. 9.4-8c and evaluating the resulting continuous function at points $2\pi u/P\Delta$. In practice, the analytic convolution is often difficult to perform, especially in two dimensions. An alternative is to perform an analytic inverse Fourier transformation of the transfer function $h_C(\omega)$ to obtain its continuous domain impulse response $h_C(x)$ and then form $b_D(u)$ from the steps of Eqs. 9.4-5 to 9.4-7. Still another alternative is to form $b_D(u)$ from $h_C(\omega)$ according to Eqs. 9.4-8a and 9.4-8b, take its discrete inverse Fourier transform, window the resulting sequence, and then form $b_D(u)$ from Eq. 9.4-7.

9.4.2. Windowing Functions

The windowing operation performed explicitly in the spatial domain according to Eq. 9.4-6 or implicitly in the Fourier domain by Eq. 9.4-8 is absolutely imperative if the wraparound error effect described in Section 9.3 is to be avoided. A common mistake in image filtering is to set the values of the discrete impulse response function arbitrarily equal to samples of the continuous impulse response function. The corresponding extended discrete impulse response function will generally possess nonzero elements in each of its J elements. That is, the length L of the discrete impulse response embedded in the extended vector of Eq. 9.4-5 will implicitly be set equal to J. Therefore, all elements of the output filtering operation will be subject to wraparound error.

A variety of window functions have been proposed for discrete linear filtering (10–12). Several of the most common are listed in Table 9.4-1 and sketched in Figure 9.4-1. Figure 9.4-2 shows plots of the transfer functions of these window functions. The window transfer functions consist of a main lobe and sidelobes whose peaks decrease in magnitude with increasing frequency. Examination of the structure of Eq. 9.4-8 indicates that the main lobe causes a loss in frequency response over the signal passband from 0 to ω_0, while the sidelobes are responsible for an aliasing error because the windowed impulse response function $b_C(\omega)$ is not bandlimited. A tapered window function reduces the magnitude of the sidelobes and consequently attenuates the aliasing error, but the main lobe becomes wider, causing the signal frequency response within the passband to be reduced. A design trade-off must be made between these complementary sources of error. Both sources of degradation can be reduced by increasing the truncation length of the windowed impulse response, but this strategy will either result in a shorter length output sequence or an increased number of computational operations.

TABLE 9.4-1. Window Functions[a]

Function	Definition
Rectangular	$w(n) = 1 \quad 0 \leq n \leq L - 1$
Barlett (triangular)	$w(n) = \begin{cases} \dfrac{2n}{L-1} & 0 \leq n - \dfrac{L-1}{2} \\ 2 - \dfrac{2}{L-1} & \dfrac{L-1}{2} \leq n \leq L - 1 \end{cases}$
Hanning	$w(n) = \dfrac{1}{2}\left(1 - \cos\left\{\dfrac{2\pi n}{L-1}\right\}\right) \quad 0 \leq n \leq L - 1$
Hamming	$w(n) = 0.54 - 0.46 \cos\left\{\dfrac{2\pi n}{L-1}\right\} \quad 0 \leq n \leq L - 1$
Blackman	$w(n) = 0.42 - 0.5 \cos\left\{\dfrac{2\pi n}{L-1}\right\} + 0.08\cos\left\{\dfrac{4\pi n}{L-1}\right\} \quad 0 \leq n \leq L - 1$
Kaiser	$\dfrac{I_0\left\{\omega_a[((L-1)/2)^2 - [n - ((L-1)/2)]^2]^{1/2}\right\}}{I_0\{\omega_a[(L-1)/2]\}} \quad 0 \leq n \leq L - 1$

[a] $I_0\{\cdot\}$ is the modified zeroth-order Bessel function of the first kind and ω_a is a design parameter.

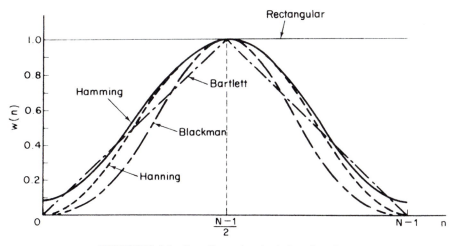

FIGURE 9.4-1. One-dimensional window functions.

9.4.3. Discrete Domain Transfer Functions

In practice, it is common to define the discrete domain transform directly in the discrete Fourier transform frequency space. The following are definitions of several widely used transfer functions for a $N \times N$ pixel image. Applications of these filters are presented in Chapter 10.

1. *Zonal low-pass filter:*

$$\mathcal{H}(u, v) = 1 \qquad 0 \leq u \leq C - 1 \qquad \text{and } 0 \leq v \leq C - 1$$

$$0 \leq u \leq C - 1 \qquad \text{and } N + 1 - C \leq v \leq N - 1$$

$$N + 1 - C \leq u \leq N - 1 \quad \text{and } 0 \leq v \leq C - 1$$

$$N + 1 - C \leq u \leq N - 1 \quad \text{and } N + 1 - C \leq v \leq N - 1 \qquad (9.4\text{-}9a)$$

$$\mathcal{H}(u, v) = 0 \qquad \text{otherwise} \qquad\qquad\qquad\qquad\qquad (9.4\text{-}9b)$$

where C is the filter cutoff frequency for $0 < C \leq 1 + N/2$. Figure 9.4-3 illustrates the low-pass filter zones.

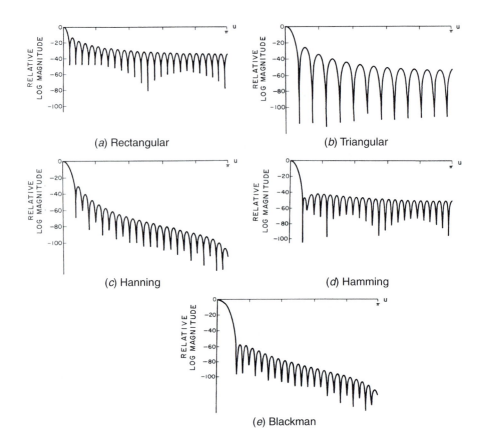

FIGURE 9.4-2. Transfer functions of one-dimensional window functions.

2. Zonal high-pass filter:

$$\mathcal{H}(0, 0) = 0 \tag{9.4-10a}$$

$$\mathcal{H}(u, v) = 0 \qquad 0 \leq u \leq C-1 \qquad \text{and } 0 \leq v \leq C-1$$

$$0 \leq u \leq C-1 \qquad \text{and } N+1-C \leq v \leq N-1$$

$$N+1-C \leq u \leq N-1 \quad \text{and } 0 \leq v \leq C-1$$

$$N+1-C \leq u \leq N-1 \quad \text{and } N+1-C \leq v \leq N-1 \tag{9.4-10b}$$

$$\mathcal{H}(u, v) = 1 \qquad \text{otherwise} \tag{9.4-10c}$$

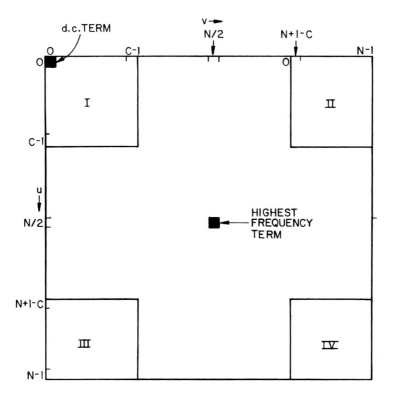

FIGURE 9.4-3. Zonal filter transfer function definition.

3. Gaussian filter:

$$\mathcal{H}(u, v) = \mathcal{G}(u, v) \quad 0 \leq u \leq N/2 \qquad \text{and } 0 \leq v \leq N/2$$

$$0 \leq u \leq N/2 \qquad \text{and } 1 + N/2 \leq v \leq N - 1$$

$$1 + N/2 \leq u \leq N - 1 \quad \text{and } 0 \leq v \leq N/2$$

$$1 + N/2 \leq u \leq N - 1 \quad \text{and } 1 + N/2 \leq v \leq N - 1 \qquad (9.4\text{-}11a)$$

where

$$\mathcal{G}(u, v) = \exp\left\{-\frac{1}{2}[(s_u u)^2 + (s_v v)^2]\right\} \qquad (9.4\text{-}11b)$$

and s_u and s_v are the Gaussian filter spread factors.

4. *Butterworth low-pass filter:*

$$\mathcal{H}(u, v) = \mathcal{B}(u, v) \quad 0 \le u \le N/2 \qquad \text{and} \ 0 \le v \le N/2$$

$$0 \le u \le N/2 \qquad \text{and} \ 1 + N/2 \le v \le N - 1$$

$$1 + N/2 \le u \le N - 1 \quad \text{and} \ 0 \le v \le N/2$$

$$1 + N/2 \le u \le N - 1 \quad \text{and} \ 1 + N/2 \le v \le N - 1 \quad (9.4\text{-}12a)$$

where

$$\mathcal{B}(u, v) = \frac{1}{1 + \left[\dfrac{(u^2 + v^2)^{1/2}}{C} \right]^{2n}} \tag{9.4-12b}$$

where the integer variable n is the order of the filter. The Butterworth low-pass filter provides an attenuation of 50% at the cutoff frequency $C = (u^2 + v^2)^{1/2}$.

5. *Butterworth high-pass filter:*

$$\mathcal{H}(u, v) = \mathcal{B}(u, v) \quad 0 \le u \le N/2 \qquad \text{and} \ 0 \le v \le N/2$$

$$0 \le u \le N/2 \qquad \text{and} \ 1 + N/2 \le v \le N - 1$$

$$1 + N/2 \le u \le N - 1 \quad \text{and} \ 0 \le v \le N/2$$

$$1 + N/2 \le u \le N - 1 \quad \text{and} \ 1 + N/2 \le v \le N - 1 \quad (9.4\text{-}13a)$$

where

$$\mathcal{B}(u, v) = \frac{1}{1 + \left[\dfrac{C}{(u^2 + v^2)^{1/2}} \right]^{2n}} \tag{9.4-13b}$$

Figure 9.4-4 shows the transfer functions of zonal and Butterworth low- and high-pass filters for a 512×512 pixel image.

9.5. SMALL GENERATING KERNEL CONVOLUTION

It is possible to perform convolution on a $N \times N$ image array $F(j, k)$ with an arbitrary $L \times L$ impulse response array $H(j, k)$ by a sequential technique called *small*

(*a*) Zonal low-pass   (*b*) Butterworth low-pass

(*c*) Zonal high-pass   (*d*) Butterworth high-pass

FIGURE 9.4-4. Zonal and Butterworth low- and high-pass transfer functions; 512×512 images; cutoff frequency = 64.

generating kernel (SGK) *convolution* (13–16). Figure 9.5-1 illustrates the decomposition process in which a $L \times L$ prototype impulse response array $H(j, k)$ is sequentially decomposed into 3×3 pixel SGKs according to the relation

$$\hat{H}(j, k) = K_1(j, k) \circledast K_2(j, k) \circledast \cdots \circledast K_Q(j, k) \qquad (9.5\text{-}1)$$

where $\hat{H}(j, k)$ is the synthesized impulse response array, the symbol $\circledast$ denotes centered two-dimensional finite-area convolution, as defined by Eq. 7.1-14, and $K_i(j, k)$ is the ith 3×3 pixel SGK of the decomposition, where $Q = (L-1)/2$. The SGK convolution technique can be extended to larger SGK kernels. Generally, the SGK synthesis of Eq. 9.5-1 is not exact. Techniques have been developed for choosing the SGKs to minimize the mean-square error between $\hat{H}(j, k)$ and $H(j, k)$ (13).

FIGURE 9.5-1. Cascade decomposition of a two-dimensional impulse response array into small generating kernels.

Two-dimensional convolution can be performed sequentially without approximation error by utilizing the singular-value decomposition technique described in Appendix A1.2 in conjunction with the SGK decimation (17–19). With this method, called *SVD/SGK convolution*, the impulse response array $H(j, k)$ is regarded as a matrix **H**. Suppose that **H** is orthogonally separable such that it can be expressed in the outer product form

$$\mathbf{H} = \mathbf{a}\mathbf{b}^T \tag{9.5-2}$$

where **a** and **b** are column and row operator vectors, respectively. Then, the two-dimensional convolution operation can be performed by first convolving the columns of $F(j, k)$ with the impulse response sequence $a(j)$ corresponding to the vector **a**, and then convolving the rows of that resulting array with the sequence $b(k)$ corresponding to the vector **b**. If **H** is not separable, the matrix can be expressed as a sum of separable matrices by the singular-value decomposition by which

$$\mathbf{H} = \sum_{i=1}^{R} \mathbf{H}_i \tag{9.5-3a}$$

$$\mathbf{H}_i = s_i \mathbf{a}_i \mathbf{b}_i^T \tag{9.5-3b}$$

where $R \geq 1$ is the rank of **H**, s_i is the ith singular value of **H**. The vectors $\mathbf{a}_i$ and $\mathbf{b}_i$ are the $L \times 1$ eigenvectors of $\mathbf{H}\mathbf{H}^T$ and $\mathbf{H}^T\mathbf{H}$, respectively.

Each eigenvector $\mathbf{a}_i$ and $\mathbf{b}_i$ of Eq. 9.5-3 can be considered to be a one-dimensional sequence, which can be decimated by a small generating kernel expansion as

$$a_i(j) = c_i[a_{i1}(j) \circledast \cdots \circledast a_{iq}(j) \circledast \cdots \circledast a_{iQ}(j)] \tag{9.5-4a}$$

$$b_i(k) = r_i[b_{i1}(k) \circledast \cdots \circledast b_{iq}(k) \circledast \cdots \circledast b_{iQ}(k)] \tag{9.5-4b}$$

where $a_{iq}(j)$ and $b_{iq}(k)$ are 3×1 impulse response sequences corresponding to the ith singular-value channel and the qth SGK expansion. The terms c_i and r_i are column and row gain constants. They are equal to the sum of the elements of their respective sequences if the sum is nonzero, and equal to the sum of the magnitudes

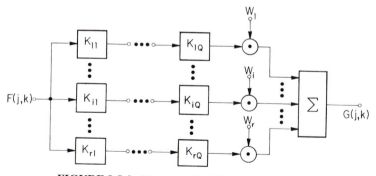

FIGURE 9.5-2. Nonseparable SVD/SGK expansion.

otherwise. The former case applies for a unit-gain filter impulse response, while the latter case applies for a differentiating filter.

As a result of the linearity of the SVD expansion of Eq. 9.5-3b, the large size impulse response array $H_i(j, k)$ corresponding to the matrix $\mathbf{H}_i$ of Eq. 9.5-3a can be synthesized by sequential 3×3 convolutions according to the relation

$$H_i(j, k) = r_i c_i [K_{i1}(j, k) \circledast \ ... \ \circledast K_{iq}(j, k) \circledast \ ... \ \circledast K_{iQ}(j, k)] \qquad (9.5\text{-}5)$$

where $K_{iq}(j, k)$ is the qth SGK of the ith SVD channel. Each $K_{iq}(j, k)$ is formed by an outer product expansion of a pair of the $a_{iq}(j)$ and $b_{iq}(k)$ terms of Eq. 9.5-4. The ordering is important only for low-precision computation when roundoff error becomes a consideration. Figure 9.5-2 is the flowchart for SVD/SGK convolution. The weighting terms in the figure are

$$W_i = s_i r_i c_i \qquad (9.5\text{-}6)$$

Reference 19 describes the design procedure for computing the $K_{iq}(j, k)$.

REFERENCES

1. W. K. Pratt, "Generalized Wiener Filtering Computation Techniques," *IEEE Trans. Computers*, **C-21**, 7, July 1972, 636–641.

2. T. G. Stockham, Jr., "High Speed Convolution and Correlation," *Proc. Spring Joint Computer Conference*, 1966, 229–233.

3. W. M. Gentleman and G. Sande, "Fast Fourier Transforms for Fun and Profit," *Proc. Fall Joint Computer Conference*, 1966, 563–578.

4. W. K. Pratt, "Vector Formulation of Two-Dimensional Signal Processing Operations," *Computer Graphics and Image Processing*, **4**, 1, March 1975, 1–24.

5. B. R. Hunt, "A Matrix Theory Proof of the Discrete Convolution Theorem," *IEEE Trans. Audio and Electroacoustics*, **AU-19**, 4, December 1973, 285–288.

6. W. K. Pratt, "Transform Domain Signal Processing Techniques," *Proc. National Electronics Conference*, Chicago, 1974.

7. H. D. Helms, "Fast Fourier Transform Method of Computing Difference Equations and Simulating Filters," *IEEE Trans. Audio and Electroacoustics*, **AU-15**, 2, June 1967, 85–90.

8. M. P. Ekstrom and V. R. Algazi, "Optimum Design of Two-Dimensional Nonrecursive Digital Filters," *Proc. 4th Asilomar Conference on Circuits and Systems*, Pacific Grove, CA, November 1970.

9. B. R. Hunt, "Computational Considerations in Digital Image Enhancement," *Proc. Conference on Two-Dimensional Signal Processing*, University of Missouri, Columbia, MO, October 1971.

10. A. V. Oppenheim and R. W. Schafer, *Digital Signal Processing*, Prentice Hall, Englewood Cliffs, NJ, 1975.

11. R. B. Blackman and J. W. Tukey, *The Measurement of Power Spectra*, Dover Publications, New York, 1958.

12. J. F. Kaiser, "Digital Filters", Chapter 7 in *Systems Analysis by Digital Computer*, F. F. Kuo and J. F. Kaiser, Eds., Wiley, New York, 1966.

13. J. F. Abramatic and O. D. Faugeras, "Design of Two-Dimensional FIR Filters from Small Generating Kernels," *Proc. IEEE Conference on Pattern Recognition and Image Processing*, Chicago, May 1978.

14. W. K. Pratt, J. F. Abramatic, and O. D. Faugeras, "Method and Apparatus for Improved Digital Image Processing," U.S. patent 4,330,833, May 18, 1982.

15. J. F. Abramatic and O. D. Faugeras, "Sequential Convolution Techniques for Image Filtering," *IEEE Trans. Acoustics, Speech, and Signal Processing*, **ASSP-30**, 1, February 1982, 1–10.

16. J. F. Abramatic and O. D. Faugeras, "Correction to Sequential Convolution Techniques for Image Filtering," *IEEE Trans. Acoustics, Speech, and Signal Processing*, **ASSP-30**, 2, April 1982, 346.

17. W. K. Pratt, "Intelligent Image Processing Display Terminal," *Proc. SPIE*, **199**, August 1979, 189–194.

18. J. F. Abramatic and S. U. Lee, "Singular Value Decomposition of 2-D Impulse Responses," *Proc. International Conference on Acoustics, Speech, and Signal Processing*, Denver, CO, April 1980, 749–752.

19. S. U. Lee, "Design of SVD/SGK Convolution Filters for Image Processing," Report USCIPI 950, University Southern California, Image Processing Institute, January 1980.

PART 4

IMAGE IMPROVEMENT

The use of digital processing techniques for image improvement has received much interest with the publicity given to applications in space imagery and medical research. Other applications include image improvement for photographic surveys and industrial radiographic analysis.

Image improvement is a term coined to denote three types of image manipulation processes: image enhancement, image restoration, and geometrical image modification. Image enhancement entails operations that improve the appearance to a human viewer, or operations to convert an image to a format better suited to machine processing. Image restoration has commonly been defined as the modification of an observed image in order to compensate for defects in the imaging system that produced the observed image. Geometrical image modification includes image magnification, minification, rotation, and nonlinear spatial warping.

Chapter 10 describes several techniques of monochrome and color image enhancement. The chapters that follow develop models for image formation and restoration, and present methods of point and spatial image restoration. The final chapter of this part considers geometrical image modification.

10

IMAGE ENHANCEMENT

Image enhancement processes consist of a collection of techniques that seek to improve the visual appearance of an image or to convert the image to a form better suited for analysis by a human or a machine. In an image enhancement system, there is no conscious effort to improve the fidelity of a reproduced image with regard to some ideal form of the image, as is done in image restoration. Actually, there is some evidence to indicate that often a distorted image, for example, an image with amplitude overshoot and undershoot about its object edges, is more subjectively pleasing than a perfectly reproduced original.

For image analysis purposes, the definition of image enhancement stops short of information extraction. As an example, an image enhancement system might emphasize the edge outline of objects in an image by high-frequency filtering. This edge-enhanced image would then serve as an input to a machine that would trace the outline of the edges, and perhaps make measurements of the shape and size of the outline. In this application, the image enhancement processor would emphasize salient features of the original image and simplify the processing task of a data-extraction machine.

There is no general unifying theory of image enhancement at present because there is no general standard of image quality that can serve as a design criterion for an image enhancement processor. Consideration is given here to a variety of techniques that have proved useful for human observation improvement and image analysis.

10.1. CONTRAST MANIPULATION

One of the most common defects of photographic or electronic images is poor contrast resulting from a reduced, and perhaps nonlinear, image amplitude range. Image

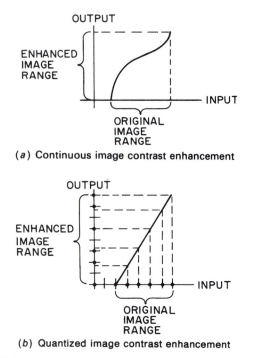

(a) Continuous image contrast enhancement

(b) Quantized image contrast enhancement

FIGURE 10.1-1. Continuous and quantized image contrast enhancement.

contrast can often be improved by amplitude rescaling of each pixel (1,2). Figure 10.1-1a illustrates a transfer function for contrast enhancement of a typical continuous amplitude low-contrast image. For continuous amplitude images, the transfer function operator can be implemented by photographic techniques, but it is often difficult to realize an arbitrary transfer function accurately. For quantized amplitude images, implementation of the transfer function is a relatively simple task. However, in the design of the transfer function operator, consideration must be given to the effects of amplitude quantization. With reference to Figure 10.1-1b, suppose that an original image is quantized to J levels, but it occupies a smaller range. The output image is also assumed to be restricted to J levels, and the mapping is linear. In the mapping strategy indicated in Figure 10.1-1b, the output level chosen is that level closest to the exact mapping of an input level. It is obvious from the diagram that the output image will have unoccupied levels within its range, and some of the gray scale transitions will be larger than in the original image. The latter effect may result in noticeable gray scale contouring. If the output image is quantized to more levels than the input image, it is possible to approach a linear placement of output levels, and hence, decrease the gray scale contouring effect.

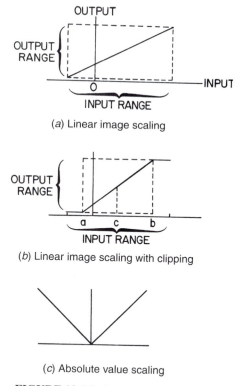

(a) Linear image scaling

(b) Linear image scaling with clipping

(c) Absolute value scaling

FIGURE 10.1-2. Image scaling methods.

10.1.1. Amplitude Scaling

A digitally processed image may occupy a range different from the range of the original image. In fact, the numerical range of the processed image may encompass negative values, which cannot be mapped directly into a light intensity range. Figure 10.1-2 illustrates several possibilities of scaling an output image back into the domain of values occupied by the original image. By the first technique, the processed image is linearly mapped over its entire range, while by the second technique, the extreme amplitude values of the processed image are clipped to maximum and minimum   limits. The second technique is often subjectively preferable, especially for images in which a relatively small number of pixels exceed the limits. Contrast enhancement algorithms often possess an option to clip a fixed percentage of the amplitude values on each end of the amplitude scale. In medical image enhancement applications, the contrast modification operation shown in Figure 10.2-2b, for $a \geq 0$, is called a *window-level transformation*. The window value is the width of the linear slope, $b - a$; the level is located at the midpoint c of the slope line. The third technique of amplitude scaling, shown in Figure 10.1-2c, utilizes an absolute value transformation for visualizing an image with negatively valued pixels. This is a

(*a*) Linear, full range, − 0.147 to 0.169

(*b*) Clipping, 0.000 to 0.169   (*c*) Absolute value, 0.000 to 0.169

FIGURE 10.1-3. Image scaling of the Q component of the YIQ representation of the `dolls_gamma` color image.

useful transformation for systems that utilize the two's complement numbering convention for amplitude representation. In such systems, if the amplitude of a pixel overshoots +1.0 (maximum luminance white) by a small amount, it wraps around by the same amount to −1.0, which is also maximum luminance white. Similarly, pixel undershoots remain near black.

Figure 10.1-3 illustrates the amplitude scaling of the Q component of the YIQ transformation, shown in Figure 3.5-14, of a monochrome image containing negative pixels. Figure 10.1-3*a* presents the result of amplitude scaling with the linear function of Figure 10.1-2*a* over the amplitude range of the image. In this example, the most negative pixels are mapped to black (0.0), and the most positive pixels are mapped to white (1.0). Amplitude scaling in which negative value pixels are clipped to zero is shown in Figure 10.1-3*b*. The black regions of the image correspond to

(*a*) Original

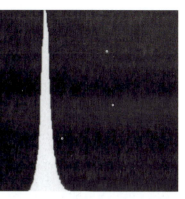

(*b*) Original histogram

(*c*) Min. clip = 0.17, max. clip = 0.64

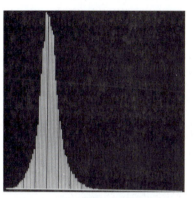

(*d*) Enhancement histogram

(*e*) Min. clip = 0.24, max. clip = 0.35

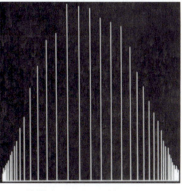

(*f*) Enhancement histogram

FIGURE 10.1-4. Window-level contrast stretching of an earth satellite image.

negative pixel values of the Q component. Absolute value scaling is presented in Figure 10.1-3c.

Figure 10.1-4 shows examples of contrast stretching of a poorly digitized original satellite image along with gray scale histograms of the original and enhanced pictures. In Figure 10.1-4c, the clip levels are set at the histogram limits of the original, while in Figure 10.1-4e, the clip levels truncate 5% of the original image upper and lower level amplitudes. It is readily apparent from the histogram of Figure 10.1-4f that the contrast-stretched image of Figure 10.1-4e has many unoccupied amplitude levels. Gray scale contouring is at the threshold of visibility.

10.1.2. Contrast Modification

Section 10.1.1 dealt with amplitude scaling of images that do not properly utilize the dynamic range of a display; they may lie partly outside the dynamic range or occupy only a portion of the dynamic range. In this section, attention is directed to point transformations that modify the contrast of an image within a display's dynamic range.

Figure 10.1-5a contains an original image of a jet aircraft that has been digitized to 256 gray levels and numerically scaled over the range of 0.0 (black) to 1.0 (white).

(a) Original

(b) Original histogram

(c) Transfer function

(d) Contrast stretched

FIGURE 10.1-5. Window-level contrast stretching of the `jet_mon` image.

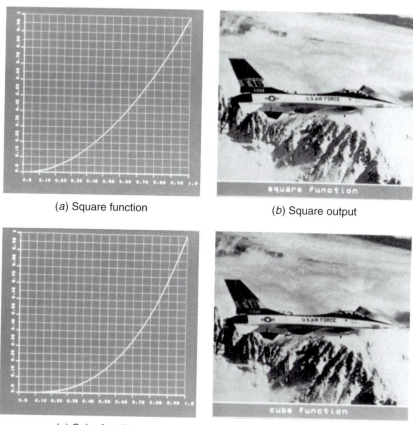

(a) Square function

(b) Square output

(c) Cube function

(d) Cube output

FIGURE 10.1-6. Square and cube contrast modification of the jet_mon image.

The histogram of the image is shown in Figure 10.1-5b. Examination of the histogram of the image reveals that the image contains relatively few low- or high-amplitude pixels. Consequently, applying the window-level contrast stretching function of Figure 10.1-5c results in the image of Figure 10.1-5d, which possesses better visual contrast but does not exhibit noticeable visual clipping.

Consideration will now be given to several nonlinear point transformations, some of which will be seen to improve visual contrast, while others clearly impair visual contrast.

Figures 10.1-6 and 10.1-7 provide examples of power law point transformations in which the processed image is defined by

$$G(j, k) = [F(j, k)]^{p}$$

(10.1-1)

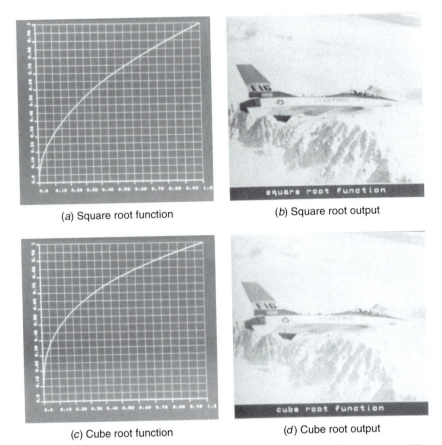

(a) Square root function

(b) Square root output

(c) Cube root function

(d) Cube root output

FIGURE 10.1-7. Square root and cube root contrast modification of the `jet_mon` image.

where $0.0 \le F(j, k) \le 1.0$ represents the original image and p is the power law variable. It is important that the amplitude limits of Eq. 10.1-1 be observed; processing of the integer code (e.g., 0 to 255) by Eq. 10.1-1 will give erroneous results. The square function provides the best visual result. The rubber band transfer function shown in Figure 10.1-8a provides a simple piecewise linear approximation to the power law curves. It is often useful in interactive enhancement machines in which the inflection point is interactively placed.

The Gaussian error function behaves like a square function for low-amplitude pixels and like a square root function for high- amplitude pixels. It is defined as

$$G(j, k) = \frac{\text{erf}\left\{\dfrac{F(j, k) - 0.5}{a\sqrt{2}}\right\} + \dfrac{0.5}{a\sqrt{2}}}{2 \; \text{erf}\left\{\dfrac{0.5}{a\sqrt{2}}\right\}} \tag{10.1-2a}$$

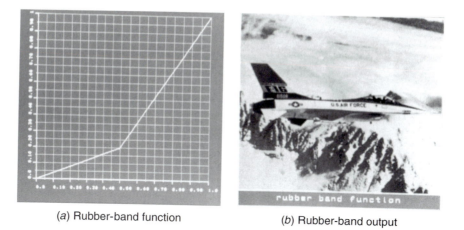

| (a) Rubber-band function | (b) Rubber-band output |

FIGURE 10.1-8. Rubber-band contrast modification of the `jet_mon` image.

where

$$\text{erf}\{x\} = \frac{2}{\sqrt{\pi}} \int_0^x \exp\{-y^2\} \, dy \qquad (10.1\text{-}2b)$$

and a is the standard deviation of the Gaussian distribution.

The logarithm function is useful for scaling image arrays with a very wide dynamic range. The logarithmic point transformation is given by

$$G(j, k) = \frac{\log_e\{1.0 + aF(j, k)\}}{\log_e\{2.0\}} \qquad (10.1\text{-}3)$$

under the assumption that $0.0 \leq F(j, k) \leq 1.0$, where a is a positive scaling factor. Figure 8.2-4 illustrates the logarithmic transformation applied to an array of Fourier transform coefficients.

There are applications in image processing in which monotonically decreasing and nonmonotonic amplitude scaling is useful. For example, contrast reverse and contrast inverse transfer functions, as illustrated in Figure 10.1-9, are often helpful in visualizing detail in dark areas of an image. The reverse function is defined as

$$G(j, k) = 1.0 - F(j, k) \qquad (10.1\text{-}4)$$

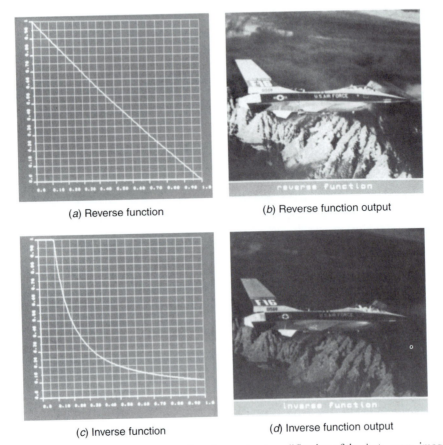

(a) Reverse function	(b) Reverse function output
(c) Inverse function	(d) Inverse function output

FIGURE 10.1-9. Reverse and inverse function contrast modification of the jet_mon image.

where $0.0 \le F(j, k) \le 1.0$ The inverse function

$$G(j, k) = \begin{cases} 1.0 & \text{for } 0.0 \le F(j, k) < 0.1 \qquad (10.1\text{-}5a) \\[2mm] \dfrac{0.1}{F(j, k)} & \text{for } 0.1 \le F(j, k) \le 1.0 \qquad (10.1\text{-}5b) \end{cases}$$

is clipped at the 10% input amplitude level to maintain the output amplitude within the range of unity.

Amplitude-level slicing, as illustrated in Figure 10.1-10, is a useful interactive tool for visually analyzing the spatial distribution of pixels of certain amplitude within an image. With the function of Figure 10.1-10a, all pixels within the amplitude passband are rendered maximum white in the output, and pixels outside the passband are rendered black. Pixels outside the amplitude passband are displayed in their original state with the function of Figure 10.1-10b.

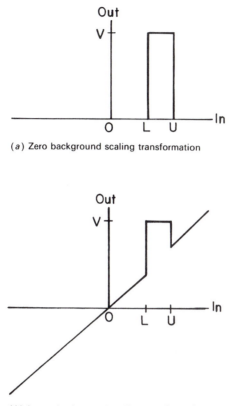

(*a*) Zero background scaling transformation

(*b*) Image background scaling transformation

FIGURE 10.1-10. Level slicing contrast modification functions.

10.2. HISTOGRAM MODIFICATION

The luminance histogram of a typical natural scene that has been linearly quantized is usually highly skewed toward the darker levels; a majority of the pixels possess a luminance less than the average. In such images, detail in the darker regions is often not perceptible. One means of enhancing these types of images is a technique called *histogram modification,* in which the original image is rescaled so that the histogram of the enhanced image follows some desired form. Andrews, Hall, and others (3–5) have produced enhanced imagery by a *histogram equalization* process for which the histogram of the enhanced image is forced to be uniform. Frei (6) has explored the use of histogram modification procedures that produce enhanced images possessing exponential or hyperbolic-shaped histograms. Ketcham (7) and Hummel (8) have demonstrated improved results by an adaptive histogram modification procedure.

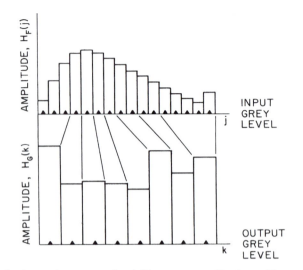

FIGURE 10.2-1. Approximate gray level histogram equalization with unequal number of quantization levels.

10.2.1. Nonadaptive Histogram Modification

Figure 10.2-1 gives an example of histogram equalization. In the figure, $H_F(c)$ for $c = 1, 2,..., C$, represents the fractional number of pixels in an input image whose amplitude is quantized to the cth reconstruction level. Histogram equalization seeks to produce an output image field G by point rescaling such that the normalized gray-level histogram $H_G(d) = 1/D$ for $d = 1, 2,..., D$. In the example of Figure 10.2-1, the number of output levels is set at one-half of the number of input levels. The scaling algorithm is developed as follows. The average value of the histogram is computed. Then, starting at the lowest gray level of the original, the pixels in the quantization bins are combined until the sum is closest to the average. All of these pixels are then rescaled to the new first reconstruction level at the midpoint of the enhanced image first quantization bin. The process is repeated for higher-value gray levels. If the number of reconstruction levels of the original image is large, it is possible to rescale the gray levels so that the enhanced image histogram is almost constant. It should be noted that the number of reconstruction levels of the enhanced image must be less than the number of levels of the original image to provide proper gray scale redistribution if all pixels in each quantization level are to be treated similarly. This process results in a somewhat larger quantization error. It is possible to perform the gray scale histogram equalization process with the same number of gray levels for the original and enhanced images, and still achieve a constant histogram of the enhanced image, by randomly redistributing pixels from input to output quantization bins.

The histogram modification process can be considered to be a monotonic point transformation $g_d = T\{f_c\}$ for which the input amplitude variable $f_1 \leq f_c \leq f_C$ is mapped into an output variable $g_1 \leq g_d \leq g_D$ such that the output probability distribution $P_R\{g_d = b_d\}$ follows some desired form for a given input probability distribution $P_R\{f_c = a_c\}$ where a_c and b_d are reconstruction values of the cth and dth levels. Clearly, the input and output probability distributions must each sum to unity. Thus,

$$\sum_{c=1}^{C} P_R\{f_c = a_c\} = 1 \qquad (10.2\text{-}1a)$$

$$\sum_{d=1}^{D} P_R\{g_d = b_d\} = 1 \qquad (10.2\text{-}1b)$$

Furthermore, the cumulative distributions must equate for any input index c. That is, the probability that pixels in the input image have an amplitude less than or equal to a_c must be equal to the probability that pixels in the output image have amplitude less than or equal to b_d, where $b_d = T\{a_c\}$ because the transformation is monotonic. Hence

$$\sum_{n=1}^{d} P_R\{g_n = b_n\} = \sum_{m=1}^{c} P_R\{f_m = a_m\} \qquad (10.2\text{-}2)$$

The summation on the right is the cumulative probability distribution of the input image. For a given image, the cumulative distribution is replaced by the cumulative histogram to yield the relationship

$$\sum_{n=1}^{d} P_R\{g_n = b_n\} = \sum_{m=1}^{c} H_F(m) \qquad (10.2\text{-}3)$$

Equation 10.2-3 now must be inverted to obtain a solution for g_d in terms of f_c. In general, this is a difficult or impossible task to perform analytically, but certainly possible by numerical methods. The resulting solution is simply a table that indicates the output image level for each input image level.

The histogram transformation can be obtained in approximate form by replacing the discrete probability distributions of Eq. 10.2-2 by continuous probability densities. The resulting approximation is

$$\int_{g_{min}}^{g} p_g(g)\,dg = \int_{f_{min}}^{f} pf(f)\,df \qquad (10.2\text{-}4)$$

TABLE 10.2-1. Histogram Modification Transfer Functions

Output Probability Density Model	Transfer Function[a]
Uniform	
$p_g(g) = \dfrac{1}{g_{max} - g_{min}}$, $\quad g_{min} \le g \le g_{max}$	$g = (g_{max} - g_{min})P_f(f) + g_{min}$
Exponential	
$p_g(g) = \alpha\exp\{-\alpha(g - g_{min})\}$ $\quad g \le g_{min}$	$g = g_{min} - \dfrac{1}{\alpha}\ln\{1 - P_f(f)\}$
Rayleigh	
$p_g(g) = \dfrac{g - g_{min}}{\alpha^2}\exp\left\{-\dfrac{(g-g_{min})^2}{2\alpha^2}\right\}$ $\quad g \ge g_{min}$	$g = g_{min} + \left[2\alpha^2\ln\left\{\dfrac{1}{1 - P_f(f)}\right\}\right]^{1/2}$
Hyperbolic (Cube root)	
$p_g(g) = \dfrac{1}{3}\dfrac{g^{-2/3}}{g_{max}^{1/3} - g_{min}^{1/3}}$	$g = \left[g_{max}^{1/3} - g_{min}^{1/3}[P_f(f)] + g_{max}^{1/3}\right]^3$
Hyperbolic (Logarithmic)	
$p_g(g) = \dfrac{1}{g[\ln\{g_{max}\} - \ln\{g_{min}\}]}$	$g = g_{min}\left(\dfrac{g_{max}}{g_{min}}\right)^{P_f(f)}$

[a]The cumulative probability distribution $P_f(f)$, of the input image is approximated by its cumulative histogram:

$$p_f(f) \approx \sum_{m=0}^{j} H_F(m)$$

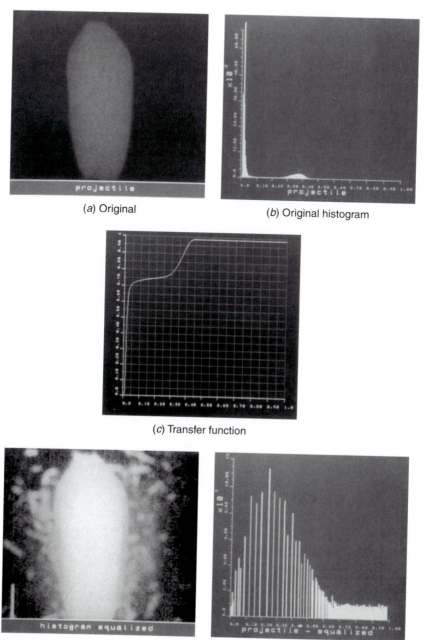

(a) Original

(b) Original histogram

(c) Transfer function

(d) Enhanced

(e) Enhanced histogram

FIGURE 10.2-2. Histogram equalization of the `projectile` image.

where $p_f(f)$ and $p_g(g)$ are the probability densities of f and g, respectively. The integral on the right is the cumulative distribution function $P_f(f)$ of the input variable f. Hence,

$$\int_{g_{min}}^{g} p_g(g)\, dg = P_f(f) \tag{10.2-5}$$

In the special case, for which the output density is forced to be the uniform density,

$$p_g(g) = \frac{1}{g_{max} - g_{min}} \tag{10.2-6}$$

for $g_{min} \le g \le g_{max}$, the histogram equalization transfer function becomes

$$g = (g_{max} - g_{min})P_f(f) + g_{min} \tag{10.2-7}$$

Table 10.2-1 lists several output image histograms and their corresponding transfer functions.

Figure 10.2-2 provides an example of histogram equalization for an x-ray of a projectile. The original image and its histogram are shown in Figure 10.2-2a and b, respectively. The transfer function of Figure 10.2-2c is equivalent to the cumulative histogram of the original image. In the histogram equalized result of Figure 10.2-2, ablating material from the projectile, not seen in the original, is clearly visible. The histogram of the enhanced image appears peaked, but close examination reveals that many gray level output values are unoccupied. If the high occupancy gray levels were to be averaged with their unoccupied neighbors, the resulting histogram would be much more uniform.

Histogram equalization usually performs best on images with detail hidden in dark regions. Good-quality originals are often degraded by histogram equalization. As an example, Figure 10.2-3 shows the result of histogram equalization on the jet image.

Frei (6) has suggested the *histogram hyperbolization* procedure listed in Table 10.2-1 and described in Figure 10.2-4. With this method, the input image histogram is modified by a transfer function such that the output image probability density is of hyperbolic form. Then the resulting gray scale probability density following the assumed logarithmic or cube root response of the photoreceptors of the eye model will be uniform. In essence, histogram equalization is performed after the cones of the retina.

10.2.2. Adaptive Histogram Modification

The histogram modification methods discussed in Section 10.2.1 involve application of the same transformation or mapping function to each pixel in an image. The mapping function is based on the histogram of the entire image. This process can be

(a) Original

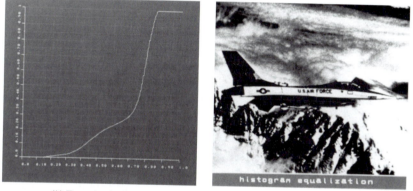

(b) Transfer function   (c) Histogram equalized

FIGURE 10.2-3. Histogram equalization of the jet_mon image.

made spatially adaptive by applying histogram modification to each pixel based on the histogram of pixels within a moving window neighborhood. This technique is obviously computationally intensive, as it requires histogram generation, mapping function computation, and mapping function application at each pixel.

Pizer et al. (9) have proposed an adaptive histogram equalization technique in which histograms are generated only at a rectangular grid of points and the mappings at each pixel are generated by interpolating mappings of the four nearest grid points. Figure 10.2-5 illustrates the geometry. A histogram is computed at each grid point in a window about the grid point. The window dimension can be smaller or larger than the grid spacing. Let M_{00}, M_{01}, M_{10}, M_{11} denote the histogram modification mappings generated at four neighboring grid points. The mapping to be applied at pixel $F(j, k)$ is determined by a bilinear interpolation of the mappings of the four nearest grid points as given by

$$M = a[bM_{00} + (1-b)M_{10}] + (1-a)[bM_{01} + (1-b)M_{11}] \qquad (10.2\text{-}8a)$$

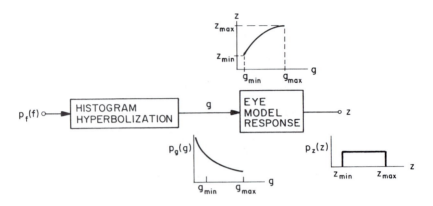

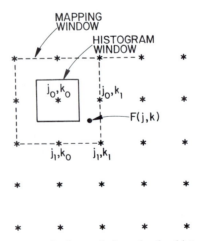

FIGURE 10.2-4. Histogram hyperbolization.

where

$$a = \frac{k - k_0}{k_1 - k_0} \tag{10.2-8b}$$

$$b = \frac{j - j_0}{j_1 - j_0} \tag{10.2-8c}$$

Pixels in the border region of the grid points are handled as special cases of Eq. 10.2-8. Equation 10.2-8 is best suited for general-purpose computer calculation.

FIGURE 10.2-5. Array geometry for interpolative adaptive histogram modification. ∗ Grid point; • pixel to be computed.

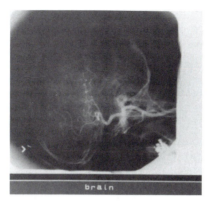

(*a*) Original

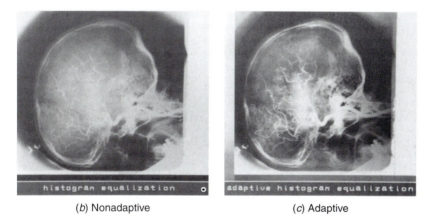

(*b*) Nonadaptive   (*c*) Adaptive

FIGURE 10.2-6. Nonadaptive and adaptive histogram equalization of the `brainscan` image.

For parallel processors, it is often more efficient to use the histogram generated in the histogram window of Figure 10.2-5 and apply the resultant mapping function to all pixels in the mapping window of the figure. This process is then repeated at all grid points. At each pixel coordinate (j, k), the four histogram modified pixels obtained from the four overlapped mappings are combined by bilinear interpolation. Figure 10.2-6 presents a comparison between nonadaptive and adaptive histogram equalization of a monochrome image. In the adaptive histogram equalization example, the histogram window is 64×64.

10.3. NOISE CLEANING

An image may be subject to noise and interference from several sources, including electrical sensor noise, photographic grain noise, and channel errors. These noise effects can be reduced by classical statistical filtering techniques to be discussed in Chapter 12. Another approach, discussed in this section, is the application of ad hoc *noise cleaning* techniques.

Image noise arising from a noisy sensor or channel transmission errors usually appears as discrete isolated pixel variations that are not spatially correlated. Pixels that are in error often appear visually to be markedly different from their neighbors. This observation is the basis of many noise cleaning algorithms (10–13). In this section we describe several linear and nonlinear techniques that have proved useful for noise reduction.

Figure 10.3-1 shows two test images, which will be used to evaluate noise cleaning techniques. Figure 10.3-1*b* has been obtained by adding uniformly distributed noise to the original image of Figure 10.3-1*a*. In the impulse noise example of Figure 10.3-1*c*, maximum-amplitude pixels replace original image pixels in a spatially random manner.

(*a*) Original

(*b*) Original with uniform noise   (*c*) Original with impulse noise

FIGURE 10.3-1. Noisy test images derived from the peppers_mon image.

10.3.1. Linear Noise Cleaning

Noise added to an image generally has a higher-spatial-frequency spectrum than the normal image components because of its spatial decorrelatedness. Hence, simple low-pass filtering can be effective for noise cleaning. Consideration will now be given to convolution and Fourier domain methods of noise cleaning.

Spatial Domain Processing. Following the techniques outlined in Chapter 7, a spatially filtered output image $G(j, k)$ can be formed by discrete convolution of an input image $F(j, k)$ with a $L \times L$ impulse response array $H(j, k)$ according to the relation

$$G(j, k) = \sum\sum F(m, n)H(m + j + C, n + k + C) \qquad (10.13\text{-}1)$$

where $C = (L + 1)/2$. Equation 10.3-1 utilizes the centered convolution notation developed by Eq. 7.1-14, whereby the input and output arrays are centered with respect to one another, with the outer boundary of $G(j, k)$ of width $(L - 1)/2$ pixels set to zero.

For noise cleaning, H should be of low-pass form, with all positive elements. Several common 3×3 pixel impulse response arrays of low-pass form are listed below.

Mask 1:
$$\mathbf{H} = \frac{1}{9}\begin{bmatrix} 1 & 1 & 1 \\ 1 & 1 & 1 \\ 1 & 1 & 1 \end{bmatrix} \qquad (10.3\text{-}2a)$$

Mask 2:
$$\mathbf{H} = \frac{1}{10}\begin{bmatrix} 1 & 1 & 1 \\ 1 & 2 & 1 \\ 1 & 1 & 1 \end{bmatrix} \qquad (10.3\text{-}2b)$$

Mask 3:
$$\mathbf{H} = \frac{1}{16}\begin{bmatrix} 1 & 2 & 1 \\ 2 & 4 & 2 \\ 1 & 2 & 1 \end{bmatrix} \qquad (10.3\text{-}2c)$$

These arrays, called *noise cleaning masks*, are normalized to unit weighting so that the noise-cleaning process does not introduce an amplitude bias in the processed image. The effect of noise cleaning with the arrays on the uniform noise and impulse noise test images is shown in Figure 10.3-2. Mask 1 and 2 of Eq. 10.3-2 are special cases of a 3×3 parametric low-pass filter whose impulse response is defined as

$$\mathbf{H} = \frac{1}{b + 2}\begin{bmatrix} 1 & b & 1 \\ b & b^2 & b \\ 1 & b & 1 \end{bmatrix} \qquad (10.3\text{-}3)$$

(a) Uniform noise, mask 1

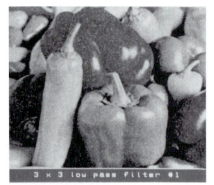

(b) Impulse noise, mask 1

(c) Uniform noise, mask 2

(d) Impulse noise, mask 2

(e) Uniform noise, mask 3

(f) Impulse noise, mask 3

FIGURE 10.3-2. Noise cleaning with 3×3 low-pass impulse response arrays on the noisy test images.

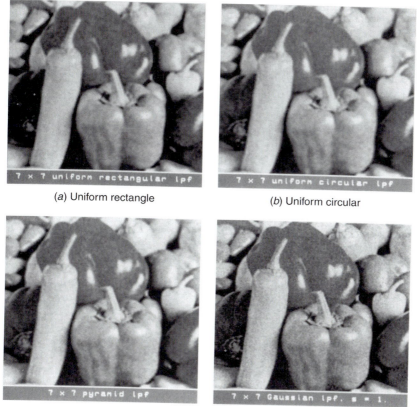

(a) Uniform rectangle   (b) Uniform circular

(c) Pyramid   (d) Gaussian, $s = 1.0$

FIGURE 10.3-3. Noise cleaning with 7×7 impulse response arrays on the noisy test image with uniform noise.

The concept of low-pass filtering noise cleaning can be extended to larger impulse response arrays. Figures 10.3-3 and 10.3-4 present noise cleaning results for several 7×7 impulse response arrays for uniform and impulse noise. As expected, use of a larger impulse response array provides more noise smoothing, but at the expense of the loss of fine image detail.

Fourier Domain Processing. It is possible to perform linear noise cleaning in the Fourier domain (13) using the techniques outlined in Section 9.3. Properly executed, there is no difference in results between convolution and Fourier filtering; the choice is a matter of implementation considerations.

High-frequency noise effects can be reduced by Fourier domain filtering with a zonal low-pass filter with a transfer function defined by Eq. 9.3-9. The sharp cutoff characteristic of the zonal low-pass filter leads to *ringing* artifacts in a filtered image. This deleterious effect can be eliminated by the use of a smooth cutoff filter,

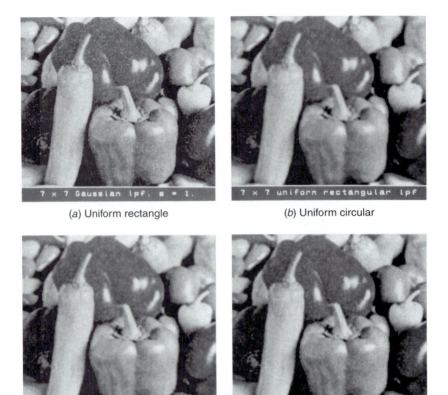

(a) Uniform rectangle

(b) Uniform circular

(c) Pyramid

(d) Gaussian, $s = 1.0$

FIGURE 10.3-4. Noise cleaning with 7×7 impulse response arrays on the noisy test image with impulse noise.

such as the Butterworth low-pass filter whose transfer function is specified by Eq. 9.4-12. Figure 10.3-5 shows the results of zonal and Butterworth low-pass filtering of noisy images.

Unlike convolution, Fourier domain processing, often provides quantitative and intuitive insight into the nature of the noise process, which is useful in designing noise cleaning spatial filters. As an example, Figure 10.3-6a shows an original image subject to periodic interference. Its two-dimensional Fourier transform, shown in Figure 10.3-6b, exhibits a strong response at the two points in the Fourier plane corresponding to the frequency response of the interference. When multiplied point by point with the Fourier transform of the original image, the bandstop filter of Figure 10.3-6c attenuates the interference energy in the Fourier domain. Figure 10.3-6d shows the noise-cleaned result obtained by taking an inverse Fourier transform of the product.

(a) Uniform noise, zonal

(b) Impulse noise, zonal

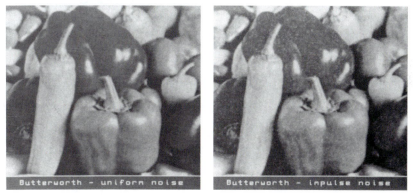

(c) Uniform noise, Butterworth

(d) Impulse noise, Butterworth

FIGURE 10.3-5. Noise cleaning with zonal and Butterworth low-pass filtering on the noisy test images; cutoff frequency = 64.

Homomorphic Filtering. Homomorphic filtering (14) is a useful technique for image enhancement when an image is subject to multiplicative noise or interference. Figure 10.3-7 describes the process. The input image $F(j, k)$ is assumed to be modeled as the product of a noise-free image $S(j, k)$ and an illumination interference array $I(j, k)$. Thus,

$$F(j, k) = I(j, k)S(j, k) \tag{10.3-4}$$

Ideally, $I(j, k)$ would be a constant for all (j, k). Taking the logarithm of Eq. 10.3-4 yields the additive linear result

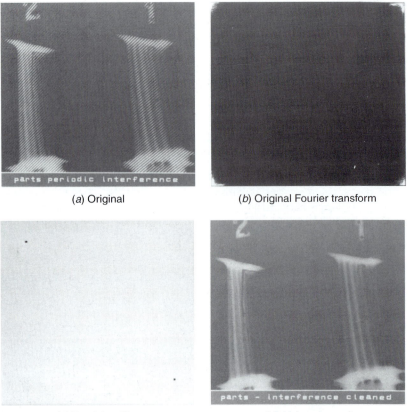

(a) Original   (b) Original Fourier transform

(c) Bandstop filter   (d) Noise cleaned

FIGURE 10.3-6. Noise cleaning with Fourier domain band stop filtering on the `parts` image with periodic interference.

$$\log\{F(j,k)\} = \log\{I(j,k)\} + \log\{S(j,k)\} \tag{10.3-5}$$

Conventional linear filtering techniques can now be applied to reduce the log interference component. Exponentiation after filtering completes the enhancement process. Figure 10.3-8 provides an example of homomorphic filtering. In this example, the illumination field $I(j,k)$ increases from left to right from a value of 0.1 to 1.0.

FIGURE 10.3-7. Homomorphic filtering.

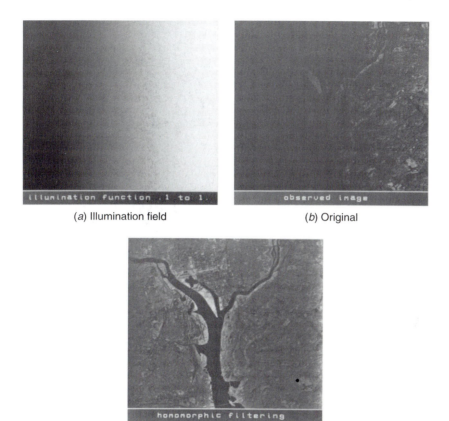

(*a*) Illumination field   (*b*) Original

(*c*) Homomorphic filtering

FIGURE 10.3-8. Homomorphic filtering on the `washington_ir` image with a Butterworth high-pass filter; cutoff frequency = 4.

Therefore, the observed image appears quite dim on its left side. Homomorphic filtering (Figure 10.3-8*c*) compensates for the nonuniform illumination.

10.3.2. Nonlinear Noise Cleaning

The linear processing techniques described previously perform reasonably well on images with continuous noise, such as additive uniform or Gaussian distributed noise. However, they tend to provide too much smoothing for impulselike noise. Nonlinear techniques often provide a better trade-off between noise smoothing and the retention of fine image detail. Several nonlinear techniques are presented below. Mastin (15) has performed subjective testing of several of these operators.

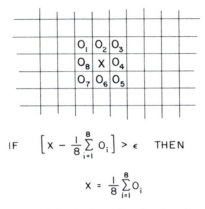

$$\text{IF} \quad \left[x - \frac{1}{8} \sum_{i=1}^{8} O_i \right] > \epsilon \quad \text{THEN}$$

$$x = \frac{1}{8} \sum_{i=1}^{8} O_i$$

FIGURE 10.3-9. Outlier noise cleaning algorithm.

Outlier. Figure 10.3-9 describes a simple *outlier* noise cleaning technique in which each pixel is compared to the average of its eight neighbors. If the magnitude of the difference is greater than some threshold level, the pixel is judged to be noisy, and it is replaced by its neighborhood average. The eight-neighbor average can be computed by convolution of the observed image with the impulse response array

$$\mathbf{H} = \frac{1}{8} \begin{bmatrix} 1 & 1 & 1 \\ 1 & 0 & 1 \\ 1 & 1 & 1 \end{bmatrix} \tag{10.3-6}$$

Figure 10.3-10 presents the results of outlier noise cleaning for a threshold level of 10%.

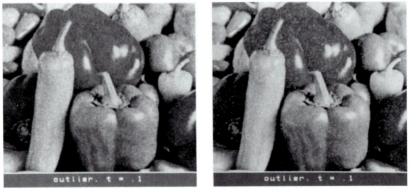

(a) Uniform noise   (b) Impulse noise

FIGURE 10.3-10. Noise cleaning with the outlier algorithm on the noisy test images.

The outlier operator can be extended straightforwardly to larger windows. Davis and Rosenfeld (16) have suggested a variant of the outlier technique in which the center pixel in a window is replaced by the average of its k neighbors whose amplitudes are closest to the center pixel.

Median Filter. *Median filtering* is a nonlinear signal processing technique developed by Tukey (17) that is useful for noise suppression in images. In one-dimensional form, the median filter consists of a sliding window encompassing an odd number of pixels. The center pixel in the window is replaced by the median of the pixels in the window. The median of a discrete sequence $a_1, a_2,..., a_N$ for N odd is that member of the sequence for which $(N - 1)/2$ elements are smaller or equal in value and $(N - 1)/2$ elements are larger or equal in value. For example, if the values of the pixels within a window are 0.1, 0.2, 0.9, 0.4, 0.5, the center pixel would be replaced by the value 0.4, which is the median value of the sorted sequence 0.1, 0.2, 0.4, 0.5, 0.9. In this example, if the value 0.9 were a noise spike in a monotonically increasing sequence, the median filter would result in a considerable improvement. On the other hand, the value 0.9 might represent a valid signal pulse for a wide-bandwidth sensor, and the resultant image would suffer some loss of resolution. Thus, in some cases the median filter will provide noise suppression, while in other cases it will cause signal suppression.

Figure 10.3-11 illustrates some examples of the operation of a median filter and a mean (smoothing) filter for a discrete step function, ramp function, pulse function, and a triangle function with a window of five pixels. It is seen from these examples that the median filter has the usually desirable property of not affecting step functions or ramp functions. Pulse functions, whose periods are less than one-half the window width, are suppressed. But the peak of the triangle is flattened.

Operation of the median filter can be analyzed to a limited extent. It can be shown that the median of the product of a constant K and a sequence $f(j)$ is

$$\text{MED}\{K[f(j)]\} = K[\text{MED}\{f(j)\}] \qquad (10.3\text{-}7)$$

However, for two arbitrary sequences $f(j)$ and $g(j)$, it does not follow that the median of the sum of the sequences is equal to the sum of their medians. That is, in general,

$$\text{MED}\{f(j) + g(j)\} \neq \text{MED}\{f(j)\} + \text{MED}\{g(j)\} \qquad (10.3\text{-}8)$$

The sequences 0.1, 0.2, 0.3, 0.4, 0.5 and 0.1, 0.2, 0.3, 0.2, 0.1 are examples for which the additive linearity property does not hold.

There are various strategies for application of the median filter for noise suppression. One method would be to try a median filter with a window of length 3. If there is no significant signal loss, the window length could be increased to 5 for median

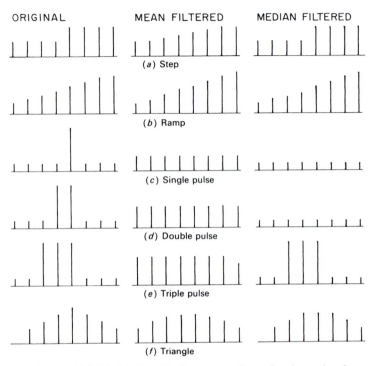

FIGURE 10.3-11. Median filtering on one-dimensional test signals.

filtering of the original. The process would be terminated when the median filter begins to do more harm than good. It is also possible to perform cascaded median filtering on a signal using a fixed-or variable-length window. In general, regions that are unchanged by a single pass of the filter will remain unchanged in subsequent passes. Regions in which the signal period is lower than one-half the window width will be continually altered by each successive pass. Usually, the process will continue until the resultant period is greater than one-half the window width, but it can be shown that some sequences will never converge (18).

The concept of the median filter can be extended easily to two dimensions by utilizing a two-dimensional window of some desired shape such as a rectangle or discrete approximation to a circle. It is obvious that a two-dimensional $L \times L$ median filter will provide a greater degree of noise suppression than sequential processing with $L \times 1$ median filters, but two-dimensional processing also results in greater signal suppression. Figure 10.3-12 illustrates the effect of two-dimensional median filtering of a spatial peg function with a 3×3 square filter and a 5×5 plus sign–shaped filter. In this example, the square median has deleted the corners of the peg, but the plus median has not affected the corners.

Figures 10.3-13 and 10.3-14 show results of plus sign shaped median filtering on the noisy test images of Figure 10.3-1 for impulse and uniform noise, respectively.

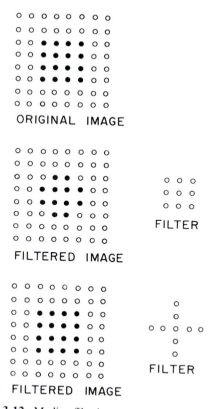

ORIGINAL IMAGE

FILTERED IMAGE

FILTER

FILTERED IMAGE

FILTER

FIGURE 10.3-12. Median filtering on two-dimensional test signals.

In the impulse noise example, application of the 3×3 median significantly reduces the noise effect, but some residual noise remains. Applying two 3×3 median filters in cascade provides further improvement. The 5×5 median filter removes almost all of the impulse noise. There is no visible impulse noise in the 7×7 median filter result, but the image has become somewhat blurred. In the case of uniform noise, median filtering provides little visual improvement.

Huang et al. (19) and Astola and Campbell (20) have developed fast median filtering algorithms. The latter can be generalized to implement any rank ordering.

Pseudomedian Filter. Median filtering is computationally intensive; the number of operations grows exponentially with window size. Pratt et al. (21) have proposed a computationally simpler operator, called the *pseudomedian filter*, which possesses many of the properties of the median filter.

Let $\{S_L\}$ denote a sequence of elements s_1, s_2,..., s_L. The pseudomedian of the sequence is

(a) 3×3 median filter

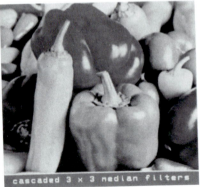

(b) 3×3 cascaded median filter

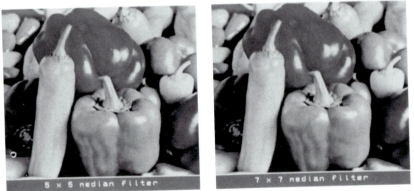

(c) 5×5 median filter

(d) 7×7 median filter

FIGURE 10.3-13. Median filtering on the noisy test image with uniform noise.

$$\mathrm{PMED}\{S_L\} = (1/2)\mathrm{MAXIMIN}\{S_L\} + (1/2)\mathrm{MINIMAX}\{S_L\} \qquad (10.3\text{-}9)$$

where for $M = (L + 1)/2$

$$\mathrm{MAXIMIN}\{S_L\} = \mathrm{MAX}\{[\mathrm{MIN}(s_1, ..., s_M)], [\mathrm{MIN}(s_2, ..., s_{M+1})]$$

$$..., [\mathrm{MIN}(s_{L-M+1}, ..., s_L)]\} \qquad (10.3\text{-}10a)$$

$$\mathrm{MINIMAX}\{S_L\} = \mathrm{MIN}\{[\mathrm{MAX}(s_1, ..., s_M)], [\mathrm{MAX}(s_2, ..., s_{M+1})]$$

$$..., [\mathrm{MAX}(s_{L-M+1}, ..., s_L)]\} \qquad (10.3\text{-}10b)$$

(*a*) 3 × 3 median filter

(*b*) 5 × 5 median filter

(*c*) 7 × 7 median filter

FIGURE 10.3-14. Median filtering on the noisy test image with uniform noise.

Operationally, the sequence of L elements is decomposed into subsequences of M elements, each of which is slid to the right by one element in relation to its predecessor, and the appropriate MAX and MIN operations are computed. As will be demonstrated, the MAXIMIN and MINIMAX operators are, by themselves, useful operators. It should be noted that it is possible to recursively decompose the MAX and MIN functions on long sequences into sliding functions of length 2 and 3 for pipeline computation (21).

The one-dimensional pseudomedian concept can be extended in a variety of ways. One approach is to compute the MAX and MIN functions over rectangular windows. As with the median filter, this approach tends to over smooth an image. A plus-shape pseudomedian generally provides better subjective results. Consider a plus-shaped window containing the following two-dimensional set elements $\{S_E\}$

$$y_1$$

.

.

.

$$x_1 \; \cdots \; x_M \; \cdots \; x_C$$

.

.

.

$$y_R$$

Let the sequences $\{X_C\}$ and $\{Y_R\}$ denote the elements along the horizontal and vertical axes of the window, respectively. Note that the element x_M is common to both sequences. Then the plus-shaped pseudomedian can be defined as

$$\text{PMED}\{S_E\} = (1/2)\text{MAX}[\text{MAXIMIN}\{X_C\}, \text{MAXIMIN}\{Y_R\}]$$

$$+ (1/2)\, \text{MIN}\, [\text{MINIMAX}\{X_C\}, \text{MINIMAX}\{Y_R\}]$$

$$(10.3\text{-}11)$$

The MAXIMIN operator in one- or two-dimensional form is useful for removing bright impulse noise but has little or no effect on dark impulse noise. Conversely, the MINIMAX operator does a good job in removing dark, but not bright, impulse noise. A logical conclusion is to cascade the operators.

Figure 10.3-16 shows the results of MAXIMIN, MINIMAX, and pseudomedian filtering on an image subjected to salt and pepper noise. As observed, the MAXIMIN operator reduces the salt noise, while the MINIMAX operator reduces the pepper noise. The pseudomedian provides attenuation for both types of noise. The cascade MINIMAX and MAXIMIN operators, in either order, show excellent results.

Wavelet De-noising. Section 8.4-3 introduced wavelet transforms. The usefulness of wavelet transforms for image coding derives from the property that most of the energy of a transformed image is concentrated in the trend transform components rather than the fluctuation components (22). The fluctuation components may be grossly quantized without serious image degradation. This energy compaction property can also be exploited for noise removal. The concept, called *wavelet de-noising* (22,23), is quite simple. The wavelet transform coefficients are thresholded such that the presumably noisy, low-amplitude coefficients are set to zero.

(a) Original

(b) MAXIMIN

(c) MINIMAX

(d) Pseudomedian

(e) MINIMAX of MAXIMIN

(f) MAXIMIN of MINIMAX

FIGURE 10.3-15. 5×5 plus-shape MINIMAX, MAXIMIN, and pseudomedian filtering on the noisy test images.

10.4. EDGE CRISPENING

Psychophysical experiments indicate that a photograph or visual signal with accentuated or *crispened* edges is often more subjectively pleasing than an exact photometric reproduction. *Edge crispening* can be accomplished in a variety of ways.

10.4.1. Linear Edge Crispening

Edge crispening can be performed by discrete convolution, as defined by Eq. 10.3-1, in which the impulse response array H is of high-pass form. Several common 3×3 high-pass masks are given below (24–26).

Mask 1:

$$H = \begin{bmatrix} 0 & -1 & 0 \\ -1 & 5 & -1 \\ 0 & -1 & 0 \end{bmatrix} \qquad (10.4\text{-}1a)$$

Mask 2:

$$H = \begin{bmatrix} -1 & -1 & -1 \\ -1 & 9 & -1 \\ -1 & -1 & -1 \end{bmatrix} \qquad (10.4\text{-}1b)$$

Mask 3:

$$H = \begin{bmatrix} 1 & -2 & 1 \\ -2 & 5 & -2 \\ 1 & -2 & 1 \end{bmatrix} \qquad (10.3\text{-}1c)$$

These masks possess the property that the sum of their elements is unity, to avoid amplitude bias in the processed image. Figure 10.4-1 provides examples of edge crispening on a monochrome image with the masks of Eq. 10.4-1. Mask 2 appears to provide the best visual results.

To obtain edge crispening on electronically scanned images, the scanner signal can be passed through an electrical filter with a high-frequency bandpass characteristic. Another possibility for scanned images is the technique of *unsharp masking* (27,28). In this process, the image is effectively scanned with two overlapping apertures, one at normal resolution and the other at a lower spatial resolution, which upon sampling produces normal and low-resolution images $F(j, k)$ and $F_L(j, k)$, respectively. An unsharp masked image

$$G(j, k) = \frac{c}{2c - 1} F(j, k) - \frac{1 - c}{2c - 1} F_L(j, k) \qquad (10.4\text{-}2)$$

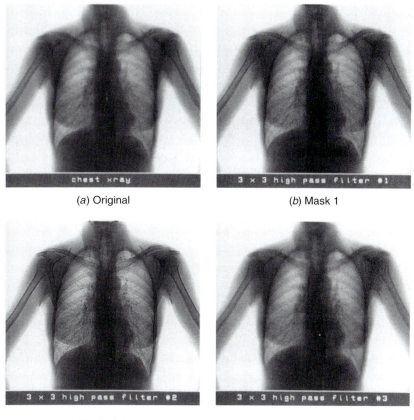

(*a*) Original   (*b*) Mask 1

(*c*) Mask 2   (*d*) Mask 3

FIGURE 10.4-1. Edge crispening with 3×3 masks on the chest_xray image.

is then generated by forming the weighted difference between the normal and low-resolution images, where c is a weighting constant. Typically, c is in the range 3/5 to 5/6, so that the ratio of normal to low-resolution components in the masked image is from 1.5:1 to 5:1. Figure 10.4-2 illustrates typical scan signals obtained when scanning over an object edge. The masked signal has a longer-duration edge gradient as well as an overshoot and undershoot, as compared to the original signal. Subjectively, the apparent sharpness of the original image is improved. Figure 10.4-3 presents examples of unsharp masking in which the low-resolution image is obtained by convolution with a uniform $L \times L$ impulse response array. The sharpening effect is stronger as L increases and c decreases.

Linear edge crispening can be performed by Fourier domain filtering. A zonal high-pass filter with a transfer function given by Eq. 9.4-10 suppresses all spatial frequencies below the cutoff frequency except for the dc component, which is necessary to maintain the average amplitude of the filtered image. Figure 10.4-4 shows

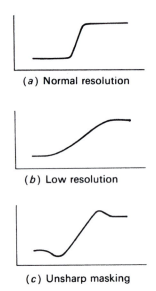

(*a*) Normal resolution

(*b*) Low resolution

(*c*) Unsharp masking

FIGURE 10.4-2. Waveforms in an unsharp masking image enhancement system.

the result of zonal high-pass filtering of an image. Zonal high-pass filtering often causes ringing in a filtered image. Such ringing can be reduced significantly by utilization of a high-pass filter with a smooth cutoff response. One such filter is the Butterworth high-pass filter, whose transfer function is defined by Eq. 9.4-13.

Figure 10.4-4 shows the results of zonal and Butterworth high-pass filtering. In both examples, the filtered images are biased to a midgray level for display.

10.4.2. Statistical Differencing

Another form of edge crispening, called *statistical differencing* (29, p. 100), involves the generation of an image by dividing each pixel value by its estimated standard deviation $D(j, k)$ according to the basic relation

$$G(j, k) = \frac{F(j, k)}{D(j, k)} \tag{10.4-3}$$

where the estimated standard deviation

$$D(j, k) = \frac{1}{W} \left[\sum_{m = j - w}^{j + w} \sum_{n = k - w}^{k + w} [F(m, n) - M(m, n)]^2 \right]^{1/2} \tag{10.4-4}$$

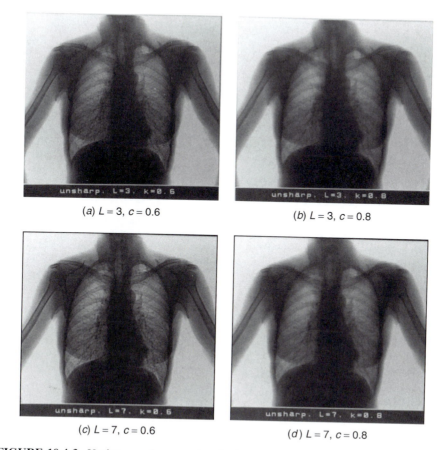

(a) L = 3, c = 0.6

(b) L = 3, c = 0.8

(c) L = 7, c = 0.6

(d) L = 7, c = 0.8

FIGURE 10.4-3. Unsharp mask processing for $L \times L$ uniform low-pass convolution on the chest_xray image.

is computed at each pixel over some $W \times W$ neighborhood where $W = 2w + 1$. The function $M(j, k)$ is the estimated mean value of the original image at point (j, k), which is computed as

$$M(j, k) = \frac{1}{W^2} \sum_{m = j - w}^{j + w} \sum_{n = k - w}^{k + w} F(m, n) \qquad (10.4\text{-}5)$$

The enhanced image $G(j, k)$ is increased in amplitude with respect to the original at pixels that deviate significantly from their neighbors, and is decreased in relative amplitude elsewhere. The process is analogous to automatic gain control for an audio signal.

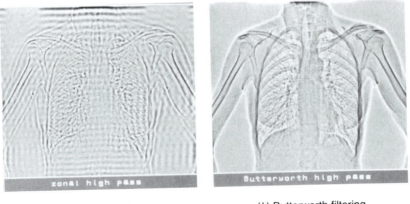

<table>
<tr><td>(a) Zonal filtering</td><td>(b) Butterworth filtering</td></tr>
</table>

FIGURE 10.4-4. Zonal and Butterworth high-pass filtering on the `chest_xray` image; cutoff frequency = 32.

Wallis (30) has suggested a generalization of the statistical differencing operator in which the enhanced image is forced to a form with desired first- and second-order moments. The *Wallis operator* is defined by

$$G(j, k) = [F(j, k) - M(j, k)] \frac{A_{max} D_d}{A_{max} D(j, k) + D_d} + [p M_d + (1 - p) M(j, k)] \qquad (10.4\text{-}6)$$

where M_d and D_d represent desired average mean and standard deviation factors, A_{max} is a maximum gain factor that prevents overly large output values when $D(j, k)$ is small and $0.0 \le p \le 1.0$ is a mean proportionality factor controlling the background flatness of the enhanced image.

The Wallis operator can be expressed in a more general form as

$$G(j, k) = [F(j, k) - M(j, k)] A(j, k) + B(j, k) \qquad (10.4\text{-}7)$$

where $A(j, k)$ is a spatially dependent gain factor and $B(j, k)$ is a spatially dependent background factor. These gain and background factors can be derived directly from Eq. 10.4-4, or they can be specified in some other manner. For the Wallis operator, it is convenient to specify the desired average standard deviation D_d such that the spatial gain ranges between maximum A_{max} and minimum A_{min} limits. This can be accomplished by setting D_d to the value

(*a*) Original

(*b*) Mean, 0.00 to 0.98

(*c*) Standard deviation, 0.01 to 0.26

(*d*) Background, 0.09 to 0.88

(*e*) Spatial gain, 0.75 to 2.35

(*f*) Wallis enhancement, – 0.07 to 1.12

FIGURE 10.4-5. Wallis statistical differencing on the `bridge` image for $M_d = 0.45$, $D_d = 0.28$, $p = 0.20$, $A_{max} = 2.50$, $A_{min} = 0.75$ using a 9×9 pyramid array.

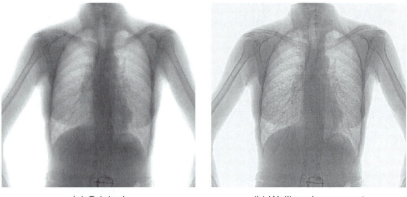

| (a) Original | (b) Wallis enhancement |

FIGURE 10.4-6. Wallis statistical differencing on the `chest_xray` image for $M_d = 0.64$, $D_d = 0.22$, $p = 0.20$, $A_{max} = 2.50$, $A_{min} = 0.75$ using a 11×11 pyramid array.

$$D_d = \frac{A_{min}A_{max}D_{max}}{A_{max} - A_{min}} \tag{10.4-8}$$

where D_{max} is the maximum value of $D(j, k)$. The summations of Eqs. 10.4-4 and 10.4-5 can be implemented by convolutions with a uniform impulse array. But, overshoot and undershoot effects may occur. Better results are usually obtained with a pyramid or Gaussian-shaped array.

Figure 10.4-5 shows the mean, standard deviation, spatial gain, and Wallis statistical differencing result on a monochrome image. Figure 10.4-6 presents a medical imaging example.

10.5. COLOR IMAGE ENHANCEMENT

The image enhancement techniques discussed previously have all been applied to monochrome images. This section considers the enhancement of natural color images and introduces the pseudocolor and false color image enhancement methods. In the literature, the terms pseudocolor and false color have often been used improperly. Pseudocolor produces a color image from a monochrome image, while false color produces an enhanced color image from an original natural color image or from multispectral image bands.

10.5.1. Natural Color Image Enhancement

The monochrome image enhancement methods described previously can be applied to natural color images by processing each color component individually. However, care must be taken to avoid changing the average value of the processed image components. Otherwise, the processed color image may exhibit deleterious shifts in hue and saturation.

Typically, color images are processed in the *RGB* color space. For some image enhancement algorithms, there are computational advantages to processing in a luma-chroma space, such as *YIQ*, or a lightness-chrominance space, such as $L*u*v*$. As an example, if the objective is to perform edge crispening of a color image, it is usually only necessary to apply the enhancement method to the luma or lightness component. Because of the high-spatial-frequency response limitations of human vision, edge crispening of the chroma or chrominance components may not be perceptible.

Faugeras (31) has investigated color image enhancement in a perceptual space based on a color vision model similar to the model presented in Figure 2.5-3. The procedure is to transform the *RGB* tristimulus value original images according to the color vision model to produce a set of three perceptual space images that, ideally, are perceptually independent. Then, an image enhancement method is applied independently to the perceptual space images. Finally, the enhanced perceptual space images are subjected to steps that invert the color vision model and produce an enhanced color image represented in *RGB* color space.

10.5.2. Pseudocolor

Pseudocolor (32–34) is a color mapping of a monochrome image array which is intended to enhance the detectability of detail within the image. The pseudocolor mapping of an array $F(j, k)$ is defined as

$$R(j, k) = O_R\{F(j, k)\} \tag{10.5-1a}$$

$$G(j, k) = O_G\{F(j, k)\} \tag{10.5-1b}$$

$$B(j, k) = O_B\{F(j, k)\} \tag{10.5-1c}$$

where $R(j, k)$, $G(j, k)$, $B(j, k)$ are display color components and $O_R\{F(j, k)\}$, $O_G\{F(j, k)\}$, $O_B\{F(j, k)\}$ are linear or nonlinear functional operators. This mapping defines a path in three-dimensional color space parametrically in terms of the array $F(j, k)$. Figure 10.5-1 illustrates the *RGB* color space and two color mappings that originate at black and terminate at white. Mapping *A* represents the *achromatic* path through all shades of gray; it is the normal representation of a monochrome image. Mapping *B* is a spiral path through color space.

Another class of pseudocolor mappings includes those mappings that exclude all shades of gray. Mapping *C*, which follows the edges of the *RGB* color cube, is such an example. This mapping follows the perimeter of the gamut of reproducible colors as depicted by the uniform chromaticity scale (UCS) chromaticity chart shown in

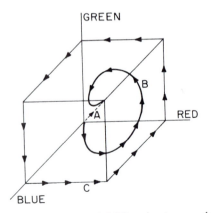

FIGURE 10.5-1. Black-to-white and *RGB* perimeter pseudocolor mappings.

Figure 10.5-2. The luminances of the colors red, green, blue, cyan, magenta, and yellow that lie along the perimeter of reproducible colors are noted in the figure. It is seen that the luminance of the pseudocolor scale varies between a minimum of 0.114 for blue to a maximum of 0.886 for yellow. A maximum luminance of unity is reached only for white. In some applications it may be desirable to fix the luminance of all displayed colors so that discrimination along the pseudocolor scale is by hue and saturation attributes of a color only. Loci of constant luminance are plotted in Figure 10.5-2.

Figure 10.5-2 also includes bounds for displayed colors of constant luminance. For example, if the *RGB* perimeter path is followed, the maximum luminance of any color must be limited to 0.114, the luminance of blue. At a luminance of 0.2, the *RGB* perimeter path can be followed except for the region around saturated blue. At higher luminance levels, the gamut of constant luminance colors becomes severely limited. Figure 10.5-2b is a plot of the 0.5 luminance locus. Inscribed within this locus is the locus of those colors of largest constant saturation. A pseudocolor scale along this path would have the property that all points differ only in hue.

With a given pseudocolor path in color space, it is necessary to choose the scaling between the data plane variable and the incremental path distance. On the UCS chromaticity chart, incremental distances are subjectively almost equally noticeable. Therefore, it is reasonable to subdivide geometrically the path length into equal increments. Figure 10.5-3 shows examples of pseudocoloring of a gray scale chart image and a seismic image.

10.5.3. False Color

False color is a point-by-point mapping of an original color image, described by its three primary colors, or of a set of multispectral image planes of a scene, to a color

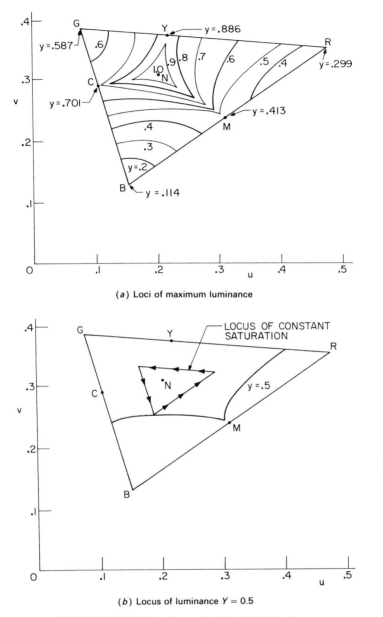

(*a*) Loci of maximum luminance

(*b*) Locus of luminance *Y* = 0.5

FIGURE 10.5-2. Luminance loci for NTSC colors.

space defined by display tristimulus values that are linear or nonlinear functions of the original image pixel values (35,36). A common intent is to provide a displayed image with objects possessing different or false colors from what might be expected.

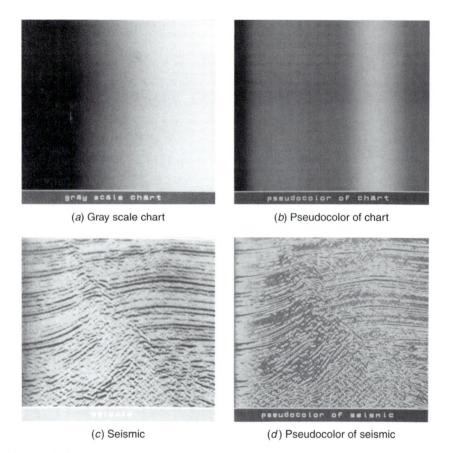

(a) Gray scale chart

(b) Pseudocolor of chart

(c) Seismic

(d) Pseudocolor of seismic

FIGURE 10.5-3. Pseudocoloring of the `gray_chart` and `seismic` images. See insert for a color representation of this figure.

For example, blue sky in a normal scene might be converted to appear red, and green grass transformed to blue. One possible reason for such a color mapping is to place normal objects in a strange color world so that a human observer will pay more attention to the objects than if they were colored normally.

Another reason for false color mappings is the attempt to color a normal scene to match the color sensitivity of a human viewer. For example, it is known that the luminance response of cones in the retina peaks in the green region of the visible spectrum. Thus, if a normally red object is false colored to appear green, it may become more easily detectable. Another psychophysical property of color vision that can be exploited is the contrast sensitivity of the eye to changes in blue light. In some situation it may be worthwhile to map the normal colors of objects with fine detail into shades of blue.

A third application of false color is to produce a natural color representation of a set of multispectral images of a scene. Some of the multispectral images may even be obtained from sensors whose wavelength response is outside the visible wavelength range, for example, infrared or ultraviolet.

In a false color mapping, the red, green, and blue display color components are related to natural or multispectral images F_i by

$$R_D = O_R\{F_1, F_2, ...\} \tag{10.5-2a}$$

$$G_D = O_G\{F_1, F_2, ...\} \tag{10.5-2b}$$

$$B_D = O_B\{F_1, F_2, ...\} \tag{10.5-2c}$$

where $O_R\{\cdot\}$, $O_R\{\cdot\}$, $O_B\{\cdot\}$ are general functional operators. As a simple example, the set of red, green, and blue sensor tristimulus values ($R_S = F_1$, $G_S = F_2$, $B_S = F_3$) may be interchanged according to the relation

$$\begin{bmatrix} R_D \\ G_D \\ B_D \end{bmatrix} = \begin{bmatrix} 0 & 1 & 0 \\ 0 & 0 & 1 \\ 1 & 0 & 0 \end{bmatrix} \begin{bmatrix} R_S \\ G_S \\ B_S \end{bmatrix} \tag{10.5-3}$$

Green objects in the original will appear red in the display, blue objects will appear green, and red objects will appear blue. A general linear false color mapping of natural color images can be defined as

$$\begin{bmatrix} R_D \\ G_D \\ B_D \end{bmatrix} = \begin{bmatrix} m_{11} & m_{12} & m_{13} \\ m_{21} & m_{22} & m_{21} \\ m_{23} & m_{32} & m_{33} \end{bmatrix} \begin{bmatrix} R_S \\ G_S \\ B_S \end{bmatrix} \tag{10.5-4}$$

This color mapping should be recognized as a linear coordinate conversion of colors reproduced by the primaries of the original image to a new set of primaries. Figure 10.5-4 provides examples of false color mappings of a pair of images.

10.6. MULTISPECTRAL IMAGE ENHANCEMENT

Enhancement procedures are often performed on multispectral image bands of a scene in order to accentuate salient features to assist in subsequent human interpretation or machine analysis (35,37). These procedures include individual image band

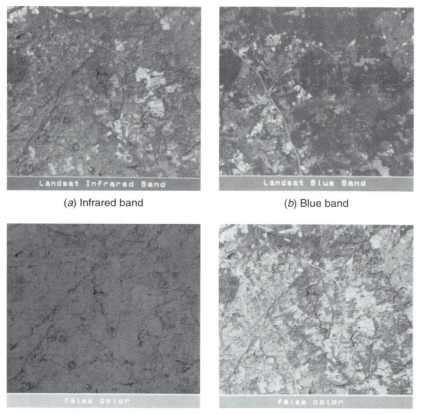

(a) Infrared band   (b) Blue band

(c) R = infrared, G = 0, B = blue   (d) R = infrared, G = 1/2 [infrared + blue], B = blue

FIGURE 10.5-4. False coloring of multispectral images. See insert for a color representation of this figure.

enhancement techniques, such as contrast stretching, noise cleaning, and edge crispening, as described earlier. Other methods, considered in this section, involve the joint processing of multispectral image bands.

Multispectral image bands can be subtracted in pairs according to the relation

$$D_{m,n}(j, k) = F_m(j, k) - F_n(j, k) \qquad (10.6\text{-}1)$$

in order to accentuate reflectivity variations between the multispectral bands. An associated advantage is the removal of any unknown but common bias components that may exist. Another simple but highly effective means of multispectral image enhancement is the formation of ratios of the image bands. The ratio image between the mth and nth multispectral bands is defined as

$$R_{m,n}(j, k) = \frac{F_m(j, k)}{F_n(j, k)} \qquad (10.6\text{-}2)$$

It is assumed that the image bands are adjusted to have nonzero pixel values. In many multispectral imaging systems, the image band $F_n(j, k)$ can be modeled by the product of an object reflectivity function $R_n(j, k)$ and an illumination function $I(j, k)$ that is identical for all multispectral bands. Ratioing of such imagery provides an automatic compensation of the illumination factor. The ratio $F_m(j, k)/[F_n(j, k) \pm \Delta(j, k)]$, for which $\Delta(j, k)$ represents a quantization level uncertainty, can vary considerably if $F_n(j, k)$ is small. This variation can be reduced significantly by forming the logarithm of the ratios defined by (24)

$$L_{m,n}(j, k) = \log\{R_{m,n}(j, k)\} = \log\{F_m(j, k)\} - \log\{F_n(j, k)\} \qquad (10.6\text{-}3)$$

There are a total of $N(N-1)$ different difference or ratio pairs that may be formed from N multispectral bands. To reduce the number of combinations to be considered, the differences or ratios are often formed with respect to an average image field:

$$A(j, k) = \frac{1}{N} \sum_{n=1}^{N} F_n(j, k) \qquad (10.6\text{-}4)$$

Unitary transforms between multispectral planes have also been employed as a means of enhancement. For N image bands, a $N \times 1$ vector

$$\mathbf{x} = \begin{bmatrix} F_1(j, k) \\ F_2(j, k) \\ \cdot \\ \cdot \\ \cdot \\ F_N(j, k) \end{bmatrix} \qquad (10.6\text{-}5)$$

is formed at each coordinate (j, k). Then, a transformation

$$\mathbf{y} = \mathbf{A}\mathbf{x} \qquad (10.6\text{-}6)$$

is formed where $\mathbf{A}$ is a $N \times N$ unitary matrix. A common transformation is the principal components decomposition, described in Section 5.8, in which the rows of the matrix $\mathbf{A}$ are composed of the eigenvectors of the covariance matrix $\mathbf{K_x}$ between the bands. The matrix $\mathbf{A}$ performs a diagonalization of the covariance matrix $\mathbf{K_x}$ such that the covariance matrix of the transformed imagery bands

$$\mathbf{K_y} = \mathbf{AK_x A}^T = \Lambda \tag{10.6-7}$$

is a diagonal matrix Λ whose elements are the eigenvalues of $\mathbf{K_x}$ arranged in descending value. The principal components decomposition, therefore, results in a set of decorrelated data arrays whose energies are ranged in amplitude. This process, of course, requires knowledge of the covariance matrix between the multispectral bands. The covariance matrix must be either modeled, estimated, or measured. If the covariance matrix is highly nonstationary, the principal components method becomes difficult to utilize.

Figure 10.6-1 contains a set of four multispectral images, and Figure 10.6-2 exhibits their corresponding log ratios (37). Principal components bands of these multispectral images are illustrated in Figure 10.6-3 (37).

(*a*) Band 4 (green)   (*b*) Band 5 (red)

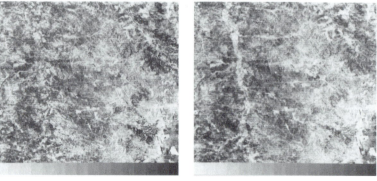

(*c*) Band 6 (infrared 1)   (*d*) Band 7 (infrared 2)

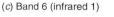

FIGURE 10.6-1. Multispectral images.

(a) $\dfrac{\text{Band 4}}{\text{Band 5}}$

(b) $\dfrac{\text{Band 4}}{\text{Band 6}}$

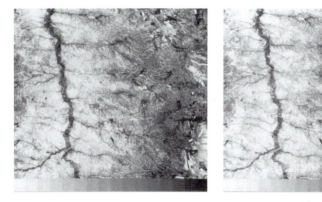

(c) $\dfrac{\text{Band 4}}{\text{Band 7}}$

(d) $\dfrac{\text{Band 5}}{\text{Band 6}}$

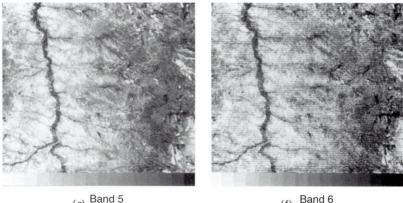

(e) $\dfrac{\text{Band 5}}{\text{Band 7}}$

(f) $\dfrac{\text{Band 6}}{\text{Band 7}}$

FIGURE 10.6-2. Logarithmic ratios of multispectral images.

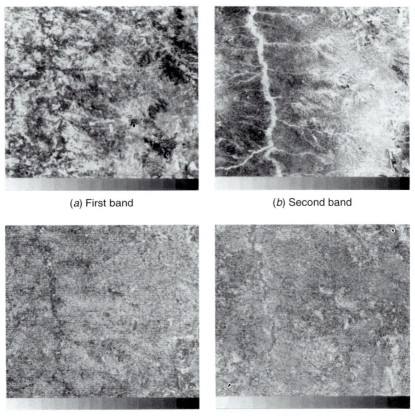

(a) First band   (b) Second band

(c) Third band   (d) Fourth band

FIGURE 10.6-3. Principal components of multispectral images.

REFERENCES

1. R. Nathan, "Picture Enhancement for the Moon, Mars, and Man," in *Pictorial Pattern Recognition*, G. C. Cheng, ed., Thompson, Washington DC, 1968, 239–235.

2. F. Billingsley, "Applications of Digital Image Processing," *Applied Optics*, **9**, 2, February 1970, 289–299.

3. H. C. Andrews, A. G. Tescher, and R. P. Kruger, "Image Processing by Digital Computer," *IEEE Spectrum*, **9**, 7, July 1972, 20–32.

4. E. L. Hall et al., "A Survey of Preprocessing and Feature Extraction Techniques for Radiographic Images," *IEEE Trans. Computers*, **C-20**, 9, September 1971, 1032–1044.

5. E. L. Hall, "Almost Uniform Distribution for Computer Image Enhancement," *IEEE Trans. Computers*, **C-23**, 2, February 1974, 207–208.

6. W. Frei, "Image Enhancement by Histogram Hyperbolization," *Computer Graphics and Image Processing*, **6**, 3, June 1977, 286–294.

7. D. J. Ketcham, "Real Time Image Enhancement Technique," *Proc. SPIE/OSA Conference on Image Processing*, Pacific Grove, CA, **74**, February 1976, 120–125.

8. R. A. Hummel, "Image Enhancement by Histogram Transformation," *Computer Graphics and Image Processing*, **6**, 2, 1977, 184–195.

9. S. M. Pizer et al., "Adaptive Histogram Equalization and Its Variations," *Computer Vision, Graphics, and Image Processing*. **39**, 3, September 1987, 355–368.

10. G. P. Dineen, "Programming Pattern Recognition," *Proc. Western Joint Computer Conference*, March 1955, 94–100.

11. R. E. Graham, "Snow Removal: A Noise Stripping Process for Picture Signals," *IRE Trans. Information Theory*, **IT-8**, 1, February 1962, 129–144.

12. A. Rosenfeld, C. M. Park, and J. P. Strong, "Noise Cleaning in Digital Pictures," *Proc. EASCON Convention Record*, October 1969, 264–273.

13. R. Nathan, "Spatial Frequency Filtering," in *Picture Processing and Psychopictorics*, B. S. Lipkin and A. Rosenfeld, Eds., Academic Press, New York, 1970, 151–164.

14. A. V. Oppenheim, R. W. Schaefer, and T. G. Stockham, Jr., "Nonlinear Filtering of Multiplied and Convolved Signals," *Proc. IEEE*, **56**, 8, August 1968, 1264–1291.

15. G. A. Mastin, "Adaptive Filters for Digital Image Noise Smoothing: An Evaluation," *Computer Vision, Graphics, and Image Processing*, **31**, 1, July 1985, 103–121.

16. L. S. Davis and A. Rosenfeld, "Noise Cleaning by Iterated Local Averaging," *IEEE Trans. Systems, Man and Cybernetics*, **SMC-7**, 1978, 705–710.

17. J. W. Tukey, *Exploratory Data Analysis*, Addison-Wesley, Reading, MA, 1971.

18. T. A. Nodes and N. C. Gallagher, Jr., "Median Filters: Some Manipulations and Their Properties," *IEEE Trans. Acoustics, Speech, and Signal Processing*, **ASSP-30**, 5, October 1982, 739–746.

19. T. S. Huang, G. J. Yang, and G. Y. Tang, "A Fast Two-Dimensional Median Filtering Algorithm," *IEEE Trans. Acoustics, Speech, and Signal Processing*, **ASSP-27**, 1, February 1979, 13–18.

20. J. T. Astola and T. G. Campbell, "On Computation of the Running Median," *IEEE Trans. Acoustics, Speech, and Signal Processing*, **37**, 4, April 1989, 572–574.

21. W. K. Pratt, T. J. Cooper, and I. Kabir, "Pseudomedian Filter," *Proc. SPIE Conference*, Los Angeles, January 1984.

22. J. S. Walker, *A Primer on Wavelets and Their Scientific Applications*, Chapman & Hall CRC Press, Boca Raton, FL, 1999.

23. S. Mallat, *A Wavelet Tour of Signal Processing*, Academic Press, New York, 1998.

24. L. G. Roberts, "Machine Perception of Three-Dimensional Solids," in *Optical and Electro-Optical Information Processing*, J. T. Tippett et al., Eds., MIT Press, Cambridge, MA, 1965.

25. J. M. S. Prewitt, "Object Enhancement and Extraction," in *Picture Processing and Psychopictorics*, B. S. Lipkin and A. Rosenfeld, eds., Academic Press, New York, 1970, 75–150.

26. A. Arcese, P. H. Mengert, and E. W. Trombini, "Image Detection Through Bipolar Correlation," *IEEE Trans. Information Theory*, **IT-16**, 5, September 1970, 534–541.

27. W. F. Schreiber, "Wirephoto Quality Improvement by Unsharp Masking," *J. Pattern Recognition*, **2**, 1970, 111–121.

28. J-S. Lee, "Digital Image Enhancement and Noise Filtering by Use of Local Statistics," *IEEE Trans. Pattern Analysis and Machine Intelligence*, **PAMI-2**, 2, March 1980, 165–168.

29. A. Rosenfeld, *Picture Processing by Computer*, Academic Press, New York, 1969.

30. R. H. Wallis, "An Approach for the Space Variant Restoration and Enhancement of Images," *Proc. Symposium on Current Mathematical Problems in Image Science*, Monterey, CA, November 1976.

31. O. D. Faugeras, "Digital Color Image Processing Within the Framework of a Human Visual Model," *IEEE Trans. Acoustics, Speech, and Signal Processing*, **ASSP-27**, 4, August 1979, 380–393.

32. C. Gazley, J. E. Reibert, and R. H. Stratton, "Computer Works a New Trick in Seeing Pseudo Color Processing," *Aeronautics and Astronautics*, **4**, April 1967, 56.

33. L. W. Nichols and J. Lamar, "Conversion of Infrared Images to Visible in Color," *Applied Optics*, **7**, 9, September 1968, 1757.

34. E. R. Kreins and L. J. Allison, "Color Enhancement of Nimbus High Resolution Infrared Radiometer Data," *Applied Optics*, **9**, 3, March 1970, 681.

35. A. F. H. Goetz et al., "Application of ERTS Images and Image Processing to Regional Geologic Problems and Geologic Mapping in Northern Arizona," Technical Report 32-1597, Jet Propulsion Laboratory, Pasadena, CA, May 1975.

36. W. Find, "Image Coloration as an Interpretation Aid," *Proc. SPIE/OSA Conference on Image Processing*, Pacific Grove, CA, February 1976, 74, 209–215.

37. G. S. Robinson and W. Frei, "Final Research Report on Computer Processing of ERTS Images," Report USCIPI 640, University of Southern California, Image Processing Institute, Los Angeles, September 1975.

11

IMAGE RESTORATION MODELS

Image restoration may be viewed as an estimation process in which operations are performed on an observed or measured image field to estimate the ideal image field that would be observed if no image degradation were present in an imaging system. Mathematical models are described in this chapter for image degradation in general classes of imaging systems. These models are then utilized in subsequent chapters as a basis for the development of image restoration techniques.

11.1. GENERAL IMAGE RESTORATION MODELS

In order effectively to design a digital image restoration system, it is necessary quantitatively to characterize the image degradation effects of the physical imaging system, the image digitizer, and the image display. Basically, the procedure is to model the image degradation effects and then perform operations to undo the model to obtain a restored image. It should be emphasized that accurate image modeling is often the key to effective image restoration. There are two basic approaches to the modeling of image degradation effects: a priori modeling and a posteriori modeling. In the former case, measurements are made on the physical imaging system, digitizer, and display to determine their response for an arbitrary image field. In some instances it will be possible to model the system response deterministically, while in other situations it will only be possible to determine the system response in a stochastic sense. The a posteriori modeling approach is to develop the model for the image degradations based on measurements of a particular image to be restored. Basically, these two approaches differ only in the manner in which information is gathered to describe the character of the image degradation.

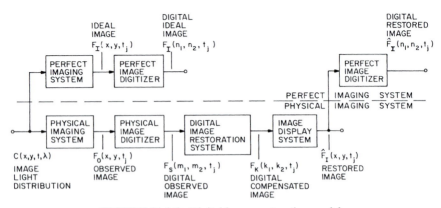

FIGURE 11.1-1. Digital image restoration model.

Figure 11.1-1 shows a general model of a digital imaging system and restoration process. In the model, a continuous image light distribution $C(x, y, t, \lambda)$ dependent on spatial coordinates (x, y), time (t), and spectral wavelength (λ) is assumed to exist as the driving force of a physical imaging system subject to point and spatial degradation effects and corrupted by deterministic and stochastic disturbances. Potential degradations include diffraction in the optical system, sensor nonlineari-ties, optical system aberrations, film nonlinearities, atmospheric turbulence effects, image motion blur, and geometric distortion. Noise disturbances may be caused by electronic imaging sensors or film granularity. In this model, the physical imaging system produces a set of output image fields $F_O^{(i)}(x, y, t_j)$ at time instant t_j described by the general relation

$$F_O^{(i)}(x, y, t_j) = O_P\{C(x, y, t, \lambda)\} \qquad (11.1-1)$$

where $O_P\{\cdot\}$ represents a general operator that is dependent on the space coordi-nates (x, y), the time history (t), the wavelength (λ), and the amplitude of the light distribution (C). For a monochrome imaging system, there will only be a single out-put field, while for a natural color imaging system, $F_O^{(i)}(x, y, t_j)$ may denote the red, green, and blue tristimulus bands for $i = 1, 2, 3$, respectively. Multispectral imagery may also involve several output bands of data.

In the general model of Figure 11.1-1, each observed image field $F_O^{(i)}(x, y, t_j)$ is digitized, following the techniques outlined in Part 3, to produce an array of image samples $F_S^{(i)}(m_1, m_2, t_j)$ at each time instant t_j. The output samples of the digitizer are related to the input observed field by

$$F_S^{(i)}(m_1, m_2, t_j) = O_G\{F_O^{(i)}(x, y, t_j)\} \qquad (11.1-2)$$

where $O_G\{\cdot\}$ is an operator modeling the image digitization process.

A digital image restoration system that follows produces an output array $F_K^{(i)}(k_1, k_2, t_j)$ by the transformation

$$F_K^{(i)}(k_1, k_2, t_j) = O_R\{F_S^{(i)}(m_1, m_2, t_j)\} \qquad (11.1\text{-}3)$$

where $O_R\{\cdot\}$ represents the designed restoration operator. Next, the output samples of the digital restoration system are interpolated by the image display system to produce a continuous image estimate $\hat{F}_I^{(i)}(x, y, t_j)$. This operation is governed by the relation

$$\hat{F}_I^{(i)}(x, y, t_j) = O_D\{F_K^{(i)}(k_1, k_2, t_j)\} \qquad (11.1\text{-}4)$$

where $O_D\{\cdot\}$ models the display transformation.

The function of the digital image restoration system is to compensate for degradations of the physical imaging system, the digitizer, and the image display system to produce an estimate of a hypothetical ideal image field $F_I^{(i)}(x, y, t_j)$ that would be displayed if all physical elements were perfect. The perfect imaging system would produce an ideal image field modeled by

$$F_I^{(i)}(x, y, t_j) = O_I\left\{\int_0^\infty \int_{t_j - T}^{t_j} C(x, y, t, \lambda) U_i(t, \lambda)\, dt\, d\lambda\right\} \qquad (11.1\text{-}5)$$

where $U_i(t, \lambda)$ is a desired temporal and spectral response function, T is the observation period, and $O_I\{\cdot\}$ is a desired point and spatial response function.

Usually, it will not be possible to restore perfectly the observed image such that the output image field is identical to the ideal image field. The design objective of the image restoration processor is to minimize some error measure between $F_I^{(i)}(x, y, t_j)$ and $\hat{F}_I^{(i)}(x, y, t_j)$. The discussion here is limited, for the most part, to a consideration of techniques that minimize the mean-square error between the ideal and estimated image fields as defined by

$$\mathcal{E}_i = E\left\{[F_I^{(i)}(x, y, t_j) - \hat{F}_I^{(i)}(x, y, t_j)]^2\right\} \qquad (11.1\text{-}6)$$

where $E\{\cdot\}$ denotes the expectation operator. Often, it will be desirable to place side constraints on the error minimization, for example, to require that the image estimate be strictly positive if it is to represent light intensities that are positive.

Because the restoration process is to be performed digitally, it is often more convenient to restrict the error measure to discrete points on the ideal and estimated image fields. These discrete arrays are obtained by mathematical models of perfect image digitizers that produce the arrays

$$F_I^{(i)}(n_1, n_2, t_j) = F_I^{(i)}(x, y, t_j)\delta(x - n_1\Delta, y - n_2\Delta) \qquad (11.1\text{-}7a)$$

$$\hat{F}_I^{(i)}(n_1, n_2, t_j) = \hat{F}_I^{(i)}(x, y, t_j)\delta(x - n_1\Delta, y - n_2\Delta) \qquad (11.1\text{-}7b)$$

It is assumed that continuous image fields are sampled at a spatial period Δ satisfying the Nyquist criterion. Also, quantization error is assumed negligible. It should be noted that the processes indicated by the blocks of Figure 11.1-1 above the dashed division line represent mathematical modeling and are not physical operations performed on physical image fields and arrays. With this discretization of the continuous ideal and estimated image fields, the corresponding mean-square restoration error becomes

$$\mathcal{E}_i = E\left\{ [F_I^{(i)}(n_1, n_2, t_j) - \hat{F}_I^{(i)}(n_1, n_2, t_j)]^2 \right\} \qquad (11.1\text{-}8)$$

With the relationships of Figure 11.1-1 quantitatively established, the restoration problem may be formulated as follows:

Given the sampled observation $F_S^{(i)}(m_1, m_2, t_j)$ expressed in terms of the image light distribution $C(x, y, t, \lambda)$, determine the transfer function $O_K\{\cdot\}$ that minimizes the error measure between $F_I^{(i)}(x, y, t_j)$ and $\hat{F}_I^{(i)}(x, y, t_j)$ subject to desired constraints.

There are no general solutions for the restoration problem as formulated above because of the complexity of the physical imaging system. To proceed further, it is necessary to be more specific about the type of degradation and the method of restoration. The following sections describe models for the elements of the generalized imaging system of Figure 11.1-1.

11.2. OPTICAL SYSTEMS MODELS

One of the major advances in the field of optics during the past 40 years has been the application of system concepts to optical imaging. Imaging devices consisting of lenses, mirrors, prisms, and so on, can be considered to provide a deterministic transformation of an input spatial light distribution to some output spatial light distribution. Also, the system concept can be extended to encompass the spatial propagation of light through free space or some dielectric medium.

In the study of geometric optics, it is assumed that light rays always travel in a straight-line path in a homogeneous medium. By this assumption, a bundle of rays passing through a clear aperture onto a screen produces a geometric light projection of the aperture. However, if the light distribution at the region between the light and

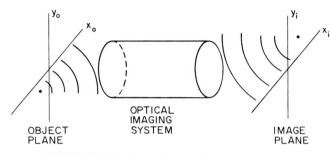

FIGURE 11.2-1. Generalized optical imaging system.

dark areas on the screen is examined in detail, it is found that the boundary is not sharp. This effect is more pronounced as the aperture size is decreased. For a pinhole aperture, the entire screen appears diffusely illuminated. From a simplistic viewpoint, the aperture causes a bending of rays called *diffraction*. Diffraction of light can be quantitatively characterized by considering light as electromagnetic radiation that satisfies Maxwell's equations. The formulation of a complete theory of optical imaging from the basic electromagnetic principles of diffraction theory is a complex and lengthy task. In the following, only the key points of the formulation are presented; details may be found in References 1 to 3.

Figure 11.2-1 is a diagram of a generalized optical imaging system. A point in the object plane at coordinate (x_o, y_o) of intensity $I_o(x_o, y_o)$ radiates energy toward an imaging system characterized by an entrance pupil, exit pupil, and intervening system transformation. Electromagnetic waves emanating from the optical system are focused to a point (x_i, y_i) on the image plane producing an intensity $I_i(x_i, y_i)$. The imaging system is said to be *diffraction limited* if the light distribution at the image plane produced by a point-source object consists of a converging spherical wave whose extent is limited only by the exit pupil. If the wavefront of the electromagnetic radiation emanating from the exit pupil is not spherical, the optical system is said to possess *aberrations*.

In most optical image formation systems, the optical radiation emitted by an object arises from light transmitted or reflected from an incoherent light source. The image radiation can often be regarded as quasimonochromatic in the sense that the spectral bandwidth of the image radiation detected at the image plane is small with respect to the center wavelength of the radiation. Under these joint assumptions, the imaging system of Figure 11.2-1 will respond as a linear system in terms of the intensity of its input and output fields. The relationship between the image intensity and object intensity for the optical system can then be represented by the superposition integral equation

$$I_i(x_i, y_i) = \int_{-\infty}^{\infty} \int_{-\infty}^{\infty} H(x_i, y_i; x_o, y_o) I_o(x_o, y_o) \, dx_o \, dy_o \qquad (11.2\text{-}1)$$

where $H(x_i, y_i; x_o, y_o)$ represents the image intensity response to a point source of light. Often, the intensity impulse response is space invariant and the input–output relationship is given by the convolution equation

$$I_i(x_i, y_i) = \int_{-\infty}^{\infty} \int_{-\infty}^{\infty} H(x_i - x_o, y_i - y_o) I_o(x_o, y_o) \, dx_o \, dy_o \qquad (11.2\text{-}2)$$

In this case, the normalized Fourier transforms

$$\mathcal{I}_o(\omega_x, \omega_y) = \frac{\int_{-\infty}^{\infty} \int_{-\infty}^{\infty} I_o(x_o, y_o) \exp\{-i(\omega_x x_o + \omega_y y_o)\} \, dx_o \, dy_o}{\int_{-\infty}^{\infty} \int_{-\infty}^{\infty} I_o(x_o, y_o) \, dx_o \, dy_o} \qquad (11.2\text{-}3a)$$

$$\mathcal{I}_i(\omega_x, \omega_y) = \frac{\int_{-\infty}^{\infty} \int_{-\infty}^{\infty} I_i(x_i, y_i) \exp\{-i(\omega_x x_i + \omega_y y_i)\} \, dx_i \, dy_i}{\int_{-\infty}^{\infty} \int_{-\infty}^{\infty} I_i(x_i, y_i) \, dx_i \, dy_i} \qquad (11.2\text{-}3b)$$

of the object and image intensity fields are related by

$$\mathcal{I}_o(\omega_x, \omega_y) = \mathcal{H}(\omega_x, \omega_y) \mathcal{I}_i(\omega_x, \omega_y) \qquad (11.2\text{-}4)$$

where $\mathcal{H}(\omega_x, \omega_y)$, which is called the *optical transfer function* (OTF), is defined by

$$\mathcal{H}(\omega_x, \omega_y) = \frac{\int_{-\infty}^{\infty} \int_{-\infty}^{\infty} H(x, y) \exp\{-i(\omega_x x + \omega_y y)\} \, dx \, dy}{\int_{-\infty}^{\infty} \int_{-\infty}^{\infty} H(x, y) \, dx \, dy} \qquad (11.2\text{-}5)$$

The absolute value $|\mathcal{H}(\omega_x, \omega_y)|$ of the OTF is known as the *modulation transfer function* (MTF) of the optical system.

The most common optical image formation system is a circular thin lens. Figure 11.2-2 illustrates the OTF for such a lens as a function of its degree of misfocus (1, p. 486; 4). For extreme misfocus, the OTF will actually become negative at some spatial frequencies. In this state, the lens will cause a contrast reversal: Dark objects will appear light, and vice versa.

Earth's atmosphere acts as an imaging system for optical radiation transversing a path through the atmosphere. Normally, the index of refraction of the atmosphere remains relatively constant over the optical extent of an object, but in some instances atmospheric turbulence can produce a spatially variable index of

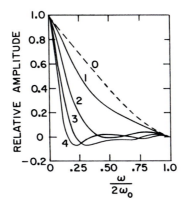

FIGURE 11.2-2. Cross section of transfer function of a lens. Numbers indicate degree of misfocus.

refraction that leads to an effective blurring of any imaged object. An equivalent impulse response

$$H(x, y) = K_1 \exp\left\{-(K_2 x^2 + K_3 y^2)^{5/6}\right\}$$

(11.2-6)

where the K_n are constants, has been predicted and verified mathematically by experimentation (5) for long-exposure image formation. For convenience in analysis, the function 5/6 is often replaced by unity to obtain a Gaussian-shaped impulse response model of the form

$$H(x, y) = K \exp\left\{-\left(\frac{x^2}{2b_x^2} + \frac{y^2}{2b_y^2}\right)\right\}$$

(11.2-7)

where K is an amplitude scaling constant and b_x and b_y are blur-spread factors.

Under the assumption that the impulse response of a physical imaging system is independent of spectral wavelength and time, the observed image field can be modeled by the superposition integral equation

$$F_O^{(i)}(x, y, t_j) = O_C\left\{\int_{-\infty}^{\infty}\int_{-\infty}^{\infty} C(\alpha, \beta, t, \lambda)H(x, y; \alpha, \beta)\, d\alpha\, d\beta\right\}$$

(11.2-8)

where $O_C\{\cdot\}$ is an operator that models the spectral and temporal characteristics of the physical imaging system. If the impulse response is spatially invariant, the model reduces to the convolution integral equation

$$F_O^{(i)}(x, y, t_j) = O_C\left\{\int_{-\infty}^{\infty}\int_{-\infty}^{\infty} C(\alpha, \beta, t, \lambda)H(x - \alpha, y - \beta)\, d\alpha\, d\beta\right\} \qquad (11.2\text{-}9)$$

11.3. PHOTOGRAPHIC PROCESS MODELS

There are many different types of materials and chemical processes that have been utilized for photographic image recording. No attempt is made here either to survey the field of photography or to deeply investigate the physics of photography. References 6 to 8 contain such discussions. Rather, the attempt here is to develop mathematical models of the photographic process in order to characterize quantitatively the photographic components of an imaging system.

11.3.1. Monochromatic Photography

The most common material for photographic image recording is silver halide emulsion, depicted in Figure 11.3-1. In this material, silver halide grains are suspended in a transparent layer of gelatin that is deposited on a glass, acetate, or paper backing. If the backing is transparent, a transparency can be produced, and if the backing is a white paper, a reflection print can be obtained. When light strikes a grain, an electrochemical conversion process occurs, and part of the grain is converted to metallic silver. A development center is then said to exist in the grain. In the development process, a chemical developing agent causes grains with partial silver content to be converted entirely to metallic silver. Next, the film is fixed by chemically removing unexposed grains.

The photographic process described above is called a *non reversal process*. It produces a negative image in the sense that the silver density is inversely proportional to the exposing light. A positive reflection print of an image can be obtained in a two-stage process with nonreversal materials. First, a negative transparency is produced, and then the negative transparency is illuminated to expose negative reflection print paper. The resulting silver density on the developed paper is then proportional to the light intensity that exposed the negative transparency.

A positive transparency of an image can be obtained with a reversal type of film. This film is exposed and undergoes a first development similar to that of a nonreversal film. At this stage in the photographic process, all grains that have been exposed

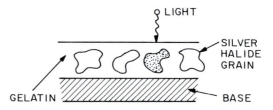

FIGURE 11.3-1. Cross section of silver halide emulsion.

to light are converted completely to metallic silver. In the next step, the metallic silver grains are chemically removed. The film is then uniformly exposed to light, or alternatively, a chemical process is performed to expose the remaining silver halide grains. Then the exposed grains are developed and fixed to produce a positive transparency whose density is proportional to the original light exposure.

The relationships between light intensity exposing a film and the density of silver grains in a transparency or print can be described quantitatively by sensitometric measurements. Through sensitometry, a model is sought that will predict the spectral light distribution passing through an illuminated transparency or reflected from a print as a function of the spectral light distribution of the exposing light and certain physical parameters of the photographic process. The first stage of the photographic process, that of exposing the silver halide grains, can be modeled to a first-order approximation by the integral equation

$$X(C) = k_x \int C(\lambda)L(\lambda)\, d\lambda \qquad (11.3\text{-}1)$$

where $X(C)$ is the integrated exposure, $C(\lambda)$ represents the spectral energy distribution of the exposing light, $L(\lambda)$ denotes the spectral sensitivity of the film or paper plus any spectral losses resulting from filters or optical elements, and k_x is an exposure constant that is controllable by an aperture or exposure time setting. Equation 11.3-1 assumes a fixed exposure time. Ideally, if the exposure time were to be increased by a certain factor, the exposure would be increased by the same factor. Unfortunately, this relationship does not hold exactly. The departure from linearity is called a *reciprocity failure* of the film. Another anomaly in exposure prediction is the intermittency effect, in which the exposures for a constant intensity light and for an intermittently flashed light differ even though the incident energy is the same for both sources. Thus, if Eq. 11.3-1 is to be utilized as an exposure model, it is necessary to observe its limitations: The equation is strictly valid only for a fixed exposure time and constant-intensity illumination.

The transmittance $\tau(\lambda)$ of a developed reversal or non-reversal transparency as a function of wavelength can be ideally related to the density of silver grains by the exponential law of absorption as given by

$$\tau(\lambda) = \exp\{-d_e D(\lambda)\} \qquad (11.3\text{-}2)$$

where $D(\lambda)$ represents the characteristic density as a function of wavelength for a reference exposure value, and d_e is a variable proportional to the actual exposure. For monochrome transparencies, the characteristic density function $D(\lambda)$ is reasonably constant over the visible region. As Eq. 11.3-2 indicates, high silver densities result in low transmittances, and vice versa. It is common practice to change the proportionality constant of Eq. 11.3-2 so that measurements are made in exponent ten units. Thus, the transparency transmittance can be equivalently written as

$$\tau(\lambda) = 10^{-d_x D(\lambda)} \tag{11.3-3}$$

where d_x is the density variable, inversely proportional to exposure, for exponent 10 units. From Eq. 11.3-3, it is seen that the photographic density is logarithmically related to the transmittance. Thus,

$$d_x D(\lambda) = -\log_{10} \tau(\lambda) \tag{11.3-4}$$

The reflectivity $r_o(\lambda)$ of a photographic print as a function of wavelength is also inversely proportional to its silver density, and follows the exponential law of absorption of Eq. 11.3-2. Thus, from Eqs. 11.3-3 and 11.3-4, one obtains directly

$$r_o(\lambda) = 10^{-d_x D(\lambda)} \tag{11.3-5}$$

$$d_x D(\lambda) = -\log_{10} r_o(\lambda) \tag{11.3-6}$$

where d_x is an appropriately evaluated variable proportional to the exposure of the photographic paper.

The relational model between photographic density and transmittance or reflectivity is straightforward and reasonably accurate. The major problem is the next step of modeling the relationship between the exposure $X(C)$ and the density variable d_x. Figure 11.3-2a shows a typical curve of the transmittance of a nonreversal transparency

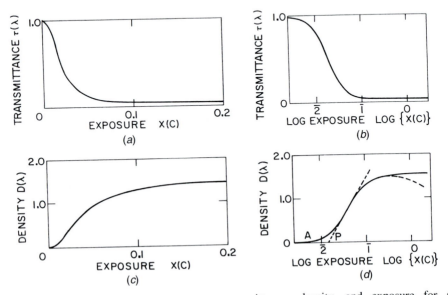

FIGURE 11.3-2. Relationships between transmittance, density, and exposure for a nonreversal film.

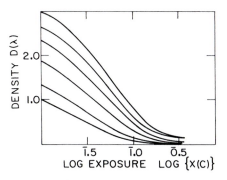

FIGURE 11.3-3. H & D curves for a reversal film as a function of development time.

as a function of exposure. It is to be noted that the curve is highly nonlinear except for a relatively narrow region in the lower exposure range. In Figure 11.3-2b, the curve of Figure 11.3-2a has been replotted as transmittance versus the logarithm of exposure. An approximate linear relationship is found to exist between transmittance and the logarithm of exposure, but operation in this exposure region is usually of little use in imaging systems. The parameter of interest in photography is the photographic density variable d_x, which is plotted as a function of exposure and logarithm of exposure in Figure 11.3-2c and 11.3-2d. The plot of density versus logarithm of exposure is known as the *H & D curve* after Hurter and Driffield, who performed fundamental investigations of the relationships between density and exposure. Figure 11.3-3 is a plot of the H & D curve for a reversal type of film. In Figure 11.3-2d, the central portion of the curve, which is approximately linear, has been approximated by the line defined by

$$d_x = \gamma[\log_{10} X(C) - K_F] \qquad (11.3\text{-}7)$$

where γ represents the slope of the line and K_F denotes the intercept of the line with the log exposure axis. The slope of the curve γ (*gamma,*) is a measure of the contrast of the film, while the factor K_F is a measure of the film speed; that is, a measure of the base exposure required to produce a negative in the linear region of the H & D curve. If the exposure is restricted to the linear portion of the H & D curve, substitution of Eq. 11.3-7 into Eq. 11.3-3 yields a transmittance function

$$\tau(\lambda) = K_\tau(\lambda)[X(C)]^{-\gamma D(\lambda)} \qquad (11.3\text{-}8a)$$

where

$$K_\tau(\lambda) \equiv 10^{\gamma K_F D(\lambda)} \qquad (11.3\text{-}8b)$$

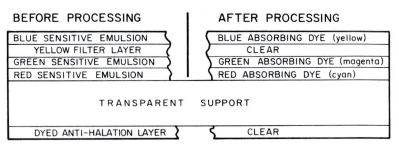

FIGURE 11.3-4. Color film integral tripack.

With the exposure model of Eq. 11.3-1, the transmittance or reflection models of Eqs. 11.3-3 and 11.3-5, and the H & D curve, or its linearized model of Eq. 11.3-7, it is possible mathematically to model the monochrome photographic process.

11.3.2. Color Photography

Modern color photography systems utilize an integral tripack film, as illustrated in Figure 11.3-4, to produce positive or negative transparencies. In a cross section of this film, the first layer is a silver halide emulsion sensitive to blue light. A yellow filter following the blue emulsion prevents blue light from passing through to the green and red silver emulsions that follow in consecutive layers and are naturally sensitive to blue light. A transparent base supports the emulsion layers. Upon development, the blue emulsion layer is converted into a yellow dye transparency whose dye concentration is proportional to the blue exposure for a negative transparency and inversely proportional for a positive transparency. Similarly, the green and blue emulsion layers become magenta and cyan dye layers, respectively. Color prints can be obtained by a variety of processes (7). The most common technique is to produce a positive print from a color negative transparency onto nonreversal color paper.

In the establishment of a mathematical model of the color photographic process, each emulsion layer can be considered to react to light as does an emulsion layer of a monochrome photographic material. To a first approximation, this assumption is correct. However, there are often significant interactions between the emulsion and dye layers, Each emulsion layer possesses a characteristic sensitivity, as shown by the typical curves of Figure 11.3-5. The integrated exposures of the layers are given by

$$X_R(C) = d_R \int C(\lambda) L_R(\lambda)\, d\lambda \tag{11.3-9a}$$

$$X_G(C) = d_G \int C(\lambda) L_G(\lambda)\, d\lambda \tag{11.3-9b}$$

$$X_B(C) = d_B \int C(\lambda) L_B(\lambda)\, d\lambda \tag{11.3-9c}$$

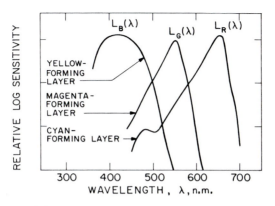

FIGURE 11.3-5. Spectral sensitivities of typical film layer emulsions.

where d_R, d_G, d_B are proportionality constants whose values are adjusted so that the exposures are equal for a reference white illumination and so that the film is not saturated. In the chemical development process of the film, a positive transparency is produced with three absorptive dye layers of cyan, magenta, and yellow dyes.

The transmittance $\tau_T(\lambda)$ of the developed transparency is the product of the transmittance of the cyan $\tau_{TC}(\lambda)$, the magenta $\tau_{TM}(\lambda)$, and the yellow $\tau_{TY}(\lambda)$ dyes. Hence,

$$\tau_T(\lambda) = \tau_{TC}(\lambda)\tau_{TM}(\lambda)\tau_{TY}(\lambda) \tag{11.3-10}$$

The transmittance of each dye is a function of its spectral absorption characteristic and its concentration. This functional dependence is conveniently expressed in terms of the relative density of each dye as

$$\tau_{TC}(\lambda) = 10^{-cD_{NC}(\lambda)} \tag{11.3-11a}$$

$$\tau_{TM}(\lambda) = 10^{-mD_{NM}(\lambda)} \tag{11.3-11b}$$

$$\tau_{TY}(\lambda) = 10^{-yD_{NY}(\lambda)} \tag{11.3-11c}$$

where c, m, y represent the relative amounts of the cyan, magenta, and yellow dyes, and $D_{NC}(\lambda)$, $D_{NM}(\lambda)$, $D_{NY}(\lambda)$ denote the spectral densities of unit amounts of the dyes. For unit amounts of the dyes, the transparency transmittance is

$$\tau_{TN}(\lambda) = 10^{-D_{TN}(\lambda)} \tag{11.3-12a}$$

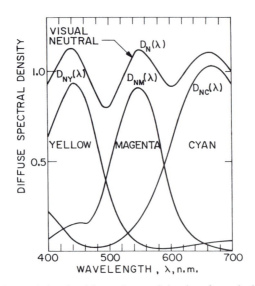

FIGURE 11.3-6. Spectral dye densities and neutral density of a typical reversal color film.

where

$$D_{TN}(\lambda) = D_{NC}(\lambda) + D_{NM}(\lambda) + D_{NY}(\lambda) \qquad (11.3\text{-}12b)$$

Such a transparency appears to be a neutral gray when illuminated by a reference white light. Figure 11.3-6 illustrates the typical dye densities and neutral density for a reversal film.

The relationship between the exposure values and dye layer densities is, in general, quite complex. For example, the amount of cyan dye produced is a nonlinear function not only of the red exposure, but is also dependent to a smaller extent on the green and blue exposures. Similar relationships hold for the amounts of magenta and yellow dyes produced by their exposures. Often, these *interimage effects* can be neglected, and it can be assumed that the cyan dye is produced only by the red exposure, the magenta dye by the green exposure, and the blue dye by the yellow exposure. For this assumption, the dye density–exposure relationship can be characterized by the Hurter–Driffield plot of equivalent neutral density versus the logarithm of exposure for each dye. Figure 11.3-7 shows a typical H & D curve for a reversal film. In the central portion of each H & D curve, the density versus exposure characteristic can be modeled as

$$c = \gamma_C \log_{10} X_R + K_{FC} \qquad (11.3\text{-}13a)$$

$$m = \gamma_M \log_{10} X_G + K_{FM} \qquad (11.3\text{-}13b)$$

$$y = \gamma_Y \log_{10} X_B + K_{FY} \qquad (11.3\text{-}13c)$$

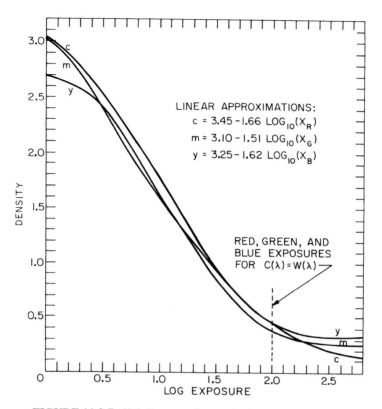

LINEAR APPROXIMATIONS:
$$c = 3.45 - 1.66 \, LOG_{10}(X_R)$$
$$m = 3.10 - 1.51 \, LOG_{10}(X_G)$$
$$y = 3.25 - 1.62 \, LOG_{10}(X_B)$$

RED, GREEN, AND BLUE EXPOSURES FOR $C(\lambda) = W(\lambda)$ →

FIGURE 11.3-7. H & D curves for a typical reversal color film.

where γ_C, γ_M, γ_Y, representing the slopes of the curves in the linear region, are called *dye layer gammas*.

The spectral energy distribution of light passing through a developed transparency is the product of the transparency transmittance and the incident illumination spectral energy distribution $E(\lambda)$ as given by

$$C_T(\lambda) = E(\lambda) 10^{-[cD_{NC}(\lambda) + mD_{NM}(\lambda) + yD_{NY}(\lambda)]} \tag{11.3-14}$$

Figure 11.3-8 is a block diagram of the complete color film recording and reproduction process. The original light with distribution $C(\lambda)$ and the light passing through the transparency $C_T(\lambda)$ at a given resolution element are rarely identical. That is, a spectral match is usually not achieved in the photographic process. Furthermore, the lights C and C_T usually do not even provide a colorimetric match.

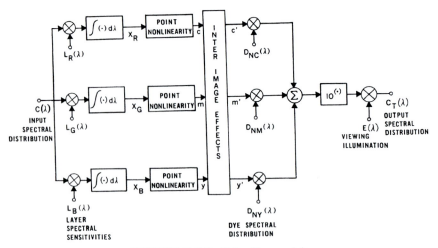

FIGURE 11.3-8. Color film model.

11.4. DISCRETE IMAGE RESTORATION MODELS

This chapter began with an introduction to a general model of an imaging system and a digital restoration process. Next, typical components of the imaging system were described and modeled within the context of the general model. Now, the discussion turns to the development of several discrete image restoration models. In the development of these models, it is assumed that the spectral wavelength response and temporal response characteristics of the physical imaging system can be separated from the spatial and point characteristics. The following discussion considers only spatial and point characteristics.

After each element of the digital image restoration system of Figure 11.1-1 is modeled, following the techniques described previously, the restoration system may be conceptually distilled to three equations:

Observed image:

$$F_S(m_1, m_2) = O_M\{F_I(n_1, n_2), N_1(m_1, m_2), \ldots, N_N(m_1, m_2)\} \qquad (11.4\text{-}1a)$$

Compensated image:

$$F_K(k_1, k_2) = O_R\{F_S(m_1, m_2)\} \qquad (11.4\text{-}1b)$$

Restored image:

$$\hat{F}_I(n_1, n_2) = O_D\{F_K(k_1, k_2)\} \qquad (11.4\text{-}1c)$$

where F_S represents an array of observed image samples, F_I and $\hat{F}_I$ are arrays of ideal image points and estimates, respectively, F_K is an array of compensated image points from the digital restoration system, N_i denotes arrays of noise samples from various system elements, and $O_M\{\cdot\}$, $O_R\{\cdot\}$, $O_D\{\cdot\}$ represent general transfer functions of the imaging system, restoration processor, and display system, respectively. Vector-space equivalents of Eq. 11.4-1 can be formed for purposes of analysis by column scanning of the arrays of Eq. 11.4-1. These relationships are given by

$$\mathbf{f}_S = O_M\{\mathbf{f}_I, \mathbf{n}_1, ..., \mathbf{n}_N\} \tag{11.4-2a}$$

$$\mathbf{f}_K = O_R\{\mathbf{f}_S\} \tag{11.4-2b}$$

$$\hat{\mathbf{f}}_I = O_D\{\mathbf{f}_K\} \tag{11.4-2c}$$

Several estimation approaches to the solution of 11.4-1 or 11.4-2 are described in the following chapters. Unfortunately, general solutions have not been found; recourse must be made to specific solutions for less general models.

The most common digital restoration model is that of Figure 11.4-1a, in which a continuous image field is subjected to a linear blur, the electrical sensor responds nonlinearly to its input intensity, and the sensor amplifier introduces additive Gaussian noise independent of the image field. The physical image digitizer that follows may also introduce an effective blurring of the sampled image as the result of sampling with extended pulses. In this model, display degradation is ignored.

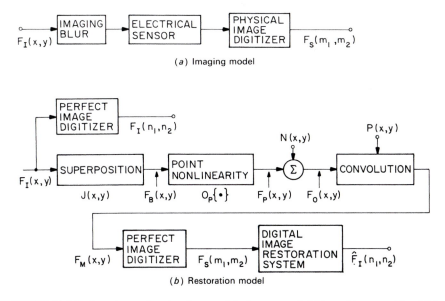

FIGURE 11.4-1. Imaging and restoration models for a sampled blurred image with additive noise.

Figure 11.4-1b shows a restoration model for the imaging system. It is assumed that the imaging blur can be modeled as a superposition operation with an impulse response $J(x, y)$ that may be space variant. The sensor is assumed to respond nonlinearly to the input field $F_B(x, y)$ on a point-by-point basis, and its output is subject to an additive noise field $N(x, y)$. The effect of sampling with extended sampling pulses, which are assumed symmetric, can be modeled as a convolution of $F_O(x, y)$ with each pulse $P(x, y)$ followed by perfect sampling.

The objective of the restoration is to produce an array of samples $\hat{F}_I(n_1, n_2)$ that are estimates of points on the ideal input image field $F_I(x, y)$ obtained by a perfect image digitizer sampling at a spatial period ΔI. To produce a digital restoration model, it is necessary quantitatively to relate the physical image samples $F_S(m_1, m_2)$ to the ideal image points $F_I(n_1, n_2)$ following the techniques outlined in Section 7.2. This is accomplished by truncating the sampling pulse equivalent impulse response $P(x, y)$ to some spatial limits $\pm T_P$, and then extracting points from the continuous observed field $F_O(x, y)$ at a grid spacing ΔP. The discrete representation must then be carried one step further by relating points on the observed image field $F_O(x, y)$ to points on the image field $F_P(x, y)$ and the noise field $N(x, y)$. The final step in the development of the discrete restoration model involves discretization of the superposition operation with $J(x, y)$. There are two potential sources of error in this modeling process: truncation of the impulse responses $J(x, y)$ and $P(x, y)$, and quadrature integration errors. Both sources of error can be made negligibly small by choosing the truncation limits T_B and T_P large and by choosing the quadrature spacings ΔI and ΔP small. This, of course, increases the sizes of the arrays, and eventually, the amount of storage and processing required. Actually, as is subsequently shown, the numerical stability of the restoration estimate may be impaired by improving the accuracy of the discretization process!

The relative dimensions of the various arrays of the restoration model are important. Figure 11.4-2 shows the nested nature of the arrays. The image array observed, $F_O(k_1, k_2)$, is smaller than the ideal image array, $F_I(n_1, n_2)$, by the half-width of the truncated impulse response $J(x, y)$. Similarly, the array of physical sample points $F_S(m_1, m_2)$ is smaller than the array of image points observed, $F_O(k_1, k_2)$, by the half-width of the truncated impulse response $P(x, y)$.

It is convenient to form vector equivalents of the various arrays of the restoration model in order to utilize the formal structure of vector algebra in the subsequent restoration analysis. Again, following the techniques of Section 7.2, the arrays are reindexed so that the first element appears in the upper-left corner of each array. Next, the vector relationships between the stages of the model are obtained by column scanning of the arrays to give

$$\mathbf{f}_S = \mathbf{B}_P \mathbf{f}_O \tag{11.4-3a}$$

$$\mathbf{f}_O = \mathbf{f}_P + \mathbf{n} \tag{11.4-3b}$$

$$\mathbf{f}_P = O_P\{\mathbf{f}_B\} \tag{11.4-3c}$$

$$\mathbf{f}_B = \mathbf{B}_B \mathbf{f}_I \tag{11.4-3d}$$

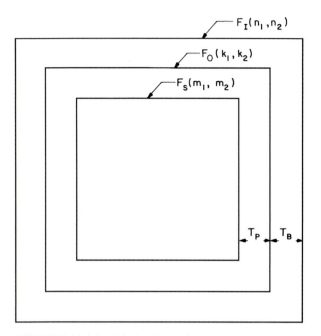

FIGURE 11.4-2. Relationships of sampled image arrays.

where the blur matrix $\mathbf{B}_P$ contains samples of $P(x, y)$ and $\mathbf{B}_B$ contains samples of $J(x, y)$. The nonlinear operation of Eq. 1 1.4-3c is defined as a point-by-point nonlinear transformation. That is,

$$f_P(i) = O_P\{f_B(i)\} \tag{11.4-4}$$

Equations 11.4-3a to 11.4-3d can be combined to yield a single equation for the observed physical image samples in terms of points on the ideal image:

$$\mathbf{f}_S = \mathbf{B}_P O_P\{\mathbf{B}_B \mathbf{f}_I\} + \mathbf{B}_P \mathbf{n} \tag{11.4-5}$$

Several special cases of Eq. 11.4-5 will now be defined. First, if the point nonlinearity is absent,

$$\mathbf{f}_S = \mathbf{B}\mathbf{f}_I + \mathbf{n}_B \tag{11.4-6}$$

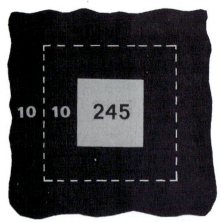

(a) Original

(b) Impulse response

(c) Observation

FIGURE 11.4-3. Image arrays for underdetermined model.

where $\mathbf{B} = \mathbf{B}_P\mathbf{B}_B$ and $\mathbf{n}_B = \mathbf{B}_P\mathbf{n}$. This is the classical discrete model consisting of a set of linear equations with measurement uncertainty. Another case that will be defined for later discussion occurs when the spatial blur of the physical image digitizer is negligible. In this case,

$$\mathbf{f}_S = O_P\{\mathbf{B}\mathbf{f}_I\} + \mathbf{n} \tag{11.4-7}$$

where $\mathbf{B} = \mathbf{B}_B$ is defined by Eq. 7.2-15.

Chapter 12 contains results for several image restoration experiments based on the restoration model defined by Eq. 11.4-6. An artificial image has been generated for these computer simulation experiments (9). The original image used for the analysis of underdetermined restoration techniques, shown in Figure 11.4-3a, consists of a 4×4 pixel square of intensity 245 placed against an extended background of intensity

10 referenced to an intensity scale of 0 to 255. All images are zoomed for display purposes. The Gaussian-shaped impulse response function is defined as

$$H(l_1, l_2) = K \exp\left\{-\left(\frac{l_1}{2b_C^2} + \frac{l_2}{2b_R^2}\right)\right\} \tag{11.4-8}$$

over a 5×5 point array where K is an amplitude scaling constant and b_C and b_R are blur-spread constants.

In the computer simulation restoration experiments, the observed blurred image model has been obtained by multiplying the column-scanned original image of Figure 11.4-3a by the blur matrix **B**. Next, additive white Gaussian observation noise has been simulated by adding output variables from an appropriate random number generator to the blurred images. For display, all image points restored are clipped to the intensity range 0 to 255.

REFERENCES

1.   M. Born and E. Wolf, *Principles of Optics*, 7th ed., Pergamon Press, New York, 1999.

2.   J. W. Goodman, *Introduction to Fourier Optics*, 2nd ed., McGraw-Hill, New York, 1996.

3.   E. L. O'Neill and E. H. O'Neill, *Introduction to Statistical Optics*, reprint ed., Addison-Wesley, Reading, MA, 1992.

4.   H. H. Hopkins, *Proc. Royal Society, A*, **231**, 1184, July 1955, 98.

5.   R. E. Hufnagel and N. R. Stanley, "Modulation Transfer Function Associated with Image Transmission Through Turbulent Media," *J. Optical Society of America*, **54**, 1, January 1964, 52–61.

6.   K. Henney and B. Dudley, *Handbook of Photography*, McGraw-Hill, New York, 1939.

7.   R. M. Evans, W. T. Hanson, and W. L. Brewer, *Principles of Color Photography*, Wiley, New York, 1953.

8.   C. E. Mees, *The Theory of Photographic Process*, Macmillan, New York, 1966.

9.   N. D. A. Mascarenhas and W. K. Pratt, "Digital Image Restoration Under a Regression Model," *IEEE Trans. Circuits and Systems*, **CAS-22**, 3, March 1975, 252–266.

12

POINT AND SPATIAL IMAGE
RESTORATION TECHNIQUES

A common defect in imaging systems is unwanted nonlinearities in the sensor and display systems. Post processing correction of sensor signals and pre-processing correction of display signals can reduce such degradations substantially (1). Such point restoration processing is usually relatively simple to implement. One of the most common image restoration tasks is that of spatial image restoration to compensate for image blur and to diminish noise effects. References 2 to 6 contain surveys of spatial image restoration methods.

12.1. SENSOR AND DISPLAY POINT NONLINEARITY CORRECTION

This section considers methods for compensation of point nonlinearities of sensors and displays.

12.1.1. Sensor Point Nonlinearity Correction

In imaging systems in which the source degradation can be separated into cascaded spatial and point effects, it is often possible directly to compensate for the point degradation (7). Consider a physical imaging system that produces an observed image field $F_O(x, y)$ according to the separable model

$$F_O(x, y) = O_Q\{O_D\{C(x, y, \lambda)\}\}$$

(12.1-1)

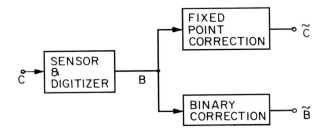

FIGURE 12.1-1. Point luminance correction for an image sensor.

where $C(x, y, \lambda)$ is the spectral energy distribution of the input light field, $O_Q\{\cdot\}$ represents the point amplitude response of the sensor and $O_D\{\cdot\}$ denotes the spatial and wavelength responses. Sensor luminance correction can then be accomplished by passing the observed image through a correction system with a point restoration operator $O_R\{\cdot\}$ ideally chosen such that

$$O_R\{O_Q\{\cdot\}\} = 1 \tag{12.1-2}$$

For continuous images in optical form, it may be difficult to implement a desired point restoration operator if the operator is nonlinear. Compensation for images in analog electrical form can be accomplished with a nonlinear amplifier, while digital image compensation can be performed by arithmetic operators or by a table look-up procedure.

Figure 12.1-1 is a block diagram that illustrates the point luminance correction methodology. The sensor input is a point light distribution function C that is converted to a binary number B for eventual entry into a computer or digital processor. In some imaging applications, processing will be performed directly on the binary representation, while in other applications, it will be preferable to convert to a real fixed-point computer number linearly proportional to the sensor input luminance. In the former case, the binary correction unit will produce a binary number $\tilde{B}$ that is designed to be linearly proportional to C, and in the latter case, the fixed-point correction unit will produce a fixed-point number $\tilde{C}$ that is designed to be equal to C.

A typical measured response B versus sensor input luminance level C is shown in Figure 12.1-2a, while Figure 12.1-2b shows the corresponding compensated response that is desired. The measured response can be obtained by scanning a gray scale test chart of known luminance values and observing the digitized binary value B at each step. Repeated measurements should be made to reduce the effects of noise and measurement errors. For calibration purposes, it is convenient to regard the binary-coded luminance as a fixed-point binary number. As an example, if the luminance range is sliced to 4096 levels and coded with 12 bits, the binary representation would be

$$B = b_8 \, b_7 \, b_6 \, b_5 \, b_4 \, b_3 \, b_2 \, b_1 . \; b_{-1} \, b_{-2} \, b_{-3} \, b_{-4} \tag{12.1-3}$$

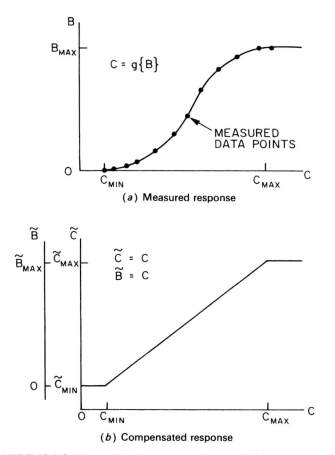

FIGURE 12.1-2. Measured and compensated sensor luminance response.

The whole-number part in this example ranges from 0 to 255, and the fractional part divides each integer step into 16 subdivisions. In this format, the scanner can produce output levels over the range

$$255.9375 \leq B \leq 0.0 \tag{12.1-4}$$

After the measured gray scale data points of Figure 12.1-2a have been obtained, a smooth analytic curve

$$C = g\{B\} \tag{12.1-5}$$

is fitted to the data. The desired luminance response in real number and binary number forms is

$$\tilde{C} = C \qquad\qquad (12.1\text{-}6a)$$

$$\tilde{B} = B_{max} \frac{C - C_{min}}{C_{max} - C_{min}} \qquad\qquad (12.1\text{-}6b)$$

Hence, the required compensation relationships are

$$\tilde{C} = g\{B\} \qquad\qquad (12.1\text{-}7a)$$

$$\tilde{B} = B_{max} \frac{g\{B\} - C_{min}}{C_{max} - C_{min}} \qquad\qquad (12.1\text{-}7b)$$

The limits of the luminance function are commonly normalized to the range 0.0 to 1.0.

To improve the accuracy of the calibration procedure, it is first wise to perform a rough calibration and then repeat the procedure as often as required to refine the correction curve. It should be observed that because B is a binary number, the corrected luminance value $\tilde{C}$ will be a quantized real number. Furthermore, the corrected binary coded luminance $\tilde{B}$ will be subject to binary roundoff of the right-hand side of Eq. 12.1-7b. As a consequence of the nonlinearity of the fitted curve $C = g\{B\}$ and the amplitude quantization inherent to the digitizer, it is possible that some of the corrected binary-coded luminance values may be unoccupied. In other words, the image histogram of $\tilde{B}$ may possess gaps. To minimize this effect, the number of output levels can be limited to less than the number of input levels. For example, B may be coded to 12 bits and $\tilde{B}$ coded to only 8 bits. Another alternative is to add pseudorandom noise to $\tilde{B}$ to smooth out the occupancy levels.

Many image scanning devices exhibit a variable spatial nonlinear point luminance response. Conceptually, the point correction techniques described previously could be performed at each pixel value using the measured calibrated curve at that point. Such a process, however, would be mechanically prohibitive. An alternative approach, called *gain correction*, that is often successful is to model the variable spatial response by some smooth normalized two-dimensional curve $G(j, k)$ over the sensor surface. Then, the corrected spatial response can be obtained by the operation

$$\tilde{F}(j, k) = \frac{F(j, k)}{G(j, k)} \qquad\qquad (12.1\text{-}8)$$

where $F(j, k)$ and $\tilde{F}(j, k)$ represent the raw and corrected sensor responses, respectively.

Figure 12.1-3 provides an example of adaptive gain correction of a charge coupled device (CCD) camera. Figure 12.1-3a is an image of a spatially flat light box surface obtained with the CCD camera. A line profile plot of a diagonal line through the original image is presented in Figure 12.1-3b. Figure 12.3-3c is the gain-corrected original, in which $G(j, k)$ is obtained by Fourier domain low-pass filtering of

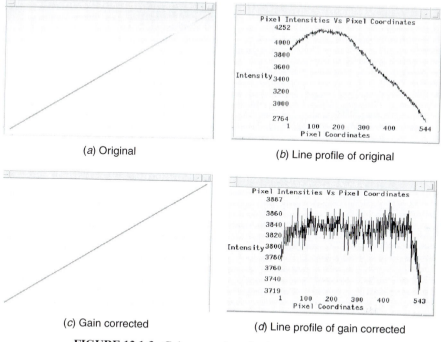

(a) Original

(b) Line profile of original

(c) Gain corrected

(d) Line profile of gain corrected

FIGURE 12.1-3. Gain correction of a CCD camera image.

the original image. The line profile plot of Figure 12.1-3d shows the "flattened" result.

12.1.2. Display Point Nonlinearity Correction

Correction of an image display for point luminance nonlinearities is identical in principle to the correction of point luminance nonlinearities of an image sensor. The procedure illustrated in Figure 12.1-4 involves distortion of the binary coded image luminance variable B to form a corrected binary coded luminance function $\tilde{B}$ so that the displayed luminance $\tilde{C}$ will be linearly proportional to B. In this formulation, the display may include a photographic record of a displayed light field. The desired overall response is

$$\tilde{C} = B \frac{\tilde{C}_{max} - \tilde{C}_{min}}{B_{max}} + \tilde{C}_{min} \tag{12.1-9}$$

Normally, the maximum and minimum limits of the displayed luminance function $\tilde{C}$ are not absolute quantities, but rather are transmissivities or reflectivities

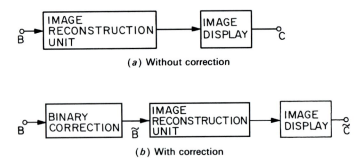

FIGURE 12.1-4. Point luminance correction of an image display.

normalized over a unit range. The measured response of the display and image reconstruction system is modeled by the nonlinear function

$$C = f\{B\} \tag{12.1-10}$$

Therefore, the desired linear response can be obtained by setting

$$\tilde{B} = g\left\{ B \frac{\tilde{C}_{max} - \tilde{C}_{min}}{B_{max}} + \tilde{C}_{min} \right\} \tag{12.1-11}$$

where $g\{\cdot\}$ is the inverse function of $f\{\cdot\}$.

The experimental procedure for determining the correction function $g\{\cdot\}$ will be described for the common example of producing a photographic print from an image display. The first step involves the generation of a digital gray scale step chart over the full range of the binary number B. Usually, about 16 equally spaced levels of B are sufficient. Next, the reflective luminance must be measured over each step of the developed print to produce a plot such as in Figure 12.1-5. The data points are then fitted by the smooth analytic curve $B = g\{C\}$, which forms the desired transformation of Eq. 12.1-10. It is important that enough bits be allocated to B so that the discrete mapping $g\{\cdot\}$ can be approximated to sufficient accuracy. Also, the number of bits allocated to $\tilde{B}$ must be sufficient to prevent gray scale contouring as the result of the nonlinear spacing of display levels. A 10-bit representation of B and an 8-bit representation of $\tilde{B}$ should be adequate in most applications.

Image display devices such as cathode ray tube displays often exhibit spatial luminance variation. Typically, a displayed image is brighter at the center of the display screen than at its periphery. Correction techniques, as described by Eq. 12.1-8, can be utilized for compensation of spatial luminance variations.

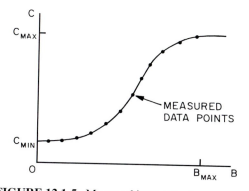

FIGURE 12.1-5. Measured image display response.

12.2. CONTINUOUS IMAGE SPATIAL FILTERING RESTORATION

For the class of imaging systems in which the spatial degradation can be modeled by a linear-shift-invariant impulse response and the noise is additive, restoration of continuous images can be performed by linear filtering techniques. Figure 12.2-1 contains a block diagram for the analysis of such techniques. An ideal image $F_I(x, y)$ passes through a linear spatial degradation system with an impulse response $H_D(x, y)$ and is combined with additive noise $N(x, y)$. The noise is assumed to be uncorrelated with the ideal image. The image field observed can be represented by the convolution operation as

$$F_O(x, y) = \int_{-\infty}^{\infty}\int_{-\infty}^{\infty} F_I(\alpha, \beta) H_D(x - \alpha, y - \beta)\, d\alpha\, d\beta + N(x, y) \qquad (12.2\text{-}1a)$$

or

$$F_O(x, y) = F_I(x, y) \circledast H_D(x, y) + N(x, y) \qquad (12.2\text{-}1b)$$

The restoration system consists of a linear-shift-invariant filter defined by the impulse response $H_R(x, y)$. After restoration with this filter, the reconstructed image becomes

$$\hat{F}_I(x, y) = \int_{-\infty}^{\infty}\int_{-\infty}^{\infty} F_O(\alpha, \beta) H_R(x - \alpha, y - \beta)\, d\alpha\, d\beta \qquad (12.2\text{-}2a)$$

or

$$\hat{F}_I(x, y) = F_O(x, y) \circledast H_R(x, y) \qquad (12.2\text{-}2b)$$

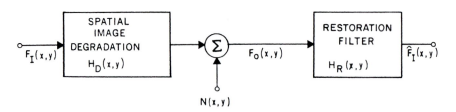

FIGURE 12.2-1. Continuous image restoration model.

Substitution of Eq. 12.2-1*b* into Eq. 12.2-2*b* yields

$$\hat{F}_I(x, y) = [F_I(x, y) \circledast H_D(x, y) + N(x, y)] \circledast H_R(x, y) \qquad (12.2\text{-}3)$$

It is analytically convenient to consider the reconstructed image in the Fourier transform domain. By the Fourier transform convolution theorem,

$$\hat{\mathcal{F}}_I(\omega_x, \omega_y) = [\mathcal{F}_I(\omega_x, \omega_y)\mathcal{H}_D(\omega_x, \omega_y) + \mathcal{N}(\omega_x, \omega_y)]\mathcal{H}_R(\omega_x, \omega_y) \qquad (12.2\text{-}4)$$

where $\mathcal{F}_I(\omega_x, \omega_y)$, $\hat{\mathcal{F}}_I(\omega_x, \omega_y)$, $\mathcal{N}(\omega_x, \omega_y)$, $\mathcal{H}_D(\omega_x, \omega_y)$, $\mathcal{H}_R(\omega_x, \omega_y)$ are the two-dimensional Fourier transforms of $F_I(x, y)$, $\hat{F}_I(x, y)$, $N(x, y)$, $H_D(x, y)$, $H_R(x, y)$, respectively.

The following sections describe various types of continuous image restoration filters.

12.2.1. Inverse Filter

The earliest attempts at image restoration were based on the concept of inverse filtering, in which the transfer function of the degrading system is inverted to yield a restored image (8–12). If the restoration *inverse filter* transfer function is chosen so that

$$\mathcal{H}_R(\omega_x, \omega_y) = \frac{1}{\mathcal{H}_D(\omega_x, \omega_y)} \qquad (12.2\text{-}5)$$

then the spectrum of the reconstructed image becomes

$$\hat{\mathcal{F}}_I(\omega_x, \omega_y) = \mathcal{F}_I(\omega_x, \omega_y) + \frac{\mathcal{N}(\omega_x, \omega_y)}{\mathcal{H}_D(\omega_x, \omega_y)} \qquad (12.2\text{-}6)$$

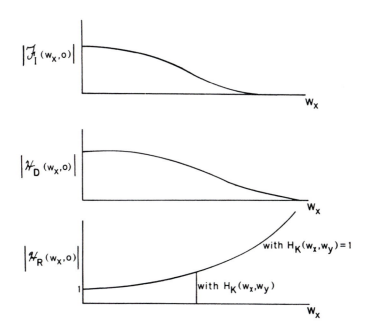

FIGURE 12.2-2. Typical spectra of an inverse filtering image restoration system.

Upon inverse Fourier transformation, the restored image field

$$\hat{F}_I(x, y) = F_I(x, y) + \frac{1}{4\pi^2} \int_{-\infty}^{\infty}\int_{-\infty}^{\infty} \frac{\mathcal{N}(\omega_x, \omega_y)}{\mathcal{H}_D(\omega_x, \omega_y)} \exp\{i(\omega_x x + \omega_y y)\} \, d\omega_x \, d\omega_y$$

$$(12.2\text{-}7)$$

is obtained. In the absence of source noise, a perfect reconstruction results, but if source noise is present, there will be an additive reconstruction error whose value can become quite large at spatial frequencies for which $\mathcal{H}_D(\omega_x, \omega_y)$ is small. Typically, $\mathcal{H}_D(\omega_x, \omega_y)$ and $\mathcal{F}_I(\omega_x, \omega_y)$ are small at high spatial frequencies, hence image quality becomes severely impaired in high-detail regions of the reconstructed image. Figure 12.2-2 shows typical frequency spectra involved in inverse filtering.

The presence of noise may severely affect the uniqueness of a restoration estimate. That is, small changes in $N(x, y)$ may radically change the value of the estimate $\hat{F}_I(x, y)$. For example, consider the dither function $Z(x, y)$ added to an ideal image to produce a perturbed image

$$F_Z(x, y) = F_I(x, y) + Z(x, y) \qquad (12.2\text{-}8)$$

There may be many dither functions for which

$$\left| \int_{-\infty}^{\infty} \int_{-\infty}^{\infty} Z(\alpha, \beta) H_D(x - \alpha, y - \beta) \, d\alpha \, d\beta \right| < |N(x, y)| \qquad (12.2\text{-}9)$$

For such functions, the perturbed image field $F_Z(x, y)$ may satisfy the convolution integral of Eq. 12.2-1 to within the accuracy of the observed image field. Specifically, it can be shown that if the dither function is a high-frequency sinusoid of arbitrary amplitude, then in the limit

$$\lim_{n \to \infty} \left\{ \int_{-\infty}^{\infty} \int_{-\infty}^{\infty} \sin\{n(\alpha + \beta)\} H_D(x - \alpha, y - \beta) \, d\alpha \, d\beta \right\} = 0 \qquad (12.2\text{-}10)$$

For image restoration, this fact is particularly disturbing, for two reasons. High-frequency signal components may be present in an ideal image, yet their presence may be masked by observation noise. Conversely, a small amount of observation noise may lead to a reconstruction of $F_I(x, y)$ that contains very large amplitude high-frequency components. If relatively small perturbations $N(x, y)$ in the observation result in large dither functions for a particular degradation impulse response, the convolution integral of Eq. 12.2-1 is said to be unstable or *ill conditioned*. This potential instability is dependent on the structure of the degradation impulse response function.

There have been several ad hoc proposals to alleviate noise problems inherent to inverse filtering. One approach (10) is to choose a restoration filter with a transfer function

$$\mathcal{H}_R(\omega_x, \omega_y) = \frac{\mathcal{H}_K(\omega_x, \omega_y)}{\mathcal{H}_D(\omega_x, \omega_y)} \qquad (12.2\text{-}11)$$

where $\mathcal{H}_K(\omega_x, \omega_y)$ has a value of unity at spatial frequencies for which the expected magnitude of the ideal image spectrum is greater than the expected magnitude of the noise spectrum, and zero elsewhere. The reconstructed image spectrum is then

$$\hat{\mathcal{F}}_I(\omega_x, \omega_y) = \mathcal{F}_I(\omega_x, \omega_y) \mathcal{H}_K(\omega_x, \omega_y) + \frac{\mathcal{N}(\omega_x, \omega_y) \mathcal{H}_K(\omega_x, \omega_y)}{\mathcal{H}_D(\omega_x, \omega_y)} \qquad (12.2\text{-}12)$$

The result is a compromise between noise suppression and loss of high-frequency image detail.

Another fundamental difficulty with inverse filtering is that the transfer function of the degradation may have zeros in its passband. At such points in the frequency spectrum, the inverse filter is not physically realizable, and therefore the filter must be approximated by a large value response at such points.

12.2.2. Wiener Filter

It should not be surprising that inverse filtering performs poorly in the presence of noise because the filter design ignores the noise process. Improved restoration quality is possible with Wiener filtering techniques, which incorporate a priori statistical knowledge of the noise field (13–17).

In the general derivation of the *Wiener filter*, it is assumed that the ideal image $F_I(x, y)$ and the observed image $F_O(x, y)$ of Figure 12.2-1 are samples of two-dimensional, continuous stochastic fields with zero-value spatial means. The impulse response of the restoration filter is chosen to minimize the mean-square restoration error

$$\mathcal{E} = E\{[F_I(x, y) - \hat{F}_I(x, y)]^2\} \tag{12.2-13}$$

The mean-square error is minimized when the following *orthogonality condition* is met (13):

$$E\{[F_I(x, y) - \hat{F}_I(x, y)]F_O(x', y')\} = 0 \tag{12.2-14}$$

for all image coordinate pairs (x, y) and (x', y'). Upon substitution of Eq. 12.2-2a for the restored image and some linear algebraic manipulation, one obtains

$$E\{F_I(x, y)F_O(x, y)\} = \int_{-\infty}^{\infty}\int_{-\infty}^{\infty} E\{F_O(\alpha, \beta)F_O(x', y')\}H_R(x - \alpha, y - \beta)\, d\alpha\, d\beta \tag{12.2-15}$$

Under the assumption that the ideal image and observed image are jointly stationary, the expectation terms can be expressed as covariance functions, as in Eq. 1.4-8. This yields

$$K_{F_I F_O}(x - x', y - y') = \int_{-\infty}^{\infty}\int_{-\infty}^{\infty} K_{F_O F_O}(\alpha - x', \beta - y')H_R(x - \alpha, y - \beta)\, d\alpha\, d\beta \tag{12.2-16}$$

Then, taking the two-dimensional Fourier transform of both sides of Eq. 12.2-16 and solving for $\mathcal{H}_R(\omega_x, \omega_y)$, the following general expression for the Wiener filter transfer function is obtained:

$$\mathcal{H}_R(\omega_x, \omega_y) = \frac{\mathcal{W}_{F_I F_O}(\omega_x, \omega_y)}{\mathcal{W}_{F_O F_O}(\omega_x, \omega_y)} \tag{12.2-17}$$

In the special case of the additive noise model of Figure 12.2-1:

$$W_{F_I F_O}(\omega_x, \omega_y) = \mathcal{H}_D^*(\omega_x, \omega_y) W_{F_I}(\omega_x, \omega_y) \tag{12.2-18a}$$

$$W_{F_O F_O}(\omega_x, \omega_y) = |\mathcal{H}_D(\omega_x, \omega_y)|^2 W_{F_I}(\omega_x, \omega_y) + W_N(\omega_x, \omega_y) \tag{12.2-18b}$$

This leads to the additive noise Wiener filter

$$\mathcal{H}_R(\omega_x, \omega_y) = \frac{\mathcal{H}_D^*(\omega_x, \omega_y) W_{F_I}(\omega_x, \omega_y)}{|\mathcal{H}_D(\omega_x, \omega_y)|^2 W_{F_I}(\omega_x, \omega_y) + W_N(\omega_x, \omega_y)} \tag{12.2-19a}$$

or

$$\mathcal{H}_R(\omega_x, \omega_y) = \frac{\mathcal{H}_D^*(\omega_x, \omega_y)}{|\mathcal{H}_D(\omega_x, \omega_y)|^2 + W_N(\omega_x, \omega_y)/W_{F_I}(\omega_x, \omega_y)} \tag{12.2-19b}$$

In the latter formulation, the transfer function of the restoration filter can be expressed in terms of the signal-to-noise power ratio

$$\mathrm{SNR}(\omega_x, \omega_y) \equiv \frac{W_{F_I}(\omega_x, \omega_y)}{W_N(\omega_x, \omega_y)} \tag{12.2-20}$$

at each spatial frequency. Figure 12.2-3 shows cross-sectional sketches of a typical ideal image spectrum, noise spectrum, blur transfer function, and the resulting Wiener filter transfer function. As noted from the figure, this version of the Wiener filter acts as a bandpass filter. It performs as an inverse filter at low spatial frequencies, and as a smooth rolloff low-pass filter at high spatial frequencies.

Equation 12.2-19 is valid when the ideal image and observed image stochastic processes are zero mean. In this case, the reconstructed image Fourier transform is

$$\hat{\mathcal{F}}_I(\omega_x, \omega_y) = \mathcal{H}_R(\omega_x, \omega_y) \mathcal{F}_O(\omega_x, \omega_y) \tag{12.2-21}$$

If the ideal image and observed image means are nonzero, the proper form of the reconstructed image Fourier transform is

$$\hat{\mathcal{F}}_I(\omega_x, \omega_y) = \mathcal{H}_R(\omega_x, \omega_y)[\mathcal{F}_O(\omega_x, \omega_y) - \mathcal{M}_O(\omega_x, \omega_y)] + \mathcal{M}_I(\omega_x, \omega_y) \tag{12.2-22a}$$

where

$$\mathcal{M}_O(\omega_x, \omega_y) = \mathcal{H}_D(\omega_x, \omega_y) \mathcal{M}_I(\omega_x, \omega_y) + \mathcal{M}_N(\omega_x, \omega_y) \tag{12.2-22b}$$

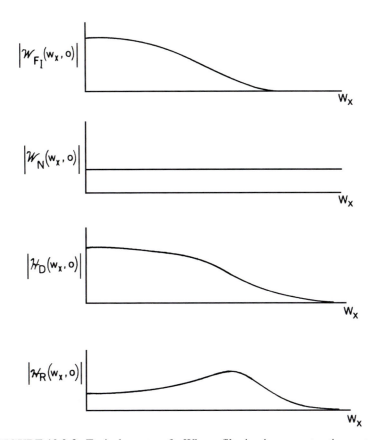

FIGURE 12.2-3. Typical spectra of a Wiener filtering image restoration system.

and $M_I(\omega_x, \omega_y)$ and $M_N(\omega_x, \omega_y)$ are the two-dimensional Fourier transforms of the means of the ideal image and noise, respectively. It should be noted that Eq. 12.2-22 accommodates spatially varying mean models. In practice, it is common to estimate the mean of the observed image by its spatial average $M_O(x, y)$ and apply the Wiener filter of Eq. 12.2-19 to the observed image difference $F_O(x, y) - M_O(x, y)$, and then add back the ideal image mean $M_I(x, y)$ to the Wiener filter result.

It is useful to investigate special cases of Eq. 12.2-19. If the ideal image is assumed to be uncorrelated with unit energy, $W_{F_I}(\omega_x, \omega_y) = 1$ and the Wiener filter becomes

$$\mathcal{H}_R(\omega_x, \omega_y) = \frac{\mathcal{H}_D^*(\omega_x, \omega_y)}{\left|\mathcal{H}_D(\omega_x, \omega_y)\right|^2 + W_N(\omega_x, \omega_y)} \qquad (12.2\text{-}23)$$

This version of the Wiener filter provides less noise smoothing than does the general case of Eq. 12.2-19. If there is no blurring of the ideal image, $\mathcal{H}_D(\omega_x, \omega_y) = 1$ and the Wiener filter becomes a noise smoothing filter with a transfer function

$$\mathcal{H}_R(\omega_x, \omega_y) = \frac{1}{1 + \mathcal{W}_N(\omega_x, \omega_y)} \tag{12.2-24}$$

In many imaging systems, the impulse response of the blur may not be fixed; rather, it changes shape in a random manner. A practical example is the blur caused by imaging through a turbulent atmosphere. Obviously, a Wiener filter applied to this problem would perform better if it could dynamically adapt to the changing blur impulse response. If this is not possible, a design improvement in the Wiener filter can be obtained by considering the impulse response to be a sample of a two-dimensional stochastic process with a known mean shape and with a random perturbation about the mean modeled by a known power spectral density. Transfer functions for this type of restoration filter have been developed by Slepian (18).

12.2.3. Parametric Estimation Filters

Several variations of the Wiener filter have been developed for image restoration. Some techniques are ad hoc, while others have a quantitative basis.

Cole (19) has proposed a restoration filter with a transfer function

$$\mathcal{H}_R(\omega_x, \omega_y) = \left[\frac{\mathcal{W}_{F_I}(\omega_x, \omega_y)}{|\mathcal{H}_D(\omega_x, \omega_y)|^2 \mathcal{W}_{F_I}(\omega_x, \omega_y) + \mathcal{W}_N(\omega_x, \omega_y)} \right]^{1/2} \tag{12.2-25}$$

The power spectrum of the filter output is

$$\mathcal{W}_{\hat{F}_I}(\omega_x, \omega_y) = |\mathcal{H}_R(\omega_x, \omega_y)|^2 \mathcal{W}_{F_O}(\omega_x, \omega_y) \tag{12.2-26}$$

where $\mathcal{W}_{F_O}(\omega_x, \omega_y)$ represents the power spectrum of the observation, which is related to the power spectrum of the ideal image by

$$\mathcal{W}_{F_O}(\omega_x, \omega_y) = |\mathcal{H}_D(\omega_x, \omega_y)|^2 \mathcal{W}_{F_I}(\omega_x, \omega_y) + \mathcal{W}_N(\omega_x, \omega_y) \tag{12.2-27}$$

Thus, it is easily seen that the power spectrum of the reconstructed image is identical to the power spectrum of the ideal image field. That is,

$$\mathcal{W}_{\hat{F}_I}(\omega_x, \omega_y) = \mathcal{W}_{F_I}(\omega_x, \omega_y) \tag{12.2-28}$$

For this reason, the restoration filter defined by Eq. 12.2-25 is called the image *power-spectrum filter*. In contrast, the power spectrum for the reconstructed image as obtained by the Wiener filter of Eq. 12.2-19 is

$$W_{\hat{F}_I}(\omega_x, \omega_y) = \frac{|\mathcal{H}_D(\omega_x, \omega_y)|^2 [W_{F_I}(\omega_x, \omega_y)]^2}{|\mathcal{H}_D(\omega_x, \omega_y)|^2 W_{F_I}(\omega_x, \omega_y) + W_N(\omega_x, \omega_y)} \tag{12.2-29}$$

In this case, the power spectra of the reconstructed and ideal images become identical only for a noise-free observation. Although equivalence of the power spectra of the ideal and reconstructed images appears to be an attractive feature of the image power-spectrum filter, it should be realized that it is more important that the Fourier spectra (Fourier transforms) of the ideal and reconstructed images be identical because their Fourier transform pairs are unique, but power-spectra transform pairs are not necessarily unique. Furthermore, the Wiener filter provides a minimum mean-square error estimate, while the image power-spectrum filter may result in a large residual mean-square error.

Cole (19) has also introduced a *geometrical mean filter*, defined by the transfer function

$$\mathcal{H}_R(\omega_x, \omega_y) = [\mathcal{H}_D(\omega_x, \omega_y)]^{-S} \left[\frac{\mathcal{H}_D^*(\omega_x, \omega_y) W_{F_I}(\omega_x, \omega_y)}{|\mathcal{H}_D(\omega_x, \omega_y)|^2 W_{F_I}(\omega_x, \omega_y) + W_N(\omega_x, \omega_y)} \right]^{1-S} \tag{12.2-30}$$

where $0 \leq S \leq 1$ is a design parameter. If $S = 1/2$ and $\mathcal{H}_D = \mathcal{H}_D^*$, the geometrical mean filter reduces to the image power-spectrum filter as given in Eq. 12.2-25.

Hunt (20) has developed another parametric restoration filter, called the *constrained least-squares filter*, whose transfer function is of the form

$$\mathcal{H}_R(\omega_x, \omega_y) = \frac{\mathcal{H}_D^*(\omega_x, \omega_y)}{|\mathcal{H}_D(\omega_x, \omega_y)|^2 + \gamma |C(\omega_x, \omega_y)|^2} \tag{12.2-31}$$

where γ is a design constant and $C(\omega_x, \omega_y)$ is a design spectral variable. If $\gamma = 1$ and $|C(\omega_x, \omega_y)|^2$ is set equal to the reciprocal of the spectral signal-to-noise power ratio of Eq. 12.2-20, the constrained least-squares filter becomes equivalent to the Wiener filter of Eq. 12.2-19b. The spectral variable can also be used to minimize higher-order derivatives of the estimate.

12.2.4. Application to Discrete Images

The inverse filtering, Wiener filtering, and parametric estimation filtering techniques developed for continuous image fields are often applied to the restoration of

(*a*) Original

(*b*) Blurred, $b = 2.0$ (*c*) Blurred with noise, SNR $= 10.0$

FIGURE 12.2-4. Blurred test images.

discrete images. The common procedure has been to replace each of the continuous spectral functions involved in the filtering operation by its discrete two-dimensional Fourier transform counterpart. However, care must be taken in this conversion process so that the discrete filtering operation is an accurate representation of the continuous convolution process and that the discrete form of the restoration filter impulse response accurately models the appropriate continuous filter impulse response.

Figures 12.2-4 to 12.2-7 present examples of continuous image spatial filtering techniques by discrete Fourier transform filtering. The original image of Figure 12.2-4*a* has been blurred with a Gaussian-shaped impulse response with $b = 2.0$ to obtain the blurred image of Figure 12.2-4*b*. White Gaussian noise has been added to the blurred image to give the noisy blurred image of Figure l2.2-4*c*, which has a signal-to-noise ratio of 10.0.

Figure 12.2-5 shows the results of inverse filter image restoration of the blurred and noisy-blurred images. In Figure 12.2-5*a*, the inverse filter transfer function follows Eq. 12.2-5 (i.e., no high-frequency cutoff). The restored image for the noise-free observation is corrupted completely by the effects of computational error. The computation was performed using 32-bit floating-point arithmetic. In Figure 12.2-5*c* the inverse filter restoration is performed with a circular cutoff inverse filter as defined by Eq. 12.2-11 with $C = 200$ for the 512×512 pixel noise-free observation. Some faint artifacts are visible in the restoration. In Figure 12.2-5*e* the cutoff frequency is reduced to $C = 150$. The restored image appears relatively sharp and free of artifacts. Figure 12.2-5*b*, *d*, and *f* show the result of inverse filtering on the noisy-blurred observed image with varying cutoff frequencies. These restorations illustrate the trade-off between the level of artifacts and the degree of deblurring.

Figure 12.2-6 shows the results of Wiener filter image restoration. In all cases, the noise power spectral density is white and the signal power spectral density is circularly symmetric Markovian with a correlation factor ρ. For the noise-free observation, the Wiener filter provides restorations that are free of artifacts but only slightly sharper than the blurred observation. For the noisy observation, the restoration artifacts are less noticeable than for an inverse filter.

Figure 12.2-7 presents restorations using the power spectrum filter. For a noise-free observation, the power spectrum filter gives a restoration of similar quality to an inverse filter with a low cutoff frequency. For a noisy observation, the power spectrum filter restorations appear to be grainier than for the Wiener filter.

The continuous image field restoration techniques derived in this section are advantageous in that they are relatively simple to understand and to implement using Fourier domain processing. However, these techniques face several important limitations. First, there is no provision for aliasing error effects caused by physical undersampling of the observed image. Second, the formulation inherently assumes that the quadrature spacing of the convolution integral is the same as the physical sampling. Third, the methods only permit restoration for linear, space-invariant degradation. Fourth, and perhaps most important, it is difficult to analyze the effects of numerical errors in the restoration process and to develop methods of combatting such errors. For these reasons, it is necessary to turn to the discrete model of a sampled blurred image developed in Section 7.2 and then reformulate the restoration problem on a firm numeric basic. This is the subject of the remaining sections of the chapter.

12.3. PSEUDOINVERSE SPATIAL IMAGE RESTORATION

The matrix pseudoinverse defined in Chapter 5 can be used for spatial image restoration of digital images when it is possible to model the spatial degradation as a vector-space operation on a vector of ideal image points yielding a vector of physical observed samples obtained from the degraded image (21–23).

(a) Noise-free, no cutoff

(b) Noisy, $C = 100$

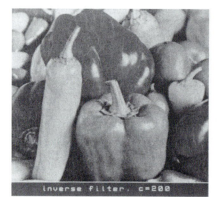

(c) Noise-free, $C = 200$

(d) Noisy, $C = 75$

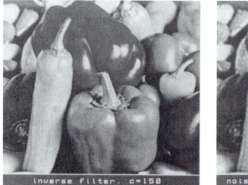

(e) Noise-free, $C = 150$

(f) Noisy, $C = 50$

FIGURE 12.2-5. Inverse filter image restoration on the blurred test images.

(a) Noise-free, $\rho = 0.9$

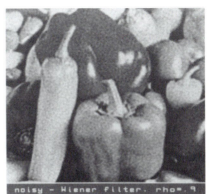

(b) Noisy, $\rho = 0.9$

(c) Noise-free, $\rho = 0.5$

(d) Noisy, $\rho = 0.5$

(e) Noise-free, $\rho = 0.0$

(f) Noisy, $\rho = 0.0$

FIGURE 12.2-6. Wiener filter image restoration on the blurred test images; SNR = 10.0.

(a) Noise-free, $\rho = 0.5$ (b) Noisy, $\rho = 0.5$

(c) Noisy, $\rho = 0.5$ (d) Noisy, $\rho = 0.0$

FIGURE 12.2-7. Power spectrum filter image restoration on the blurred test images; SNR = 10.0.

12.3.1. Pseudoinverse: Image Blur

The first application of the pseudoinverse to be considered is that of the restoration of a blurred image described by the vector-space model

$$\mathbf{g} = \mathbf{Bf} \tag{12.3-1}$$

as derived in Eq. 11.5-6, where $\mathbf{g}$ is a $P \times 1$ vector $(P = M^2)$ containing the $M \times M$ physical samples of the blurred image, $\mathbf{f}$ is a $Q \times 1$ vector $(Q = N^2)$ containing $N \times N$ points of the ideal image and $\mathbf{B}$ is the $P \times Q$ matrix whose elements are points on the impulse function. If the physical sample period and the quadrature representation period are identical, P will be smaller than Q, and the system of equations will be *underdetermined*. By oversampling the blurred image, it is possible to force $P > Q$ or even $P = Q$. In either case, the system of equations is called *overdetermined*. An overdetermined set of equations can also be obtained if some of the elements of the ideal image vector can be specified through a priori knowledge. For example, if the ideal image is known to contain a limited size object against a black background (zero luminance), the elements of **f** beyond the limits may be set to zero.

In discrete form, the restoration problem reduces to finding a solution $\hat{\mathbf{f}}$ to Eq. 12.3-1 in the sense that

$$\mathbf{B}\hat{\mathbf{f}} = \mathbf{g} \qquad (12.3\text{-}2)$$

Because the vector **g** is determined by physical sampling and the elements of **B** are specified independently by system modeling, there is no guarantee that a $\hat{\mathbf{f}}$ even exists to satisfy Eq. 12.3-2. If there is a solution, the system of equations is said to be *consistent*; otherwise, the system of equations is *inconsistent*.

In Appendix 1 it is shown that inconsistency in the set of equations of Eq. 12.3-1 can be characterized as

$$\mathbf{g} = \mathbf{Bf} + \mathbf{e}\{\mathbf{f}\} \qquad (12.3\text{-}3)$$

where $\mathbf{e}\{\mathbf{f}\}$ is a vector of remainder elements whose value depends on **f**. If the set of equations is inconsistent, a solution of the form

$$\hat{\mathbf{f}} = \mathbf{Wg} \qquad (12.3\text{-}4)$$

is sought for which the linear operator **W** minimizes the least-squares modeling error

$$\mathcal{E}_M = [\mathbf{e}\{\hat{\mathbf{f}}\}]^T [\mathbf{e}\{\hat{\mathbf{f}}\}] = [\mathbf{g} - \mathbf{B}\hat{\mathbf{f}}]^T [\mathbf{g} - \mathbf{B}\hat{\mathbf{f}}] \qquad (12.3\text{-}5)$$

This error is shown, in Appendix 1, to be minimized when the operator $\mathbf{W} = \mathbf{B}^{\$}$ is set equal to the least-squares inverse of **B**. The least-squares inverse is not necessarily unique. It is also proved in Appendix 1 that the generalized inverse operator $\mathbf{W} = \mathbf{B}^-$, which is a special case of the least-squares inverse, is unique, minimizes the least-squares modeling error, and simultaneously provides a minimum norm estimate. That is, the sum of the squares of $\hat{\mathbf{f}}$ is a minimum for all possible minimum least-square error estimates. For the restoration of image blur, the generalized inverse provides a lowest-intensity restored image.

If Eq. 12.3-1 represents a consistent set of equations, one or more solutions may exist for Eq. 12.3-2. The solution commonly chosen is the estimate that minimizes the least-squares estimation error defined in the equivalent forms

$$\mathcal{E}_E = (\mathbf{f} - \hat{\mathbf{f}})^T (\mathbf{f} - \hat{\mathbf{f}}) \tag{12.3-6a}$$

$$\mathcal{E}_E = \text{tr}\{(\mathbf{f} - \hat{\mathbf{f}})(\mathbf{f} - \hat{\mathbf{f}})^T\} \tag{12.3-6b}$$

In Appendix 1 it is proved that the estimation error is minimum for a generalized inverse $(\mathbf{W} = \mathbf{B}^-)$ estimate. The resultant residual estimation error then becomes

$$\mathcal{E}_E = \mathbf{f}^T [\mathbf{I} - [\mathbf{B}^- \ \mathbf{B}]] \mathbf{f} \tag{12.3-7a}$$

or

$$\mathcal{E}_E = \text{tr}\{\mathbf{f}\mathbf{f}^T [\mathbf{I} - [\mathbf{B}^- \ \mathbf{B}]]\} \tag{12.3-7b}$$

The estimate is perfect, of course, if $\mathbf{B}^- \mathbf{B} = \mathbf{I}$.

Thus, it is seen that the generalized inverse is an optimal solution, in the sense defined previously, for both consistent and inconsistent sets of equations modeling image blur. From Eq. 5.5-5, the generalized inverse has been found to be algebraically equivalent to

$$\mathbf{B}^- = [\mathbf{B}^T \mathbf{B}]^{-1} \mathbf{B}^T \tag{12.3-8a}$$

if the $P \times Q$ matrix $\mathbf{B}$ is of rank Q. If $\mathbf{B}$ is of rank P, then

$$\mathbf{B}^- = \mathbf{B}^T [\mathbf{B}^T \mathbf{B}]^{-1} \tag{12.3-8b}$$

For a consistent set of equations and a rank Q generalized inverse, the estimate

$$\hat{\mathbf{f}} = \mathbf{B}^- \mathbf{g} = \mathbf{B}^- \mathbf{B}\mathbf{f} = [[\mathbf{B}^T \mathbf{B}]^{-1} \mathbf{B}^T] \mathbf{B}\mathbf{f} = \mathbf{f} \tag{12.3-9}$$

is obviously perfect. However, in all other cases, a residual estimation error may occur. Clearly, it would be desirable to deal with an overdetermined blur matrix of rank Q in order to achieve a perfect estimate. Unfortunately, this situation is rarely achieved in image restoration. Oversampling the blurred image can produce an overdetermined set of equations ($P > Q$), but the rank of the blur matrix is likely to be much less than Q because the rows of the blur matrix will become more linearly dependent with finer sampling.

A major problem in application of the generalized inverse to image restoration is dimensionality. The generalized inverse is a $Q \times P$ matrix where P is equal to the number of pixel observations and Q is equal to the number of pixels to be estimated in an image. It is usually not computationally feasible to use the generalized inverse operator, defined by Eq. 12.3-8, over large images because of difficulties in reliably computing the generalized inverse and the large number of vector multiplications associated with Eq. 12.3-4. Computational savings can be realized if the blur matrix **B** is separable such that

$$\mathbf{B} = \mathbf{B}_C \otimes \mathbf{B}_R \qquad (12.3\text{-}10)$$

where $\mathbf{B}_C$ and $\mathbf{B}_R$ are column and row blur operators. In this case, the generalized inverse is separable in the sense that

$$\mathbf{B}^- = \mathbf{B}_C^- \otimes \mathbf{B}_R^- \qquad (12.3\text{-}11)$$

where $\mathbf{B}_C^-$ and $\mathbf{B}_R^-$ are generalized inverses of $\mathbf{B}_C$ and $\mathbf{B}_R$, respectively. Thus, when the blur matrix is of separable form, it becomes possible to form the estimate of the image by sequentially applying the generalized inverse of the row blur matrix to each row of the observed image array and then using the column generalized inverse operator on each column of the array.

Pseudoinverse restoration of large images can be accomplished in an approximate fashion by a block mode restoration process, similar to the block mode filtering technique of Section 9.3, in which the blurred image is partitioned into small blocks that are restored individually. It is wise to overlap the blocks and accept only the pixel estimates in the center of each restored block because these pixels exhibit the least uncertainty. Section 12.3.3 describes an efficient computational algorithm for pseudoinverse restoration for space-invariant blur.

Figure 12.3-1a shows a blurred image based on the model of Figure 11.5-3. Figure 12.3-1b shows a restored image using generalized inverse image restoration. In this example, the observation is noise free and the blur impulse response function is Gaussian shaped, as defined in Eq. 11.5-8, with $b_R = b_C = 1.2$. Only the center 8×8 region of the 12×12 blurred picture is displayed, zoomed to an image size of 256×256 pixels. The restored image appears to be visually improved compared to the blurred image, but the restoration is not identical to the original unblurred image of Figure 11.5-3a. The figure also gives the percentage least-squares error (PLSE) as defined in Appendix 3, between the blurred image and the original unblurred image, and between the restored image and the original. The restored image has less error than the blurred image.

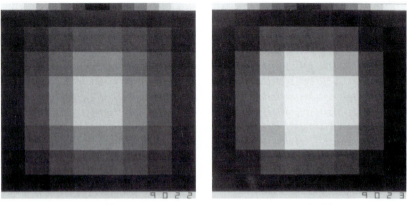

(a) Blurred, PLSE = 4.97%   (b) Restored, PLSE = 1.41%

FIGURE 12.3-1. Pseudoinverse image restoration for test image blurred with Gaussian shape impulse response. $M = 8$, $N = 12$, $L = 5$; $b_R = b_C = 1.2$; noise-free observation.

12.3.2. Pseudoinverse: Image Blur Plus Additive Noise

In many imaging systems, an ideal image is subject to both blur and additive noise; the resulting vector-space model takes the form

$$\mathbf{g} = \mathbf{Bf} + \mathbf{n} \qquad (12.3\text{-}12)$$

where $\mathbf{g}$ and $\mathbf{n}$ are $P \times 1$ vectors of the observed image field and noise field, respectively, $\mathbf{f}$ is a $Q \times 1$ vector of ideal image points, and $\mathbf{B}$ is a $P \times Q$ blur matrix. The vector $\mathbf{n}$ is composed of two additive components: samples of an additive external noise process and elements of the vector difference $(\mathbf{g} - \mathbf{Bf})$ arising from modeling errors in the formulation of $\mathbf{B}$. As a result of the noise contribution, there may be no vector solutions $\hat{\mathbf{f}}$ that satisfy Eq. 12.3-12. However, as indicated in Appendix 1, the generalized inverse $\mathbf{B}^-$ can be utilized to determine a least-squares error, minimum norm estimate. In the absence of modeling error, the estimate

$$\hat{\mathbf{f}} = \mathbf{B}^- \mathbf{g} = \mathbf{B}^- \mathbf{Bf} + \mathbf{B}^- \mathbf{n} \qquad (12.3\text{-}13)$$

differs from the ideal image because of the additive noise contribution $\mathbf{B}^- \mathbf{n}$. Also, for the underdetermined model, $\mathbf{B}^- \mathbf{B}$ will not be an identity matrix. If $\mathbf{B}$ is an overdetermined rank Q matrix, as defined in Eq. 12.3-8a, then $\mathbf{B}^- \mathbf{B} = \mathbf{I}$, and the resulting estimate is equal to the original image vector $\mathbf{f}$ plus a perturbation vector $\Delta \mathbf{f} = \mathbf{B}^- \mathbf{n}$. The perturbation error in the estimate can be measured as the ratio of the vector norm of the perturbation to the vector norm of the estimate. It can be shown (24, p. 52) that the relative error is subject to the bound

$$\frac{||\Delta \mathbf{f}||}{||\mathbf{f}||} < ||\mathbf{B}^-|| \cdot ||\mathbf{B}|| \frac{||\mathbf{n}||}{||\mathbf{g}||} \qquad (12.3\text{-}14)$$

The product $||\mathbf{B}^-|| \cdot ||\mathbf{B}||$, which is called the *condition number* $C\{\mathbf{B}\}$ of $\mathbf{B}$, determines the relative error in the estimate in terms of the ratio of the vector norm of the noise to the vector norm of the observation. The condition number can be computed directly or found in terms of the ratio

$$C\{\mathbf{B}\} = \frac{W_1}{W_N} \qquad (12.3\text{-}15)$$

of the largest W_1 to smallest W_N singular values of $\mathbf{B}$. The noise perturbation error for the underdetermined matrix $\mathbf{B}$ is also governed by Eqs. 12.3-14 and 12.3-15 if W_N is defined to be the smallest nonzero singular value of $\mathbf{B}$ (25, p. 41). Obviously, the larger the condition number of the blur matrix, the greater will be the sensitivity to noise perturbations.

Figure 12.3-2 contains image restoration examples for a Gaussian-shaped blur function for several values of the blur standard deviation and a noise variance of 10.0 on an amplitude scale of 0.0 to 255.0. As expected, observation noise degrades the restoration. Also as expected, the restoration for a moderate degree of blur is worse than the restoration for less blur. However, this trend does not continue; the restoration for severe blur is actually better in a subjective sense than for moderate blur. This seemingly anomalous behavior, which results from spatial truncation of the point-spread function, can be explained in terms of the condition number of the blur matrix. Figure 12.3-3 is a plot of the condition number of the blur matrix of the previous examples as a function of the blur coefficient (21). For small amounts of blur, the condition number is low. A maximum is attained for moderate blur, followed by a decrease in the curve for increasing values of the blur coefficient. The curve tends to stabilize as the blur coefficient approaches infinity. This curve provides an explanation for the previous experimental results. In the restoration operation, the blur impulse response is spatially truncated over a square region of 5×5 quadrature points. As the blur coefficient increases, for fixed M and N, the blur impulse response becomes increasingly wider, and its tails become truncated to a greater extent. In the limit, the nonzero elements in the blur matrix become constant values, and the condition number assumes a constant level. For small values of the blur coefficient, the truncation effect is negligible, and the condition number curve follows an ascending path toward infinity with the asymptotic value obtained for a smoothly represented blur impulse response. As the blur factor increases, the number of nonzero elements in the blur matrix increases, and the condition number stabilizes to a constant value. In effect, a trade-off exists between numerical errors caused by ill-conditioning and modeling accuracy. Although this conclusion

Blurred   Restored

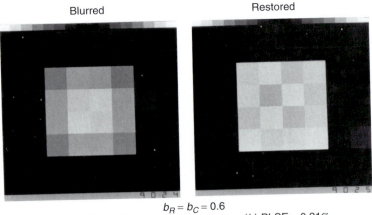

$b_R = b_C = 0.6$

(a) PLSE = 1.30%   (b) PLSE = 0.21%

$b_R = b_C = 1.2$

(c) PLSE = 4.91%   (d) PLSE = 2695.81%

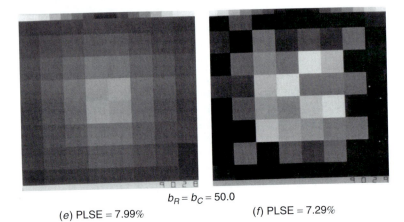

$b_R = b_C = 50.0$

(e) PLSE = 7.99%   (f) PLSE = 7.29%

FIGURE 12.3-2. Pseudoinverse image restoration for test image blurred with Gaussian shape impulse response. $M = 8$, $N = 12$, $L = 5$; noisy observation, Var = 10.0.

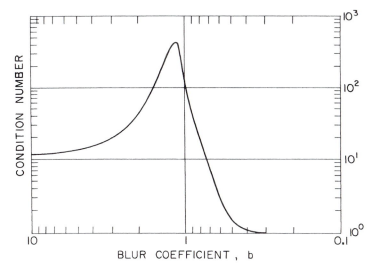

FIGURE 12.3-3. Condition number curve.

is formulated on the basis of a particular degradation model, the inference seems to be more general because the inverse of the integral operator that describes the blur is unbounded. Therefore, the closer the discrete model follows the continuous model, the greater the degree of ill-conditioning. A move in the opposite direction reduces singularity but imposes modeling errors. This inevitable dilemma can only be broken with the intervention of correct a priori knowledge about the original image.

12.3.3. Pseudoinverse Computational Algorithms

Efficient computational algorithms have been developed by Pratt and Davarian (22) for pseudoinverse image restoration for space-invariant blur. To simplify the explanation of these algorithms, consideration will initially be limited to a one-dimensional example.

Let the $N \times 1$ vector $\mathbf{f}_T$ and the $M \times 1$ vector $\mathbf{g}_T$ be formed by selecting the center portions of $\mathbf{f}$ and $\mathbf{g}$, respectively. The truncated vectors are obtained by dropping $L - 1$ elements at each end of the appropriate vector. Figure 12.3-4a illustrates the relationships of all vectors for $N = 9$ original vector points, $M = 7$ observations and an impulse response of length $L = 3$.

The elements $\mathbf{f}_T$ and $\mathbf{g}_T$ are entries in the adjoint model

$$\mathbf{q}_E = \mathbf{C}\mathbf{f}_E + \mathbf{n}_E \qquad (12.3\text{-}16a)$$

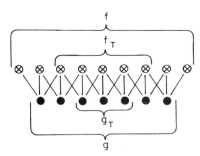

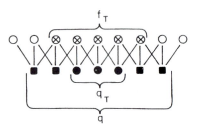

(b) Discrete convolution

(a) Sampled continuous convolution

FIGURE 12.3-4. One-dimensional sampled continuous convolution and discrete convolution.

where the extended vectors $\mathbf{q}_E$, $\mathbf{f}_E$ and $\mathbf{n}_E$ are defined in correspondence with

$$
\begin{bmatrix} \mathbf{g} \\ \mathbf{0} \end{bmatrix} = \begin{bmatrix} \mathbf{C} \end{bmatrix} \begin{bmatrix} \mathbf{f}_T \\ \mathbf{0} \end{bmatrix} + \begin{bmatrix} \mathbf{n}_T \\ \mathbf{0} \end{bmatrix} \tag{12.3-16b}
$$

where $\mathbf{g}$ is a $M \times 1$ vector, $\mathbf{f}_T$ and $\mathbf{n}_T$ are $K \times 1$ vectors, and $\mathbf{C}$ is a $J \times J$ matrix. As noted in Figure 12.3-4b, the vector $\mathbf{q}$ is identical to the image observation $\mathbf{g}$ over its $R = M - 2(L - 1)$ center elements. The outer elements of $\mathbf{q}$ can be approximated by

$$
\mathbf{q} \approx \tilde{\mathbf{q}} = \mathbf{Eg} \tag{12.3-17}
$$

where $\mathbf{E}$, called an *extraction weighting matrix*, is defined as

$$
\mathbf{E} = \begin{bmatrix} \mathbf{a} & \mathbf{0} & \mathbf{0} \\ \mathbf{0} & \mathbf{I} & \mathbf{0} \\ \mathbf{0} & \mathbf{0} & \mathbf{b} \end{bmatrix} \tag{12.3-18}
$$

where $\mathbf{a}$ and $\mathbf{b}$ are $L \times L$ submatrices, which perform a windowing function similar to that described in Section 9.4.2 (22).

Combining Eqs. 12.3-17 and 12.3-18, an estimate of $\mathbf{f}_T$ can be obtained from

$$
\hat{\mathbf{f}}_E = \mathbf{C}^{-1} \hat{\mathbf{q}}_E \tag{12.3-19}
$$

(a) Original image vectors, **f**

(b) Truncated image vectors, $\mathbf{f}_T$

(c) Observation vectors, **g**

(d) Windowed observation vectors, **q**

(e) Restoration without windowing, $\hat{\mathbf{f}}_T$

(f) Restoration with windowing, $\hat{\mathbf{f}}_T$

FIGURE 12.3-5. Pseudoinverse image restoration for small degree of horizontal blur, $b_R = 1.5$.

Equation 12.3-19 can be solved efficiently using Fourier domain convolution techniques, as described in Section 9.3. Computation of the pseudoinverse by Fourier processing requires on the order of $J^2(1 + 4 \log_2 J)$ operations in two dimensions; spatial domain computation requires about $M^2 N^2$ operations. As an example, for $M = 256$ and $L = 17$, the computational savings are nearly 1750:1 (22).

Figure 12.3-5 is a computer simulation example of the operation of the pseudoinverse image restoration algorithm for one-dimensional blur of an image. In the first step of the simulation, the center K pixels of the original image are extracted to form the set of truncated image vectors $\mathbf{f}_T$ shown in Figure 12.3-5b. Next, the truncated image vectors are subjected to a simulated blur with a Gaussian-shaped impulse response with $b_R = 1.5$ to produce the observation of Figure 12.3-5c. Figure 12.3-5d shows the result of the extraction operation on the observation. Restoration results without and with the extraction weighting operator $\mathbf{E}$ are presented in Figure 12.3-5e and f, respectively. These results graphically illustrate the importance of the

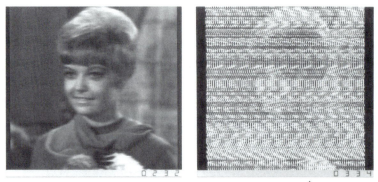

(a) Observation, $\mathbf{g}$ (b) Restoration, $\hat{\mathbf{f}}_T$

Gaussian blur, $b_R = 2.0$

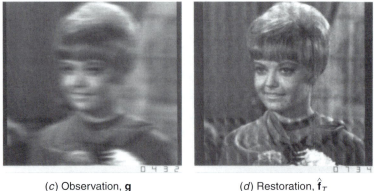

(c) Observation, $\mathbf{g}$ (d) Restoration, $\hat{\mathbf{f}}_T$

Uniform motion blur, $L = 15.0$

FIGURE 12.3-6. Pseudoinverse image restoration for moderate and high degrees of horizontal blur.

extraction operation. Without weighting, errors at the observation boundary completely destroy the estimate in the boundary region, but with weighting the restoration is subjectively satisfying, and the restoration error is significantly reduced. Figure 12.3-6 shows simulation results for the experiment of Figure 12.3-5 when the degree of blur is increased by setting $b_R = 2.0$. The higher degree of blur greatly increases the ill-conditioning of the blur matrix, and the residual error in formation of the modified observation after weighting leads to the disappointing estimate of Figure 12.3-6b. Figure 12.3-6c and d illustrate the restoration improvement obtained with the pseudoinverse algorithm for horizontal image motion blur. In this example, the blur impulse response is constant, and the corresponding blur matrix is better conditioned than the blur matrix for Gaussian image blur.

12.4. SVD PSEUDOINVERSE SPATIAL IMAGE RESTORATION

In Appendix 1 it is shown that any matrix can be decomposed into a series of eigenmatrices by the technique of singular value decomposition. For image restoration, this concept has been extended (26–29) to the eigendecomposition of blur matrices in the imaging model

$$\mathbf{g} = \mathbf{Bf} + \mathbf{n} \tag{12.4-1}$$

From Eq. A1.2-3, the blur matrix $\mathbf{B}$ may be expressed as

$$\mathbf{B} = \mathbf{U}\boldsymbol{\Lambda}^{1/2}\mathbf{V}^{T} \tag{12.4-2}$$

where the $P \times P$ matrix $\mathbf{U}$ and the $Q \times Q$ matrix $\mathbf{V}$ are unitary matrices composed of the eigenvectors of $\mathbf{BB}^{T}$ and $\mathbf{B}^{T}\mathbf{B}$, respectively and $\boldsymbol{\Lambda}$ is a $P \times Q$ matrix whose diagonal terms $\lambda(i)$ contain the eigenvalues of $\mathbf{BB}^{T}$ and $\mathbf{B}^{T}\mathbf{B}$. As a consequence of the orthogonality of $\mathbf{U}$ and $\mathbf{V}$, it is possible to express the blur matrix in the series form

$$\mathbf{B} = \sum_{i=1}^{R} [\lambda(i)]^{1/2}\mathbf{u}_i\mathbf{v}_i^{T} \tag{12.4-3}$$

where $\mathbf{u}_i$ and $\mathbf{v}_i$ are the ith columns of $\mathbf{U}$ and $\mathbf{V}$, respectively, and R is the rank of the matrix $\mathbf{B}$.

From Eq. 12.4-2, because $\mathbf{U}$ and $\mathbf{V}$ are unitary matrices, the generalized inverse of $\mathbf{B}$ is

$$\mathbf{B}^{-} = \mathbf{V}\boldsymbol{\Lambda}^{1/2}\mathbf{U}^{T} = \sum_{i=1}^{R} [\lambda(i)]^{-1/2}\mathbf{v}_i\mathbf{u}_i^{T} \tag{12.4-4}$$

Figure 12.4-1 shows an example of the SVD decomposition of a blur matrix. The generalized inverse estimate can then be expressed as

(a) Blur matrix, *B*

(b) $u_1 v_1^T$, $\lambda(1) = 0.871$ *(c)* $u_2 v_2^T$, $\lambda(2) = 0.573$

(d) $u_3 v_3^T$, $\lambda(3) = 0.285$ *(e)* $u_4 v_4^T$, $\lambda(4) = 0.108$

(f) $u_5 v_5^T$, $\lambda(5) = 0.034$ *(g)* $u_6 v_6^T$, $\lambda(6) = 0.014$

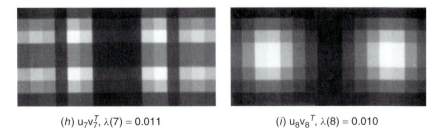

(h) $u_7 v_7^T$, $\lambda(7) = 0.011$ *(i)* $u_8 v_8^T$, $\lambda(8) = 0.010$

FIGURE 12.4-1. SVD decomposition of a blur matrix for $b_R = 2.0$, $M = 8$, $N = 16$, $L = 9$.

$$\hat{\mathbf{f}} = \mathbf{B}^{-}\mathbf{g} = \mathbf{V}\mathbf{\Lambda}^{1/2}\mathbf{U}^{T}\mathbf{g} \qquad (12.4\text{-}5a)$$

or, equivalently,

$$\hat{\mathbf{f}} = \sum_{i=1}^{R} [\lambda(i)]^{-1/2}\mathbf{v}_{i}\mathbf{u}_{i}^{T}\mathbf{g} = \sum_{i=1}^{R} [\lambda(i)]^{-1/2}[\mathbf{u}_{i}^{T}\mathbf{g}]\mathbf{v}_{i} \qquad (12.4\text{-}5b)$$

recognizing the fact that the inner product $\mathbf{u}_{i}^{T}\mathbf{g}$ is a scalar. Equation 12.4-5 provides the basis for sequential estimation; the kth estimate of $\mathbf{f}$ in a sequence of estimates is equal to

$$\hat{\mathbf{f}}_{k} = \hat{\mathbf{f}}_{k-1} + [\lambda(k)]^{-1/2}[\mathbf{u}_{k}^{T}\mathbf{g}]\mathbf{v}_{k} \qquad (12.4\text{-}6)$$

One of the principal advantages of the sequential formulation is that problems of ill-conditioning generally occur only for higher-order singular values. Thus, it is possible interactively to terminate the expansion before numerical problems occur.

Figure 12.4-2 shows an example of sequential SVD restoration for the underdetermined model example of Figure 11.5-3 with a poorly conditioned Gaussian blur matrix. A one-step pseudoinverse would have resulted in the final image estimate that is totally overwhelmed by numerical errors. The sixth step, which is the best subjective restoration, offers a considerable improvement over the blurred original, but the lowest least-squares error occurs for three singular values.

The major limitation of the SVD image restoration method formulation in Eqs. 12.4-5 and 12.4-6 is computational. The eigenvectors $\mathbf{u}_{i}$ and $\mathbf{v}_{i}$ must first be determined for the matrix $\mathbf{B}\mathbf{B}^{T}$ and $\mathbf{B}^{T}\mathbf{B}$. Then the vector computations of Eq 12.4-5 or 12.4-6 must be performed. Even if $\mathbf{B}$ is direct-product separable, permitting separable row and column SVD pseudoinversion, the computational task is staggering in the general case.

The pseudoinverse computational algorithm described in the preceding section can be adapted for SVD image restoration in the special case of space-invariant blur (23). From the adjoint model of Eq. 12.3-16 given by

$$\mathbf{q}_{E} = \mathbf{C}\mathbf{f}_{E} + \mathbf{n}_{E} \qquad (12.4\text{-}7)$$

the circulant matrix $\mathbf{C}$ can be expanded in SVD form as

$$\mathbf{C} = \mathbf{X}\mathbf{\Delta}^{1/2}\mathbf{Y}*^{T} \qquad (12.4\text{-}8)$$

where $\mathbf{X}$ and $\mathbf{Y}$ are unitary matrices defined by

(*a*) 8 singular values PLSE = 2695.81%   (*b*) 7 singular values PLSE = 148.93%

(*c*) 6 singular values PLSE = 6.88%   (*d*) 5 singular values PLSE = 3.31%

(*e*) 4 singular values PLSE = 3.06%   (*f*) 3 singular values PLSE = 3.05%

(*g*) 2 singular values PLSE = 9.52%   (*h*) 1 singular value PLSE = 9.52%

FIGURE 12.4-2. SVD restoration for test image blurred with a Gaussian-shaped impulse response. $b_R = b_C = 1.2$, $M = 8$, $N = 12$, $L = 5$; noisy observation, Var = 10.0.

$$\mathbf{X}[\mathbf{CC}^T]\mathbf{X}*^T = \Delta \tag{12.4-9a}$$

$$\mathbf{Y}[\mathbf{C}^T\mathbf{C}]\mathbf{Y}*^T = \Delta \tag{12.4-9b}$$

Because $\mathbf{C}$ is circulant, $\mathbf{CC}^T$ is also circulant. Therefore $\mathbf{X}$ and $\mathbf{Y}$ must be equivalent to the Fourier transform matrix $\mathbf{A}$ or $\mathbf{A}^{-1}$ because the Fourier matrix produces a diagonalization of a circulant matrix. For purposes of standardization, let $\mathbf{X} = \mathbf{Y} = \mathbf{A}^{-1}$. As a consequence, the eigenvectors $\mathbf{x}_i = \mathbf{y}_i$, which are rows of $\mathbf{X}$ and $\mathbf{Y}$, are actually the complex exponential basis functions

$$x_k^*(j) = \exp\left\{\frac{2\pi i}{J}(k-1)(j-1)\right\} \tag{12.4-10}$$

of a Fourier transform for $1 \leq j, k \leq J$. Furthermore,

$$\Delta = \mathcal{C}\mathcal{C}*^T \tag{12.4-11}$$

where $\mathcal{C}$ is the Fourier domain circular area convolution matrix. Then, in correspondence with Eq. 12.4-5

$$\hat{\mathbf{f}}_E = \mathbf{A}^{-1}\mathbf{\Lambda}^{-1/2}\mathbf{A}\tilde{\mathbf{q}}_E \tag{12.4-12}$$

where $\tilde{\mathbf{q}}_E$ is the modified blurred image observation of Eqs. 12.3-19 and 12.3-20. Equation 12.4-12 should be recognized as being a Fourier domain pseudoinverse estimate. Sequential SVD restoration, analogous to the procedure of Eq. 12.4-6, can be obtained by replacing the SVD pseudoinverse matrix $\mathbf{\Lambda}^{-1/2}$ of Eq. 12.4-12 by the operator

$$\mathbf{\Lambda}_T^{-1/2} = \begin{bmatrix} [\Delta_T(1)]^{-1/2} & & & & & 0 \\ & [\Delta_T(2)]^{-1/2} & & & & \cdot \\ & & \cdot & & & \cdot \\ & & \cdots & & & \cdot \\ \cdot & & & [\Delta_T(T)]^{-1/2} & & \cdot \\ \cdot & & & & 0 & \cdot \\ \cdot & & & & & \cdot \\ & & & & \cdots & \\ 0 & & & & & 0 \end{bmatrix} \tag{12.4-13}$$

(*a*) Blurred observation   (*b*) Restoration, $T = 58$

(*c*) Restoration, $T = 60$

FIGURE 12.4-3. Sequential SVD pseudoinverse image restoration for horizontal Gaussian blur, $b_R = 3.0$, $L = 23$, $J = 256$.

Complete truncation of the high-frequency terms to avoid ill-conditioning effects may not be necessary in all situations. As an alternative to truncation, the diagonal zero elements can be replaced by $[\Delta_T(T)]^{-1/2}$ or perhaps by some sequence that declines in value as a function of frequency. This concept is actually analogous to the truncated inverse filtering technique defined by Eq. 12.2-11 for continuous image fields.

Figure 12.4-3 shows an example of SVD pseudoinverse image restoration for one-dimensional Gaussian image blur with $b_R = 3.0$. It should be noted that the restoration attempt with the standard pseudoinverse shown in Figure 12.3-6*b* was subject to severe ill-conditioning errors at a blur spread of $b_R = 2.0$.

12.5. STATISTICAL ESTIMATION SPATIAL IMAGE RESTORATION

A fundamental limitation of pseudoinverse restoration techniques is that observation noise may lead to severe numerical instability and render the image estimate unusable. This problem can be alleviated in some instances by statistical restoration techniques that incorporate some a priori statistical knowledge of the observation noise (21).

12.5.1. Regression Spatial Image Restoration

Consider the vector-space model

$$\mathbf{g} = \mathbf{Bf} + \mathbf{n} \tag{12.5-1}$$

for a blurred image plus additive noise in which $\mathbf{B}$ is a $P \times Q$ blur matrix and the noise is assumed to be zero mean with known covariance matrix $\mathbf{K_n}$. The regression method seeks to form an estimate

$$\hat{\mathbf{f}} = \mathbf{Wg} \tag{12.5-2}$$

where $\mathbf{W}$ is a restoration matrix that minimizes the weighted error measure

$$\Theta\{\hat{\mathbf{f}}\} = [\mathbf{g} - \mathbf{B}\hat{\mathbf{f}}]^T \mathbf{K_n}^{-1} [\mathbf{g} - \mathbf{B}\hat{\mathbf{f}}] \tag{12.5-3}$$

Minimization of the restoration error can be accomplished by the classical method of setting the partial derivative of $\Theta\{\hat{\mathbf{f}}\}$ with respect to $\hat{\mathbf{f}}$ to zero. In the underdetermined case, for which $P < Q$, it can be shown (30) that the minimum norm estimate regression operator is

$$\mathbf{W} = [\mathbf{K}^{-1}\mathbf{B}]^{-} \mathbf{K}^{-1} \tag{12.5-4}$$

where $\mathbf{K}$ is a matrix obtained from the spectral factorization

$$\mathbf{K_n} = \mathbf{KK}^T \tag{12.5-5}$$

of the noise covariance matrix $\mathbf{K_n}$. For white noise, $\mathbf{K} = \sigma_n^2 \mathbf{I}$, and the regression operator assumes the form of a rank P generalized inverse for an underdetermined system as given by Eq. 12.3-8b.

12.5.2. Wiener Estimation Spatial Image Restoration

With the regression technique of spatial image restoration, the noise field is modeled as a sample of a two-dimensional random process with a known mean and covariance function. Wiener estimation techniques assume, in addition, that the ideal image is also a sample of a two-dimensional random process with known first and second moments (21,22,31).

Wiener Estimation: General Case. Consider the general discrete model of Figure 12.5-1 in which a $Q \times 1$ image vector $\mathbf{f}$ is subject to some unspecified type of point and spatial degradation resulting in the $P \times 1$ vector of observations $\mathbf{g}$. An estimate of $\mathbf{f}$ is formed by the linear operation

$$\hat{\mathbf{f}} = \mathbf{Wg} + \mathbf{b} \qquad (12.5\text{-}6)$$

where $\mathbf{W}$ is a $Q \times P$ restoration matrix and $\mathbf{b}$ is a $Q \times 1$ bias vector. The objective of Wiener estimation is to choose $\mathbf{W}$ and $\mathbf{b}$ to minimize the mean-square restoration error, which may be defined as

$$\mathcal{E} = E\{[\mathbf{f} - \hat{\mathbf{f}}]^T [\mathbf{f} - \hat{\mathbf{f}}]\} \qquad (12.5\text{-}7a)$$

or

$$\mathcal{E} = \mathrm{tr}\{E\{[\mathbf{f} - \hat{\mathbf{f}}][\mathbf{f} - \hat{\mathbf{f}}]^T\}\} \qquad (12.5\text{-}7b)$$

Equation 12.5-7a expresses the error in inner-product form as the sum of the squares of the elements of the error vector $[\mathbf{f} - \hat{\mathbf{f}}]$, while Eq. 12.5-7b forms the covariance matrix of the error, and then sums together its variance terms (diagonal elements) by the trace operation. Minimization of Eq. 12.5-7 in either of its forms can be accomplished by differentiation of $\mathcal{E}$ with respect to $\hat{\mathbf{f}}$. An alternative approach,

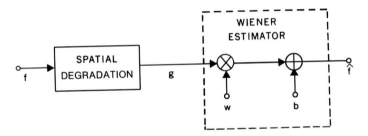

FIGURE 12.5-1. Wiener estimation for spatial image restoration.

which is of quite general utility, is to employ the *orthogonality principle* (32, p. 219) to determine the values of **W** and **b** that minimize the mean-square error. In the context of image restoration, the orthogonality principle specifies two necessary and sufficient conditions for the minimization of the mean-square restoration error:

1. The expected value of the image estimate must equal the expected value of the image

$$E\{\hat{\mathbf{f}}\} = E\{\mathbf{f}\} \qquad (12.5\text{-}8)$$

2. The restoration error must be orthogonal to the observation about its mean

$$E\{[\mathbf{f} - \hat{\mathbf{f}}][\mathbf{g} - E\{\mathbf{g}\}]^{T}\} = \mathbf{0} \qquad (12.5\text{-}9)$$

From condition 1, one obtains

$$\mathbf{b} = E\{\mathbf{f}\} - \mathbf{W}E\{\mathbf{g}\} \qquad (12.5\text{-}10)$$

and from condition 2

$$E\{[\mathbf{W} + \mathbf{b} - \mathbf{f}][\mathbf{g} - E\{\mathbf{g}\}]^{T}\} = \mathbf{0} \qquad (12.5\text{-}11)$$

Upon substitution for the bias vector **b** from Eq. 12.5-10 and simplification, Eq. 12.5-11 yields

$$\mathbf{W} = \mathbf{K}_{\mathbf{fg}}[\mathbf{K}_{\mathbf{gg}}]^{-1} \qquad (12.5\text{-}12)$$

where $\mathbf{K}_{\mathbf{gg}}$ is the $P \times P$ covariance matrix of the observation vector (assumed nonsingular) and $\mathbf{K}_{\mathbf{fg}}$ is the $Q \times P$ cross-covariance matrix between the image and observation vectors. Thus, the optimal bias vector **b** and restoration matrix **W** may be directly determined in terms of the first and second joint moments of the ideal image and observation vectors. It should be noted that these solutions apply for nonlinear and space-variant degradations. Subsequent sections describe applications of Wiener estimation to specific restoration models.

Wiener Estimation: Image Blur with Additive Noise. For the discrete model for a blurred image subjective to additive noise given by

$$\mathbf{g} = \mathbf{Bf} + \mathbf{n} \qquad (12.5\text{-}13)$$

the Wiener estimator is composed of a bias term

$$\mathbf{b} = E\{\mathbf{f}\} - \mathbf{W}E\{\mathbf{g}\} = E\{\mathbf{f}\} - \mathbf{W}BE\{\mathbf{f}\} + \mathbf{W}E\{\mathbf{n}\} \tag{12.5-14}$$

and a matrix operator

$$\mathbf{W} = \mathbf{K}_{\mathbf{fg}}[\mathbf{K}_{\mathbf{gg}}]^{-1} = \mathbf{K}_{\mathbf{f}}\mathbf{B}^T[\mathbf{B}\mathbf{K}_{\mathbf{f}}\mathbf{B}^T + \mathbf{K}_{\mathbf{n}}]^{-1} \tag{12.5-15}$$

If the ideal image field is assumed uncorrelated, $\mathbf{K}_{\mathbf{f}} = \sigma_f^2\mathbf{I}$ where σ_f^2 represents the image energy. Equation 12.5-15 then reduces to

$$\mathbf{W} = \sigma_f^2\mathbf{B}^T[\sigma_f^2\mathbf{B}\mathbf{B}^T + \mathbf{K}_{\mathbf{n}}]^{-1} \tag{12.5-16}$$

For a white-noise process with energy σ_n^2, the Wiener filter matrix becomes

$$\mathbf{W} = \mathbf{B}^T\left(\mathbf{B}\mathbf{B}^T + \frac{\sigma_n^2}{\sigma_f^2}\mathbf{I}\right) \tag{12.5-17}$$

As the ratio of image energy to noise energy (σ_f^2/σ_n^2) approaches infinity, the Wiener estimator of Eq. 12.5-17 becomes equivalent to the generalized inverse estimator.

Figure 12.5-2 shows restoration examples for the model of Figure 11.5-3 for a Gaussian-shaped blur function. Wiener restorations of large size images are given in Figure 12.5-3 using a fast computational algorithm developed by Pratt and Davarian (22). In the example of Figure 12.5-3a illustrating horizontal image motion blur, the impulse response is of rectangular shape of length $L = 11$. The center pixels have been restored and replaced within the context of the blurred image to show the visual restoration improvement. The noise level and blur impulse response of the electron microscope original image of Figure 12.5-3c were estimated directly from the photographic transparency using techniques to be described in Section 12.7. The parameters were then utilized to restore the center pixel region, which was then replaced in the context of the blurred original.

12.6. CONSTRAINED IMAGE RESTORATION

The previously described image restoration techniques have treated images as arrays of numbers. They have not considered that a restored natural image should be subject to physical constraints. A restored natural image should be spatially smooth and strictly positive in amplitude.

Blurred   Restored

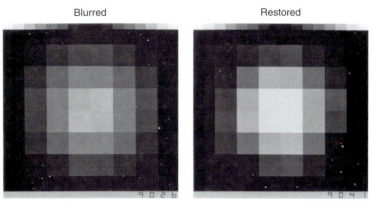

$b_R = b_C = 1.2$, Var = 10.0, $\rho = 0.75$, SNR = 200.0

(a) PLSE = 4.91%   (b) PLSE = 3.71%

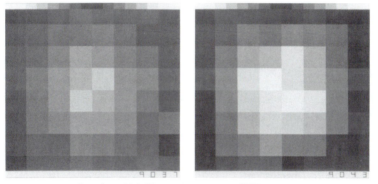

$b_R = b_C = 50.0$, Var = 10.0, $\rho = 0.75$, SNR = 200.0

(c) PLSE = 7.99%   (d) PLSE = 4.20%

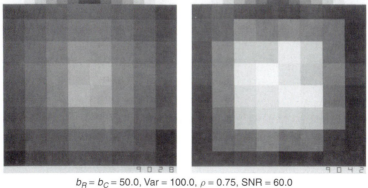

$b_R = b_C = 50.0$, Var = 100.0, $\rho = 0.75$, SNR = 60.0

(e) PLSE = 7.93%   (f) PLSE = 4.74%

FIGURE 12.5-2. Wiener estimation for test image blurred with Gaussian-shaped impulse response. $M = 8$, $N = 12$, $L = 5$.

(*a*) Observation   (*b*) Restoration

(*c*) Observation   (*d*) Restoration

FIGURE 12.5-3. Wiener image restoration.

12.6.1. Smoothing Methods

Smoothing and regularization techniques (33–35) have been used in an attempt to overcome the ill-conditioning problems associated with image restoration. Basically, these methods attempt to force smoothness on the solution of a least-squares error problem.

Two formulations of these methods are considered (21). The first formulation consists of finding the minimum of $\hat{\mathbf{f}}^T \mathbf{S} \hat{\mathbf{f}}$ subject to the equality constraint

$$[\mathbf{g} - \mathbf{B}\hat{\mathbf{f}}]^T \mathbf{M}[\mathbf{g} - \mathbf{B}\hat{\mathbf{f}}] = e \qquad (12.6\text{-}1)$$

where $\mathbf{S}$ is a smoothing matrix, $\mathbf{M}$ is an error-weighting matrix, and e denotes a residual scalar estimation error. The error-weighting matrix is often chosen to be equal to the inverse of the observation noise covariance matrix, $\mathbf{M} = \mathbf{K_n}^{-1}$. The Lagrangian estimate satisfying Eq. 12.6-1 is (19)

$$\hat{\mathbf{f}} = \mathbf{S}^{-1}\mathbf{B}^T\left[\mathbf{BS}^{-1}\mathbf{B}^T + \frac{1}{\gamma}\mathbf{M}^{-1}\right]^{-1}\mathbf{g} \qquad (12.6\text{-}2)$$

In Eq. 12.6-2, the Lagrangian factor γ is chosen so that Eq. 12.6-1 is satisfied; that is, the compromise between residual error and smoothness of the estimator is deemed satisfactory.

Now consider the second formulation, which involves solving an equality-constrained least-squares problem by minimizing the left-hand side of Eq. 12.6-1 such that

$$\hat{\mathbf{f}}^T\mathbf{S}\hat{\mathbf{f}} = d \qquad (12.6\text{-}3)$$

where the scalar d represents a fixed degree of smoothing. In this case, the optimal solution for an underdetermined nonsingular system is found to be

$$\hat{\mathbf{f}} = \mathbf{S}^{-1}\mathbf{B}^T[\mathbf{BS}^{-1}\mathbf{B}^T + \gamma\mathbf{M}^{-1}]^{-1}\mathbf{g} \qquad (12.6\text{-}4)$$

A comparison of Eqs. 12.6-2 and 12.6-4 reveals that the two inverse problems are solved by the same expression, the only difference being the Lagrange multipliers, which are inverses of one another. The smoothing estimates of Eq. 12.6-4 are closely related to the regression and Wiener estimates derived previously. If $\gamma = 0$, $\mathbf{S} = \mathbf{I}$ and $\mathbf{M} = \mathbf{K_n}^{-1}$ where $\mathbf{K_n}$ is the observation noise covariance matrix, then the smoothing and regression estimates become equivalent. Substitution of $\gamma = 1$, $\mathbf{S} = \mathbf{K_f}^{-1}$ and $\mathbf{M} = \mathbf{K_n}^{-1}$ where $\mathbf{K_f}$ is the image covariance matrix results in equivalence to the Wiener estimator. These equivalences account for the relative smoothness of the estimates obtained with regression and Wiener restoration as compared to pseudoinverse restoration. A problem that occurs with the smoothing and regularizing techniques is that even though the variance of a solution can be calculated, its bias can only be determined as a function of $\mathbf{f}$.

12.6.2. Constrained Restoration Techniques

Equality and inequality constraints have been suggested (21) as a means of improving restoration performance for ill-conditioned restoration models. Examples of constraints include the specification of individual pixel values, of ratios of the values of some pixels, or the sum of part or all of the pixels, or amplitude limits of pixel values.

Quite often a priori information is available in the form of inequality constraints involving pixel values. The physics of the image formation process requires that pixel values be non-negative quantities. Furthermore, an upper bound on these values is often known because images are digitized with a finite number of bits assigned to each pixel. Amplitude constraints are also inherently introduced by the need to "fit" a restored image to the dynamic range of a display. One approach is linearly to rescale the restored image to the display image. This procedure is usually undesirable because only a few out-of-range pixels will cause the contrast of all other pixels to be reduced. Also, the average luminance of a restored image is usually affected by rescaling. Another common display method involves clipping of all pixel values exceeding the display limits. Although this procedure is subjectively preferable to rescaling, bias errors may be introduced.

If a priori pixel amplitude limits are established for image restoration, it is best to incorporate these limits directly in the restoration process rather than arbitrarily invoke the limits on the restored image. Several techniques of inequality constrained restoration have been proposed.

Consider the general case of constrained restoration in which the vector estimate $\hat{\mathbf{f}}$ is subject to the inequality constraint

$$\mathbf{l} \leq \hat{\mathbf{f}} \leq \mathbf{u} \tag{12.6-5}$$

where $\mathbf{u}$ and $\mathbf{l}$ are vectors containing upper and lower limits of the pixel estimate, respectively. For least-squares restoration, the quadratic error must be minimized subject to the constraint of Eq. 12.6-5. Under this framework, restoration reduces to the solution of a quadratic programming problem (21). In the case of an absolute error measure, the restoration task can be formulated as a linear programming problem (36,37). The a priori knowledge involving the inequality constraints may substantially reduce pixel uncertainty in the restored image; however, as in the case of equality constraints, an unknown amount of bias may be introduced.

Figure 12.6-1 is an example of image restoration for the Gaussian blur model of Chapter 11 by pseudoinverse restoration and with inequality constrained (21) in which the scaled luminance of each pixel of the restored image has been limited to the range of 0 to 255. The improvement obtained by the constraint is substantial. Unfortunately, the quadratic programming solution employed in this example requires a considerable amount of computation. A brute-force extension of the procedure does not appear feasible.

Several other methods have been proposed for constrained image restoration. One simple approach, based on the concept of homomorphic filtering, is to take the logarithm of each observation. Exponentiation of the corresponding estimates automatically yields a strictly positive result. Burg (38), Edward and Fitelson (39), and Frieden (6,40,41) have developed restoration methods providing a positivity constraint, which are based on a maximum entropy principle originally employed to estimate a probability density from observation of its moments. Huang et al. (42) have introduced a projection method of constrained image restoration in which the set of equations $\mathbf{g} = \mathbf{Bf}$ are iteratively solved by numerical means. At each stage of the solution the intermediate estimates are amplitude clipped to conform to amplitude limits.

(*a*) Blurred observation

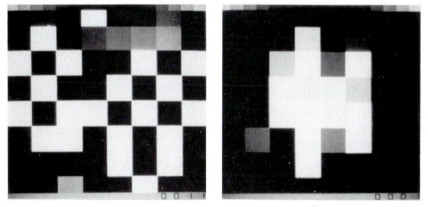

(*b*) Unconstrained restoration   (*c*) Constrained restoration

FIGURE 12.6-1. Comparison of unconstrained and inequality constrained image restoration for a test image blurred with Gaussian-shaped impulse response. $b_R = b_C = 1.2$, $M = 12$, $N = 8$, $L = 5$; noisy observation, Var = 10.0.

12.7. BLIND IMAGE RESTORATION

Most image restoration techniques are based on some a priori knowledge of the image degradation; the point luminance and spatial impulse responses of the system degradation are assumed known. In many applications, such information is simply not available. The degradation may be difficult to measure or may be time varying in an unpredictable manner. In such cases, information about the degradation must be extracted from the observed image either explicitly or implicitly. This task is called *blind image restoration* (5,19,43). Discussion here is limited to blind image restoration methods for blurred images subject to additive noise.

There are two major approaches to blind image restoration: direct measurement and indirect estimation. With the former approach, the blur impulse response and noise level are first estimated from an image to be restored, and then these parameters are utilized in the restoration. Indirect estimation techniques employ temporal or spatial averaging to either obtain a restoration or to determine key elements of a restoration algorithm.

12.7.1. Direct Measurement Methods

Direct measurement blind restoration of a blurred noisy image usually requires measurement of the blur impulse response and noise power spectrum or covariance function of the observed image. The blur impulse response is usually measured by isolating the image of a suspected object within a picture. By definition, the blur impulse response is the image of a point-source object. Therefore, a point source in the observed scene yields a direct indication of the impulse response. The image of a suspected sharp edge can also be utilized to derive the blur impulse response. Averaging several parallel line scans normal to the edge will significantly reduce noise effects. The noise covariance function of an observed image can be estimated by measuring the image covariance over a region of relatively constant background luminance. References 5, 44, and 45 provide further details on direct measurement methods.

12.7.2. Indirect Estimation Methods

Temporal redundancy of scenes in real-time television systems can be exploited to perform blind restoration indirectly. As an illustration, consider the ith observed image frame

$$G_i(x, y) = F_I(x, y) + N_i(x, y) \tag{12.7-1}$$

of a television system in which $F_I(x, y)$ is an ideal image and $N_i(x, y)$ is an additive noise field independent of the ideal image. If the ideal image remains constant over a sequence of M frames, then temporal summation of the observed images yields the relation

$$F_I(x, y) = \frac{1}{M} \sum_{i=1}^{M} G_i(x, y) - \frac{1}{M} \sum_{i=1}^{M} N_i(x, y) \tag{12.7-2}$$

The value of the noise term on the right will tend toward its ensemble average $E\{N(x, y)\}$ for M large. In the common case of zero-mean white Gaussian noise, the

(*a*) `dolls_linear`

(*b*) `dolls_gamma`

Color photographs of the `dolls_linear` and the `dolls_gamma` color images. See pages 74 and 80 for discussion of these images.

(*a*) Gray scale chart

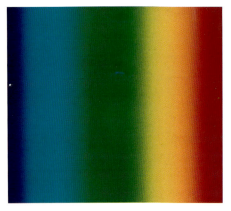

(*b*) Pseudocolor of chart

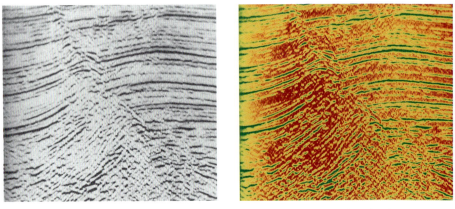

(*c*) Seismic

(*d*) Pseudocolor of seismic

Figure 10.5-3. Pseudocoloring of the `gray_chart` and `seismic` images. See page 288 for discussion of this figure.

(*a*) Infrared band

(*b*) Blue band

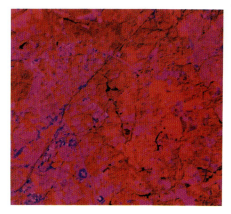

(*c*) R = infrared, G = 0, B = blue

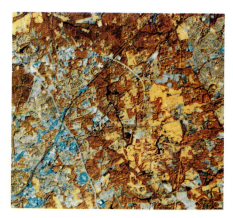

(*d*) R = infrared, G = 1/2 [infrared + blue],
B = blue

Figure 10.5-4. False coloring of multispectral images. See page 290 for discussion of this figure.

(a) Color representation

(b) Red component

(c) Green component

(d) Blue component

Figure 15.6-1. The `peppers_gamma` color image and its *RGB* color components. See page 502 for discussion of this figure.

(*a*) Noise-free original

(*b*) Noisy image 1

(*c*) Noisy image 2

(*d*) Temporal average

FIGURE 12.7-1 Temporal averaging of a sequence of eight noisy images. SNR = 10.0.

ensemble average is zero at all (*x*, y), and it is reasonable to form the estimate as

$$\hat{F}_I(x, y) = \frac{1}{M} \sum_{i=1}^{M} G_i(x, y) \tag{12.7-3}$$

Figure 12.7-1 presents a computer-simulated example of temporal averaging of a sequence of noisy images. In this example the original image is unchanged in the sequence. Each image observed is subjected to a different additive random noise pattern.

The concept of temporal averaging is also useful for image deblurring. Consider an imaging system in which sequential frames contain a relatively stationary object degraded by a different linear-shift invariant impulse response $H_i(x, y)$ over each frame. This type of imaging would be encountered, for example, when photographing distant objects through a turbulent atmosphere if the object does not move significantly between frames. By taking a short exposure at each frame, the atmospheric turbulence is "frozen" in space at each frame interval. For this type of object, the degraded image at the ith frame interval is given by

$$G_i(x, y) = F_I(x, y) \circledast H_i(x, y) \tag{12.7-4}$$

for $i = 1, 2, ..., M$. The Fourier spectra of the degraded images are then

$$\mathcal{G}_i(\omega_x, \omega_y) = \mathcal{F}_I(\omega_x, \omega_y)\mathcal{H}_i(\omega_x, \omega_y) \tag{12.7-5}$$

On taking the logarithm of the degraded image spectra

$$\ln\{\mathcal{G}_i(\omega_x, \omega_y)\} = \ln\{\mathcal{F}_I(\omega_x, \omega_y)\} + \ln\{\mathcal{H}_i(\omega_x, \omega_y)\} \tag{12.7-6}$$

the spectra of the ideal image and the degradation transfer function are found to separate additively. It is now possible to apply any of the common methods of statistical estimation of a signal in the presence of additive noise. If the degradation impulse responses are uncorrelated between frames, it is worthwhile to form the sum

$$\sum_{i=1}^{M} \ln\{\mathcal{G}_i(\omega_x, \omega_y)\} = M \ln\{\mathcal{F}_I(\omega_x, \omega_y)\} + \sum_{i=1}^{M} \ln\{\mathcal{H}_i(\omega_x, \omega_y)\} \tag{12.7-7}$$

because for large M the latter summation approaches the constant value

$$\mathcal{H}_M(\omega_x, \omega_y) = \lim_{M \to \infty}\left\{\sum_{i=1}^{M} \ln\{\mathcal{H}_i(\omega_x, \omega_y)\}\right\} \tag{12.7-8}$$

The term $\mathcal{H}_M(\omega_x, \omega_y)$ may be viewed as the average logarithm transfer function of the atmospheric turbulence. An image estimate can be expressed as

$$\hat{\mathcal{F}}_I(\omega_x, \omega_y) = \exp\left\{-\frac{\mathcal{H}_M(\omega_x, \omega_y)}{M}\right\}\prod_{i=1}^{M}[\mathcal{G}_i(\omega_x, \omega_y)]^{1/M} \tag{12.7-9}$$

An inverse Fourier transform then yields the spatial domain estimate. In any practical imaging system, Eq. 12.7-4 must be modified by the addition of a noise component $N_i(x, y)$. This noise component unfortunately invalidates the separation step of Eq. 12.7-6, and therefore destroys the remainder of the derivation. One possible ad hoc solution to this problem would be to perform noise smoothing or filtering on each observed image field and then utilize the resulting estimates as assumed noise-less observations in Eq. 12.7-9. Alternatively, the blind restoration technique of Stockham et al. (43) developed for nonstationary speech signals may be adapted to the multiple-frame image restoration problem.

REFERENCES

1. D. A. O'Handley and W. B. Green, "Recent Developments in Digital Image Processing at the Image Processing Laboratory at the Jet Propulsion Laboratory," *Proc. IEEE*, **60**, 7, July 1972, 821–828.

2. M. M. Sondhi, "Image Restoration: The Removal of Spatially Invariant Degradations," *Proc. IEEE*, **60**, 7, July 1972, 842–853.

3. H. C. Andrews, "Digital Image Restoration: A Survey," *IEEE Computer*, **7**, 5, May 1974, 36–45.

4. B. R. Hunt, "Digital Image Processing," *Proc. IEEE*, **63**, 4, April 1975, 693–708.

5. H. C. Andrews and B. R. Hunt, *Digital Image Restoration*, Prentice Hall, Englewood Cliffs, NJ, 1977.

6. B. R. Frieden, "Image Enhancement and Restoration," in *Picture Processing and Digital Filtering*, T. S. Huang, Ed., Springer-Verlag, New York, 1975.

7. T. G. Stockham, Jr., "A–D and D–A Converters: Their Effect on Digital Audio Fidelity," in *Digital Signal Processing*, L. R. Rabiner and C. M. Rader, Eds., IEEE Press, New York, 1972, 484–496.

8. A. Marechal, P. Croce, and K. Dietzel, "Amelioration du contrast des details des images photographiques par filtrage des fréquencies spatiales," *Optica Acta*, **5**, 1958, 256–262.

9. J. Tsujiuchi, "Correction of Optical Images by Compensation of Aberrations and by Spatial Frequency Filtering," in *Progress in Optics*, Vol. **2**, E. Wolf, Ed., Wiley, New York, 1963, 131–180.

10. J. L. Harris, Sr., "Image Evaluation and Restoration," *J. Optical Society of America*, **56**, 5, May 1966, 569–574.

11. B. L. McGlamery, "Restoration of Turbulence-Degraded Images," *J. Optical Society of America*, **57**, 3, March 1967, 293–297.

12. P. F. Mueller and G. O. Reynolds, "Image Restoration by Removal of Random Media Degradations," *J. Optical Society of America*, **57**, 11, November 1967, 1338–1344.

13. C. W. Helstrom, "Image Restoration by the Method of Least Squares," *J. Optical Society of America*, **57**, 3, March 1967, 297–303.

14. J. L. Harris, Sr., "Potential and Limitations of Techniques for Processing Linear Motion-Degraded Imagery," in *Evaluation of Motion Degraded Images*, US Government Printing Office, Washington DC, 1968, 131–138.

15. J. L. Homer, "Optical Spatial Filtering with the Least-Mean-Square-Error Filter," *J. Optical Society of America*, **51**, 5, May 1969, 553–558.

16. J. L. Homer, "Optical Restoration of Images Blurred by Atmospheric Turbulence Using Optimum Filter Theory," *Applied Optics*, **9**, 1, January 1970, 167–171.

17. B. L. Lewis and D. J. Sakrison, "Computer Enhancement of Scanning Electron Micrographs," *IEEE Trans. Circuits and Systems*, **CAS-22**, 3, March 1975, 267–278.

18. D. Slepian, "Restoration of Photographs Blurred by Image Motion," *Bell System Technical J.*, **XLVI**, 10, December 1967, 2353–2362.

19. E. R. Cole, "The Removal of Unknown Image Blurs by Homomorphic Filtering," Ph.D. dissertation, Department of Electrical Engineering, University of Utah, Salt Lake City, UT June 1973.

20. B. R. Hunt, "The Application of Constrained Least Squares Estimation to Image Restoration by Digital Computer," *IEEE Trans. Computers*, **C-23**, 9, September 1973, 805–812.

21. N. D. A. Mascarenhas and W. K. Pratt, "Digital Image Restoration Under a Regression Model," *IEEE Trans. Circuits and Systems*, **CAS-22**, 3, March 1975, 252–266.

22. W. K. Pratt and F. Davarian, "Fast Computational Techniques for Pseudoinverse and Wiener Image Restoration," *IEEE Trans. Computers*, **C-26**, 6, June 1977, 571–580.

23. W. K. Pratt, "Pseudoinverse Image Restoration Computational Algorithms," in *Optical Information Processing* Vol. **2**, G. W. Stroke, Y. Nesterikhin, and E. S. Barrekette, Eds., Plenum Press, New York, 1977.

24. B. W. Rust and W. R. Burrus, *Mathematical Programming and the Numerical Solution of Linear Equations*, American Elsevier, New York, 1972.

25. A. Albert, *Regression and the Moore–Penrose Pseudoinverse*, Academic Press, New York, 1972.

26. H. C. Andrews and C. L. Patterson, "Outer Product Expansions and Their Uses in Digital Image Processing," *American Mathematical. Monthly*, **1**, 82, January 1975, 1–13.

27. H. C. Andrews and C. L. Patterson, "Outer Product Expansions and Their Uses in Digital Image Processing," *IEEE Trans. Computers*, **C-25**, 2, February 1976, 140–148.

28. T. S. Huang and P. M. Narendra, "Image Restoration by Singular Value Decomposition," *Applied Optics*, **14**, 9, September 1975, 2213–2216.

29. H. C. Andrews and C. L. Patterson, "Singular Value Decompositions and Digital Image Processing," *IEEE Trans. Acoustics, Speech, and Signal Processing*, **ASSP-24**, 1, February 1976, 26–53.

30. T. O. Lewis and P. L. Odell, *Estimation in Linear Models*, Prentice Hall, Englewood Cliffs, NJ, 1971.

31. W. K. Pratt, "Generalized Wiener Filter Computation Techniques," *IEEE Trans. Computers*, **C-21**, 7, July 1972, 636–641.

32. A. Papoulis, *Probability Random Variables and Stochastic Processes*, 3rd Ed., McGraw-Hill, New York, 1991.

33. S. Twomey, "On the Numerical Solution of Fredholm Integral Equations of the First Kind by the Inversion of the Linear System Produced by Quadrature," *J. Association for Computing Machinery*, **10**, 1963, 97–101.

34. D. L. Phillips, "A Technique for the Numerical Solution of Certain Integral Equations of the First Kind," *J. Association for Computing Machinery*, **9**, 1964, 84-97.

35. A. N. Tikonov, "Regularization of Incorrectly Posed Problems," *Soviet Mathematics*, **4**, 6, 1963, 1624–1627.

36. E. B. Barrett and R. N. Devich, "Linear Programming Compensation for Space-Variant Image Degradation," *Proc. SPIE/OSA Conference on Image Processing*, J. C. Urbach, Ed., Pacific Grove, CA, February 1976, **74**, 152–158.

37. D. P. MacAdam, "Digital Image Restoration by Constrained Deconvolution," *J. Optical Society of America*, **60**, 12, December 1970, 1617–1627.

38. J. P. Burg, "Maximum Entropy Spectral Analysis," 37th Annual Society of Exploration Geophysicists Meeting, Oklahoma City, OK, 1967.

39. J. A. Edward and M. M. Fitelson, "Notes on Maximum Entropy Processing," *IEEE Trans. Information Theory*, **IT-19**, 2, March 1973, 232–234.

40. B. R. Frieden, "Restoring with Maximum Likelihood and Maximum Entropy," *J. Optical Society America*, **62**, 4, April 1972, 511–518.

41. B. R. Frieden, "Maximum Entropy Restorations of Garrymede," *in Proc. SPIE/OSA Conference on Image Processing*, J. C. Urbach, Ed., Pacific Grove, CA, February 1976, **74**, 160–165.

42. T. S. Huang, D. S. Baker, and S. P. Berger, "Iterative Image Restoration," *Applied Optics*, **14**, 5, May 1975, 1165–1168.

43. T. G. Stockham, Jr., T. M. Cannon, and P. B. Ingebretsen, "Blind Deconvolution Through Digital Signal Processing," *Proc. IEEE*, **63**, 4, April 1975, 678–692.

44. A. Papoulis, "Approximations of Point Spreads for Deconvolution," *J. Optical Society of America*, **62**, 1, January 1972, 77–80.

45. B. Tatian, "Asymptotic Expansions for Correcting Truncation Error in Transfer-Function Calculations," *J. Optical Society of America*, **61**, 9, September 1971, 1214–1224.

13

GEOMETRICAL IMAGE MODIFICATION

One of the most common image processing operations is geometrical modification in which an image is spatially translated, scaled, rotated, nonlinearly warped, or viewed from a different perspective.

13.1. TRANSLATION, MINIFICATION, MAGNIFICATION, AND ROTATION

Image translation, scaling, and rotation can be analyzed from a unified standpoint. Let $G(j, k)$ for $1 \leq j \leq J$ and $1 \leq k \leq K$ denote a discrete output image that is created by geometrical modification of a discrete input image $F(p, q)$ for $1 \leq p \leq P$ and $1 \leq q \leq Q$. In this derivation, the input and output images may be different in size. Geometrical image transformations are usually based on a Cartesian coordinate system representation in which the origin $(0, 0)$ is the lower left corner of an image, while for a discrete image, typically, the upper left corner unit dimension pixel at indices $(1, 1)$ serves as the address origin. The relationships between the Cartesian coordinate representations and the discrete image arrays of the input and output images are illustrated in Figure 13.1-1. The output image array indices are related to their Cartesian coordinates by

$$x_k = k - \tfrac{1}{2} \qquad \qquad (13.1\text{-}1a)$$

$$y_k = J + \tfrac{1}{2} - j \qquad \qquad (13.1\text{-}1b)$$

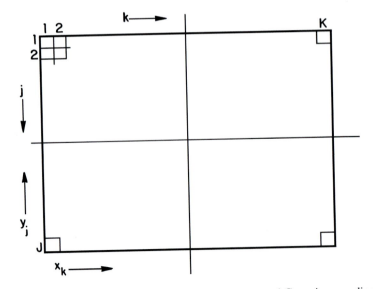

FIGURE 13.1-1. Relationship between discrete image array and Cartesian coordinate representation.

Similarly, the input array relationship is given by

$$u_q = q - \frac{1}{2} \tag{13.1-2a}$$

$$v_p = P + \frac{1}{2} - p \tag{13.1-2b}$$

13.1.1. Translation

Translation of $F(p, q)$ with respect to its Cartesian origin to produce $G(j, k)$ involves the computation of the relative offset addresses of the two images. The translation address relationships are

$$x_k = u_q + t_x \tag{13.1-3a}$$

$$y_j = v_p + t_y \tag{13.1-3b}$$

where t_x and t_y are translation offset constants. There are two approaches to this computation for discrete images: forward and reverse address computation. In the forward approach, u_q and v_p are computed for each input pixel (p, q) and substituted into Eq. 13.1-3 to obtain x_k and y_j. Next, the output array addresses (j, k) are computed by inverting Eq. 13.1-1. The composite computation reduces to

$$j' = p - (P - J) - t_y \tag{13.1-4a}$$

$$k' = q + t_x \tag{13.1-4b}$$

where the prime superscripts denote that j' and k' are not integers unless t_x and t_y are integers. If j' and k' are rounded to their nearest integer values, data voids can occur in the output image. The reverse computation approach involves calculation of the input image addresses for integer output image addresses. The composite address computation becomes

$$p' = j + (P - J) + t_y \tag{13.1-5a}$$

$$q' = k - t_x \tag{13.1-5b}$$

where again, the prime superscripts indicate that p' and q' are not necessarily integers. If they are not integers, it becomes necessary to interpolate pixel amplitudes of $F(p, q)$ to generate a resampled pixel estimate $\hat{F}(p, q)$, which is transferred to $G(j, k)$. The geometrical resampling process is discussed in Section 13.5.

13.1.2. Scaling

Spatial size scaling of an image can be obtained by modifying the Cartesian coordinates of the input image according to the relations

$$x_k = s_x u_q \tag{13.1-6a}$$

$$y_j = s_y v_p \tag{13.1-6b}$$

where s_x and s_y are positive-valued scaling constants, but not necessarily integer valued. If s_x and s_y are each greater than unity, the address computation of Eq. 13.1-6 will lead to magnification. Conversely, if s_x and s_y are each less than unity, minification results. The reverse address relations for the input image address are found to be

$$p' = (1/s_y)(j + J - \tfrac{1}{2}) + P + \tfrac{1}{2} \tag{13.1-7a}$$

$$q' = (1/s_x)(k - \tfrac{1}{2}) + \tfrac{1}{2} \tag{13.1-7b}$$

As with generalized translation, it is necessary to interpolate $F(p, q)$ to obtain $G(j, k)$.

13.1.3. Rotation

Rotation of an input image about its Cartesian origin can be accomplished by the address computation

$$x_k = u_q \cos \theta - v_p \sin \theta \qquad (13.1\text{-}8a)$$

$$y_j = u_q \sin \theta + v_p \cos \theta \qquad (13.1\text{-}8b)$$

where θ is the counterclockwise angle of rotation with respect to the horizontal axis of the input image. Again, interpolation is required to obtain $G(j, k)$. Rotation of an input image about an arbitrary pivot point can be accomplished by translating the origin of the image to the pivot point, performing the rotation, and then translating back by the first translation offset. Equation 13.1-8 must be inverted and substitutions made for the Cartesian coordinates in terms of the array indices in order to obtain the reverse address indices (p', q'). This task is straightforward but results in a messy expression. A more elegant approach is to formulate the address computation as a vector-space manipulation.

13.1.4. Generalized Linear Geometrical Transformations

The vector-space representations for translation, scaling, and rotation are given below.

Translation:

$$\begin{bmatrix} x_k \\ y_j \end{bmatrix} = \begin{bmatrix} u_q \\ v_p \end{bmatrix} + \begin{bmatrix} t_x \\ t_y \end{bmatrix} \qquad (13.1\text{-}9)$$

Scaling:

$$\begin{bmatrix} x_k \\ y_j \end{bmatrix} = \begin{bmatrix} s_x & 0 \\ 0 & s_y \end{bmatrix} \begin{bmatrix} u_q \\ v_p \end{bmatrix} \qquad (13.1\text{-}10)$$

Rotation:

$$\begin{bmatrix} x_k \\ y_j \end{bmatrix} = \begin{bmatrix} \cos \theta & -\sin \theta \\ \sin \theta & \cos \theta \end{bmatrix} \begin{bmatrix} u_q \\ v_p \end{bmatrix} \qquad (13.1\text{-}11)$$

Now, consider a compound geometrical modification consisting of translation, followed by scaling followed by rotation. The address computations for this compound operation can be expressed as

$$
\begin{bmatrix} x_k \\ y_j \end{bmatrix} = \begin{bmatrix} \cos\theta & -\sin\theta \\ \sin\theta & \cos\theta \end{bmatrix} \begin{bmatrix} s_x & 0 \\ 0 & s_y \end{bmatrix} \begin{bmatrix} u_q \\ v_p \end{bmatrix} + \begin{bmatrix} \cos\theta & -\sin\theta \\ \sin\theta & \cos\theta \end{bmatrix} \begin{bmatrix} s_x & 0 \\ 0 & s_y \end{bmatrix} \begin{bmatrix} t_x \\ t_y \end{bmatrix} \quad (13.1\text{-}12a)
$$

or upon consolidation

$$
\begin{bmatrix} x_k \\ y_j \end{bmatrix} = \begin{bmatrix} s_x\cos\theta & -s_y\sin\theta \\ s_x\sin\theta & s_y\cos\theta \end{bmatrix} \begin{bmatrix} u_q \\ v_p \end{bmatrix} + \begin{bmatrix} s_x t_x\cos\theta - s_y t_y\sin\theta \\ s_x t_x\sin\theta + s_y t_y\cos\theta \end{bmatrix} \quad (13.1\text{-}12b)
$$

Equation 13.1-12b is, of course, linear. It can be expressed as

$$
\begin{bmatrix} x_k \\ y_j \end{bmatrix} = \begin{bmatrix} c_0 & c_1 \\ d_0 & d_1 \end{bmatrix} \begin{bmatrix} u_q \\ v_p \end{bmatrix} + \begin{bmatrix} c_2 \\ d_2 \end{bmatrix} \quad (13.1\text{-}13a)
$$

in one-to-one correspondence with Eq. 13.1-12b. Equation 13.1-13a can be rewritten in the more compact form

$$
\begin{bmatrix} x_k \\ y_j \end{bmatrix} = \begin{bmatrix} c_0 & c_1 & c_2 \\ d_0 & d_1 & d_2 \end{bmatrix} \begin{bmatrix} u_q \\ v_p \\ 1 \end{bmatrix} \quad (13.1\text{-}13b)
$$

As a consequence, the three address calculations can be obtained as a single linear address computation. It should be noted, however, that the three address calculations are not commutative. Performing rotation followed by minification followed by translation results in a mathematical transformation different than Eq. 13.1-12. The overall results can be made identical by proper choice of the individual transformation parameters.

To obtain the reverse address calculation, it is necessary to invert Eq. 13.1-13b to solve for (u_q, v_p) in terms of (x_k, y_j). Because the matrix in Eq. 13.1-13b is not square, it does not possess an inverse. Although it is possible to obtain (u_q, v_p) by a pseudoinverse operation, it is convenient to augment the rectangular matrix as follows:

$$\begin{bmatrix} x_k \\ y_j \\ 1 \end{bmatrix} = \begin{bmatrix} c_0 & c_1 & c_2 \\ d_0 & d_1 & d_2 \\ 0 & 0 & 1 \end{bmatrix} \begin{bmatrix} u_q \\ v_p \\ 1 \end{bmatrix} \tag{13.1-14}$$

This three-dimensional vector representation of a two-dimensional vector is a special case of a *homogeneous coordinates* representation (1–3).

The use of homogeneous coordinates enables a simple formulation of concatenated operators. For example, consider the rotation of an image by an angle θ about a pivot point (x_c, y_c) in the image. This can be accomplished by

$$\begin{bmatrix} x_k \\ y_j \\ 1 \end{bmatrix} = \begin{bmatrix} 1 & 0 & x_c \\ 0 & 1 & y_c \\ 0 & 0 & 1 \end{bmatrix} \begin{bmatrix} \cos\theta & -\sin\theta & 0 \\ \sin\theta & \cos\theta & 0 \\ 0 & 0 & 1 \end{bmatrix} \begin{bmatrix} 1 & 0 & -x_c \\ 0 & 1 & -y_c \\ 0 & 0 & 1 \end{bmatrix} \begin{bmatrix} u_q \\ v_p \\ 1 \end{bmatrix} \tag{13.1-15}$$

which reduces to a single 3×3 transformation:

$$\begin{bmatrix} x_k \\ y_j \\ 1 \end{bmatrix} = \begin{bmatrix} \cos\theta & -\sin\theta & -x_c\cos\theta + y_c\sin\theta + x_c \\ \sin\theta & \cos\theta & -x_c\sin\theta - y_c\cos\theta + y_c \\ 0 & 0 & 1 \end{bmatrix} \begin{bmatrix} u_q \\ v_p \\ 1 \end{bmatrix} \tag{13.1-16}$$

The reverse address computation for the special case of Eq. 13.1-16, or the more general case of Eq. 13.1-13, can be obtained by inverting the 3×3 transformation matrices by numerical methods. Another approach, which is more computationally efficient, is to initially develop the homogeneous transformation matrix in reverse order as

$$\begin{bmatrix} u_q \\ v_p \\ 1 \end{bmatrix} = \begin{bmatrix} a_0 & a_1 & a_2 \\ b_0 & b_1 & b_2 \\ 0 & 0 & 1 \end{bmatrix} \begin{bmatrix} x_k \\ y_j \\ 1 \end{bmatrix} \tag{13.1-17}$$

where for translation

$$a_0 = 1 \tag{13.1-18a}$$

$$a_1 = 0 \tag{13.1-18b}$$

$$a_2 = -t_x \tag{13.1-18c}$$

$$b_0 = 0 \tag{13.1-18d}$$

$$b_1 = 1 \tag{13.1-18e}$$

$$b_2 = -t_y \tag{13.1-18f}$$

and for scaling

$$a_0 = 1/s_x \qquad (13.1\text{-}19a)$$

$$a_1 = 0 \qquad (13.1\text{-}19b)$$

$$a_2 = 0 \qquad (13.1\text{-}19c)$$

$$b_0 = 0 \qquad (13.1\text{-}19d)$$

$$b_1 = 1/s_y \qquad (13.1\text{-}19e)$$

$$b_2 = 0 \qquad (13.1\text{-}19f)$$

and for rotation

$$a_0 = \cos\theta \qquad (13.1\text{-}20a)$$

$$a_1 = \sin\theta \qquad (13.1\text{-}20b)$$

$$a_2 = 0 \qquad (13.1\text{-}20c)$$

$$b_0 = -\sin\theta \qquad (13.1\text{-}20d)$$

$$b_1 = \cos\theta \qquad (13.1\text{-}20e)$$

$$b_2 = 0 \qquad (13.1\text{-}20f)$$

Address computation for a rectangular destination array $G(j, k)$ from a rectangular source array $F(p, q)$ of the same size results in two types of ambiguity: some pixels of $F(p, q)$ will map outside of $G(j, k)$; and some pixels of $G(j, k)$ will not be mappable from $F(p, q)$ because they will lie outside its limits. As an example, Figure 13.1-2 illustrates rotation of an image by $45°$ about its center. If the desire of the mapping is to produce a complete destination array $G(j, k)$, it is necessary to access a sufficiently large source image $F(p, q)$ to prevent mapping voids in $G(j, k)$. This is accomplished in Figure 13.1-2d by embedding the original image of Figure 13.1-2a in a zero background that is sufficiently large to encompass the rotated original.

13.1.5. Affine Transformation

The geometrical operations of translation, size scaling, and rotation are special cases of a geometrical operator called an *affine transformation*. It is defined by Eq. 13.1-13b, in which the constants c_i and d_i are general weighting factors. The affine transformation is not only useful as a generalization of translation, scaling, and rotation. It provides a means of image shearing in which the rows or columns are successively uniformly translated with respect to one another. Figure 13.1-3

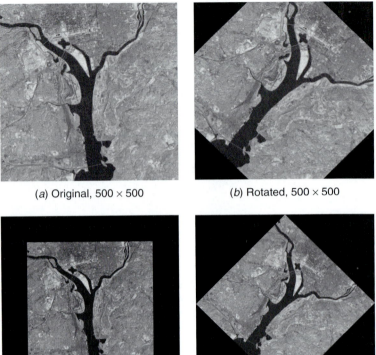

(a) Original, 500 × 500   (b) Rotated, 500 × 500

(c) Original, 708 × 708   (d) Rotated, 708 × 708

FIGURE 13.1-2. Image rotation by 45° on the `washington_ir` image about its center.

illustrates image shearing of rows of an image. In this example, $c_0 = d_1 = 1.0$, $c_1 = 0.1$, $d_0 = 0.0$, and $c_2 = d_2 = 0.0$.

13.1.6. Separable Translation, Scaling, and Rotation

The address mapping computations for translation and scaling are separable in the sense that the horizontal output image coordinate x_k depends only on u_q, and y_j depends only on v_p. Consequently, it is possible to perform these operations separably in two passes. In the first pass, a one-dimensional address translation is performed independently on each row of an input image to produce an intermediate array $I(p, k)$. In the second pass, columns of the intermediate array are processed independently to produce the final result $G(j, k)$.

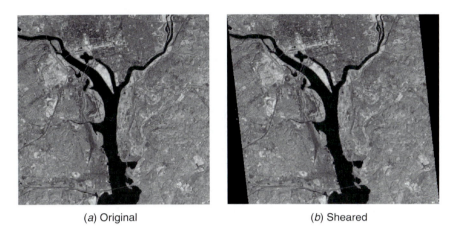

| (a) Original | (b) Sheared |

FIGURE 13.1-3. Horizontal image shearing on the `washington_ir` image.

Referring to Eq. 13.1-8, it is observed that the address computation for rotation is of a form such that x_k is a function of both u_q and v_p; and similarly for y_j. One might then conclude that rotation cannot be achieved by separable row and column processing, but Catmull and Smith (4) have demonstrated otherwise. In the first pass of the Catmull and Smith procedure, each row of $F(p, q)$ is mapped into the corresponding row of the intermediate array $I(p, k)$ using the standard row address computation of Eq. 13.1-8a. Thus

$$x_k = u_q \cos \theta - v_p \sin \theta \qquad (13.1-21)$$

Then, each column of $I(p, k)$ is processed to obtain the corresponding column of $G(j, k)$ using the address computation

$$y_j = \frac{x_k \sin \theta + v_p}{\cos \theta} \qquad (13.1-22)$$

Substitution of Eq. 13.1-21 into Eq. 13.1-22 yields the proper composite y-axis transformation of Eq. 13.1-8b. The "secret" of this separable rotation procedure is the ability to invert Eq. 13.1-21 to obtain an analytic expression for u_q in terms of x_k. In this case,

$$u_q = \frac{x_k + v_p \sin \theta}{\cos \theta} \qquad (13.1-23)$$

when substituted into Eq. 13.1-21, gives the intermediate column warping function of Eq. 13.1-22.

The Catmull and Smith two-pass algorithm can be expressed in vector-space form as

$$\begin{bmatrix} x_k \\ y_j \end{bmatrix} = \begin{bmatrix} 1 & 0 \\ \tan\theta & \dfrac{1}{\cos\theta} \end{bmatrix} \begin{bmatrix} \cos\theta & -\sin\theta \\ 0 & 1 \end{bmatrix} \begin{bmatrix} u_q \\ v_p \end{bmatrix} \qquad (13.1\text{-}24)$$

The separable processing procedure must be used with caution. In the special case of a rotation of 90°, all of the rows of $F(p, q)$ are mapped into a single column of $I(p, k)$, and hence the second pass cannot be executed. This problem can be avoided by processing the columns of $F(p, q)$ in the first pass. In general, the best overall results are obtained by minimizing the amount of spatial pixel movement. For example, if the rotation angle is + 80°, the original should be rotated by +90° by conventional row–column swapping methods, and then that intermediate image should be rotated by −10° using the separable method.

Figure 13.14 provides an example of separable rotation of an image by 45°. Figure 13.1-4a is the original, Figure 13.1-4b shows the result of the first pass and Figure 13.1-4c presents the final result.

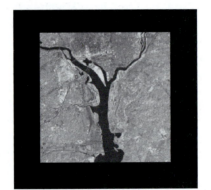

(*a*) Original

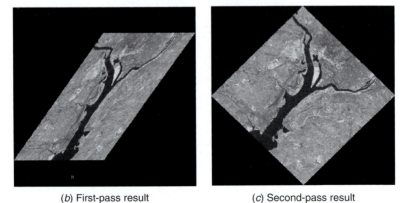

(*b*) First-pass result   (*c*) Second-pass result

FIGURE 13.1-4. Separable two-pass image rotation on the `washington_ir` image.

Separable, two-pass rotation offers the advantage of simpler computation compared to one-pass rotation, but there are some disadvantages to two-pass rotation. Two-pass rotation causes loss of high spatial frequencies of an image because of the intermediate scaling step (5), as seen in Figure 13.1-4b. Also, there is the potential of increased aliasing error (5,6), as discussed in Section 13.5.

Several authors (5,7,8) have proposed a three-pass rotation procedure in which there is no scaling step and hence no loss of high-spatial-frequency content with proper interpolation. The vector-space representation of this procedure is given by

$$\begin{bmatrix} x_k \\ y_j \end{bmatrix} = \begin{bmatrix} 1 & -\tan(\theta/2) \\ 0 & 1 \end{bmatrix} \begin{bmatrix} 1 & 0 \\ \sin\theta & 1 \end{bmatrix} \begin{bmatrix} 1 & -\tan(\theta/2) \\ 0 & 1 \end{bmatrix} \begin{bmatrix} u_q \\ v_p \end{bmatrix} \qquad (13.1\text{-}25)$$

This transformation is a series of image shearing operations without scaling. Figure 13.1-5 illustrates three-pass rotation for rotation by 45°.

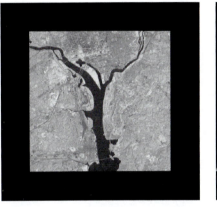

(*a*) Original

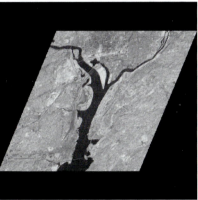

(*b*) First-pass result

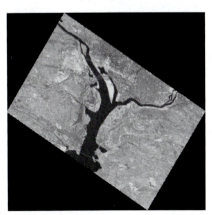

(*c*) Second-pass result

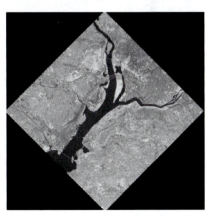

(*d*) Third-pass result

FIGURE 13.1-5. Separable three-pass image rotation on the `washington_ir` image.

13.2 SPATIAL WARPING

The address computation procedures described in the preceding section can be extended to provide nonlinear spatial warping of an image. In the literature, this process is often called *rubber-sheet stretching* (9,10). Let

$$x = X(u, v) \tag{13.2-1a}$$

$$y = Y(u, v) \tag{13.2-1b}$$

denote the generalized forward address mapping functions from an input image to an output image. The corresponding generalized reverse address mapping functions are given by

$$u = U(x, y) \tag{13.2-2a}$$

$$v = V(x, y) \tag{13.2-2b}$$

For notational simplicity, the (j, k) and (p, q) subscripts have been dropped from these and subsequent expressions. Consideration is given next to some examples and applications of spatial warping.

13.2.1. Polynomial Warping

The reverse address computation procedure given by the linear mapping of Eq. 13.1-17 can be extended to higher dimensions. A second-order polynomial warp address mapping can be expressed as

$$u = a_0 + a_1 x + a_2 y + a_3 x^2 + a_4 xy + a_5 y^2 \tag{13.2-3a}$$

$$v = b_0 + b_1 x + b_2 y + b_3 x^2 + b_4 xy + b_5 y^2 \tag{13.2-3b}$$

In vector notation,

$$\begin{bmatrix} u \\ v \end{bmatrix} = \begin{bmatrix} a_0 & a_1 & a_2 & a_3 & a_4 & a_5 \\ b_0 & b_1 & b_2 & b_3 & b_4 & b_5 \end{bmatrix} \begin{bmatrix} 1 \\ x \\ y \\ x^2 \\ xy \\ y^2 \end{bmatrix} \tag{13.2-3c}$$

For first-order address mapping, the weighting coefficients (a_i, b_i) can easily be related to the physical mapping as described in Section 13.1. There is no simple physical

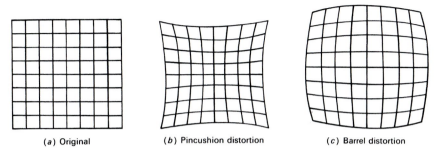

(*a*) Original   (*b*) Pincushion distortion   (*c*) Barrel distortion

FIGURE 13.2-1. Geometric distortion.

counterpart for second address mapping. Typically, second-order and higher-order address mapping are performed to compensate for spatial distortion caused by a physical imaging system. For example, Figure 13.2-1 illustrates the effects of imaging a rectangular grid with an electronic camera that is subject to nonlinear pincushion or barrel distortion. Figure 13.2-2 presents a generalization of the problem. An ideal image $F(j, k)$ is subject to an unknown physical spatial distortion. The observed image is measured over a rectangular array $O(p, q)$. The objective is to perform a spatial correction warp to produce a corrected image array $\hat{F}(j, k)$. Assume that the address mapping from the ideal image space to the observation space is given by

$$u = O_u\{x, y\} \tag{13.2-4a}$$

$$v = O_v\{x, y\} \tag{13.2-4b}$$

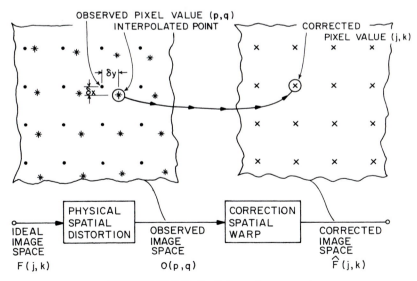

FIGURE 13.2-2. Spatial warping concept.

where $O_u\{x,y\}$ and $O_v\{x,y\}$ are physical mapping functions. If these mapping functions are known, then Eq. 13.2-4 can, in principle, be inverted to obtain the proper corrective spatial warp mapping. If the physical mapping functions are not known, Eq. 13.2-3 can be considered as an estimate of the physical mapping functions based on the weighting coefficients (a_i, b_i). These polynomial weighting coefficients are normally chosen to minimize the mean-square error between a set of observation coordinates (u_m, v_m) and the polynomial estimates (u, v) for a set $(1 \le m \le M)$ of known data points (x_m, y_m) called *control points*. It is convenient to arrange the observation space coordinates into the vectors

$$\mathbf{u}^T = [u_1, u_2, ..., u_M] \tag{13.2-5a}$$

$$\mathbf{v}^T = [v_1, v_2, ..., v_M] \tag{13.2-5b}$$

Similarly, let the second-order polynomial coefficients be expressed in vector form as

$$\mathbf{a}^T = [a_0, a_1, ..., a_5] \tag{13.2-6a}$$

$$\mathbf{b}^T = [b_0, b_1, ..., b_5] \tag{13.2-6b}$$

The mean-square estimation error can be expressed in the compact form

$$\mathcal{E} = (\mathbf{u} - \mathbf{Aa})^T(\mathbf{u} - \mathbf{Aa}) + (\mathbf{v} - \mathbf{Ab})^T(\mathbf{v} - \mathbf{Ab}) \tag{13.2-7}$$

where

$$\mathbf{A} = \begin{bmatrix} 1 & x_1 & y_1 & x_1^2 & x_1 y_1 & y_1^2 \\ 1 & x_2 & y_2 & x_2^2 & x_2 y_2 & y_2^2 \\ & & & & & \\ 1 & x_M & y_M & x_M^2 & x_M y_M & y_M^2 \end{bmatrix} \tag{13.2-8}$$

From Appendix 1, it has been determined that the error will be minimum if

$$\mathbf{a} = \mathbf{A}^- \mathbf{u} \tag{13.2-9a}$$

$$\mathbf{b} = \mathbf{A}^- \mathbf{v} \tag{13.2-9b}$$

where $\mathbf{A}^-$ is the generalized inverse of $\mathbf{A}$. If the number of control points is chosen greater than the number of polynomial coefficients, then

$$\mathbf{A}^- = [\mathbf{A}^T\mathbf{A}]^{-1}\mathbf{A} \tag{13.2-10}$$

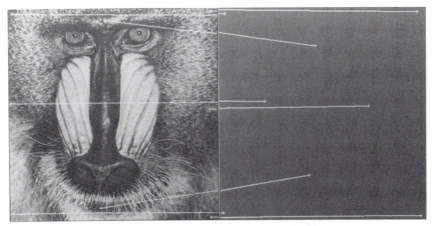

(a) Source control points   (b) Destination control points

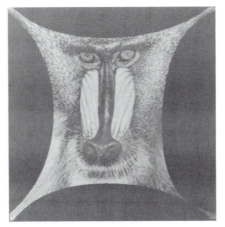

(c) Warped

FIGURE 13.2-3. Second-order polynomial spatial warping on the `mandrill_mon` image.

provided that the control points are not linearly related. Following this procedure, the polynomial coefficients (a_i, b_i) can easily be computed, and the address mapping of Eq. 13.2-1 can be obtained for all (j, k) pixels in the corrected image. Of course, proper interpolation is necessary.

Equation 13.2-3 can be extended to provide a higher-order approximation to the physical mapping of Eq. 13.2-3. However, practical problems arise in computing the pseudoinverse accurately for higher-order polynomials. For most applications, second-order polynomial computation suffices. Figure 13.2-3 presents an example of second-order polynomial warping of an image. In this example, the mapping of control points is indicated by the graphics overlay.

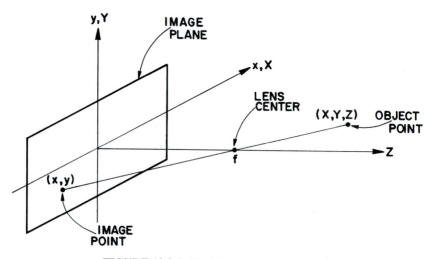

FIGURE 13.3-1. Basic imaging system model.

13.3. PERSPECTIVE TRANSFORMATION

Most two-dimensional images are views of three-dimensional scenes from the physical perspective of a camera imaging the scene. It is often desirable to modify an observed image so as to simulate an alternative viewpoint. This can be accomplished by use of a *perspective transformation*.

Figure 13.3-1 shows a simple model of an imaging system that projects points of light in three-dimensional object space to points of light in a two-dimensional image plane through a lens focused for distant objects. Let (X, Y, Z) be the continuous domain coordinate of an object point in the scene, and let (x, y) be the continuous domain-projected coordinate in the image plane. The image plane is assumed to be at the center of the coordinate system. The lens is located at a distance f to the right of the image plane, where f is the focal length of the lens. By use of similar triangles, it is easy to establish that

$$x = \frac{fX}{f-Z} \tag{13.3-1a}$$

$$y = \frac{fY}{f-Z} \tag{13.3-1b}$$

Thus the projected point (x, y) is related nonlinearly to the object point (X, Y, Z). This relationship can be simplified by utilization of homogeneous coordinates, as introduced to the image processing community by Roberts (1).

Let

$$\mathbf{v} = \begin{bmatrix} X \\ Y \\ Z \end{bmatrix} \tag{13.3-2}$$

be a vector containing the object point coordinates. The homogeneous vector $\tilde{\mathbf{v}}$ corresponding to $\mathbf{v}$ is

$$\tilde{\mathbf{v}} = \begin{bmatrix} sX \\ sY \\ sZ \\ s \end{bmatrix}$$

(13.3-3)

where s is a scaling constant. The Cartesian vector $\mathbf{v}$ can be generated from the homogeneous vector $\tilde{\mathbf{v}}$ by dividing each of the first three components by the fourth. The utility of this representation will soon become evident.

Consider the following perspective transformation matrix:

$$\mathbf{P} = \begin{bmatrix} 1 & 0 & 0 & 0 \\ 0 & 1 & 0 & 0 \\ 0 & 0 & 1 & 0 \\ 0 & 0 & -1/f & 1 \end{bmatrix}$$

(13.3-4)

This is a modification of the Roberts (1) definition to account for a different labeling of the axes and the use of column rather than row vectors. Forming the vector product

$$\tilde{\mathbf{w}} = \mathbf{P}\tilde{\mathbf{v}}$$

(13.3-5a)

yields

$$\tilde{\mathbf{w}} = \begin{bmatrix} sX \\ sY \\ sZ \\ s - sZ/f \end{bmatrix}$$

(13.3-5b)

The corresponding image plane coordinates are obtained by normalization of $\tilde{\mathbf{w}}$ to obtain

$$\mathbf{w} = \begin{bmatrix} \dfrac{fX}{f-Z} \\[2mm] \dfrac{fY}{f-Z} \\[2mm] \dfrac{fZ}{f-Z} \end{bmatrix}$$

(13.3-6)

It should be observed that the first two elements of $\mathbf{w}$ correspond to the imaging relationships of Eq. 13.3-1.

It is possible to project a specific image point (x_i, y_i) back into three-dimensional object space through an inverse perspective transformation

$$\tilde{\mathbf{v}} = \mathbf{P}^{-1}\tilde{\mathbf{w}} \tag{13.3-7a}$$

where

$$\mathbf{P}^{-1} = \begin{bmatrix} 1 & 0 & 0 & 0 \\ 0 & 1 & 0 & 0 \\ 0 & 0 & 1 & 0 \\ 0 & 0 & 1/f & 1 \end{bmatrix} \tag{13.3-7b}$$

and

$$\tilde{\mathbf{w}} = \begin{bmatrix} sx_i \\ sy_i \\ sz_i \\ s \end{bmatrix} \tag{13.3-7c}$$

In Eq. 13.3-7c, z_i is regarded as a free variable. Performing the inverse perspective transformation yields the homogeneous vector

$$\tilde{\mathbf{w}} = \begin{bmatrix} sx_i \\ sy_i \\ sz_i \\ s + sz_i/f \end{bmatrix} \tag{13.3-8}$$

The corresponding Cartesian coordinate vector is

$$\mathbf{w} = \begin{bmatrix} \dfrac{fx_i}{f - z_i} \\ \dfrac{fy_i}{f - z_i} \\ \dfrac{fz_i}{f - z_i} \end{bmatrix} \tag{13.3-9}$$

or equivalently,

$$x = \frac{fx_i}{f - z_i} \qquad (13.3\text{-}10a)$$

$$y = \frac{fy_i}{f - z_i} \qquad (13.3\text{-}10b)$$

$$z = \frac{fz_i}{f - z_i} \qquad (13.3\text{-}10c)$$

Equation 13.3-10 illustrates the many-to-one nature of the perspective transformation. Choosing various values of the free variable z_i results in various solutions for (X, Y, Z), all of which lie along a line from (x_i, y_i) in the image plane through the lens center. Solving for the free variable z_i in Eq. 13.3-10c and substituting into Eqs. 13.3-10a and 13.3-10b gives

$$X = \frac{x_i}{f}(f - Z) \qquad (13.3\text{-}11a)$$

$$Y = \frac{y_i}{f}(f - Z) \qquad (13.3\text{-}11b)$$

The meaning of this result is that because of the nature of the many-to-one perspective transformation, it is necessary to specify one of the object coordinates, say Z, in order to determine the other two from the image plane coordinates (x_i, y_i). Practical utilization of the perspective transformation is considered in the next section.

13.4. CAMERA IMAGING MODEL

The imaging model utilized in the preceding section to derive the perspective transformation assumed, for notational simplicity, that the center of the image plane was coincident with the center of the world reference coordinate system. In this section, the imaging model is generalized to handle physical cameras used in practical imaging geometries (11). This leads to two important results: a derivation of the fundamental relationship between an object and image point; and a means of changing a camera perspective by digital image processing.

Figure 13.4-1 shows an electronic camera in world coordinate space. This camera is physically supported by a gimbal that permits panning about an angle θ (horizontal movement in this geometry) and tilting about an angle ϕ (vertical movement). The gimbal center is at the coordinate (X_G, Y_G, Z_G) in the world coordinate system. The gimbal center and image plane center are offset by a vector with coordinates (X_o, Y_o, Z_o).

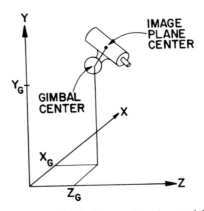

FIGURE 13.4-1. Camera imaging model.

If the camera were to be located at the center of the world coordinate origin, not panned nor tilted with respect to the reference axes, and if the camera image plane was not offset with respect to the gimbal, the homogeneous image model would be as derived in Section 13.3; that is

$$\tilde{\mathbf{w}} = \mathbf{P}\tilde{\mathbf{v}} \tag{13.4-1}$$

where $\tilde{\mathbf{v}}$ is the homogeneous vector of the world coordinates of an object point, $\tilde{\mathbf{w}}$ is the homogeneous vector of the image plane coordinates, and $\mathbf{P}$ is the perspective transformation matrix defined by Eq. 13.3-4. The camera imaging model can easily be derived by modifying Eq. 13.4-1 sequentially using a three-dimensional extension of translation and rotation concepts presented in Section 13.1.

The offset of the camera to location (X_G, Y_G, Z_G) can be accommodated by the translation operation

$$\tilde{\mathbf{w}} = \mathbf{P}\mathbf{T}_G\tilde{\mathbf{v}} \tag{13.4-2}$$

where

$$\mathbf{T}_G = \begin{bmatrix} 1 & 0 & 0 & -X_G \\ 0 & 1 & 0 & -Y_G \\ 0 & 0 & 1 & -Z_G \\ 0 & 0 & 0 & 1 \end{bmatrix} \tag{13.4-3}$$

Pan and tilt are modeled by a rotation transformation

$$\tilde{\mathbf{w}} = \mathbf{PRT}_G\tilde{\mathbf{v}} \tag{13.4-4}$$

where $\mathbf{R} = \mathbf{R}_\phi\mathbf{R}_\theta$ and

$$\mathbf{R}_\theta = \begin{bmatrix} \cos\theta & -\sin\theta & 0 & 0 \\ \sin\theta & \cos\theta & 0 & 0 \\ 0 & 0 & 1 & 0 \\ 0 & 0 & 0 & 1 \end{bmatrix} \tag{13.4-5}$$

and

$$\mathbf{R}_\phi = \begin{bmatrix} 1 & 0 & 0 & 0 \\ 0 & \cos\phi & -\sin\phi & 0 \\ 0 & \sin\phi & \cos\phi & 0 \\ 0 & 0 & 0 & 1 \end{bmatrix} \tag{13.4-6}$$

The composite rotation matrix then becomes

$$\mathbf{R} = \begin{bmatrix} \cos\theta & -\sin\theta & 0 & 0 \\ \cos\phi\sin\theta & \cos\phi\cos\theta & -\sin\phi & 0 \\ \sin\phi\sin\theta & \sin\phi\cos\theta & \cos\phi & 0 \\ 0 & 0 & 0 & 1 \end{bmatrix} \tag{13.4-7}$$

Finally, the camera-to-gimbal offset is modeled as

$$\tilde{\mathbf{w}} = \mathbf{PT}_C\mathbf{RT}_G\tilde{\mathbf{v}} \tag{13.4-8}$$

where

$$\mathbf{T}_C = \begin{bmatrix} 1 & 0 & 0 & -X_o \\ 0 & 1 & 0 & -Y_o \\ 0 & 0 & 1 & -Z_o \\ 0 & 0 & 0 & 1 \end{bmatrix} \tag{13.4-9}$$

Equation 13.4-8 is the final result giving the complete camera imaging model transformation between an object and an image point. The explicit relationship between an object point (X, Y, Z) and its image plane projection (x, y) can be obtained by performing the matrix multiplications analytically and then forming the Cartesian coordinates by dividing the first two components of $\tilde{\mathbf{w}}$ by the fourth. Upon performing these operations, one obtains

$$x = \frac{f[(X-X_G)\cos\theta - (Y-Y_G)\sin\theta - X_0]}{-(X-X_G)\sin\theta\sin\phi - (Y-Y_G)\cos\theta\sin\phi - (Z-Z_G)\cos\phi + Z_0 + f} \qquad (13.4\text{-}10a)$$

$$y = \frac{f[(X-X_G)\sin\theta\cos\phi + (Y-Y_G)\cos\theta\cos\phi - (Z-Z_G)\sin\phi - Y_0]}{-(X-X_G)\sin\theta\sin\phi - (Y-Y_G)\cos\theta\sin\phi - (Z-Z_G)\cos\phi + Z_0 + f} \qquad (13.4\text{-}10b)$$

Equation 13.4-10 can be used to predict the spatial extent of the image of a physical scene on an imaging sensor.

Another important application of the camera imaging model is to form an image by postprocessing such that the image appears to have been taken by a camera at a different physical perspective. Suppose that two images defined by $\tilde{\mathbf{w}}_1$ and $\tilde{\mathbf{w}}_2$ are formed by taking two views of the same object with the same camera. The resulting camera model relationships are then

$$\tilde{\mathbf{w}}_1 = \mathbf{PT}_C\mathbf{R}_1\mathbf{T}_{G1}\tilde{\mathbf{v}} \qquad (13.4\text{-}11a)$$

$$\tilde{\mathbf{w}}_2 = \mathbf{PT}_C\mathbf{R}_2\mathbf{T}_{G2}\tilde{\mathbf{v}} \qquad (13.4\text{-}11b)$$

Because the camera is identical for the two images, the matrices $\mathbf{P}$ and $\mathbf{T}_C$ are invariant in Eq. 13.4-11. It is now possible to perform an inverse computation of Eq. 13.4-11b to obtain

$$\tilde{\mathbf{v}} = [\mathbf{T}_{G1}]^{-1}[\mathbf{R}_1]^{-1}[\mathbf{T}_C]^{-1}[\mathbf{P}]^{-1}\tilde{\mathbf{w}}_1 \qquad (13.4\text{-}12)$$

and by substitution into Eq. 13.4-11b, it is possible to relate the image plane coordinates of the image of the second view to that obtained in the first view. Thus

$$\tilde{\mathbf{w}}_2 = \mathbf{PT}_C\mathbf{R}_2\mathbf{T}_{G2}[\mathbf{T}_{G1}]^{-1}[\mathbf{R}_1]^{-1}[\mathbf{T}_C]^{-1}[\mathbf{P}]^{-1}\tilde{\mathbf{w}}_1 \qquad (13.4\text{-}13)$$

As a consequence, an artificial image of the second view can be generated by performing the matrix multiplications of Eq. 13.4-13 mathematically on the physical image of the first view. Does this always work? No, there are limitations. First, if some portion of a physical scene were not "seen" by the physical camera, perhaps it was occluded by structures within the scene, then no amount of processing will recreate the missing data. Second, the processed image may suffer severe degradations resulting from undersampling if the two camera aspects are radically different. Nevertheless, this technique has valuable applications.

13.5. GEOMETRICAL IMAGE RESAMPLING

As noted in the preceding sections of this chapter, the reverse address computation process usually results in an address result lying between known pixel values of an input image. Thus it is necessary to estimate the unknown pixel amplitude from its known neighbors. This process is related to the image reconstruction task, as described in Chapter 4, in which a space-continuous display is generated from an array of image samples. However, the geometrical resampling process is usually not spatially regular. Furthermore, the process is discrete to discrete; only one output pixel is produced for each input address.

In this section, consideration is given to the general geometrical resampling process in which output pixels are estimated by interpolation of input pixels. The special, but common case, of image magnification by an integer zooming factor is also discussed. In this case, it is possible to perform pixel estimation by convolution.

13.5.1. Interpolation Methods

The simplest form of resampling interpolation is to choose the amplitude of an output image pixel to be the amplitude of the input pixel nearest to the reverse address. This process, called *nearest-neighbor interpolation*, can result in a spatial offset error by as much as $1/\sqrt{2}$ pixel units. The resampling interpolation error can be significantly reduced by utilizing all four nearest neighbors in the interpolation. A common approach, called *bilinear interpolation*, is to interpolate linearly along each row of an image and then interpolate that result linearly in the columnar direction. Figure 13.5-1 illustrates the process. The estimated pixel is easily found to be

$$F(p', q') = (1 - a)[(1 - b)F(p, q) + bF(p, q + 1)]$$

$$+ a[(1 - b)F(p + 1, q) + bF(p + 1, q + 1)] \qquad (13.5-1)$$

Although the horizontal and vertical interpolation operations are each linear, in general, their sequential application results in a nonlinear surface fit between the four neighboring pixels.

The expression for bilinear interpolation of Eq. 13.5-1 can be generalized for any interpolation function $R\{x\}$ that is zero-valued outside the range of ±1 sample spacing. With this generalization, interpolation can be considered as the summing of four weighted interpolation functions as given by

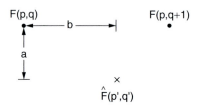

FIGURE 13.5-1. Bilinear interpolation.

$$F(p', q') = F(p, q)R\{-a\}R\{b\} + F(p, q + 1)R\{-a\}R\{-(1 - b)\}$$

$$+ F(p + 1, q)R\{1 - a\}R\{b\} + F(p + 1, q + 1)R\{1 - a\}R\{-(1 - b)\}$$

$$(13.5\text{-}2)$$

In the special case of linear interpolation, $R\{x\} = R_1\{x\}$, where $R_1\{x\}$ is defined in Eq. 4.3-2. Making this substitution, it is found that Eq. 13.5-2 is equivalent to the bilinear interpolation expression of Eq. 13.5-1.

Typically, for reasons of computational complexity, resampling interpolation is limited to a 4×4 pixel neighborhood. Figure 13.5-2 defines a generalized bicubic interpolation neighborhood in which the pixel $F(p, q)$ is the nearest neighbor to the pixel to be interpolated. The interpolated pixel may be expressed in the compact form

$$F(p', q') = \sum_{m = -1}^{2} \sum_{n = -1}^{2} F(p + m, q + n)R_C\{(m - a)\}R_C\{-(n - b)\} \quad (13.5\text{-}3)$$

where $R_C(x)$ denotes a bicubic interpolation function such as a cubic B-spline or cubic interpolation function, as defined in Section 4.3-2.

13.5.2. Convolution Methods

When an image is to be magnified by an integer zoom factor, pixel estimation can be implemented efficiently by convolution (12). As an example, consider image magnification by a factor of 2:1. This operation can be accomplished in two stages. First, the input image is transferred to an array in which rows and columns of zeros are interleaved with the input image data as follows:

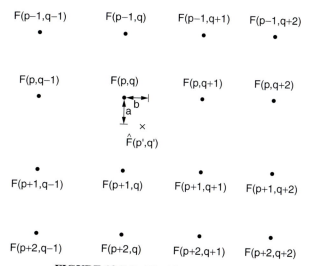

F(p–1,q–1)   F(p–1,q)   F(p–1,q+1)   F(p–1,q+2)

F(p,q–1)   F(p,q)   F(p,q+1)   F(p,q+2)

$\hat{F}(p',q')$

F(p+1,q–1)   F(p+1,q)   F(p+1,q+1)   F(p+1,q+2)

F(p+2,q–1)   F(p+2,q)   F(p+2,q+1)   F(p+2,q+2)

FIGURE 13.5-2. Bicubic interpolation.

Peg

$$\begin{bmatrix} 1 & 1 \\ 1 & 1 \end{bmatrix}$$

Pyramid

$$\frac{1}{4}\begin{bmatrix} 1 & 2 & 1 \\ 2 & 4 & 2 \\ 1 & 2 & 1 \end{bmatrix}$$

Bell

$$\frac{1}{16}\begin{bmatrix} 1 & 3 & 3 & 1 \\ 3 & 9 & 9 & 3 \\ 3 & 9 & 9 & 3 \\ 1 & 3 & 3 & 1 \end{bmatrix}$$

Cubic B-spline

$$\frac{1}{64}\begin{bmatrix} 1 & 4 & 6 & 4 & 1 \\ 4 & 16 & 24 & 16 & 4 \\ 6 & 24 & 36 & 24 & 6 \\ 4 & 16 & 24 & 16 & 4 \\ 1 & 4 & 6 & 4 & 1 \end{bmatrix}$$

FIGURE 13.5-3. Interpolation kernels for 2:1 magnification.

(*a*) Original

(*b*) Zero interleaved quadrant

(*c*) Peg

(*d*) Pyramid

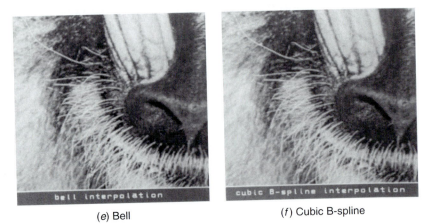

(*e*) Bell

(*f*) Cubic B-spline

FIGURE 13.5-4. Image interpolation on the mandrill_mon image for 2:1 magnification.

$$\begin{bmatrix} A & B \\ C & D \end{bmatrix}$$

$$\begin{bmatrix} A & 0 & B \\ 0 & 0 & 0 \\ C & 0 & D \end{bmatrix}$$

input image   zero-interleaved
neighborhood   neighborhood

Next, the zero-interleaved neighborhood image is convolved with one of the discrete interpolation kernels listed in Figure 13.5-3. Figure 13.5-4 presents the magnification results for several interpolation kernels. The inevitable visual trade-off between the interpolation error (the jaggy line artifacts) and the loss of high spatial frequency detail in the image is apparent from the examples.

This discrete convolution operation can easily be extended to higher-order magnification factors. For N:1 magnification, the core kernel is a $N \times N$ peg array. For large kernels it may be more computationally efficient in many cases, to perform the interpolation indirectly by Fourier domain filtering rather than by convolution (6).

REFERENCES

1. L. G. Roberts, "Machine Perception of Three-Dimensional Solids," in *Optical and Electro-Optical Information Processing*, J. T. Tippett et al., Eds., MIT Press, Cambridge, MA, 1965.

2. D. F. Rogers, *Mathematical Elements for Computer Graphics*, 2nd ed., McGraw-Hill, New York, 1989.

3. J. D. Foley et al., *Computer Graphics: Principles and Practice*, 2nd ed. in C, Addison-Wesley, Reading, MA, 1996.

4. E. Catmull and A. R. Smith, "3-D Transformation of Images in Scanline Order," *Computer Graphics, SIGGRAPH '80 Proc.*, **14**, 3, July 1980, 279–285.

5. M. Unser, P. Thevenaz, and L. Yaroslavsky, "Convolution-Based Interpolation for Fast, High-Quality Rotation of Images, *IEEE Trans. Image Processing*, **IP-4**, 10, October 1995, 1371–1381.

6. D. Fraser and R. A. Schowengerdt, "Avoidance of Additional Aliasing in Multipass Image Rotations," *IEEE Trans. Image Processing*, **IP-3**, 6, November 1994, 721–735.

7. A. W. Paeth, "A Fast Algorithm for General Raster Rotation," *in Proc. Graphics Interface '86-Vision Interface*, 1986, 77–81.

8. P. E. Danielson and M. Hammerin, "High Accuracy Rotation of Images, in *CVGIP: Graphical Models and Image Processing*, **54**, 4, July 1992, 340–344.

9. R. Bernstein, "Digital Image Processing of Earth Observation Sensor Data," *IBM J. Research and Development*, **20**, 1, 1976, 40–56.

10. D. A. O'Handley and W. B. Green, "Recent Developments in Digital Image Processing at the Image Processing Laboratory of the Jet Propulsion Laboratory," *Proc. IEEE*, **60**, 7, July 1972, 821–828.

11. K. S. Fu, R. C. Gonzalez and C. S. G. Lee, *Robotics: Control, Sensing, Vision, and Intelligence*, McGraw-Hill, New York, 1987.

12. W. K. Pratt, "Image Processing and Analysis Using Primitive Computational Elements," in *Selected Topics in Signal Processing*, S. Haykin, Ed., Prentice Hall, Englewood Cliffs, NJ, 1989.

PART 5

IMAGE ANALYSIS

Image analysis is concerned with the extraction of measurements, data or information from an image by automatic or semiautomatic methods. In the literature, this field has been called image data extraction, scene analysis, image description, automatic photo interpretation, image understanding, and a variety of other names.

Image analysis is distinguished from other types of image processing, such as coding, restoration, and enhancement, in that the ultimate product of an image analysis system is usually numerical output rather than a picture. Image analysis also diverges from classical pattern recognition in that analysis systems, by definition, are not limited to the classification of scene regions to a fixed number of categories, but rather are designed to provide a description of complex scenes whose variety may be enormously large and ill-defined in terms of a priori expectation.

14

MORPHOLOGICAL IMAGE PROCESSING

Morphological image processing is a type of processing in which the spatial form or structure of objects within an image are modified. Dilation, erosion, and skeletonization are three fundamental morphological operations. With dilation, an object grows uniformly in spatial extent, whereas with erosion an object shrinks uniformly. Skeletonization results in a stick figure representation of an object.

The basic concepts of morphological image processing trace back to the research on spatial set algebra by Minkowski (1) and the studies of Matheron (2) on topology. Serra (3–5) developed much of the early foundation of the subject. Steinberg (6,7) was a pioneer in applying morphological methods to medical and industrial vision applications. This research work led to the development of the cytocomputer for high-speed morphological image processing (8,9).

In the following sections, morphological techniques are first described for binary images. Then these morphological concepts are extended to gray scale images.

14.1. BINARY IMAGE CONNECTIVITY

Binary image morphological operations are based on the geometrical relationship or *connectivity* of pixels that are deemed to be of the same class (10,11). In the binary image of Figure 14.1-1*a*, the ring of black pixels, by all reasonable definitions of connectivity, divides the image into three segments: the white pixels exterior to the ring, the white pixels interior to the ring, and the black pixels of the ring itself. The pixels within each segment are said to be connected to one another. This concept of connectivity is easily understood for Figure 14.1-1*a*, but ambiguity arises when considering Figure 14.1-1*b*. Do the black pixels still define a ring, or do they instead form four disconnected lines? The answers to these questions depend on the definition of connectivity.

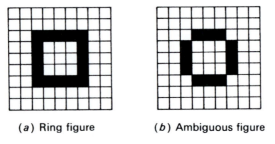

(*a*) Ring figure   (*b*) Ambiguous figure

FIGURE 14.1-1. Connectivity.

Consider the following neighborhood pixel pattern:

$$
\begin{array}{ccc}
X_3 & X_2 & X_1 \\
X_4 & X & X_0 \\
X_5 & X_6 & X_7
\end{array}
$$

in which a binary-valued pixel $F(j, k) = X$, where $X = 0$ (white) or $X = 1$ (black) is surrounded by its eight nearest neighbors $X_0, X_1, ..., X_7$. An alternative nomenclature is to label the neighbors by compass directions: north, northeast, and so on:

$$
\begin{array}{ccc}
NW & N & NE \\
W & X & E \\
SW & S & SE
\end{array}
$$

Pixel X is said to be *four-connected* to a neighbor if it is a logical 1 and if its east, north, west, or south (X_0, X_2, X_4, X_6) neighbor is a logical 1. Pixel X is said to be *eight-connected* if it is a logical 1 and if its north, northeast, etc. $(X_0, X_1, ..., X_7)$ neighbor is a logical 1.

The connectivity relationship between a center pixel and its eight neighbors can be quantified by the concept of a *pixel bond*, the sum of the bond weights between the center pixel and each of its neighbors. Each four-connected neighbor has a bond of two, and each eight-connected neighbor has a bond of one. In the following example, the pixel bond is seven.

$$
\begin{array}{ccc}
1 & 1 & 1 \\
0 & X & 0 \\
1 & 1 & 0
\end{array}
$$

0	0	0		0	0	0		0	0	0
0	1	1		0	1	1		0	1	0
0	1	0		0	0	1		0	0	0

Four-connected	Eight-connected	Isolated
$B = 4$	$B = 3$	$B = 0$

0	0	0		1	0	0		1	1	1
0	1	0		1	1	1		0	1	0
0	0	1		1	0	1		1	1	1

Spur	Bridge	H-connected
$B = 1$	$B = 7$	$B = 8$

0	0	0		0	1	1		0	1	1
0	1	1		1	1	1		0	1	1
0	1	1		1	1	1		0	1	1

Corner	Interior	Exterior
$B = 5$	$B = 1$	$B = 8$

FIGURE 14.1-2. Pixel neighborhood connectivity definitions.

Under the definition of four-connectivity, Figure 14.1-1*b* has four disconnected black line segments, but with the eight-connectivity definition, Figure 14.1-1*b* has a ring of connected black pixels. Note, however, that under eight-connectivity, all white pixels are connected together. Thus a paradox exists. If the black pixels are to be eight-connected together in a ring, one would expect a division of the white pixels into pixels that are interior and exterior to the ring. To eliminate this dilemma, eight-connectivity can be defined for the black pixels of the object, and four-connectivity can be established for the white pixels of the background. Under this definition, a string of black pixels is said to be *minimally connected* if elimination of any black pixel results in a loss of connectivity of the remaining black pixels. Figure 14.1-2 provides definitions of several other neighborhood connectivity relationships between a center black pixel and its neighboring black and white pixels.

The preceding definitions concerning connectivity have been based on a discrete image model in which a continuous image field is sampled over a rectangular array of points. Golay (12) has utilized a hexagonal grid structure. With such a structure, many of the connectivity problems associated with a rectangular grid are eliminated. In a hexagonal grid, neighboring pixels are said to be *six-connected* if they are in the same set and share a common edge boundary. Algorithms have been developed for the linking of boundary points for many feature extraction tasks (13). However, two major drawbacks have hindered wide acceptance of the hexagonal grid. First, most image scanners are inherently limited to rectangular scanning. The second problem is that the hexagonal grid is not well suited to many spatial processing operations, such as convolution and Fourier transformation.

14.2. BINARY IMAGE HIT OR MISS TRANSFORMATIONS

The two basic morphological operations, dilation and erosion, plus many variants can be defined and implemented by *hit-or-miss transformations* (3). The concept is quite simple. Conceptually, a small odd-sized mask, typically 3×3, is scanned over a binary image. If the binary-valued pattern of the mask matches the state of the pixels under the mask (hit), an output pixel in spatial correspondence to the center pixel of the mask is set to some desired binary state. For a pattern mismatch (miss), the output pixel is set to the opposite binary state. For example, to perform simple binary noise cleaning, if the isolated 3×3 pixel pattern

$$
\begin{array}{ccc}
0 & 0 & 0 \\
0 & 1 & 0 \\
0 & 0 & 0
\end{array}
$$

is encountered, the output pixel is set to zero; otherwise, the output pixel is set to the state of the input center pixel. In more complicated morphological algorithms, a large number of the $2^9 = 512$ possible mask patterns may cause hits.

It is often possible to establish simple neighborhood logical relationships that define the conditions for a hit. In the isolated pixel removal example, the defining equation for the output pixel $G(j, k)$ becomes

$$
G(j, k) = X \cap (X_0 \cup X_1 \cup \cdots \cup X_7) \tag{14.2-1}
$$

where $\cap$ denotes the intersection operation (logical AND) and $\cup$ denotes the union operation (logical OR). For complicated algorithms, the logical equation method of definition can be cumbersome. It is often simpler to regard the hit masks as a collection of binary patterns.

Hit-or-miss morphological algorithms are often implemented in digital image processing hardware by a pixel stacker followed by a look-up table (LUT), as shown in Figure 14.2-1 (14). Each pixel of the input image is a positive integer, represented by a conventional binary code, whose most significant bit is a 1 (black) or a 0 (white). The pixel stacker extracts the bits of the center pixel X and its eight neighbors and puts them in a neighborhood pixel stack. Pixel stacking can be performed by convolution with the 3×3 pixel kernel

$$
\begin{bmatrix}
2^{-4} & 2^{-3} & 2^{-2} \\
2^{-5} & 2^{0} & 2^{-1} \\
2^{-6} & 2^{-7} & 2^{-8}
\end{bmatrix}
$$

The binary number state of the neighborhood pixel stack becomes the numeric input address of the LUT whose entry is Y For isolated pixel removal, integer entry 256, corresponding to the neighborhood pixel stack state 100000000, contains $Y = 0$; all other entries contain $Y = X$.

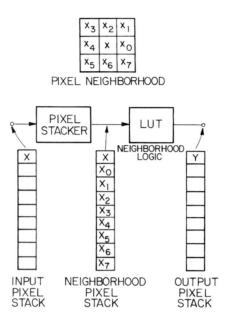

FIGURE 14.2-1. Look-up table flowchart for binary unconditional operations.

Several other 3×3 hit-or-miss operators are described in the following subsections.

14.2.1. Additive Operators

Additive hit-or-miss morphological operators cause the center pixel of a 3×3 pixel window to be converted from a logical 0 state to a logical 1 state if the neighboring pixels meet certain predetermined conditions. The basic operators are now defined.

Interior Fill. Create a black pixel if all four-connected neighbor pixels are black.

$$G(j, k) = X \cup [X_0 \cap X_2 \cap X_4 \cap X_6]$$

(14.2-2)

Diagonal Fill. Create a black pixel if creation eliminates the eight-connectivity of the background.

$$G(j, k) = X \cup [P_1 \cup P_2 \cup P_3 \cup P_4]$$

(14.2-3a)

where

$$P_1 = \bar{X} \cap X_0 \cap \bar{X}_1 \cap X_2 \tag{14.2-3b}$$

$$P_2 = \bar{X} \cap X_2 \cap \bar{X}_3 \cap X_4 \tag{14.2-3c}$$

$$P_3 = \bar{X} \cap X_4 \cap \bar{X}_5 \cap X_6 \tag{14.2-3d}$$

$$P_4 = \bar{X} \cap X_6 \cap \bar{X}_7 \cap X_0 \tag{14.2-3e}$$

In Eq. 14.2-3, the overbar denotes the logical complement of a variable.

Bridge. Create a black pixel if creation results in connectivity of previously unconnected neighboring black pixels.

$$G(j, k) = X \cup [P_1 \cup P_2 \cup \cdots \cup P_6] \tag{14.2-4a}$$

where

$$P_1 = \bar{X}_2 \cap \bar{X}_6 \cap [X_3 \cup X_4 \cup X_5] \cap [X_0 \cup X_1 \cup X_7] \cap \bar{P}_Q \tag{14.2-4b}$$

$$P_2 = \bar{X}_0 \cap \bar{X}_4 \cap [X_1 \cup X_2 \cup X_3] \cap [X_5 \cup X_6 \cup X_7] \cap \bar{P}_Q \tag{14.2-4c}$$

$$P_3 = \bar{X}_0 \cap \bar{X}_6 \cap X_7 \cap [X_2 \cup X_3 \cup X_4] \tag{14.2-4d}$$

$$P_4 = \bar{X}_0 \cap \bar{X}_2 \cap X_1 \cap [X_4 \cup X_5 \cup X_6] \tag{14.2-4e}$$

$$P_5 = \bar{X}_2 \cap \bar{X}_4 \cap X_3 \cap [X_0 \cup X_6 \cup X_7] \tag{14.2-4f}$$

$$P_6 = \bar{X}_4 \cap \bar{X}_6 \cap X_5 \cap [X_0 \cup X_1 \cup X_2] \tag{14.2-4g}$$

and

$$P_Q = L_1 \cup L_2 \cup L_3 \cup L_4 \tag{14.2-4h}$$

$$L_1 = \bar{X} \cap \bar{X}_0 \cap X_1 \cap \bar{X}_2 \cap X_3 \cap \bar{X}_4 \cap \bar{X}_5 \cap \bar{X}_6 \cap \bar{X}_7 \tag{14.2-4i}$$

$$L_2 = \bar{X} \cap \bar{X}_0 \cap \bar{X}_1 \cap \bar{X}_2 \cap X_3 \cap \bar{X}_4 \cap X_5 \cap \bar{X}_6 \cap \bar{X}_7 \tag{14.2-4j}$$

$$L_3 = \bar{X} \cap \bar{X}_0 \cap \bar{X}_1 \cap \bar{X}_2 \cap \bar{X}_3 \cap \bar{X}_4 \cap X_5 \cap \bar{X}_6 \cap X_7 \tag{14.2-4k}$$

$$L_4 = \bar{X} \cap \bar{X}_0 \cap X_1 \cap \bar{X}_2 \cap \bar{X}_3 \cap \bar{X}_4 \cap \bar{X}_5 \cap \bar{X}_6 \cap X_7 \tag{14.2-4l}$$

The following is one of 119 qualifying patterns

$$
\begin{array}{ccc}
1 & 0 & 0 \\
1 & 0 & 1 \\
0 & 0 & 1
\end{array}
$$

A pattern such as

$$
\begin{array}{ccc}
0 & 0 & 0 \\
0 & 0 & 0 \\
1 & 0 & 1
\end{array}
$$

does not qualify because the two black pixels will be connected when they are on the middle row of a subsequent observation window if they are indeed unconnected.

Eight-Neighbor Dilate. Create a black pixel if at least one eight-connected neighbor pixel is black.

$$G(j, k) = X \cup X_0 \cup \cdots \cup X_7 \tag{14.2-5}$$

This hit-or-miss definition of dilation is a special case of a generalized dilation operator that is introduced in Section 14.4. The dilate operator can be applied recursively. With each iteration, objects will grow by a single pixel width ring of exterior pixels. Figure 14.2-2 shows dilation for one and for three iterations for a binary image. In the example, the original pixels are recorded as black, the background pixels are white, and the added pixels are midgray.

Fatten. Create a black pixel if at least one eight-connected neighbor pixel is black, provided that creation does not result in a bridge between previously unconnected black pixels in a 3×3 neighborhood.

The following is an example of an input pattern in which the center pixel would be set black for the basic dilation operator, but not for the fatten operator.

$$
\begin{array}{ccc}
0 & 0 & 1 \\
1 & 0 & 0 \\
1 & 1 & 0
\end{array}
$$

There are 132 such qualifying patterns. This strategem will not prevent connection of two objects separated by two rows or columns of white pixels. A solution to this problem is considered in Section 14.3. Figure 14.2-3 provides an example of fattening.

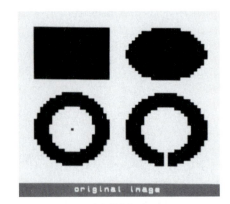

(a) Original

(b) One iteration   (c) Three iterations

FIGURE 14.2-2. Dilation of a binary image.

14.2.2. Subtractive Operators

Subtractive hit-or-miss morphological operators cause the center pixel of a 3×3 window to be converted from black to white if its neighboring pixels meet predetermined conditions. The basic subtractive operators are defined below.

Isolated Pixel Remove. Erase a black pixel with eight white neighbors.

$$G(j, k) = X \cap [X_0 \cup X_1 \cup \cdots \cup X_7] \qquad (14.2\text{-}6)$$

Spur Remove. Erase a black pixel with a single eight-connected neighbor.

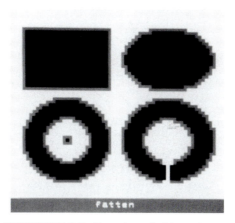

FIGURE 14.2-3. Fattening of a binary image.

The following is one of four qualifying patterns:

$$
\begin{array}{ccc}
0 & 0 & 0 \\
0 & 1 & 0 \\
1 & 0 & 0
\end{array}
$$

Interior Pixel Remove. Erase a black pixel if all four-connected neighbors are black.

$$G(j, k) = X \cap [\bar{X}_0 \cup \bar{X}_2 \cup \bar{X}_4 \cup \bar{X}_6] \tag{14.2-7}$$

There are 16 qualifying patterns.

H-Break. Erase a black pixel that is H-connected.
   There are two qualifying patterns.

$$
\begin{array}{ccc}
1 & 1 & 1 \\
0 & 1 & 0 \\
1 & 1 & 1
\end{array}
\qquad\qquad
\begin{array}{ccc}
1 & 0 & 1 \\
1 & 1 & 1 \\
1 & 0 & 1
\end{array}
$$

Eight-Neighbor Erode. Erase a black pixel if at least one eight-connected neighbor pixel is white.

$$G(j, k) = X \cap X_0 \cap \cdots \cap X_7 \tag{14.2-8}$$

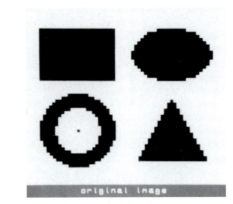

(*a*) Original

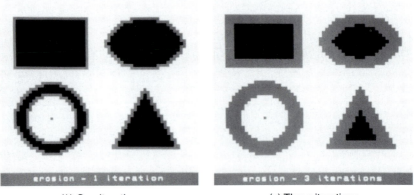

(*b*) One iteration   (*c*) Three iterations

FIGURE 14.2-4. Erosion of a binary image.

A generalized erosion operator is defined in Section 14.4. Recursive application of the erosion operator will eventually erase all black pixels. Figure 14.2-4 shows results for one and three iterations of the erode operator. The eroded pixels are midgray. It should be noted that after three iterations, the ring is totally eroded.

14.2.3. Majority Black Operator

The following is the definition of the *majority black operator:*

Majority Black. Create a black pixel if five or more pixels in a 3×3 window are black; otherwise, set the output pixel to white.

The majority black operator is useful for filling small holes in objects and closing short gaps in strokes. An example of its application to edge detection is given in Chapter 15.

14.3. BINARY IMAGE SHRINKING, THINNING, SKELETONIZING, AND THICKENING

Shrinking, thinning, skeletonizing, and thickening are forms of conditional erosion in which the erosion process is controlled to prevent total erasure and to ensure connectivity.

14.3.1. Binary Image Shrinking

The following is a definition of *shrinking:*

Shrink. Erase black pixels such that an object without holes erodes to a single pixel at or near its center of mass, and an object with holes erodes to a connected ring lying midway between each hole and its nearest outer boundary.

A 3×3 pixel object will be shrunk to a single pixel at its center. A 2×2 pixel object will be arbitrarily shrunk, by definition, to a single pixel at its lower right corner.

It is not possible to perform shrinking using single-stage 3×3 pixel hit-or-miss transforms of the type described in the previous section. The 3×3 window does not provide enough information to prevent total erasure and to ensure connectivity. A 5×5 hit-or-miss transform could provide sufficient information to perform proper shrinking. But such an approach would result in excessive computational complexity (i.e., 2^{25} possible patterns to be examined!). References 15 and 16 describe two-stage shrinking and thinning algorithms that perform a conditional marking of pixels for erasure in a first stage, and then examine neighboring marked pixels in a second stage to determine which ones can be unconditionally erased without total erasure or loss of connectivity. The following algorithm developed by Pratt and Kabir (17) is a pipeline processor version of the conditional marking scheme.

In the algorithm, two concatenated 3×3 hit-or-miss transformations are performed to obtain indirect information about pixel patterns within a 5×5 window. Figure 14.3-1 is a flowchart for the look-up table implementation of this algorithm. In the first stage, the states of nine neighboring pixels are gathered together by a pixel stacker, and a following look-up table generates a conditional mark M for possible erasures. Table 14.3-1 lists all patterns, as indicated by the letter S in the table column, which will be conditionally marked for erasure. In the second stage of the algorithm, the center pixel X and the conditional marks in a 3×3 neighborhood centered about X are examined to create an output pixel. The shrinking operation can be expressed logically as

$$G(j, k) = X \cap [\overline{M} \cup P(M, M_0, ..., M_7)] \qquad (14.3\text{-}1)$$

where $P(M, M_0, ..., M_7)$ is an erasure inhibiting logical variable, as defined in Table 14.3-2. The first four patterns of the table prevent strokes of single pixel width from being totally erased. The remaining patterns inhibit erasure that would break object connectivity. There are a total of 157 inhibiting patterns. This two-stage process must be performed iteratively until there are no further erasures.

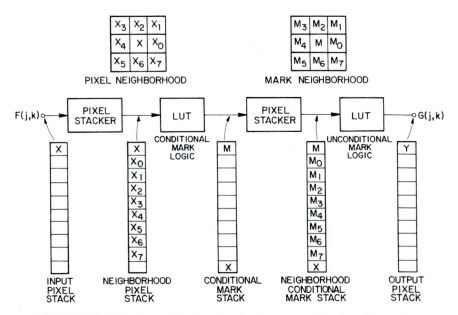

FIGURE 14.3-1. Look-up table flowchart for binary conditional mark operations.

As an example, the 2×2 square pixel object

$$
\begin{array}{cc}
1 & 1 \\
1 & 1
\end{array}
$$

results in the following intermediate array of conditional marks

$$
\begin{array}{cc}
M & M \\
M & M
\end{array}
$$

The corner cluster pattern of Table 14.3-2 gives a hit only for the lower right corner mark. The resulting output is

$$
\begin{array}{cc}
0 & 0 \\
0 & 1
\end{array}
$$

TABLE 14.3-1. Shrink, Thin, and Skeletonize Conditional Mark Patterns [*M* = 1 if hit]

Table	Bond	Pattern							
		0 0 1	1 0 0	0 0 0	0 0 0				
S	1	0 1 0	0 1 0	0 1 0	0 1 0				
		0 0 0	0 0 0	1 0 0	0 0 1				
		0 0 0	0 1 0	0 0 0	0 0 0				
S	2	0 1 1	0 1 0	1 1 0	0 1 0				
		0 0 0	0 0 0	0 0 0	0 1 0				
		0 0 1	0 1 1	1 1 0	1 0 0	0 0 0	0 0 0	0 0 0	0 0 0
S	3	0 1 1	0 1 0	0 1 0	1 1 0	1 1 0	0 1 0	0 1 0	0 1 1
		0 0 0	0 0 0	0 0 0	0 0 0	1 0 0	1 1 0	0 1 1	0 0 1
		0 1 0	0 1 0	0 0 0	0 0 0				
TK	4	0 1 1	1 1 0	1 1 0	0 1 1				
		0 0 0	0 0 0	0 1 0	0 1 0				
		0 0 1	1 1 1	1 0 0	0 0 0				
STK	4	0 1 1	0 1 0	1 1 0	0 1 0				
		0 0 1	0 0 0	1 0 0	1 1 1				
		1 1 0	0 1 0	0 1 1	0 0 1				
ST	5	0 1 1	0 1 1	1 1 0	0 1 1				
		0 0 0	0 0 1	0 0 0	0 1 0				
		0 1 1	1 1 0	0 0 0	0 0 0				
ST	5	0 1 1	1 1 0	1 1 0	0 1 1				
		0 0 0	0 0 0	1 1 0	0 1 1				
		1 1 0	0 1 1						
ST	6	0 1 1	1 1 0						
		0 0 1	1 0 0						
		1 1 1	0 1 1	1 1 1	1 1 0	1 0 0	0 0 0	0 0 0	0 0 1
STK	6	0 1 1	0 1 1	1 1 0	1 1 0	1 1 0	1 1 0	0 1 1	0 1 1
		0 0 0	0 0 1	0 0 0	1 0 0	1 1 0	1 1 1	1 1 1	0 1 1

(*Continued*)

TABLE 14.3-1 (Continued)

Table	Bond	Pattern							
		1 1 1	1 1 1	1 0 0	0 0 1				
STK	7	0 1 1	1 1 0	1 1 0	0 1 1				
		0 0 1	1 0 0	1 1 1	1 1 1				
		0 1 1	1 1 1	1 1 0	0 0 0				
STK	8	0 1 1	1 1 1	1 1 0	1 1 1				
		0 1 1	0 0 0	1 1 0	1 1 1				
		1 1 1	0 1 1	1 1 1	1 1 1	1 1 1	1 1 0	1 0 0	0 0 1
STK	9	0 1 1	0 1 1	1 1 1	1 1 1	1 1 0	1 1 0	1 1 1	1 1 1
		0 1 1	1 1 1	1 0 0	0 0 1	1 1 0	1 1 1	1 1 1	1 1 1
		1 1 1	1 1 1	1 1 1	1 0 1				
STK	10	0 1 1	1 1 1	1 1 0	1 1 1				
		1 1 1	1 0 1	1 1 1	1 1 1				
		1 1 1	1 1 1	1 1 0	0 1 1				
K	11	1 1 1	1 1 1	1 1 1	1 1 1				
		0 1 1	1 1 0	1 1 1	1 1 1				

Figure 14.3-2 shows an example of the shrinking of a binary image for four and 13 iterations of the algorithm. No further shrinking occurs for more than 13 iterations. At this point, the shrinking operation has become *idempotent* (i. e., reapplication evokes no further change. This shrinking algorithm does not shrink the symmetric original ring object to a ring that is also symmetric because of some of the conditional mark patterns of Table 14.3-2, which are necessary to ensure that objects of even dimension shrink to a single pixel. For the same reason, the shrink ring is not minimally connected.

14.3.2. Binary Image Thinning

The following is a definition of *thinning:*

Thin. Erase black pixels such that an object without holes erodes to a minimally connected stroke located equidistant from its nearest outer boundaries, and an object with holes erodes to a minimally connected ring midway between each hole and its nearest outer boundary.

TABLE 14.3-2. Shrink and Thin Unconditional Mark Patterns
$[P(M, M_0, M_1, M_2, M_3, M_4, M_5, M_6, M_7) = 1 \text{ if hit}]^a$

								Pattern

Spur   Single 4-connection

0 0 *M*	*M* 0 0	0 0 0	0 0 0				
0 *M* 0	0 *M* 0	0 *M* 0	0 *MM*				
0 0 0	0 0 0	0 *M* 0	0 0 0				

L Cluster (thin only)

0 0 *M*	0 *MM*	*MM* 0	*M* 0 0	0 0 0	0 0 0	0 0 0	0 0 0
0 *MM*	0 *M* 0	0 *M* 0	*MM* 0	*MM* 0	0 *M* 0	0 *M* 0	0 *MM*
0 0 0	0 0 0	0 0 0	0 0 0	*M* 0 0	*MM* 0	0 *MM*	0 0 *M*

4-Connected offset

0 *MM*	*MM* 0	0 *M* 0	0 0 *M*
MM 0	0 *MM*	0 *MM*	0 *MM*
0 0 0	0 0 0	0 0 *M*	0 *M* 0

Spur corner cluster

0 *A M*	*M B* 0	0 0 *M*	*M* 0 0
0 *M B*	*A M* 0	*A M* 0	0 *M B*
M 0 0	0 0 *M*	*M B* 0	0 *A M*

Corner cluster

M M D
M M D
D D D

Tee branch

D M 0	0 *M D*	0 0 *D*	*D* 0 0	*D M D*	0 *M* 0	0 *M* 0	*D M D*
M M M	*M M M*	*M M M*	*M M M*	*M M* 0	*M M* 0	0 *M M*	0 *M M*
D 0 0	0 0 *D*	0 *M D*	*D M* 0	0 *M* 0	*D M D*	*D M D*	0 *M* 0

Vee branch

M D M	*M D C*	*C B A*	*A D M*
D M D	*D M B*	*D M D*	*B M D*
A B C	*M D A*	*M D M*	*C D M*

Diagonal branch

D M 0	0 *M D*	*D* 0 *M*	*M* 0 *D*
0 *M M*	*M M* 0	*M M* 0	0 *M M*
M 0 *D*	*D* 0 *M*	0 *M D*	*D M* 0

$^a A \cup B \cup C = 1$ $D = 0 \cup 1$ $A \cup B = 1.$

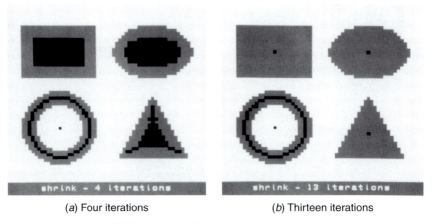

(a) Four iterations   (b) Thirteen iterations

FIGURE 14.3-2. Shrinking of a binary image.

The following is an example of the thinning of a 3×5 pixel object without holes

```
1  1  1  1  1        0  0  0  0  0
1  1  1  1  1        0  1  1  1  0
1  1  1  1  1        0  0  0  0  0
```

         before   after

A 2×5 object is thinned as follows:

```
1  1  1  1  1        0  0  0  0  0
1  1  1  1  1        0  1  1  1  1
```

         before   after

Table 14.3-1 lists the conditional mark patterns, as indicated by the letter T in the table column, for thinning by the conditional mark algorithm of Figure 14.3-1. The shrink and thin unconditional patterns are identical, as shown in Table 14.3-2.

Figure 14.3-3 contains an example of the thinning of a binary image for four and eight iterations. Figure 14.3-4 provides an example of the thinning of an image of a printed circuit board in order to locate solder pads that have been deposited improperly and that do not have holes for component leads. The pads with holes erode to a minimally connected ring, while the pads without holes erode to a point.

Thinning can be applied to the background of an image containing several objects as a means of separating the objects. Figure 14.3-5 provides an example of the process. The original image appears in Figure 14.3-5a, and the background-reversed image is Figure 14.3-5b. Figure 14.3-5c shows the effect of thinning the background. The thinned strokes that separate the original objects are minimally

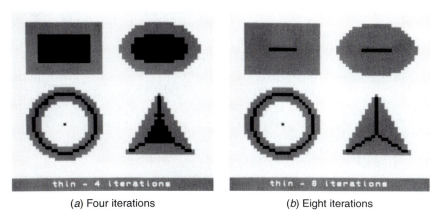

(a) Four iterations   (b) Eight iterations

FIGURE 14.3-3. Thinning of a binary image.

connected, and therefore the background of the separating strokes is eight-connected throughout the image. This is an example of the connectivity ambiguity discussed in Section 14.1. To resolve this ambiguity, a diagonal fill operation can be applied to the thinned strokes. The result, shown in Figure 14.3-5d, is called the *exothin* of the original image. The name derives from the exoskeleton, discussed in the following section.

14.3.3. Binary Image Skeletonizing

A skeleton or stick figure representation of an object can be used to describe its structure. Thinned objects sometimes have the appearance of a skeleton, but they are not always uniquely defined. For example, in Figure 14.3-3, both the rectangle and ellipse thin to a horizontal line.

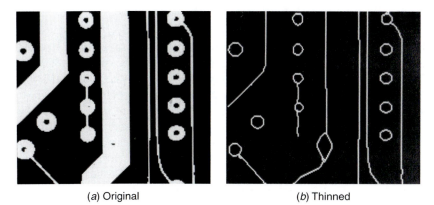

(a) Original   (b) Thinned

FIGURE 14.3-4. Thinning of a printed circuit board image.

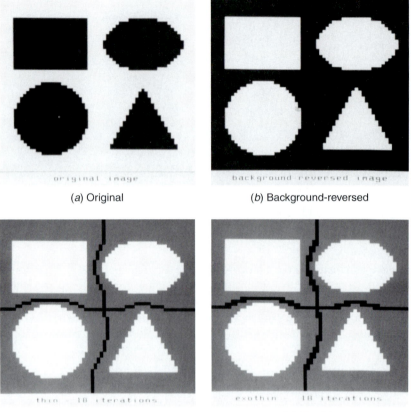

(a) Original

(b) Background-reversed

(c) Thinned background

(d) Exothin

FIGURE 14.3-5. Exothinning of a binary image.

Blum (18) has introduced a skeletonizing technique called *medial axis transformation* that produces a unique skeleton for a given object. An intuitive explanation of the medial axis transformation is based on the *prairie fire* analogy (19–22). Consider the circle and rectangle regions of Figure 14.3-6 to be composed of dry grass on a bare dirt background. If a fire were to be started simultaneously on the perimeter of the grass, the fire would proceed to burn toward the center of the regions until all the grass was consumed. In the case of the circle, the fire would burn to the center point of the circle, which is the *quench point* of the circle. For the rectangle, the fire would proceed from each side. As the fire moved simultaneously from left and top, the fire lines would meet and quench the fire. The quench points or quench lines of a figure are called its *medial axis skeleton*. More generally, the medial axis skeleton consists of the set of points that are equally distant from two closest points of an object boundary. The minimal distance function is called the *quench distance* of the object. From the medial axis skeleton of an object and its quench distance, it is

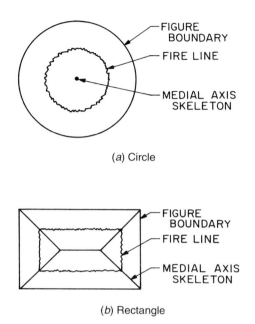

(a) Circle

(b) Rectangle

FIGURE 14.3-6. Medial axis transforms.

possible to reconstruct the object boundary. The object boundary is determined by the union of a set of circular disks formed by circumscribing a circle whose radius is the quench distance at each point of the medial axis skeleton.

A reasonably close approximation to the medial axis skeleton can be implemented by a slight variation of the conditional marking implementation shown in Figure 14.3-1. In this approach, an image is iteratively eroded using conditional and unconditional mark patterns until no further erosion occurs. The conditional mark patterns for skeletonization are listed in Table 14.3-1 under the table indicator K. Table 14.3-3 lists the unconditional mark patterns. At the conclusion of the last iteration, it is necessary to perform a single iteration of bridging as defined by Eq. 14.2-4 to restore connectivity, which will be lost whenever the following pattern is encountered:

$$1\ 1\ 1\ 1\ 1$$
$$1\ 1\ 1\ 1\ 1$$

Inhibiting the following mark pattern created by the bit pattern above:

$$M\ M$$
$$M\ M$$

will prevent elliptically shaped objects from being improperly skeletonized.

TABLE 14.3-3. Skeletonize Unconditional Mark Patterns
$[P(M, M_0, M_1, M_2, M_3, M_4, M_5, M_6, M_7) = 1$ if hit$]^a$

					Pattern						
Spur											
0	0	0	0	0	0	0	0	M	M	0	0
0	M	0	0	M	0	0	M	0	0	M	0
0	0	M	M	0	0	0	0	0	0	0	0
Single 4-connection											
0	0	0	0	0	0	0	0	0	0	M	0
0	M	0	0	M	M	M	M	0	0	M	0
0	M	0	0	0	0	0	0	0	0	0	0
L corner											
0	M	0	0	M	0	0	0	0	0	0	0
0	M	M	M	M	0	0	M	M	M	M	0
0	0	0	0	0	0	0	M	0	0	M	0
Corner cluster											
D	M	M	D	D	D	M	M	D	D	D	D
D	M	M	M	M	D	M	M	D	D	M	M
D	D	D	M	M	D	D	D	D	D	M	M
Tee branch											
D	M	D	D	M	D	D	D	D	D	M	D
M	M	M	M	M	D	M	M	M	D	M	M
D	0	0	D	M	D	D	M	D	D	M	D
Vee branch											
M	D	M	M	D	C	C	B	A	A	D	M
D	M	D	D	M	B	D	M	D	B	M	D
A	B	C	M	D	A	M	D	M	C	D	M
Digonal branch											
D	M	0	0	M	D	D	0	M	M	0	D
0	M	M	M	M	0	M	M	0	0	M	M
M	0	D	D	0	M	0	M	D	D	M	0

$^a A \cup B \cup C = 1$ $D = 0 \cup 1$.

(*a*) Four iterations

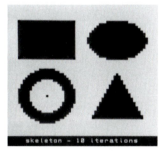

(*b*) Ten iterations

FIGURE 14.3-7. Skeletonizing of a binary image.

Figure 14.3-7 shows an example of the skeletonization of a binary image. The eroded pixels are midgray. It should observed that skeletonizing gives different results than thinning for many objects. Prewitt (23, p. 136) has coined the term *exoskeleton* for the skeleton of the background of object in a scene. The exoskeleton partitions each objects from neighboring object, as does the thinning of the background.

14.3.4. Binary Image Thickening

In Section 14.2.1, the fatten operator was introduced as a means of dilating objects such that objects separated by a single pixel stroke would not be fused. But the fatten operator does not prevent fusion of objects separated by a double width white stroke. This problem can be solved by iteratively thinning the background of an image and then performing a diagonal fill operation. This process, called *thickening*, when taken to its idempotent limit, forms the exothin of the image, as discussed in Section 14.3.2. Figure 14.3-8 provides an example of thickening. The exothin operation is repeated three times on the background reversed version of the original image. Figure 14.3-8*b* shows the final result obtained by reversing the background of the exothinned image.

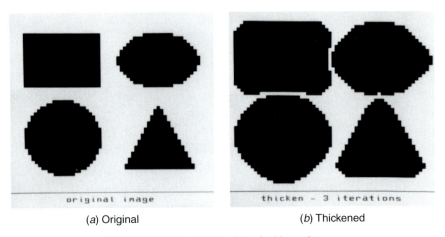

original image

(*a*) Original

thicken - 3 iterations

(*b*) Thickened

FIGURE 14.3-8. Thickening of a binary image.

14.4. BINARY IMAGE GENERALIZED DILATION AND EROSION

Dilation and erosion, as defined earlier in terms of hit-or-miss transformations, are limited to object modification by a single ring of boundary pixels during each iteration of the process. The operations can be generalized.

Before proceeding further, it is necessary to introduce some fundamental concepts of image set algebra that are the basis for defining the generalized dilation and erosions operators. Consider a binary-valued source image function $F(j, k)$. A pixel at coordinate (j, k) is a member of $F(j, k)$, as indicated by the symbol $\in$, if and only if it is a logical 1. A binary-valued image $B(j, k)$ is a subset of a binary-valued image $A(j, k)$, as indicated by $B(j, k) \subseteq A(j, k)$, if for every spatial occurrence of a logical 1 of $A(j, k)$, $B(j, k)$ is a logical 1. The complement $\bar{F}(j, k)$ of $F(j, k)$ is a binary-valued image whose pixels are in the opposite logical state of those in $F(j, k)$. Figure 14.4-1 shows an example of the complement process and other image set algebraic operations on a pair of binary images. A reflected image $\tilde{F}(j, k)$ is an image that has been flipped from left to right and from top to bottom. Figure 14.4-2 provides an example of image complementation. Translation of an image, as indicated by the function

$$G(j, k) = T_{r, c}\{F(j, k)\} \qquad (14.4\text{-}1)$$

consists of spatially offsetting $F(j, k)$ with respect to itself by r rows and c columns, where $-R \leq r \leq R$ and $-C \leq c \leq C$. Figure 14.4-2 presents an example of the translation of a binary image.

```
0 0 0 0 0 0       0 0 0 0 0 0       1 1 1 1 1 1
0 0 1 1 0 0       0 0 0 0 0 0       1 1 0 0 1 1
0 0 1 1 0 0       0 1 1 1 1 0       1 1 0 0 1 1
0 0 1 1 0 0       0 1 1 1 1 0       1 1 0 0 1 1
0 0 1 1 0 0       0 0 0 0 0 0       1 1 0 0 1 1
0 0 0 0 0 0       0 0 0 0 0 0       1 1 1 1 1 1
        A                 B                 Ā

                                      complement


0 0 0 0 0 0       0 0 0 0 0 0       0 0 0 0 0 0
0 0 1 1 0 0       0 0 0 0 0 0       0 0 1 1 0 0
0 1 1 1 1 0       0 0 1 1 0 0       0 1 0 0 1 0
0 1 1 1 1 0       0 0 1 1 0 0       0 1 0 0 1 0
0 0 1 1 0 0       0 0 0 0 0 0       0 0 1 1 0 0
0 0 0 0 0 0       0 0 0 0 0 0       0 0 0 0 0 0
      A ∪ B             A ∩ B            A XOR B
      union          intersection      exclusive-OR
        OR               AND               XOR
```

FIGURE 14.4-1. Image set algebraic operations on binary arrays.

14.4.1. Generalized Dilation

Generalized dilation is expressed symbolically as

$$G(j, k) = F(j, k) \oplus H(j, k) \tag{14.4-2}$$

where $F(j, k)$ for $1 \le j, k \le N$ is a binary-valued image and $H(j, k)$ for $1 \le j, k \le L$, where L is an odd integer, is a binary-valued array called a *structuring element*. For notational simplicity, $F(j, k)$ and $H(j, k)$ are assumed to be square arrays. Generalized dilation can be defined mathematically and implemented in several ways. The *Minkowski addition* definition (1) is

$$G(j, k) = \underbrace{\bigcup \bigcup T_{r, c} \{F(j, k)\}}_{(r, c) \in H} \tag{14.4-3}$$

0	0	0	0	0	0	0	0		0	0	0	0	0	0	0	0		0	0	0	0	0	0	0	0
0	0	0	0	0	0	0	0		0	0	0	0	0	0	0	0		0	0	0	0	0	0	0	0
0	0	1	0	0	0	0	0		0	0	0	1	1	1	0	0		0	0	0	0	0	0	0	0
0	0	1	0	0	0	0	0		0	0	0	0	0	1	0	0		0	0	0	0	1	0	0	0
0	0	1	0	0	0	0	0		0	0	0	0	0	1	0	0		0	0	0	0	1	0	0	0
0	0	1	1	1	0	0	0		0	0	0	0	0	1	0	0		0	0	0	0	1	0	0	0
0	0	0	0	0	0	0	0		0	0	0	0	0	0	0	0		0	0	0	0	1	1	1	0
0	0	0	0	0	0	0	0		0	0	0	0	0	0	0	0		0	0	0	0	0	0	0	0

Original $F(j,k)$ Reflection $\tilde{F}(j,k)$ Translation $T_{1,2}\{F(j,k)\}$

FIGURE 14.4-2. Reflection and translation of a binary array.

It states that $G(j, k)$ is formed by the union of all translates of $F(j, k)$ with respect to itself in which the translation distance is the row and column index of pixels of $H(j, k)$ that is a logical 1. Figure 14.4-3 illustrates the concept. Equation 14.4-3 results in an $M \times M$ output array $G(j, k)$ that is justified with the upper left corner of the input array $F(j, k)$. The output array is of dimension $M = N + L - 1$, where L is the size of the structuring element. In order to register the input and output images properly, $F(j, k)$ should be translated diagonally right by $Q = (L - 1)/2$ pixels. Figure 14.4-3 shows the exclusive-OR difference between $G(j, k)$ and the translate of $F(j, k)$. This operation identifies those pixels that have been added as a result of generalized dilation.

An alternative definition of generalized dilation is based on the scanning and processing of $F(j, k)$ by the structuring element $H(j, k)$. With this approach, generalized dilation is formulated as (17)

$$G(j, k) = \bigcup_m \bigcup_n F(m, n) \cap H(j - m + 1, k - n + 1) \tag{14.4-4}$$

With reference to Eq. 7.1-7, the spatial limits of the union combination are

$$\text{MAX}\{1, j - L + 1\} \le m \le \text{MIN}\{N, j\} \tag{14.4-5a}$$

$$\text{MAX}\{1, k - L + 1\} \le n \le \text{MIN}\{N, k\} \tag{14.4-5b}$$

Equation 14.4-4 provides an output array that is justified with the upper left corner of the input array. In image processing systems, it is often convenient to center the input and output images and to limit their size to the same overall dimension. This can be accomplished easily by modifying Eq. 14.4-4 to the form

$$G(j, k) = \bigcup_m \bigcup_n F(m, n) \cap H(j - m + S, k - n + S) \tag{14.4-6}$$

```
0  0  0  0  0      1  1  0
0  0  1  0  0      1  1  0
0  1  1  0  0      1  0  0
0  0  1  1  0
0  0  0  0  0
    F(j, k)         H(j, k)
```

```
0  0  0  0  0    ·  0  0  0  0  0      ·  ·  ·  ·  ·      ·  ·  ·  ·  ·  ·      ·  ·  ·  ·  ·
0  0  1  0  0    ·  0  0  1  0  0    0  0  0  0  0    ·  0  0  0  0  0      ·  ·  ·  ·  ·
0  1  1  0  0    ·  0  1  1  0  0    0  0  1  0  0    ·  0  0  1  0  0    0  0  0  0  0
0  0  1  1  0    ·  0  0  1  1  0    0  1  1  0  0    ·  0  1  1  0  0    0  0  1  0  0
0  0  0  0  0    ·  0  0  0  0  0    0  0  1  1  0    ·  0  0  1  1  0    0  1  1  0  0
                                    0  0  0  0  0    ·  0  0  0  0  0    0  0  1  1  0
                                                                        0  0  0  0  0

  $T_{0,0}\{F(j,k)\}$      $T_{0,1}\{F(j,k)\}$      $T_{1,0}\{F(j,k)\}$      $T_{1,1}\{F(j,k)\}$      $T_{2,0}\{F(j,k)\}$
```

```
0  0  0  0  0  0  0
0  0  1  1  0  0  0
0  1  1  1  0  0  0
0  1  1  1  1  0  0
0  1  1  1  1  0  0
0  0  1  1  0  0  0
0  0  0  0  0  0  0
```

$$G(j, k) = T_{0,0}\{F(j,k)\} \cup T_{0,1}\{F(j,k)\} \cup T_{1,0}\{F(j,k)\} \cup T_{1,1}\{F(j,k)\} \cup T_{2,0}\{F(j,k)\}$$

```
0  0  0  0  0  0  0
0  0  1  1  0  0  0
0  1  1  0  0  0  0
0  1  0  0  1  0  0
0  1  1  0  0  0  0
0  0  1  1  0  0  0
0  0  0  0  0  0  0
```

$$G(j, k) \text{ XOR } T_{1,1}\{F(j,k)\}$$

FIGURE 14.4-3. Generalized dilation computed by Minkowski addition.

where $S = (L-1)/2$ and, from Eq. 7.1-10, the limits of the union combination are

$$\text{MAX}\{1, j - Q\} \le m \le \text{MIN}\{N, j + Q\} \tag{14.4-7a}$$

$$\text{MAX}\{1, k - Q\} \le n \le \text{MIN}\{N, k + Q\} \tag{14.4-7b}$$

and where $Q = (L-1)/2$. Equation 14.4-6 applies for $S \le j, k \le N - Q$ and $G(j, k) = 0$ elsewhere. The Minkowski addition definition of generalized erosion given in Eq. 14.4-2 can be modified to provide a centered result by taking the translations about the center of the structuring element. In the following discussion, only the centered definitions of generalized dilation will be utilized. In the special case for which $L = 3$, Eq. 14.4-6 can be expressed explicitly as

$$(G(j, k)) =$$
$$[H(3, 3) \cap F(j-1, k-1)] \cup [H(3, 2) \cap F(j-1, k)] \cup [H(3, 1) \cap F(j-1, K+1)]$$
$$\cup [H(2, 3) \cap F(j, k-1)] \cup [H(2, 2) \cap F(j, k)] \cup [H(2, 1) \cap F(j, k+1)]$$
$$\cup [H(1, 3) \cap F(j+1, k-1)] \cup [H(1, 2) \cap F(j+1, k)] \cup [H(1, 1) \cap F(j+1, k+1)]$$

$$(14.4-8)$$

If $H(j, k) = 1$ for $1 \le j, k \le 3$, then $G(j, k)$, as computed by Eq. 14.4-8, gives the same result as hit-or-miss dilation, as defined by Eq. 14.2-5.

It is interesting to compare Eqs. 14.4-6 and 14.4-8, which define generalized dilation, and Eqs. 7.1-14 and 7.1-15, which define convolution. In the generalized dilation equation, the union operations are analogous to the summation operations of convolution, while the intersection operation is analogous to point-by-point multiplication. As with convolution, dilation can be conceived as the scanning and processing of $F(j, k)$ by $H(j, k)$ rotated by $180°$.

14.4.2. Generalized Erosion

Generalized erosion is expressed symbolically as

$$G(j, k) = F(j, k) \ominus H(j, k) \qquad (14.4-9)$$

where again $H(j, k)$ is an odd size $L \times L$ structuring element. Serra (3) has adopted, as his definition for erosion, the dual relationship of Minkowski addition given by Eq. 14.4-1, which was introduced by Hadwiger (24). By this formulation, generalized erosion is defined to be

$$G(j, k) = \underbrace{\bigcap \bigcap T_{r, c} \{F(j, k)\}}_{(r, c) \in H} \qquad (14.4-10)$$

The meaning of this relation is that erosion of $F(j, k)$ by $H(j, k)$ is the intersection of all translates of $F(j, k)$ in which the translation distance is the row and column index of pixels of $H(j, k)$ that are in the logical 1 state. Steinberg et al. (6,25) have adopted the subtly different formulation

Sternberg definition of generalized erosion:

```
1 1 1 1 1
1 1 1 1 1    1 1 1    0 0 0
1 1 0 0 0 ⊖ 1 0 0 = 1 1 0
1 1 1 1 1    1 1 1    0 0 0
1 1 1 1 1
  F(j, k)       H(j, k)      G(j, k)
```

Serra definition of generalized erosion:

```
1 1 1 1 1
1 1 1 1 1    1 1 1    0 0 0
1 1 0 0 0 ⊖ 1 0 0 = 0 0 0
1 1 1 1 1    1 1 1    0 0 0
1 1 1 1 1
  F(j, k)       H(j, k)      G(j, k)
```

FIGURE 14.4-4. Comparison of erosion results for two definitions of generalized erosion.

$$G(j, k) = \underbrace{\bigcap \bigcap T_{r, c} \{F(j, k)\}}_{(r, c) \in \tilde{H}} \qquad (14.4\text{-}11)$$

introduced by Matheron (2), in which the translates of $F(j, k)$ are governed by the reflection $\tilde{H}(j, k)$ of the structuring element rather than by $H(j, k)$ itself.

Using the Steinberg definition, $G(j, k)$ is a logical 1 if and only if the logical 1s of $H(j, k)$ form a subset of the spatially corresponding pattern of the logical 1s of $F(j, k)$ as $H(j, k)$ is scanned over $F(j, k)$. It should be noted that the logical zeros of $H(j, k)$ do not have to match the logical zeros of $F(j, k)$. With the Serra definition, the statements above hold when $F(j, k)$ is scanned and processed by the reflection of the structuring element. Figure 14.4-4 presents a comparison of the erosion results for the two definitions of erosion. Clearly, the results are inconsistent.

Pratt (26) has proposed a relation, which is the dual to the generalized dilation expression of Eq. 14.4-6, as a definition of generalized erosion. By this formulation, generalized erosion in centered form is

$$G(j, k) = \bigcap_{m} \bigcap_{n} F(m, n) \cup \overline{H}(j - m + S, k - n + S) \qquad (14.4\text{-}12)$$

where $S = (L - 1)/2$, and the limits of the intersection combination are given by Eq. 14.4-7. In the special case for which $L = 3$, Eq. 14.4-12 becomes

$G(j, k) =$

$[\overline{H}(3, 3) \cup F(j-1, k-1)] \cap [\overline{H}(3, 2) \cup F(j-1, k)] \cap [\overline{H}(3, 1) \cup F(j-1, k+1)]$

$\cup [\overline{H}(2, 3) \cup F(j, k-1)] \cap [\overline{H}(2, 2) \cup F(j, k)] \cap [\overline{H}(2, 1) \cup F(j, k+1)]$

$\cap [\overline{H}(1, 3) \cup F(j+1, k-1)] \cap [\overline{H}(1, 2) \cup F(j+1, k)] \cap [H(1, 1) \cup F(j+1, k+1)]$

$$(14.4\text{-}13)$$

If $H(j, k) = 1$ for $1 \leq j, k \leq 3$, Eq. 14.4-13 gives the same result as hit-or-miss eight-neighbor erosion as defined by Eq. 14.2-6. Pratt's definition is the same as the Serra definition. However, Eq. 14.4-12 can easily be modified by substituting the reflection $\tilde{H}(j, k)$ for $H(j, k)$ to provide equivalency with the Steinberg definition. Unfortunately, the literature utilizes both definitions, which can lead to confusion. The definition adopted in this book is that of Hadwiger, Serra, and Pratt, because the

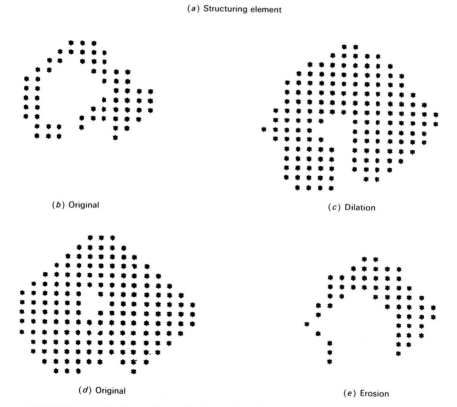

(a) Structuring element

(b) Original

(c) Dilation

(d) Original

(e) Erosion

FIGURE 14.4-5. Generalized dilation and erosion for a 5×5 structuring element.

defining relationships (Eq. 14.4-1 or 14.4-12) are duals to their counterparts for generalized dilation (Eq. 14.4-3 or 14.4-6).

Figure 14.4-5 shows examples of generalized dilation and erosion for a symmetric 5×5 structuring element.

14.4.3. Properties of Generalized Dilation and Erosion

Consideration is now given to several mathematical properties of generalized dilation and erosion. Proofs of these properties are found in Reference 25. For notational simplicity, in this subsection the spatial coordinates of a set are dropped, i.e., $A(j, k) = \mathbf{A}$. Dilation is commutative:

$$\mathbf{A} \oplus \mathbf{B} = \mathbf{B} \oplus \mathbf{A} \tag{14.4-14a}$$

But in general, erosion is not commutative:

$$\mathbf{A} \ominus \mathbf{B} \neq \mathbf{B} \ominus \mathbf{A} \tag{14.4-14b}$$

Dilation and erosion are increasing operations in the sense that if $\mathbf{A} \subseteq \mathbf{B}$, then

$$\mathbf{A} \oplus \mathbf{C} \subseteq \mathbf{B} \oplus \mathbf{C} \tag{14.4-15a}$$

$$\mathbf{A} \ominus \mathbf{C} \subseteq \mathbf{B} \ominus \mathbf{C} \tag{14.4-15b}$$

Dilation and erosion are opposite in effect; dilation of the background of an object behaves like erosion of the object. This statement can be quantified by the duality relationship

$$\overline{\mathbf{A} \ominus \mathbf{B}} = \overline{\mathbf{A}} \oplus \mathbf{B} \tag{14.4-16}$$

For the Steinberg definition of erosion, $\mathbf{B}$ on the right-hand side of Eq. 14.4-16 should be replaced by its reflection $\tilde{\mathbf{B}}$. Figure 14.4-6 contains an example of the duality relationship.

The dilation and erosion of the intersection and union of sets obey the following relations:

$$[\mathbf{A} \cap \mathbf{B}] \oplus \mathbf{C} \subseteq [\mathbf{A} \oplus \mathbf{C}] \cap [\mathbf{B} \oplus \mathbf{C}] \tag{14.4-17a}$$

$$[\mathbf{A} \cap \mathbf{B}] \ominus \mathbf{C} = [\mathbf{A} \ominus \mathbf{C}] \cap [\mathbf{B} \ominus \mathbf{C}] \tag{14.4-17b}$$

$$[\mathbf{A} \cup \mathbf{B}] \oplus \mathbf{C} = [\mathbf{A} \oplus \mathbf{C}] \cup [\mathbf{B} \oplus \mathbf{C}] \tag{14.4-17c}$$

$$[\mathbf{A} \cup \mathbf{B}] \ominus \mathbf{C} \supseteq [\mathbf{A} \ominus \mathbf{C}] \cup [\mathbf{B} \ominus \mathbf{C}] \tag{14.4-17d}$$

$$\underset{\mathbf{A}}{\begin{array}{ccccc}1&1&0&1&1\\1&1&1&1&1\\1&1&0&0&0\\1&1&1&1&1\\1&1&1&0&1\end{array}}\qquad\underset{\mathbf{B}}{\begin{array}{ccc}1&1&1\\0&0&1\\0&1&1\end{array}}\qquad\underset{\mathbf{A}\ominus\mathbf{B}}{\begin{array}{ccc}0&0&0\\1&1&0\\1&0&0\end{array}}\qquad\underset{\overline{\mathbf{A}\ominus\mathbf{B}}}{\begin{array}{ccc}1&1&1\\0&0&1\\0&1&1\end{array}}$$

$$\underset{\overline{\mathbf{A}}}{\begin{array}{ccccc}0&0&1&0&0\\0&0&0&0&0\\0&0&1&1&1\\0&0&0&0&0\\0&0&0&1&0\end{array}}\qquad\underset{\mathbf{B}}{\begin{array}{ccc}1&1&1\\0&0&1\\0&1&1\end{array}}\qquad\underset{\overline{\mathbf{A}}\oplus\mathbf{B}}{\begin{array}{ccc}1&1&1\\0&0&1\\0&1&1\end{array}}$$

FIGURE 14.4-6. Duality relationship between dilation and erosion.

The dilation and erosion of a set by the intersection of two other sets satisfy these containment relations:

$$\mathbf{A} \oplus [\mathbf{B} \cap \mathbf{C}] \subseteq [\mathbf{A} \oplus \mathbf{B}] \cap [\mathbf{A} \oplus \mathbf{C}] \tag{14.4-18a}$$

$$\mathbf{A} \ominus [\mathbf{B} \cap \mathbf{C}] \supseteq [\mathbf{A} \ominus \mathbf{B}] \cup [\mathbf{A} \ominus \mathbf{C}] \tag{14.4-18b}$$

On the other hand, dilation and erosion of a set by the union of a pair of sets are governed by the equality relations

$$\mathbf{A} \oplus [\mathbf{B} \cup \mathbf{C}] = [\mathbf{A} \oplus \mathbf{B}] \cup [\mathbf{A} \oplus \mathbf{C}] \tag{14.4-19a}$$

$$\mathbf{A} \ominus [\mathbf{B} \cup \mathbf{C}] = [\mathbf{A} \ominus \mathbf{B}] \cup [\mathbf{A} \ominus \mathbf{C}] \tag{14.4-19b}$$

The following chain rules hold for dilation and erosion.

$$\mathbf{A} \oplus [B \oplus C] = [\mathbf{A} \oplus B] \oplus \mathbf{C} \tag{14.4-20a}$$

$$\mathbf{A} \ominus [B \oplus C] = [\mathbf{A} \ominus B] \ominus \mathbf{C} \tag{14.4-20b}$$

14.4.4. Structuring Element Decomposition

Equation 14.4-20 is important because it indicates that if a $L \times L$ structuring element can be expressed as

$$H(j, k) = K_1(j, k) \oplus \cdots \oplus K_q(j, k) \oplus \cdots \oplus K_Q(j, k) \tag{14.4-21}$$

```
                                    1  1  1  1  1
         1  1  1    1  1  1         1  1  1  1  1
         1  1  1 ⊕  1  1  1  =      1  1  1  1  1
         1  1  1    1  1  1         1  1  1  1  1
                                    1  1  1  1  1


                                    0  1  1  1  0
         1  1  1    0  1  0         1  1  1  1  1
         1  1  1 ⊕  1  1  1  =      1  1  1  1  1
         1  1  1    0  1  0         1  1  1  1  1
                                    0  1  1  1  0


                                    0  0  1  0  0
         0  1  0    0  1  0         0  1  1  1  0
         1  1  1 ⊕  1  1  1  =      1  1  1  1  1
         0  1  0    0  1  0         0  1  1  1  0
                                    0  0  1  0  0


                                    0  0  1  1  1
         0  1  1    0  1  1         0  1  1  1  1
         1  1  1 ⊕  1  1  1  =      1  1  1  1  1
         1  1  0    1  1  0         1  1  1  1  0
                                    1  1  1  0  0
```

FIGURE 14.4-7. Structuring element decomposition.

where $K_q(j, k)$ is a small structuring element, it is possible to perform dilation and erosion by operating on an image sequentially. In Eq. 14.4-21, if the small structuring elements $K_q(j, k)$ are all 3×3 arrays, then $Q = (L - 1)/2$. Figure 14.4-7 gives several examples of small structuring element decomposition. Sequential small structuring element (SSE) dilation and erosion is analogous to small generating kernel (SGK) convolution as given by Eq. 9.6-1. Not every large impulse response array can be decomposed exactly into a sequence of SGK convolutions; similarly, not every large structuring element can be decomposed into a sequence of SSE dilations or erosions. Following is an example in which a 5×5 structuring element cannot be decomposed into the sequential dilation of two 3×3 SSEs. Zhuang and Haralick (27) have developed a computational search method to find a SEE decomposition into 1×2 and 2×1 elements.

```
1 1 1 1 1     1 1 1 1 0     0 0 0 0 1     0 0 0 0 0     0 0 0 0 0
1 0 0 0 1     0 0 0 0 0     0 0 0 0 1     0 0 0 0 0     1 0 0 0 0
1 0 0 0 1  =  0 0 0 0 0  ∪  0 0 0 0 1  ∪  0 0 0 0 0  ∪  1 0 0 0 0
1 0 0 0 1     0 0 0 0 0     0 0 0 0 1     0 0 0 0 0     1 0 0 0 0
1 1 1 1 1     0 0 0 0 0     0 0 0 0 1     0 1 1 1 1     1 0 0 0 0
```

```
1 1 1 1 0
0 0 0 0 0      1 1 1     1 1 0
0 0 0 0 0  =   0 0 0  ⊕  0 0 0
0 0 0 0 0      0 0 0     0 0 0
0 0 0 0 0
```

```
0 0 0 0 1
0 0 0 0 1      0 0 1     0 0 1
0 0 0 0 1  =   0 0 1  ⊕  0 0 1
0 0 0 0 1      0 0 1     0 0 0
0 0 0 0 0
```

```
0 0 0 0 0
0 0 0 0 0      0 0 0     0 0 0
0 0 0 0 0  =   0 0 0  ⊕  0 0 0
0 0 0 0 0      1 1 1     0 1 1
0 1 1 1 1
```

```
0 0 0 0 0
1 0 0 0 0      1 0 0     0 0 0
1 0 0 0 0  =   1 0 0  ⊕  1 0 0
1 0 0 0 0      1 0 0     1 0 0
1 0 0 0 0
```

FIGURE 14.4-8. Small structuring element decomposition of a 5 × 5 pixel ring.

```
1 1 1 1 1
1 0 0 0 1
1 0 0 0 1
1 0 0 0 1
1 1 1 1 1
```

For two-dimensional convolution it is possible to decompose any large impulse response array into a set of sequential SGKs that are computed in parallel and summed together using the singular-value decomposition/small generating kernel (SVD/SGK) algorithm, as illustrated by the flowchart of Figure 9.6-2. It is logical to conjecture as to whether an analog to the SVD/SGK algorithm exists for dilation and erosion. Equation 14.4-19 suggests that such an algorithm may exist. Figure 14.4-8 illustrates an SSE decomposition of the 5×5 ring example based on Eqs. 14.4-19a and 14.4-21. Unfortunately, no systematic method has yet been found to decompose an arbitrarily large structuring element.

14.5. BINARY IMAGE CLOSE AND OPEN OPERATIONS

Dilation and erosion are often applied to an image in concatenation. Dilation followed by erosion is called a *close operation*. It is expressed symbolically as

$$G(j, k) = F(j, k) \bullet H(j, k) \tag{14.5-1a}$$

where $H(j, k)$ is a $L \times L$ structuring element. In accordance with the Serra formulation of erosion, the close operation is defined as

$$G(j, k) = [F(j, k) \oplus H(j, k)] \ominus \tilde{H}(j, k) \tag{14.5-1b}$$

where it should be noted that erosion is performed with the reflection of the structuring element. Closing of an image with a compact structuring element without holes (zeros), such as a square or circle, smooths contours of objects, eliminates small holes in objects, and fuses short gaps between objects.

An *open operation*, expressed symbolically as

$$G(j, k) = F(j, k) \circ H(j, k) \tag{14.5-2a}$$

consists of erosion followed by dilation. It is defined as

$$G(j, k) = [F(j, k) \ominus \tilde{H}(j, k)] \oplus H(j, k) \tag{14.5-2b}$$

where again, the erosion is with the reflection of the structuring element. Opening of an image smooths contours of objects, eliminates small objects, and breaks narrow strokes.

The close operation tends to increase the spatial extent of an object, while the open operation decreases its spatial extent. In quantitative terms

$$F(j, k) \bullet H(j, k) \supseteq F(j, k) \tag{14.5-3a}$$

$$F(j, k) \circ H(j, k) \subseteq F(j, k) \tag{14.5-3b}$$

(*a*) Original

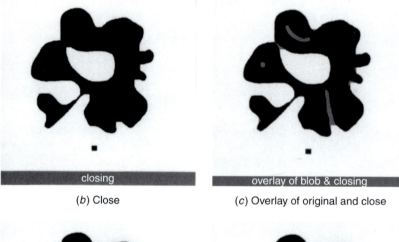

(*b*) Close

(*c*) Overlay of original and close

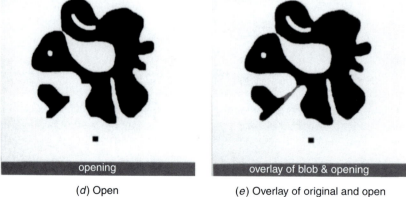

(*d*) Open

(*e*) Overlay of original and open

FIGURE 14.5-1. Close and open operations on a binary image.

It can be shown that the close and open operations are stable in the sense that (25)

$$[F(j, k) \bullet H(j, k)] \bullet H(j, k) = F(j, k) \bullet H(j, k) \qquad (14.5\text{-}4a)$$

$$[F(j, k) \circ H(j, k)] \circ H(j, k) = F(j, k) \circ H(j, k) \qquad (14.5\text{-}4b)$$

Also, it can be easily shown that the open and close operations satisfy the following duality relationship:

$$\overline{F(j, k) \bullet H(j, k)} = \overline{F(j, k)} \circ H(j, k) \qquad (14.5\text{-}5)$$

Figure 14.5-1 presents examples of the close and open operations on a binary image.

14.6. GRAY SCALE IMAGE MORPHOLOGICAL OPERATIONS

Morphological concepts can be extended to gray scale images, but the extension often leads to theoretical issues and to implementation complexities. When applied to a binary image, dilation and erosion operations cause an image to increase or decrease in spatial extent, respectively. To generalize these concepts to a gray scale image, it is assumed that the image contains visually distinct gray scale objects set against a gray background. Also, it is assumed that the objects and background are both relatively spatially smooth. Under these conditions, it is reasonable to ask: Why not just threshold the image and perform binary image morphology? The reason for not taking this approach is that the thresholding operation often introduces significant error in segmenting objects from the background. This is especially true when the gray scale image contains shading caused by nonuniform scene illumination.

14.6.1. Gray Scale Image Dilation and Erosion

Dilation or erosion of an image could, in principle, be accomplished by hit-or-miss transformations in which the quantized gray scale patterns are examined in a 3×3 window and an output pixel is generated for each pattern. This approach is, however, not computationally feasible. For example, if a look-up table implementation were to be used, the table would require 2^{72} entries for 256-level quantization of each pixel! The common alternative is to use gray scale extremum operations over a 3×3 pixel neighborhoods.

Consider a gray scale image $F(j, k)$ quantized to an arbitrary number of gray levels. According to the extremum method of gray scale image dilation, the dilation operation is defined as

$$G(j, k) = \text{MAX}\{F(j, k), F(j, k + 1), F(j - 1, k + 1), ..., F(j + 1, k + 1)\} \quad (14.6\text{-}1)$$

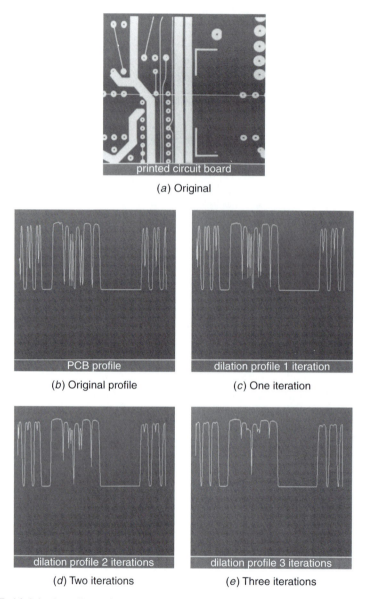

(a) Original

(b) Original profile   (c) One iteration

(d) Two iterations   (e) Three iterations

FIGURE 14.6-1. One-dimensional gray scale image dilation on a printed circuit board image.

where $\mathrm{MAX}\{S_1, ..., S_9\}$ generates the largest-amplitude pixel of the nine pixels in the neighborhood. If $F(j, k)$ is quantized to only two levels, Eq. 14.6-1 provides the same result as that using binary image dilation as defined by Eq. 14.2-5.

By the extremum method, gray scale image erosion is defined as

$$G(j, k) = \text{MIN}\{F(j, k), F(j, k + 1), F(j - 1, k + 1), ..., F(j + 1, k + 1)\} \qquad (14.6\text{-}2)$$

where $\text{MIN}\{S_1, ..., S_9\}$ generates the smallest-amplitude pixel of the nine pixels in the 3×3 pixel neighborhood. If $F(j, k)$ is binary-valued, then Eq. 14.6-2 gives the same result as hit-or-miss erosion as defined in Eq. 14.2-8.

In Chapter 10, when discussing the pseudomedian, it was shown that the MAX and MIN operations can be computed sequentially. As a consequence, Eqs. 14.6-1 and 14.6-2 can be applied iteratively to an image. For example, three iterations gives the same result as a single iteration using a 7×7 moving-window MAX or MIN operator. By selectively excluding some of the terms $S_1, ..., S_9$ of Eq. 14.6-1 or 14.6-2 during each iteration, it is possible to synthesize large nonsquare gray scale structuring elements in the same number as illustrated in Figure 14.4-7 for binary structuring elements. However, no systematic decomposition procedure has yet been developed.

Figures 14.6-1 and 14.6-2 show the amplitude profile of a row of a gray scale image of a printed circuit board (PCB) after several dilation and erosion iterations. The row selected is indicated by the white horizontal line in Figure 14.6-l*a*. In Figure 14.6-2, two-dimensional gray scale dilation and erosion are performed on the PCB image.

14.6.2. Gray Scale Image Close and Open Operators

The close and open operations introduced in Section 14.5 for binary images can easily be extended to gray scale images. Gray scale closing is realized by first performing gray scale dilation with a gray scale structuring element, then gray scale erosion with the same structuring element. Similarly, gray scale opening is accomplished by gray scale erosion followed by gray scale dilation. Figure 14.6-3 gives examples of gray scale image closing and opening.

Steinberg (28) has introduced the use of three-dimensional structuring elements for gray scale image closing and opening operations. Although the concept is well defined mathematically, it is simpler to describe in terms of a structural image model. Consider a gray scale image to be modeled as an array of closely packed square pegs, each of which is proportional in height to the amplitude of a corresponding pixel. Then a three-dimensional structuring element, for example a sphere, is placed over each peg. The bottom of the structuring element as it is translated over the peg array forms another spatially discrete surface, which is the close array of the original image. A spherical structuring element will touch pegs at peaks of the original peg array, but will not touch pegs at the bottom of steep valleys. Consequently, the close surface "fills in" dark spots in the original image. The opening of a gray scale image can be conceptualized in a similar manner. An original image is modeled as a peg array in which the height of each peg is inversely proportional to

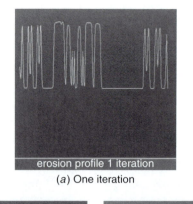

(*a*) One iteration

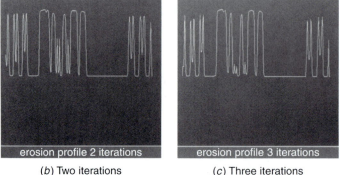

(*b*) Two iterations   (*c*) Three iterations

FIGURE 14.6-2. One-dimensional gray scale image erosion on a printed circuit board image.

the amplitude of each corresponding pixel (i.e., the gray scale is subtractively inverted). The translated structuring element then forms the open surface of the original image. For a spherical structuring element, bright spots in the original image are made darker.

14.6.3. Conditional Gray Scale Image Morphological Operators

There have been attempts to develop morphological operators for gray scale images that are analogous to binary image shrinking, thinning, skeletonizing, and thickening. The stumbling block to these extensions is the lack of a definition for connectivity of neighboring gray scale pixels. Serra (4) has proposed approaches based on topographic mapping techniques. Another approach is to iteratively perform the basic dilation and erosion operations on a gray scale image and then use a binary thresholded version of the resultant image to determine connectivity at each iteration.

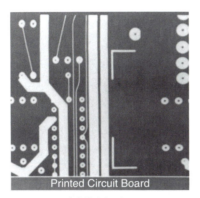

(*a*) Original

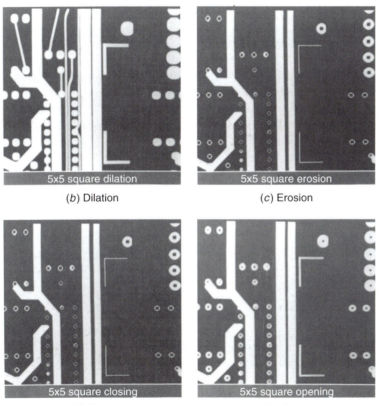

(*b*) Dilation   (*c*) Erosion

(*d*) Close   (*e*) Open

FIGURE 14.6-3. Two-dimensional gray scale image dilation, erosion, close, and open on a printed circuit board image.

REFERENCES

1. H. Minkowski, "Volumen und Oberfiläche," *Mathematische Annalen*, **57**, 1903, 447–459.

2. G. Matheron, *Random Sets and Integral Geometry*, Wiley, New York, 1975.

3. J. Serra, *Image Analysis and Mathematical Morphology*, Vol. 1, Academic Press, London, 1982.

4. J. Serra, *Image Analysis and Mathematical Morphology: Theoretical Advances*, Vol. 2, Academic Press, London, 1988.

5. J. Serra, "Introduction to Mathematical Morphology," *Computer Vision, Graphics, and Image Processing*, **35**, 3, September 1986, 283–305.

6. S. R. Steinberg, "Parallel Architectures for Image Processing," *Proc. 3rd International IEEE Compsac*, Chicago, 1981.

7. S. R. Steinberg, "Biomedical Image Processing," *IEEE Computer*, January 1983, 22–34.

8. S. R. Steinberg, "Automatic Image Processor," US patent 4,167,728.

9. R. M. Lougheed and D. L. McCubbrey, "The Cytocomputer: A Practical Pipelined Image Processor," *Proc. 7th Annual International Symposium on Computer Architecture*, 1980.

10. A. Rosenfeld, "Connectivity in Digital Pictures," *J. Association for Computing Machinery*, **17**, 1, January 1970, 146–160.

11. A. Rosenfeld, *Picture Processing by Computer*, Academic Press, New York, 1969.

12. M. J. E. Golay, "Hexagonal Pattern Transformation," *IEEE Trans. Computers*, **C-18**, 8, August 1969, 733–740.

13. K. Preston, Jr., "Feature Extraction by Golay Hexagonal Pattern Transforms," *IEEE Trans. Computers*, **C-20**, 9, September 1971, 1007–1014.

14. F. A. Gerritsen and P. W. Verbeek, "Implementation of Cellular Logic Operators Using 3×3 Convolutions and Lookup Table Hardware," *Computer Vision, Graphics, and Image Processing*, **27**, 1, 1984, 115–123.

15. A. Rosenfeld, "A Characterization of Parallel Thinning Algorithms," *Information and Control*, **29**, 1975, 286–291.

16. T. Pavlidis, "A Thinning Algorithm for Discrete Binary Images," *Computer Graphics and Image Processing*, **13**, 2, 1980, 142–157.

17. W. K. Pratt and I. Kabir, "Morphological Binary Image Processing with a Local Neighborhood Pipeline Processor," *Computer Graphics*, Tokyo, 1984.

18. H. Blum, "A Transformation for Extracting New Descriptors of Shape," in *Symposium Models for Perception of Speech and Visual Form*, W. Whaten-Dunn, Ed., MIT Press, Cambridge, MA, 1967.

19. R. O. Duda and P. E. Hart, *Pattern Classification and Scene Analysis*, Wiley-Interscience, New York, 1973.

20. L. Calabi and W. E. Harnett, "Shape Recognition, Prairie Fires, Convex Deficiencies and Skeletons," *American Mathematical Monthly*, **75**, 4, April 1968, 335–342.

21. J. C. Mott-Smith, "Medial Axis Transforms," in *Picture Processing and Psychopictorics*, B. S. Lipkin and A. Rosenfeld, Eds., Academic Press, New York, 1970.

22. C. Arcelli and G. Sanniti Di Baja, "On the Sequential Approach to Medial Line Thinning Transformation," *IEEE Trans. Systems, Man and Cybernetics*, **SMC-8**, 2, 1978, 139–144.

23. J. M. S. Prewitt, "Object Enhancement and Extraction," in *Picture Processing and Psychopictorics*, B. S. Lipkin and A. Rosenfeld, Eds., Academic Press, New York, 1970.

24. H. Hadwiger, *Vorslesunger uber Inhalt, Oberfläche und Isoperimetrie*, Springer-Verlag, Berlin, 1957.

25. R. M. Haralick, S. R. Steinberg, and X. Zhuang, "Image Analysis Using Mathematical Morphology," *IEEE Trans. Pattern Analysis and Machine Intelligence*, **PAMI-9**, 4, July 1987, 532–550.

26. W. K. Pratt, "Image Processing with Primitive Computational Elements," McMaster University, Hamilton, Ontario, Canada, 1987.

27. X. Zhuang and R. M. Haralick, "Morphological Structuring Element Decomposition," *Computer Vision, Graphics, and Image Processing*, **35**, 3, September 1986, 370–382.

28. S. R. Steinberg, "Grayscale Morphology," *Computer Vision, Graphics, and Image Processing*, **35**, 3, September 1986, 333–355.

15

EDGE DETECTION

Changes or discontinuities in an image amplitude attribute such as luminance or tristimulus value are fundamentally important primitive characteristics of an image because they often provide an indication of the physical extent of objects within the image. Local discontinuities in image luminance from one level to another are called *luminance edges*. Global luminance discontinuities, called *luminance boundary segments*, are considered in Section 17.4. In this chapter the definition of a luminance edge is limited to image amplitude discontinuities between reasonably smooth regions. Discontinuity detection between textured regions is considered in Section 17.5. This chapter also considers edge detection in color images, as well as the detection of lines and spots within an image.

15.1. EDGE, LINE, AND SPOT MODELS

Figure 15.1-1a is a sketch of a continuous domain, one-dimensional *ramp edge* modeled as a ramp increase in image amplitude from a low to a high level, or vice versa. The edge is characterized by its height, slope angle, and horizontal coordinate of the slope midpoint. An edge exists if the edge height is greater than a specified value. An ideal edge detector should produce an edge indication localized to a single pixel located at the midpoint of the slope. If the slope angle of Figure 15.1-1a is 90°, the resultant edge is called a *step edge*, as shown in Figure 15.1-1b. In a digital imaging system, step edges usually exist only for artificially generated images such as test patterns and bilevel graphics data. Digital images, resulting from digitization of optical images of real scenes, generally do not possess step edges because the anti aliasing low-pass filtering prior to digitization reduces the edge slope in the digital image caused by any sudden luminance change in the scene. The one-dimensional

443

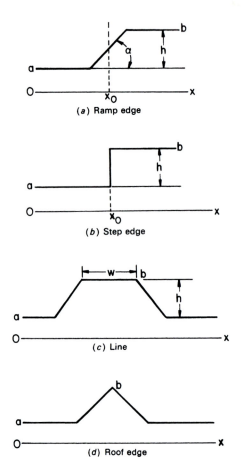

FIGURE 15.1-1. One-dimensional, continuous domain edge and line models.

profile of a *line* is shown in Figure 15.1-1c. In the limit, as the line width w approaches zero, the resultant amplitude discontinuity is called a *roof edge*.

Continuous domain, two-dimensional models of edges and lines assume that the amplitude discontinuity remains constant in a small neighborhood orthogonal to the edge or line profile. Figure 15.1-2a is a sketch of a two-dimensional edge. In addition to the edge parameters of a one-dimensional edge, the orientation of the edge slope with respect to a reference axis is also important. Figure 15.1-2b defines the edge orientation nomenclature for edges of an octagonally shaped object whose amplitude is higher than its background.

Figure 15.1-3 contains step and unit width ramp edge models in the discrete domain. The vertical ramp edge model in the figure contains a single transition pixel whose amplitude is at the midvalue of its neighbors. This edge model can be obtained by performing a 2×2 pixel moving window average on the vertical step edge

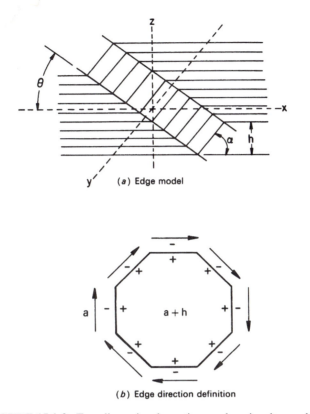

(*a*) Edge model

(*b*) Edge direction definition

FIGURE 15.1-2. Two-dimensional, continuous domain edge model.

model. The figure also contains two versions of a diagonal ramp edge. The single-pixel transition model contains a single midvalue transition pixel between the regions of high and low amplitude; the smoothed transition model is generated by a 2×2 pixel moving window average of the diagonal step edge model. Figure 15.1-3 also presents models for a discrete step and ramp corner edge. The edge location for discrete step edges is usually marked at the higher-amplitude side of an edge transition. For the single-pixel transition model and the smoothed transition vertical and corner edge models, the proper edge location is at the transition pixel. The smoothed transition diagonal ramp edge model has a pair of adjacent pixels in its transition zone. The edge is usually marked at the higher-amplitude pixel of the pair. In Figure 15.1-3 the edge pixels are italicized.

Discrete two-dimensional single-pixel line models are presented in Figure 15.1-4 for step lines and unit width ramp lines. The single-pixel transition model has a midvalue transition pixel inserted between the high value of the line plateau and the low-value background. The smoothed transition model is obtained by performing a 2×2 pixel moving window average on the step line model.

```
a a a a a b b b b b        a a a b b b b b b b        a a a a a a a a a a
a a a a a b b b b b        a a a a b b b b b b        a a a a a a a a a a
a a a a a b b b b b        a a a a a b b b b b        a a a a a b b b b b
a a a a a b b b b b        a a a a a a b b b b        a a a a a b b b b b
a a a a a b b b b b        a a a a a a a b b b        a a a a a b b b b b
      Vertical step edge            Diagonal step edge               Corner step edge
```

```
a a a a c b b b b b        a a c b b b b b b b        a a a a a a a a a a
a a a a c b b b b b        a a a c b b b b b b        a a a a c c c c c c
a a a a c b b b b b        a a a a c b b b b b        a a a a c b b b b b
a a a a c b b b b b        a a a a a c b b b b        a a a a c b b b b b
a a a a c b b b b b        a a a a a a c b b b        a a a a c b b b b b
      Vertical ramp edge            Diagonal ramp edge               Corner ramp edge
```

Single pixel transition

```
a a a a c b b b b b        a a d e b b b b b b        a a a a a a a a a a
a a a a c b b b b b        a a a d e b b b b b        a a a a d c c c c c
a a a a c b b b b b        a a a a d e b b b b        a a a a c b b b b b
a a a a c b b b b b        a a a a a d e b b b        a a a a c b b b b b
a a a a c b b b b b        a a a a a a d e b b        a a a a c b b b b b
      Vertical ramp edge            Diagonal ramp edge               Corner ramp edge
```

Smoothed transition

$$c = \frac{a+b}{2} \qquad d = \frac{3a+b}{4} \qquad e = \frac{a+3b}{4} \qquad b > a$$

FIGURE 15.1-3. Two-dimensional, discrete domain edge models.

A *spot*, which can only be defined in two dimensions, consists of a plateau of high amplitude against a lower amplitude background, or vice versa. Figure 15.1-5 presents single-pixel spot models in the discrete domain.

There are two generic approaches to the detection of edges, lines, and spots in a luminance image: differential detection and model fitting. With the differential detection approach, as illustrated in Figure 15.1-6, spatial processing is performed on an original image $F(j, k)$ to produce a differential image $G(j, k)$ with accentuated spatial amplitude changes. Next, a differential detection operation is executed to determine the pixel locations of significant differentials. The second general approach to edge, line, or spot detection involves fitting of a local region of pixel values to a model of the edge, line, or spot, as represented in Figures 15.1-1 to 15.1-5. If the fit is sufficiently close, an edge, line, or spot is said to exist, and its assigned parameters are those of the appropriate model. A binary indicator map $E(j, k)$ is often generated to indicate the position of edges, lines, or spots within an

```
a  a  a  a  a  b  b  b  b  b      a  a  a  b  b  b  b  b  b  b      a  a  a  a  a  a  a  a  a  a
a  a  a  a  a  b  b  b  b  b      a  a  a  a  b  b  b  b  b  b      a  a  a  a  a  a  a  a  a  a
a  a  a  a  a  b  b  b  b  b      a  a  a  a  a  b  b  b  b  b      a  a  a  a  a  b  b  b  b  b
a  a  a  a  a  b  b  b  b  b      a  a  a  a  a  a  b  b  b  b      a  a  a  a  a  b  b  b  b  b
a  a  a  a  a  b  b  b  b  b      a  a  a  a  a  a  a  b  b  b      a  a  a  a  a  b  b  b  b  b
```
Vertical step edge   Diagonal step edge   Corner step edge

```
a  a  a  a  c  b  b  b  b  b      a  a  c  b  b  b  b  b  b  b      a  a  a  a  a  a  a  a  a  a
a  a  a  a  c  b  b  b  b  b      a  a  a  c  b  b  b  b  b  b      a  a  a  a  c  c  c  c  c  c
a  a  a  a  c  b  b  b  b  b      a  a  a  a  c  b  b  b  b  b      a  a  a  a  c  b  b  b  b  b
a  a  a  a  c  b  b  b  b  b      a  a  a  a  a  c  b  b  b  b      a  a  a  a  c  b  b  b  b  b
a  a  a  a  c  b  b  b  b  b      a  a  a  a  a  a  c  b  b  b      a  a  a  a  c  b  b  b  b  b
```
Vertical ramp edge   Diagonal ramp edge   Corner ramp edge

Single pixel transition

```
a  a  a  a  c  b  b  b  b  b      a  a  d  e  b  b  b  b  b  b      a  a  a  a  a  a  a  a  a  a
a  a  a  a  c  b  b  b  b  b      a  a  a  d  e  b  b  b  b  b      a  a  a  a  d  c  c  c  c  c
a  a  a  a  c  b  b  b  b  b      a  a  a  a  d  e  b  b  b  b      a  a  a  a  c  b  b  b  b  b
a  a  a  a  c  b  b  b  b  b      a  a  a  a  a  d  e  b  b  b      a  a  a  a  c  b  b  b  b  b
a  a  a  a  c  b  b  b  b  b      a  a  a  a  a  a  d  e  b  b      a  a  a  a  c  b  b  b  b  b
```
Vertical ramp edge   Diagonal ramp edge   Corner ramp edge

Smoothed transition

$$c = \frac{a + b}{2} \qquad d = \frac{3a + b}{4} \qquad e = \frac{a + 3b}{4} \qquad b > a$$

FIGURE 15.1-4. Two-dimensional, discrete domain line models.

image. Typically, edge, line, and spot locations are specified by black pixels against a white background.

There are two major classes of differential edge detection: first- and second-order derivative. For the first-order class, some form of spatial first-order differentiation is performed, and the resulting edge gradient is compared to a threshold value. An edge is judged present if the gradient exceeds the threshold. For the second-order derivative class of differential edge detection, an edge is judged present if there is a significant spatial change in the polarity of the second derivative.

Sections 15.2 and 15.3 discuss the first- and second-order derivative forms of edge detection, respectively. Edge fitting methods of edge detection are considered in Section 15.4.

```
a  a  a  a  a  a  a

a  a  a  a  a  a  a

a  a  a  a  a  a  a

a  a  a  b  a  a  a

a  a  a  a  a  a  a

a  a  a  a  a  a  a

a  a  a  a  a  a  a
```

Step spot

```
a  a  a  a  a  a  a

a  a  a  a  a  a  a

a  a  c  c  c  a  a

a  a  c  b  c  a  a

a  a  c  c  c  a  a

a  a  a  a  a  a  a

a  a  a  a  a  a  a
```

Single pixel transition spot

```
a  a  a  a  a  a  a

a  a  a  a  a  a  a

a  a  a  a  a  a  a

a  a  a  d  d  a  a

a  a  a  d  d  a  a

a  a  a  a  a  a  a

a  a  a  a  a  a  a
```

Smoothed transition spot

$$c = \frac{a + b}{2} \qquad d = \frac{3a + b}{4} \qquad b > a$$

FIGURE 15.1-5. Two-dimensional, discrete domain single pixel spot models.

15.2. FIRST-ORDER DERIVATIVE EDGE DETECTION

There are two fundamental methods for generating first-order derivative edge gradients. One method involves generation of gradients in two orthogonal directions in an image; the second utilizes a set of directional derivatives.

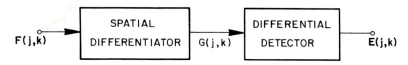

FIGURE 15.1-6. Differential edge, line, and spot detection.

15.2.1. Orthogonal Gradient Generation

An edge in a continuous domain edge segment $F(x, y)$ such as the one depicted in Figure 15.1-2a can be detected by forming the continuous one-dimensional gradient $G(x, y)$ along a line normal to the edge slope, which is at an angle θ with respect to the horizontal axis. If the gradient is sufficiently large (i.e., above some threshold value), an edge is deemed present. The gradient along the line normal to the edge slope can be computed in terms of the derivatives along orthogonal axes according to the following (1, p. 106)

$$G(x, y) = \frac{\partial F(x, y)}{\partial x} \cos \theta + \frac{\partial F(x, y)}{\partial y} \sin \theta \qquad (15.2\text{-}1)$$

Figure 15.2-1 describes the generation of an *edge gradient* $G(x, y)$ in the discrete domain in terms of a *row gradient* $G_R(j, k)$ and a *column gradient* $G_C(j, k)$. The spatial gradient amplitude is given by

$$G(j, k) = [[G_R(j, k)]^2 + [G_C(j, k)]^2]^{1/2} \qquad (15.2\text{-}2)$$

For computational efficiency, the gradient amplitude is sometimes approximated by the magnitude combination

$$G(j, k) = |G_R(j, k)| + |G_C(j, k)| \qquad (15.2\text{-}3)$$

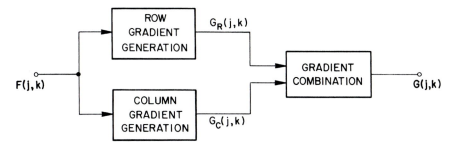

FIGURE 15.2-1. Orthogonal gradient generation.

The orientation of the spatial gradient with respect to the row axis is

$$\theta(j, k) = \arctan \left\{ \frac{G_C(j, k)}{G_R(j, k)} \right\} \qquad (15.2\text{-}4)$$

The remaining issue for discrete domain orthogonal gradient generation is to choose a good discrete approximation to the continuous differentials of Eq. 15.2-1.

The simplest method of discrete gradient generation is to form the running difference of pixels along rows and columns of the image. The row gradient is defined as

$$G_R(j, k) = F(j, k) - F(j, k - 1) \qquad (15.2\text{-}5a)$$

and the column gradient is

$$G_C(j, k) = F(j, k) - F(j + 1, k) \qquad (15.2\text{-}5b)$$

These definitions of row and column gradients, and subsequent extensions, are chosen such that G_R and G_C are positive for an edge that increases in amplitude from left to right and from bottom to top in an image.

As an example of the response of a pixel difference edge detector, the following is the row gradient along the center row of the vertical step edge model of Figure 15.1-3:

$$0 \quad 0 \quad 0 \quad 0 \quad h \quad 0 \quad 0 \quad 0 \quad 0$$

In this sequence, $h = b - a$ is the step edge height. The row gradient for the vertical ramp edge model is

$$0 \quad 0 \quad 0 \quad 0 \quad \frac{h}{2} \quad \frac{h}{2} \quad 0 \quad 0 \quad 0$$

For ramp edges, the running difference edge detector cannot localize the edge to a single pixel. Figure 15.2-2 provides examples of horizontal and vertical differencing gradients of the monochrome peppers image. In this and subsequent gradient display photographs, the gradient range has been scaled over the full contrast range of the photograph. It is visually apparent from the photograph that the running difference technique is highly susceptible to small fluctuations in image luminance and that the object boundaries are not well delineated.

(*a*) Original

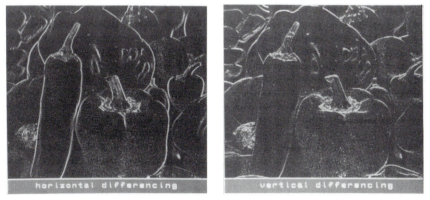

(*b*) Horizontal magnitude   (*c*) Vertical magnitude

FIGURE 15.2-2. Horizontal and vertical differencing gradients of the peppers_mon image.

Diagonal edge gradients can be obtained by forming running differences of diagonal pairs of pixels. This is the basis of the Roberts (2) cross-difference operator, which is defined in magnitude form as

$$G(j, k) = |G_1(j, k)| + |G_2(j, k)| \qquad (15.2\text{-}6a)$$

and in square-root form as

$$G(j, k) = [[G_1(j, k)]^2 + [G_2(j, k)]^2]^{1/2} \qquad (15.2\text{-}6b)$$

where

$$G_1(j, k) = F(j, k) - F(j + 1, k + 1) \qquad (15.2\text{-}6c)$$

$$G_2(j, k) = F(j, k + 1) - F(j + 1, k) \qquad (15.2\text{-}6d)$$

The edge orientation with respect to the row axis is

$$\theta(j, k) = \frac{\pi}{4} + \arctan\left\{\frac{G_2(j, k)}{G_1(j, k)}\right\} \qquad (15.2\text{-}7)$$

Figure 15.2-3 presents the edge gradients of the peppers image for the *Roberts operators*. Visually, the objects in the image appear to be slightly better distinguished with the Roberts square-root gradient than with the magnitude gradient. In Section 15.5, a quantitative evaluation of edge detectors confirms the superiority of the square-root combination technique.

The pixel difference method of gradient generation can be modified to localize the edge center of the ramp edge model of Figure 15.1-3 by forming the pixel difference separated by a null value. The row and column gradients then become

$$G_R(j, k) = F(j, k + 1) - F(j, k - 1) \qquad (15.2\text{-}8a)$$

$$G_C(j, k) = F(j - 1, k) - F(j + 1, k) \qquad (15.2\text{-}8b)$$

The row gradient response for a vertical ramp edge model is then

$$0 \ 0 \ \frac{h}{2} \ h \ \frac{h}{2} \ 0 \ 0$$

(a) Magnitude   (b) Square root

FIGURE 15.2-3. Roberts gradients of the peppers_mon image.

A_0	A_1	A_2
A_7	$F(j,k)$	A_3
A_6	A_5	A_4

FIGURE 15.2-4. Numbering convention for 3×3 edge detection operators.

Although the ramp edge is properly localized, the separated pixel difference gradient generation method remains highly sensitive to small luminance fluctuations in the image. This problem can be alleviated by using two-dimensional gradient formation operators that perform differentiation in one coordinate direction and spatial averaging in the orthogonal direction simultaneously.

Prewitt (1, p. 108) has introduced a 3×3 pixel edge gradient operator described by the pixel numbering convention of Figure 15.2-4. The *Prewitt operator* square root edge gradient is defined as

$$G(j, k) = [[G_R(j, k)]^2 + [G_C(j, k)]^2]^{1/2} \qquad (15.2\text{-}9a)$$

with

$$G_R(j, k) = \frac{1}{K+2}[(A_2 + KA_3 + A_4) - (A_0 + KA_7 + A_6)] \qquad (15.2\text{-}9b)$$

$$G_C(j, k) = \frac{1}{K+2}[(A_0 + KA_1 + A_2) - (A_6 + KA_5 + A_4)] \qquad (15.2\text{-}9c)$$

where $K = 1$. In this formulation, the row and column gradients are normalized to provide unit-gain positive and negative weighted averages about a separated edge position. The *Sobel operator* edge detector (3, p. 271) differs from the Prewitt edge detector in that the values of the north, south, east, and west pixels are doubled (i. e., $K = 2$). The motivation for this weighting is to give equal importance to each pixel in terms of its contribution to the spatial gradient. Frei and Chen (4) have proposed north, south, east, and west weightings by $K = \sqrt{2}$ so that the gradient is the same for horizontal, vertical, and diagonal edges. The edge gradient $G(j, k)$ for these three operators along a row through the single pixel transition vertical ramp edge model of Figure 15.1-3 is

$$0 \quad 0 \quad \frac{h}{2} \quad h \quad \frac{h}{2} \quad 0 \quad 0$$

Along a row through the single transition pixel diagonal ramp edge model, the gradient is

$$0 \qquad \frac{h}{\sqrt{2}\,(2+K)} \qquad \frac{h}{\sqrt{2}} \qquad \frac{\sqrt{2}\,(1+K)h}{2+K} \qquad \frac{h}{\sqrt{2}} \qquad \frac{h}{\sqrt{2}\,(2+K)} \qquad 0$$

In the *Frei–Chen operator* with $K = \sqrt{2}$, the edge gradient is the same at the edge center for the single-pixel transition vertical and diagonal ramp edge models. The Prewitt gradient for a diagonal edge is 0.94 times that of a vertical edge. The

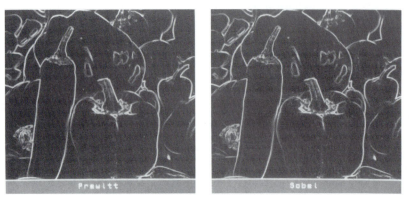

(a) Prewitt   (b) Sobel

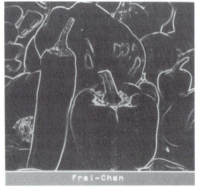

(c) Frei–Chen

FIGURE 15.2-5. Prewitt, Sobel, and Frei–Chen gradients of the `peppers_mon` image.

corresponding factor for a Sobel edge detector is 1.06. Consequently, the Prewitt operator is more sensitive to horizontal and vertical edges than to diagonal edges; the reverse is true for the Sobel operator. The gradients along a row through the smoothed transition diagonal ramp edge model are different for vertical and diagonal edges for all three of the 3×3 edge detectors. None of them are able to localize the edge to a single pixel.

Figure 15.2-5 shows examples of the Prewitt, Sobel, and Frei–Chen gradients of the peppers image. The reason that these operators visually appear to better delineate object edges than the Roberts operator is attributable to their larger size, which provides averaging of small luminance fluctuations.

The row and column gradients for all the edge detectors mentioned previously in this subsection involve a linear combination of pixels within a small neighborhood. Consequently, the row and column gradients can be computed by the convolution relationships

$$G_R(j, k) = F(j, k) \circledast H_R(j, k) \tag{15.2-10a}$$

$$G_C(j, k) = F(j, k) \circledast H_C(j, k) \tag{15.2-10b}$$

where $H_R(j, k)$ and $H_C(j, k)$ are 3×3 row and column impulse response arrays, respectively, as defined in Figure 15.2-6. It should be noted that this specification of the gradient impulse response arrays takes into account the 180° rotation of an impulse response array inherent to the definition of convolution in Eq. 7.1-14.

A limitation common to the edge gradient generation operators previously defined is their inability to detect accurately edges in high-noise environments. This problem can be alleviated by properly extending the size of the neighborhood operators over which the differential gradients are computed. As an example, a Prewitt-type 7×7 operator has a row gradient impulse response of the form

$$\mathbf{H}_R = \frac{1}{21} \begin{bmatrix} 1 & 1 & 1 & 0 & -1 & -1 & -1 \\ 1 & 1 & 1 & 0 & -1 & -1 & -1 \\ 1 & 1 & 1 & 0 & -1 & -1 & -1 \\ 1 & 1 & 1 & 0 & -1 & -1 & -1 \\ 1 & 1 & 1 & 0 & -1 & -1 & -1 \\ 1 & 1 & 1 & 0 & -1 & -1 & -1 \\ 1 & 1 & 1 & 0 & -1 & -1 & -1 \end{bmatrix} \tag{15.2-11}$$

An operator of this type is called a *boxcar operator*. Figure 15.2-7 presents the boxcar gradient of a 7×7 array.

Operator	Row gradient	Column gradient
Pixel difference	$\begin{bmatrix} 0 & 0 & 0 \\ 0 & 1 & -1 \\ 0 & 0 & 0 \end{bmatrix}$	$\begin{bmatrix} 0 & -1 & 0 \\ 0 & 1 & 0 \\ 0 & 0 & 0 \end{bmatrix}$
Separated pixel difference	$\begin{bmatrix} 0 & 0 & 0 \\ 1 & 0 & -1 \\ 0 & 0 & 0 \end{bmatrix}$	$\begin{bmatrix} 0 & -1 & 0 \\ 0 & 0 & 0 \\ 0 & 1 & 0 \end{bmatrix}$
Roberts	$\begin{bmatrix} 0 & 0 & -1 \\ 0 & 1 & 0 \\ 0 & 0 & 0 \end{bmatrix}$	$\begin{bmatrix} -1 & 0 & 0 \\ 0 & 1 & 0 \\ 0 & 0 & 0 \end{bmatrix}$
Prewitt	$\frac{1}{3}\begin{bmatrix} 1 & 0 & -1 \\ 1 & 0 & -1 \\ 1 & 0 & -1 \end{bmatrix}$	$\frac{1}{3}\begin{bmatrix} -1 & -1 & -1 \\ 0 & 0 & 0 \\ 1 & 1 & 1 \end{bmatrix}$
Sobel	$\frac{1}{4}\begin{bmatrix} 1 & 0 & -1 \\ 2 & 0 & -2 \\ 1 & 0 & -1 \end{bmatrix}$	$\frac{1}{4}\begin{bmatrix} -1 & -2 & -1 \\ 0 & 0 & 0 \\ 1 & 2 & 1 \end{bmatrix}$
Frei–Chen	$\frac{1}{2+\sqrt{2}}\begin{bmatrix} 1 & 0 & -1 \\ \sqrt{2} & 0 & -\sqrt{2} \\ 1 & 0 & -1 \end{bmatrix}$	$\frac{1}{2+\sqrt{2}}\begin{bmatrix} -1 & -\sqrt{2} & -1 \\ 0 & 0 & 0 \\ 1 & \sqrt{2} & 1 \end{bmatrix}$

FIGURE 15.2-6. Impulse response arrays for 3×3 orthogonal differential gradient edge operators.

Abdou (5) has suggested a *truncated pyramid operator* that gives a linearly decreasing weighting to pixels away from the center of an edge. The row gradient impulse response array for a 7×7 truncated pyramid operator is given by

$$\mathbf{H}_R = \frac{1}{34}\begin{bmatrix} 1 & 1 & 1 & 0 & -1 & -1 & -1 \\ 1 & 2 & 2 & 0 & -2 & -2 & -1 \\ 1 & 2 & 3 & 0 & -3 & -2 & -1 \\ 1 & 2 & 3 & 0 & -3 & -2 & -1 \\ 1 & 2 & 3 & 0 & -3 & -2 & -1 \\ 1 & 2 & 2 & 0 & -2 & -2 & -1 \\ 1 & 1 & 1 & 0 & -1 & -1 & -1 \end{bmatrix} \tag{15.2-12}$$

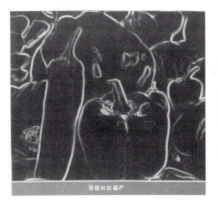

(*a*) 7 × 7 boxcar

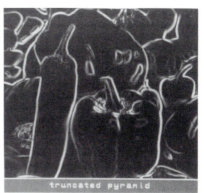

(*b*) 9 × 9 truncated pyramid

(*c*) 11 × 11 Argyle, *s* = 2.0

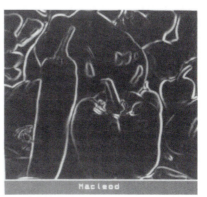

(*d*) 11 × 11 Macleod, *s* = 2.0

(*e*) 11 × 11 FDOG, *s* = 2.0

FIGURE 15.2-7. Boxcar, truncated pyramid, Argyle, Macleod, and FDOG gradients of the peppers_mon image.

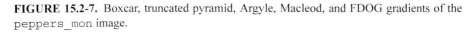

Argyle (6) and Macleod (7,8) have proposed large neighborhood Gaussian-shaped weighting functions as a means of noise suppression. Let

$$g(x, s) = [2\pi s^2]^{-1/2} \exp\{-1/2(x/s)^2\} \tag{15.2-13}$$

denote a continuous domain Gaussian function with standard deviation s. Utilizing this notation, the *Argyle operator* horizontal coordinate impulse response array can be expressed as a sampled version of the continuous domain impulse response

$$H_R(j, k) = \begin{cases} -2g(x, s)g(y, t) & \text{for } x \geq 0 \tag{15.2-14a} \\[12pt] 2g(x, s)g(y, t) & \text{for } x < 0 \tag{15.2-14b} \end{cases}$$

where s and t are spread parameters. The vertical impulse response function can be expressed similarly. The *Macleod operator* horizontal gradient impulse response function is given by

$$H_R(j, k) = [g(x + s, s) - g(x - s, s)]g(y, t) \tag{15.2-15}$$

The Argyle and Macleod operators, unlike the boxcar operator, give decreasing importance to pixels far removed from the center of the neighborhood. Figure 15.2-7 provides examples of the Argyle and Macleod gradients.

Extended-size differential gradient operators can be considered to be compound operators in which a smoothing operation is performed on a noisy image followed by a differentiation operation. The compound gradient impulse response can be written as

$$H(j, k) = H_G(j, k) \circledast H_S(j, k) \tag{15.2-16}$$

where $H_G(j, k)$ is one of the gradient impulse response operators of Figure 15.2-6 and $H_S(j, k)$ is a low-pass filter impulse response. For example, if $H_S(j, k)$ is the 3×3 Prewitt row gradient operator and $H_S(j, k) = 1/9$, for all (j, k), is a 3×3 uniform smoothing operator, the resultant 5×5 row gradient operator, after normalization to unit positive and negative gain, becomes

$$\mathbf{H}_R = \frac{1}{18} \begin{bmatrix} 1 & 1 & 0 & -1 & -1 \\ 2 & 2 & 0 & -2 & -2 \\ 3 & 3 & 0 & -3 & -3 \\ 2 & 2 & 0 & -2 & -2 \\ 1 & 1 & 0 & -1 & -1 \end{bmatrix} \tag{15.2-17}$$

The decomposition of Eq. 15.2-16 applies in both directions. By applying the SVD/ SGK decomposition of Section 9.6, it is possible, for example, to decompose a 5×5 boxcar operator into the sequential convolution of a 3×3 smoothing kernel and a 3×3 differentiating kernel.

A well-known example of a compound gradient operator is the *first derivative of Gaussian* (FDOG) *operator*, in which Gaussian-shaped smoothing is followed by differentiation (9). The FDOG continuous domain horizontal impulse response is

$$H_R(j, k) = \frac{-\partial[g(x, s)g(y, t)]}{\partial x} \tag{15.2-18a}$$

which upon differentiation yields

$$H_R(j, k) = \frac{-xg(x, s)g(y, t)}{s^2} \tag{15.2-18b}$$

Figure 15.2-7 presents an example of the FDOG gradient.

All of the differential edge enhancement operators presented previously in this subsection have been derived heuristically. Canny (9) has taken an analytic approach to the design of such operators. Canny's development is based on a one-dimensional continuous domain model of a step edge of amplitude h_E plus additive white Gaussian noise with standard deviation σ_n. It is assumed that edge detection is performed by convolving a one-dimensional continuous domain noisy edge signal $f(x)$ with an antisymmetric impulse response function $h(x)$, which is of zero amplitude outside the range $[-W, W]$. An edge is marked at the local maximum of the convolved gradient $f(x) \circledast h(x)$. The *Canny operator* impulse response $h(x)$ is chosen to satisfy the following three criteria.

1. *Good detection.* The amplitude signal-to-noise ratio (SNR) of the gradient is maximized to obtain a low probability of failure to mark real edge points and a low probability of falsely marking nonedge points. The SNR for the model is

$$\text{SNR} = \frac{h_E S(h)}{\sigma_n} \tag{15.2-19a}$$

with

$$S(h) = \frac{\int_{-W}^{0} h(x)\, dx}{\int_{-W}^{W} [h(x)]^2\, dx} \tag{15.2-19b}$$

2. *Good localization.* Edge points marked by the operator should be as close to the center of the edge as possible. The localization factor is defined as

$$\text{LOC} = \frac{h_E L(h)}{\sigma_n} \qquad (15.2\text{-}20a)$$

with

$$L(h) = \frac{h'(0)}{\int_{-W}^{W} [h'(x)]^2 \, dx} \qquad (15.2\text{-}20b)$$

where $h'(x)$ is the derivative of $h(x)$.

3. *Single response.* There should be only a single response to a true edge. The distance between peaks of the gradient when only noise is present, denoted as x_m, is set to some fraction k of the operator width factor W. Thus

$$x_m = kW \qquad (15.2\text{-}21)$$

Canny has combined these three criteria by maximizing the product $S(h)L(h)$ subject to the constraint of Eq. 15.2-21. Because of the complexity of the formulation, no analytic solution has been found, but a variational approach has been developed. Figure 15.2-8 contains plots of the Canny impulse response functions in terms of x_m.

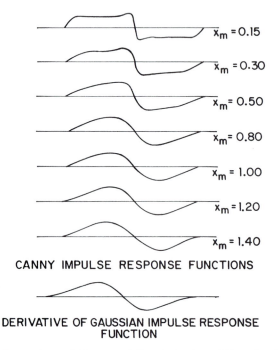

CANNY IMPULSE RESPONSE FUNCTIONS

DERIVATIVE OF GAUSSIAN IMPULSE RESPONSE FUNCTION

FIGURE 15.2-8. Comparison of Canny and first derivative of Gaussian impulse response functions.

As noted from the figure, for low values of x_m, the Canny function resembles a box-car function, while for x_m large, the Canny function is closely approximated by a FDOG impulse response function.

Discrete domain versions of the large operators defined in the continuous domain can be obtained by sampling their continuous impulse response functions over some $W \times W$ window. The window size should be chosen sufficiently large that truncation of the impulse response function does not cause high-frequency artifacts. Demigny and Kamlé (10) have developed a discrete version of Canny's criteria, which lead to the computation of discrete domain edge detector impulse response arrays.

15.2.2. Edge Template Gradient Generation

With the orthogonal differential edge enhancement techniques discussed previously, edge gradients are computed in two orthogonal directions, usually along rows and columns, and then the edge direction is inferred by computing the vector sum of the gradients. Another approach is to compute gradients in a large number of directions by convolution of an image with a set of template gradient impulse response arrays. The edge template gradient is defined as

$$G(j, k) = \text{MAX}\{|G_1(j, k)|, ..., |G_m(j, k)|, ..., |G_M(j, k)|\} \qquad (15.2\text{-}22a)$$

where

$$G_m(j, k) = F(j, k) \circledast H_m(j, k) \qquad (15.2\text{-}22b)$$

is the gradient in the mth equispaced direction obtained by convolving an image with a gradient impulse response array $H_m(j, k)$. The edge angle is determined by the direction of the largest gradient.

Figure 15.2-9 defines eight gain-normalized compass gradient impulse response arrays suggested by Prewitt (1, p. 111). The compass names indicate the slope direction of maximum response. Kirsch (11) has proposed a directional gradient defined by

$$G(j, k) = \underset{i = 0}{\overset{7}{\text{MAX}}} \left\{ |5S_i - 3T_i| \right\} \qquad (15.2\text{-}23a)$$

where

$$S_i = A_i + A_{i+1} + A_{i+2} \qquad (15.2\text{-}23b)$$

$$T_i = A_{i+3} + A_{i+4} + A_{i+5} + A_{i+5} + A_{i+6} \qquad (15.2\text{-}23c)$$

Gradient direction	Prewitt compass gradient	Kirsch	Robinson 3-level	Robinson 5-level
East H_1	$\begin{bmatrix} 1 & 1 & -1 \\ 1 & -2 & -1 \\ 1 & 1 & -1 \end{bmatrix}$	$\begin{bmatrix} 5 & -3 & -3 \\ 5 & 0 & -3 \\ 5 & -3 & -3 \end{bmatrix}$	$\begin{bmatrix} 1 & 0 & -1 \\ 1 & 0 & -1 \\ 1 & 0 & -1 \end{bmatrix}$	$\begin{bmatrix} 1 & 0 & -1 \\ 2 & 0 & -2 \\ 1 & 0 & -1 \end{bmatrix}$
Northeast H_2	$\begin{bmatrix} 1 & -1 & -1 \\ 1 & -2 & -1 \\ 1 & 1 & 1 \end{bmatrix}$	$\begin{bmatrix} -3 & -3 & -3 \\ 5 & 0 & -3 \\ 5 & 5 & -3 \end{bmatrix}$	$\begin{bmatrix} 0 & -1 & -1 \\ 1 & 0 & -1 \\ 1 & 1 & 0 \end{bmatrix}$	$\begin{bmatrix} 0 & -1 & -2 \\ 1 & 0 & -1 \\ 2 & 1 & 0 \end{bmatrix}$
North H_3	$\begin{bmatrix} -1 & -1 & -1 \\ 1 & -2 & 1 \\ 1 & 1 & 1 \end{bmatrix}$	$\begin{bmatrix} -3 & -3 & -3 \\ -3 & 0 & -3 \\ 5 & 5 & 5 \end{bmatrix}$	$\begin{bmatrix} -1 & -1 & -1 \\ 0 & 0 & 0 \\ 1 & 1 & 1 \end{bmatrix}$	$\begin{bmatrix} -1 & -2 & -1 \\ 0 & 0 & 0 \\ 1 & 2 & 1 \end{bmatrix}$
Northwest H_4	$\begin{bmatrix} -1 & -1 & 1 \\ -1 & -2 & 1 \\ 1 & 1 & 1 \end{bmatrix}$	$\begin{bmatrix} -3 & -3 & -3 \\ -3 & 0 & 5 \\ -3 & 5 & 5 \end{bmatrix}$	$\begin{bmatrix} -1 & -1 & 0 \\ -1 & 0 & 1 \\ 0 & 1 & 1 \end{bmatrix}$	$\begin{bmatrix} -2 & -1 & 0 \\ -1 & 0 & 1 \\ 0 & 1 & 2 \end{bmatrix}$
West H_5	$\begin{bmatrix} -1 & 1 & 1 \\ -1 & -2 & 1 \\ -1 & 1 & 1 \end{bmatrix}$	$\begin{bmatrix} -3 & -3 & 5 \\ -3 & 0 & 5 \\ -3 & -3 & 5 \end{bmatrix}$	$\begin{bmatrix} -1 & 0 & 1 \\ -1 & 0 & 1 \\ -1 & 0 & 1 \end{bmatrix}$	$\begin{bmatrix} -1 & 0 & 1 \\ -2 & 0 & 2 \\ -1 & 0 & 1 \end{bmatrix}$
Southwest H_6	$\begin{bmatrix} 1 & 1 & 1 \\ -1 & -2 & 1 \\ -1 & -1 & 1 \end{bmatrix}$	$\begin{bmatrix} -3 & 5 & 5 \\ -3 & 0 & 5 \\ -3 & -3 & -3 \end{bmatrix}$	$\begin{bmatrix} 0 & 1 & 1 \\ -1 & 0 & 1 \\ -1 & -1 & 0 \end{bmatrix}$	$\begin{bmatrix} 0 & 1 & 2 \\ -1 & 0 & 1 \\ -2 & -1 & 0 \end{bmatrix}$
South H_7	$\begin{bmatrix} 1 & 1 & 1 \\ 1 & -2 & 1 \\ -1 & -1 & -1 \end{bmatrix}$	$\begin{bmatrix} 5 & 5 & 5 \\ -3 & 0 & -3 \\ -3 & -3 & -3 \end{bmatrix}$	$\begin{bmatrix} 1 & 1 & 1 \\ 0 & 0 & 0 \\ -1 & -1 & -1 \end{bmatrix}$	$\begin{bmatrix} 1 & 2 & 1 \\ 0 & 0 & 0 \\ -1 & -2 & -1 \end{bmatrix}$
Southeast H_8	$\begin{bmatrix} 1 & 1 & 1 \\ 1 & -2 & -1 \\ 1 & -1 & -1 \end{bmatrix}$	$\begin{bmatrix} 5 & 5 & -3 \\ 5 & 0 & -3 \\ -3 & -3 & -3 \end{bmatrix}$	$\begin{bmatrix} 1 & 1 & 0 \\ 1 & 0 & -1 \\ 0 & -1 & -1 \end{bmatrix}$	$\begin{bmatrix} 2 & 1 & 0 \\ 1 & 0 & -1 \\ 0 & -1 & -2 \end{bmatrix}$
Scale factor	$\dfrac{1}{5}$	$\dfrac{1}{15}$	$\dfrac{1}{3}$	$\dfrac{1}{4}$

FIGURE 15.2-9. Template gradient 3×3 impulse response arrays.

The subscripts of A_i are evaluated modulo 8. It is possible to compute the Kirsch gradient by convolution as in Eq. 15.2-22b. Figure 15.2-9 specifies the gain-normalized *Kirsch operator* impulse response arrays. This figure also defines two other sets of gain-normalized impulse response arrays proposed by Robinson (12), called the *Robinson three-level operator* and the *Robinson five-level operator*, which are derived from the Prewitt and Sobel operators, respectively. Figure 15.2-10 provides a comparison of the edge gradients of the peppers image for the four 3×3 template gradient operators.

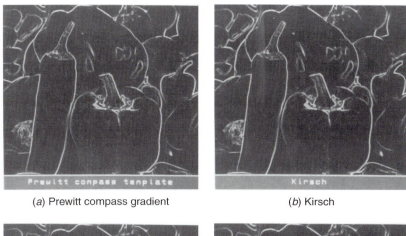

<table>
<tr><td>(a) Prewitt compass gradient</td><td>(b) Kirsch</td></tr>
</table>

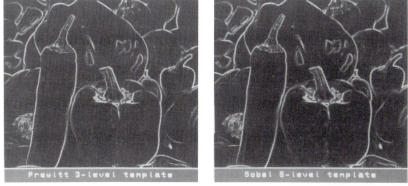

(c) Robinson three-level	(d) Robinson five-level

FIGURE 15.2-10. 3×3 template gradients of the peppers_mon image.

Nevatia and Babu (13) have developed an edge detection technique in which the gain-normalized 5×5 masks defined in Figure 15.2-11 are utilized to detect edges in 30° increments. Figure 15.2-12 shows the template gradients for the peppers image. Larger template masks will provide both a finer quantization of the edge orientation angle and a greater noise immunity, but the computational requirements increase. Paplinski (14) has developed a design procedure for n-directional template masks of arbitrary size.

15.2.3. Threshold Selection

After the edge gradient is formed for the differential edge detection methods, the gradient is compared to a threshold to determine if an edge exists. The threshold value determines the sensitivity of the edge detector. For noise-free images, the

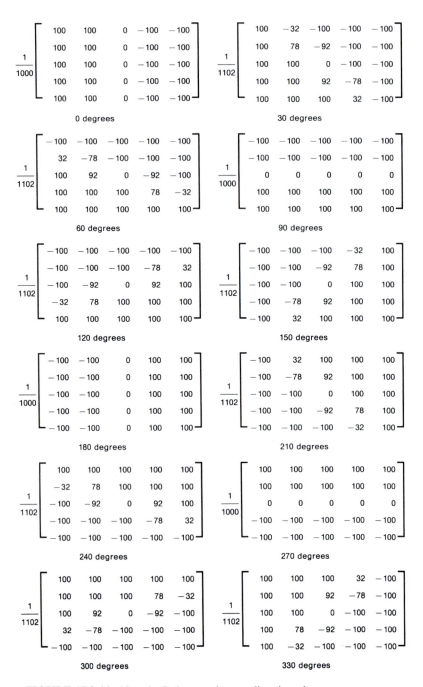

FIGURE 15.2-11. Nevatia–Babu template gradient impulse response arrays.

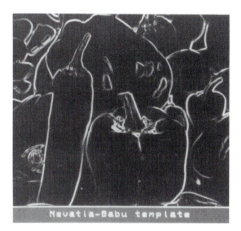

FIGURE 15.2-12.   Nevatia–Babu gradient of the `peppers_mon` image.

threshold can be chosen such that all amplitude discontinuities of a minimum contrast level are detected as *edges*, and all others are called *nonedges*. With noisy images, threshold selection becomes a trade-off between missing valid edges and creating noise-induced false edges.

Edge detection can be regarded as a hypothesis-testing problem to determine if an image region contains an edge or contains no edge (15). Let $P(\text{edge})$ and $P(\text{no-edge})$ denote the a priori probabilities of these events. Then the edge detection process can be characterized by the probability of correct edge detection,

$$P_D = \int_t^\infty p(G|\text{edge})\ dG \qquad (15.2\text{-}24a)$$

and the probability of false detection,

$$P_F = \int_t^\infty p(G|\text{no-edge})\ dG \qquad (15.2\text{-}24b)$$

where t is the edge detection threshold and $p(G|\text{edge})$ and $p(G|\text{no-edge})$ are the conditional probability densities of the edge gradient $G(j, k)$. Figure 15.2-13 is a sketch of typical edge gradient conditional densities. The probability of edge misclassification error can be expressed as

$$P_E = (1 - P_D)P(\text{edge}) + (P_F)P(\text{no-edge}) \qquad (15.2\text{-}25)$$

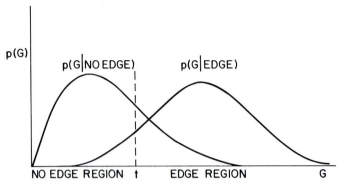

FIGURE 15.2-13. Typical edge gradient conditional probability densities.

This error will be minimum if the threshold is chosen such that an edge is deemed present when

$$\frac{p(G|\text{edge})}{p(G|\text{no–edge})} \geq \frac{P(\text{no–edge})}{P(\text{edge})} \qquad (15.2\text{-}26)$$

and the no-edge hypothesis is accepted otherwise. Equation 15.2-26 defines the well-known *maximum likelihood ratio* test associated with the Bayes minimum error decision rule of classical decision theory (16). Another common decision strategy, called the *Neyman–Pearson test*, is to choose the threshold t to minimize P_F for a fixed acceptable P_D (16).

Application of a statistical decision rule to determine the threshold value requires knowledge of the a priori edge probabilities and the conditional densities of the edge gradient. The a priori probabilities can be estimated from images of the class under analysis. Alternatively, the a priori probability ratio can be regarded as a sensitivity control factor for the edge detector. The conditional densities can be determined, in principle, for a statistical model of an ideal edge plus noise. Abdou (5) has derived these densities for 2×2 and 3×3 edge detection operators for the case of a ramp edge of width $w = 1$ and additive Gaussian noise. Henstock and Chelberg (17) have used gamma densities as models of the conditional probability densities.

There are two difficulties associated with the statistical approach of determining the optimum edge detector threshold: reliability of the stochastic edge model and analytic difficulties in deriving the edge gradient conditional densities. Another approach, developed by Abdou and Pratt (5,15), which is based on pattern recognition techniques, avoids the difficulties of the statistical method. The pattern recognition method involves creation of a large number of prototype noisy image regions, some of which contain edges and some without edges. These prototypes are then used as a training set to find the threshold that minimizes the classification error. Details of the design procedure are found in Reference 5. Table 15.2-1

TABLE 15.2-1. Threshold Levels and Associated Edge Detection Probabilities for 3 × 3 Edge Detectors as Determined by the Abdou and Pratt Pattern Recognition Design Procedure

Operator	Vertical Edge						Diagonal Edge					
	SNR = 1			SNR = 10			SNR = 1			SNR = 10		
	t_N	P_D	P_F	t_N	P_D	P_F	t_N	P_D	P_F	t_N	P_D	P_F
Roberts orthogonal gradient	1.36	0.559	0.400	0.67	0.892	0.105	1.74	0.551	0.469	0.78	0.778	0.221
Prewitt orthogonal gradient	1.16	0.608	0.384	0.66	0.912	0.480	1.19	0.593	0.387	0.64	0.931	0.064
Sobel orthogonal gradient	1.18	0.600	0.395	0.66	0.923	0.057	1.14	0.604	0.376	0.63	0.947	0.053
Prewitt compass template gradient	1.52	0.613	0.466	0.73	0.886	0.136	1.51	0.618	0.472	0.71	0.900	0.153
Kirsch template gradient	1.43	0.531	0.341	0.69	0.898	0.058	1.45	0.524	0.324	0.79	0.825	0.023
Robinson three-level template gradient	1.16	0.590	0.369	0.65	0.926	0.038	1.16	0.587	0.365	0.61	0.946	0.056
Robinson five-level template gradient	1.24	0.581	0.361	0.66	0.924	0.049	1.22	0.593	0.374	0.65	0.931	0.054

(a) Sobel, t = 0.06

(b) FDOG, t = 0.08

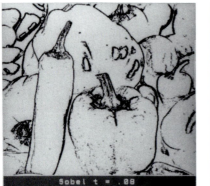

(c) Sobel, t = 0.08

(d) FDOG, t = 0.10

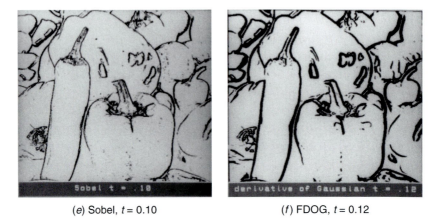

(e) Sobel, t = 0.10

(f) FDOG, t = 0.12

FIGURE 15.2-14. Threshold sensitivity of the Sobel and first derivative of Gaussian edge detectors for the peppers_mon image.

provides a tabulation of the optimum threshold for several 2×2 and 3×3 edge detectors for an experimental design with an evaluation set of 250 prototypes not in the training set (15). The table also lists the probability of correct and false edge detection as defined by Eq. 15.2-24 for theoretically derived gradient conditional densities. In the table, the threshold is normalized such that $t_N = t/G_M$, where G_M is the maximum amplitude of the gradient in the absence of noise. The power signal-to-noise ratio is defined as $\mathrm{SNR} = (h/\sigma_n)^2$, where h is the edge height and σ_n is the noise standard deviation. In most of the cases of Table 15.2-1, the optimum threshold results in approximately equal error probabilities (i.e., $P_F = 1 - P_D$). This is the same result that would be obtained by the Bayes design procedure when edges and nonedges are equally probable. The tests associated with Table 15.2-1 were conducted with relatively low signal-to-noise ratio images. Section 15.5 provides examples of such images. For high signal-to-noise ratio images, the optimum threshold is much lower. As a rule of thumb, under the condition that $P_F = 1 - P_D$, the edge detection threshold can be scaled linearly with signal-to-noise ratio. Hence, for an image with SNR = 100, the threshold is about 10% of the peak gradient value.

Figure 15.2-14 shows the effect of varying the first derivative edge detector threshold for the 3×3 Sobel and the 11×11 FDOG edge detectors for the peppers image, which is a relatively high signal-to-noise ratio image. For both edge detectors, variation of the threshold provides a trade-off between delineation of strong edges and definition of weak edges.

15.2.4. Morphological Post Processing

It is possible to improve edge delineation of first-derivative edge detectors by applying morphological operations on their edge maps. Figure 15.2-15 provides examples for the 3×3 Sobel and 11×11 FDOG edge detectors. In the Sobel example, the threshold is lowered slightly to improve the detection of weak edges. Then the morphological majority black operation is performed on the edge map to eliminate noise-induced edges. This is followed by the thinning operation to thin the edges to minimally connected lines. In the FDOG example, the majority black noise smoothing step is not necessary.

15.3. SECOND-ORDER DERIVATIVE EDGE DETECTION

Second-order derivative edge detection techniques employ some form of spatial second-order differentiation to accentuate edges. An edge is marked if a significant spatial change occurs in the second derivative. Two types of second-order derivative methods are considered: Laplacian and directed second derivative.

15.3.1. Laplacian Generation

The edge Laplacian of an image function $F(x, y)$ in the continuous domain is defined as

(*a*) Sobel, *t* = 0.07

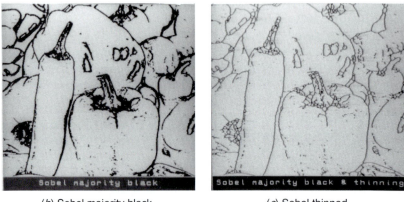

(*b*) Sobel majority black   (*c*) Sobel thinned

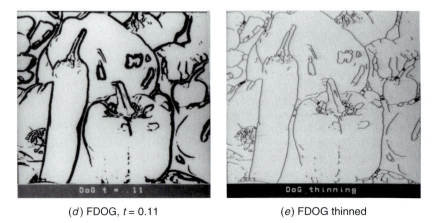

(*d*) FDOG, *t* = 0.11   (*e*) FDOG thinned

FIGURE 15.2-15. Morphological thinning of edge maps for the `peppers_mon` image.

$$G(x, y) = -\nabla^2\{F(x, y)\} \tag{15.3-1a}$$

where, from Eq. 1.2-17, the *Laplacian* is

$$\nabla^2 = \frac{\partial^2}{\partial x^2} + \frac{\partial^2}{\partial y^2} \tag{15.3-1b}$$

The Laplacian $G(x, y)$ is zero if $F(x, y)$ is constant or changing linearly in amplitude. If the rate of change of $F(x, y)$ is greater than linear, $G(x, y)$ exhibits a sign change at the point of inflection of $F(x, y)$. The zero crossing of $G(x, y)$ indicates the presence of an edge. The negative sign in the definition of Eq. 15.3-la is present so that the zero crossing of $G(x, y)$ has a positive slope for an edge whose amplitude increases from left to right or bottom to top in an image.

Torre and Poggio (18) have investigated the mathematical properties of the Laplacian of an image function. They have found that if $F(x, y)$ meets certain smoothness constraints, the zero crossings of $G(x, y)$ are closed curves.

In the discrete domain, the simplest approximation to the continuous Laplacian is to compute the difference of slopes along each axis:

$$G(j, k) = [F(j, k) - F(j, k - 1)] - [F(j, k + 1) - F(j, k)]$$

$$+ [F(j, k) - F(j + 1, k)] - [F(j - 1, k) - F(j, k)] \tag{15.3-2}$$

This four-neighbor Laplacian (1, p. 111) can be generated by the convolution operation

$$G(j, k) = F(j, k) \circledast H(j, k) \tag{15.3-3}$$

with

$$\mathbf{H} = \begin{bmatrix} 0 & 0 & 0 \\ -1 & 2 & -1 \\ 0 & 0 & 0 \end{bmatrix} + \begin{bmatrix} 0 & -1 & 0 \\ 0 & 2 & 0 \\ 0 & -1 & 0 \end{bmatrix} \tag{15.3-4a}$$

or

$$\mathbf{H} = \begin{bmatrix} 0 & -1 & 0 \\ -1 & 4 & -1 \\ 0 & -1 & 0 \end{bmatrix} \tag{15.3-4b}$$

where the two arrays of Eq. 15.3-4a correspond to the second derivatives along image rows and columns, respectively, as in the continuous Laplacian of Eq. 15.3-1b. The four-neighbor Laplacian is often normalized to provide unit-gain averages of the positive weighted and negative weighted pixels in the 3×3 pixel neighborhood. The gain-normalized four-neighbor Laplacian impulse response is defined by

$$\mathbf{H} = \frac{1}{4} \begin{bmatrix} 0 & -1 & 0 \\ -1 & 4 & -1 \\ 0 & -1 & 0 \end{bmatrix} \tag{15.3-5}$$

Prewitt (1, p. 111) has suggested an eight-neighbor Laplacian defined by the gain-normalized impulse response array

$$\mathbf{H} = \frac{1}{8} \begin{bmatrix} -1 & -1 & -1 \\ -1 & 8 & -1 \\ -1 & -1 & -1 \end{bmatrix} \tag{15.3-6}$$

This array is not separable into a sum of second derivatives, as in Eq. 15.3-4a. A separable eight-neighbor Laplacian can be obtained by the construction

$$\mathbf{H} = \begin{bmatrix} -1 & 2 & -1 \\ -1 & 2 & -1 \\ -1 & 2 & -1 \end{bmatrix} + \begin{bmatrix} -1 & -1 & -1 \\ 2 & 2 & 2 \\ -1 & -1 & -1 \end{bmatrix} \tag{15.3-7}$$

in which the difference of slopes is averaged over three rows and three columns. The gain-normalized version of the separable eight-neighbor Laplacian is given by

$$\mathbf{H} = \frac{1}{8} \begin{bmatrix} -2 & 1 & -2 \\ 1 & 4 & 1 \\ -2 & 1 & -2 \end{bmatrix} \tag{15.3-8}$$

It is instructive to examine the Laplacian response to the edge models of Figure 15.1-3. As an example, the separable eight-neighbor Laplacian corresponding to the center row of the vertical step edge model is

$$0 \quad \frac{-3h}{8} \quad \frac{3h}{8} \quad 0$$

where $h = b - a$ is the edge height. The Laplacian response of the vertical ramp edge model is

$$0 \quad \frac{-3h}{16} \quad 0 \quad \frac{3h}{16} \quad 0$$

For the vertical edge ramp edge model, the edge lies at the zero crossing pixel between the negative- and positive-value Laplacian responses. In the case of the step edge, the zero crossing lies midway between the neighboring negative and positive response pixels; the edge is correctly marked at the pixel to the right of the zero crossing. The Laplacian response for a single-transition-pixel diagonal ramp edge model is

$$0 \quad \frac{-h}{8} \quad \frac{-h}{8} \quad 0 \quad \frac{h}{8} \quad \frac{h}{8} \quad 0$$

and the edge lies at the zero crossing at the center pixel. The Laplacian response for the smoothed transition diagonal ramp edge model of Figure 15.1-3 is

$$0 \quad \frac{-h}{16} \quad \frac{-h}{8} \quad \frac{-h}{16} \quad \frac{h}{16} \quad \frac{h}{8} \quad \frac{h}{16} \quad 0$$

In this example, the zero crossing does not occur at a pixel location. The edge should be marked at the pixel to the right of the zero crossing. Figure 15.3-1 shows the Laplacian response for the two ramp corner edge models of Figure 15.1-3. The edge transition pixels are indicated by line segments in the figure. A zero crossing exists at the edge corner for the smoothed transition edge model, but not for the single-pixel transition model. The zero crossings adjacent to the edge corner do not occur at pixel samples for either of the edge models. From these examples, it can be

0	0	0	0	0	0	0	0
0	0	0	0	0	0	0	0
0	0	−1	−1	−3/2	−3/2	−3/2	−3/2
0	0	−1	1	1	0	0	0
0	0	−3/2	1	2	3/2	3/2	3/2
0	0	−3/2	0	3/2	0	0	0
0	0	−3/2	0	3/2	0	0	0
0	0	−3/2	0	3/2	0	0	0

Single pixel transition model

0	0	0	0	0	0	0	0
0	0	0	0	0	0	0	0
0	0	−1/2	−3/4	−1	−3/2	−3/2	−3/2
0	0	−3/4	0	3/4	0	0	0
0	0	−1	3/4	5/2	3/2	3/2	3/2
0	0	−3/2	0	3/2	0	0	0
0	0	−3/2	0	3/2	0	0	0
0	0	−3/2	0	3/2	0	0	0

Smoothed transition model

FIGURE 15.3-1. Separable eight-neighbor Laplacian responses for ramp corner models; all values should be scaled by $h/8$.

concluded that zero crossings of the Laplacian do not always occur at pixel samples. But for these edge models, marking an edge at a pixel with a positive response that has a neighbor with a negative response identifies the edge correctly.

Figure 15.3-2 shows the Laplacian responses of the peppers image for the three types of 3×3 Laplacians. In these photographs, negative values are depicted as dimmer than midgray and positive values are brighter than midgray.

Marr and Hildrith (19) have proposed the *Laplacian of Gaussian* (LOG) edge detection operator in which Gaussian-shaped smoothing is performed prior to application of the Laplacian. The continuous domain LOG gradient is

$$G(x, y) = -\nabla^2 \{ F(x, y) \circledast H_S(x, y) \} \tag{15.3-9a}$$

where

$$G(x, y) = g(x, s)g(y, s) \tag{15.3-9b}$$

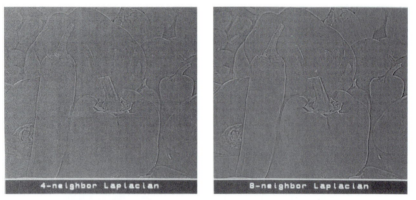

(*a*) Four-neighbor   (*b*) Eight-neighbor

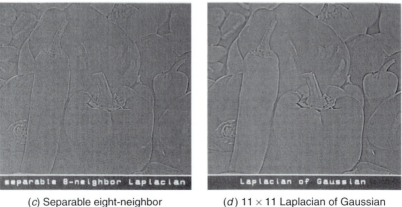

(*c*) Separable eight-neighbor   (*d*) 11 × 11 Laplacian of Gaussian

FIGURE 15.3-2. Laplacian responses of the `peppers_mon` image.

is the impulse response of the Gaussian smoothing function as defined by Eq. 15.2-13. As a result of the linearity of the second derivative operation and of the linearity of convolution, it is possible to express the LOG response as

$$G(j, k) = F(j, k) \circledast H(j, k) \tag{15.3-10a}$$

where

$$H(x, y) = -\nabla^2\{g(x, s)g(y, s)\} \tag{15.3-10b}$$

Upon differentiation, one obtains

$$H(x, y) = \frac{1}{\pi s^4}\left(1 - \frac{x^2 + y^2}{2s^2}\right)\exp\left\{-\frac{x^2 + y^2}{2s^2}\right\} \tag{15.3-11}$$

Figure 15.3-3 is a cross-sectional view of the LOG continuous domain impulse response. In the literature it is often called the *Mexican hat filter*. It can be shown (20,21) that the LOG impulse response can be expressed as

$$H(x, y) = \frac{1}{\pi s^2}\left(1 - \frac{y^2}{s^2}\right)g(x, s)g(y, s) + \frac{1}{\pi s^2}\left(1 - \frac{x^2}{s^2}\right)g(x, s)g(y, s) \tag{15.3-12}$$

Consequently, the convolution operation can be computed separably along rows and columns of an image. It is possible to approximate the LOG impulse response closely by a *difference of Gaussians* (DOG) operator. The resultant impulse response is

$$H(x, y) = g(x, s_1)g(y, s_1) - g(x, s_2)g(y, s_2) \tag{15.3-13}$$

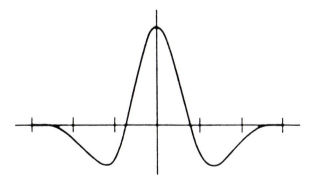

FIGURE 15.3-3. Cross section of continuous domain Laplacian of Gaussian impulse response.

where $s_1 < s_2$. Marr and Hildrith (19) have found that the ratio $s_2/s_1 = 1.6$ provides a good approximation to the LOG.

A discrete domain version of the LOG operator can be obtained by sampling the continuous domain impulse response function of Eq. 15.3-11 over a $W \times W$ window. To avoid deleterious truncation effects, the size of the array should be set such that $W = 3c$, or greater, where $c = 2\sqrt{2}\, s$ is the width of the positive center lobe of the LOG function (21). Figure 15.3-2d shows the LOG response of the peppers image for a 11×11 operator.

15.3.2. Laplacian Zero-Crossing Detection

From the discrete domain Laplacian response examples of the preceding section, it has been shown that zero crossings do not always lie at pixel sample points. In fact, for real images subject to luminance fluctuations that contain ramp edges of varying slope, zero-valued Laplacian response pixels are unlikely.

A simple approach to Laplacian zero-crossing detection in discrete domain images is to form the maximum of all positive Laplacian responses and to form the minimum of all negative-value responses in a 3×3 window, If the magnitude of the difference between the maxima and the minima exceeds a threshold, an edge is judged present.

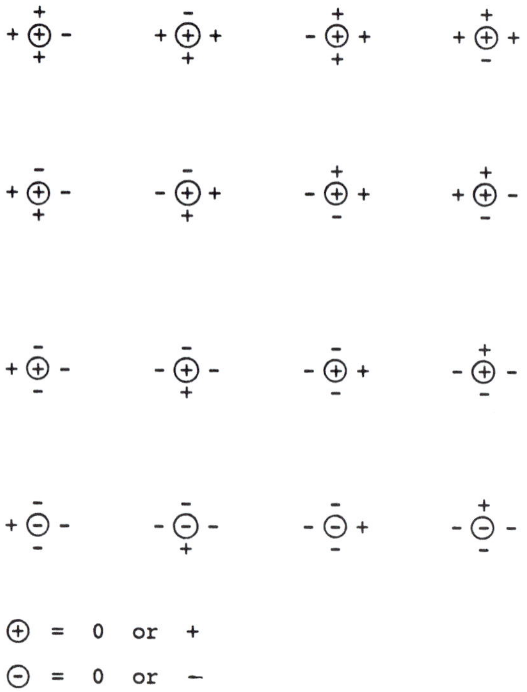

FIGURE 15.3-4. Laplacian zero-crossing patterns.

Huertas and Medioni (21) have developed a systematic method for classifying 3×3 Laplacian response patterns in order to determine edge direction. Figure 15.3-4 illustrates a somewhat simpler algorithm. In the figure, plus signs denote positive-value Laplacian responses, and negative signs denote negative Laplacian responses. The algorithm can be implemented efficiently using morphological image processing techniques.

15.3.3. Directed Second-Order Derivative Generation

Laplacian edge detection techniques employ rotationally invariant second-order differentiation to determine the existence of an edge. The direction of the edge can be ascertained during the zero-crossing detection process. An alternative approach is first to estimate the edge direction and then compute the one-dimensional second-order derivative along the edge direction. A zero crossing of the second-order derivative specifies an edge.

The directed second-order derivative of a continuous domain image $F(x, y)$ along a line at an angle θ with respect to the horizontal axis is given by

$$F''(x, y) = \frac{\partial^2 F(x, y)}{\partial x^2} \cos^2\theta + \frac{\partial^2 F(x, y)}{\partial x \partial y} \cos\theta \sin\theta + \frac{\partial^2 F(x, y)}{\partial y^2} \sin^2\theta \qquad (15.3-14)$$

It should be noted that unlike the Laplacian, the directed second-order derivative is a nonlinear operator. Convolving a smoothing function with $F(x, y)$ prior to differentiation is not equivalent to convolving the directed second derivative of $F(x, y)$ with the smoothing function.

A key factor in the utilization of the directed second-order derivative edge detection method is the ability to determine its suspected edge direction accurately. One approach is to employ some first-order derivative edge detection method to estimate the edge direction, and then compute a discrete approximation to Eq. 15.3-14. Another approach, proposed by Haralick (22), involves approximating $F(x, y)$ by a two-dimensional polynomial, from which the directed second-order derivative can be determined analytically.

As an illustration of Haralick's approximation method, called *facet modeling*, let the continuous image function $F(x, y)$ be approximated by a two-dimensional quadratic polynomial

$$\hat{F}(r, c) = k_1 + k_2 r + k_3 c + k_4 r^2 + k_5 rc + k_6 c^2 + k_7 rc^2 + k_8 r^2 c + k_9 r^2 c^2 \qquad (15.3-15)$$

about a candidate edge point (j, k) in the discrete image $F(j, k)$, where the k_n are weighting factors to be determined from the discrete image data. In this notation, the indices $-(W-1)/2 \le r, c \le (W-1)/2$ are treated as continuous variables in the row (y-coordinate) and column (x-coordinate) directions of the discrete image, but the discrete image is, of course, measurable only at integer values of r and c. From this model, the estimated edge angle is

$$\theta = \arctan\left\{\frac{k_2}{k_3}\right\}$$ (15.3-16)

In principle, any polynomial expansion can be used in the approximation. The expansion of Eq. 15.3-15 was chosen because it can be expressed in terms of a set of orthogonal polynomials. This greatly simplifies the computational task of determining the weighting factors. The quadratic expansion of Eq. 15.3-15 can be rewritten as

$$\hat{F}(r, c) = \sum_{n=1}^{N} a_n P_n(r, c)$$ (15.3-17)

where $P_n(r, c)$ denotes a set of discrete orthogonal polynomials and the a_n are weighting coefficients. Haralick (22) has used the following set of 3×3 *Chebyshev orthogonal polynomials:*

$$P_1(r, c) = 1$$ (15.3-18a)

$$P_2(r, c) = r$$ (15.3-18b)

$$P_3(r, c) = c$$ (15.3-18c)

$$P_4(r, c) = r^2 - \frac{2}{3}$$ (15.3-18d)

$$P_5(r, c) = rc$$ (15.3-18e)

$$P_6(r, c) = c^2 - \frac{2}{3}$$ (15.3-18f)

$$P_7(r, c) = c\left(r^2 - \frac{2}{3}\right)$$ (15.3-18g)

$$P_8(r, c) = r\left(c^2 - \frac{2}{3}\right)$$ (15.3-18h)

$$P_9(r, c) = \left(r^2 - \frac{2}{3}\right)\left(c^2 - \frac{2}{3}\right)$$ (15.3-18i)

defined over the (r, c) index set $\{-1, 0, 1\}$. To maintain notational consistency with the gradient techniques discussed previously, r and c are indexed in accordance with the (x, y) Cartesian coordinate system (i.e., r is incremented positively up rows and c is incremented positively left to right across columns). The polynomial coefficients k_n of Eq. 15.3-15 are related to the Chebyshev weighting coefficients by

$$k_1 = a_1 - \frac{2}{3}a_4 - \frac{2}{3}a_6 + \frac{4}{9}a_9 \tag{15.3-19a}$$

$$k_2 = a_2 - \frac{2}{3}a_7 \tag{15.3-19b}$$

$$k_3 = a_3 - \frac{2}{3}a_8 \tag{15.3-19c}$$

$$k_4 = a_4 - \frac{2}{3}a_9 \tag{15.3-19d}$$

$$k_5 = a_5 \tag{15.3-19e}$$

$$k_6 = a_6 - \frac{2}{3}a_9 \tag{15.3-19f}$$

$$k_7 = a_7 \tag{15.3-19g}$$

$$k_8 = a_8 \tag{15.3-19h}$$

$$k_9 = a_9 \tag{15.3-19i}$$

The optimum values of the set of weighting coefficients a_n that minimize the mean-square error between the image data $F(r, c)$ and its approximation $\hat{F}(r, c)$ are found to be (22)

$$a_n = \frac{\sum\sum P_n(r, c)F(r, c)}{\sum\sum [P_n(r, c)]^2} \tag{15.3-20}$$

As a consequence of the linear structure of this equation, the weighting coefficients $A_n(j, k) = a_n$ at each point in the image $F(j, k)$ can be computed by convolution of the image with a set of impulse response arrays. Hence

$$A_n(j, k) = F(j, k) \circledast H_n(j, k) \tag{15.3-21a}$$

where

$$H_n(j, k) = \frac{P_n(-j, -k)}{\sum\sum [P_n(r, c)]^2} \tag{15.3-21b}$$

Figure 15.3-5 contains the nine impulse response arrays corresponding to the 3×3 Chebyshev polynomials. The arrays H_2 and H_3, which are used to determine the edge angle, are seen from Figure 15.3-5 to be the Prewitt column and row operators, respectively. The arrays H_4 and H_6 are second derivative operators along columns

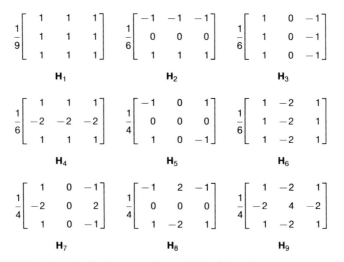

FIGURE 15.3-5. Chebyshev polynomial 3×3 impulse response arrays.

and rows, respectively, as noted in Eq. 15.3-7. Figure 15.3-6 shows the nine weighting coefficient responses for the peppers image.

The second derivative along the line normal to the edge slope can be expressed explicitly by performing second-order differentiation on Eq. 15.3-15. The result is

$$\hat{F}''(r, c) = 2k_4 \sin^2\theta + 2k_5 \sin\theta \cos\theta + 2k_6 \cos^2\theta$$

$$+ (4k_7 \sin\theta\cos\theta + 2k_8 \cos^2\theta)r + (2k_7 \sin^2\theta + 4k_8 \sin\theta\cos\theta)c$$

$$+ (2k_9 \cos^2\theta)r^2 + (8k_9 \sin\theta \cos\theta)rc + (2k_9 \sin^2\theta)c^2 \qquad (15.3-22)$$

This second derivative need only be evaluated on a line in the suspected edge direction. With the substitutions $r = \rho \sin\theta$ and $c = \rho \cos\theta$, the directed second-order derivative can be expressed as

$$\hat{F}''(\rho) = 2(k_4 \sin^2\theta + k_5 \sin\theta\cos\theta + k_6 \cos^2\theta)$$

$$+ 6 \sin\theta\cos\theta (k_7 \sin\theta + k_8 \cos\theta)\rho + 12 (k_9 \sin^2\theta\cos^2\theta)\rho^2 \quad (15.3-23)$$

The next step is to detect zero crossings of $\hat{F}''(\rho)$ in a unit pixel range $-0.5 \le \rho \le 0.5$ of the suspected edge. This can be accomplished by computing the real root (if it exists) within the range of the quadratic relation of Eq. 15.3-23.

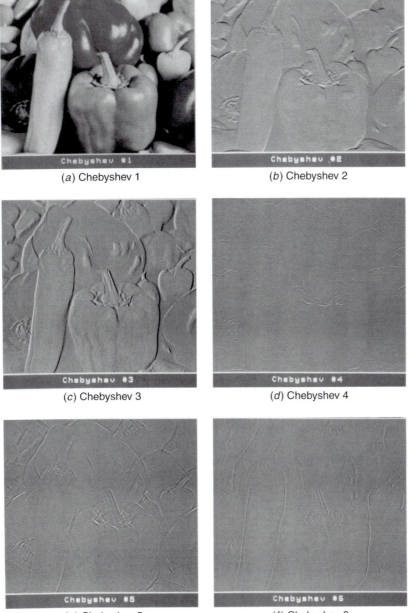

(a) Chebyshev 1

(b) Chebyshev 2

(c) Chebyshev 3

(d) Chebyshev 4

(e) Chebyshev 5

(f) Chebyshev 6

FIGURE 15.3-6. 3×3 Chebyshev polynomial responses for the `peppers_mon` image.

(*g*) Chebyshev 7

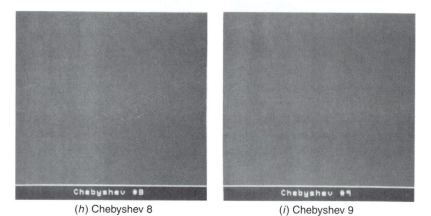

(*h*) Chebyshev 8   (*i*) Chebyshev 9

FIGURE 15.3-6 (*Continued*). 3×3 Chebyshev polynomial responses for the `peppers_mon` image.

15.4. EDGE-FITTING EDGE DETECTION

Ideal edges may be viewed as one- or two-dimensional edges of the form sketched in Figure 15.1-1. Actual image data can then be matched against, or fitted to, the ideal edge models. If the fit is sufficiently accurate at a given image location, an edge is assumed to exist with the same parameters as those of the ideal edge model.

In the one-dimensional edge-fitting case described in Figure 15.4-1, the image signal $f(x)$ is fitted to a step function

$$s(x) = \begin{cases} a & \text{for } x < x_0 & (15.4\text{-}1a) \\ \\ a + h & \text{for } x \geq x_0 & (15.4\text{-}1b) \end{cases}$$

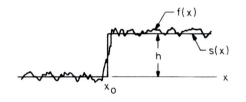

(*a*) One dimensional

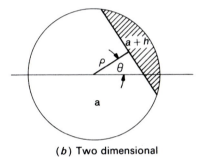

(*b*) Two dimensional

FIGURE 15.4-1. One- and two-dimensional edge fitting.

An edge is assumed present if the mean-square error

$$E = \int_{x_0 - L}^{x_0 + L} [f(x) - s(x)]^2 \, dx \qquad (15.4\text{-}2)$$

is below some threshold value. In the two-dimensional formulation, the ideal step edge is defined as

$$s(x) = \begin{cases} a & \text{for } x\cos\theta + y\sin\theta < \rho & (15.4\text{-}3a) \\ \\ a + h & \text{for } x\cos\theta + y\sin\theta \geq \rho & (15.4\text{-}3b) \end{cases}$$

where θ and ρ jointly specify the polar distance from the center of a circular test region to the normal point of the edge. The edge-fitting error is

$$E = \iint [F(x, y) - S(x, y)]^2 \, dx \, dy \qquad (15.4\text{-}4)$$

where the integration is over the circle in Figure 15.4-1.

Hueckel (23) has developed a procedure for two-dimensional edge fitting in which the pixels within the circle of Figure 15.4-1 are expanded in a set of two-dimensional basis functions by a Fourier series in polar coordinates. Let $B_i(x, y)$ represent the basis functions. Then, the weighting coefficients for the expansions of the image and the ideal step edge become

$$f_i = \iint B_i(x, y) F(x, y) \, dx \, dy \qquad (15.4\text{-}5a)$$

$$s_i = \iint B_i(x, y) S(x, y) \, dx \, dy \qquad (15.4\text{-}5b)$$

In Hueckel's algorithm, the expansion is truncated to eight terms for computational economy and to provide some noise smoothing. Minimization of the mean-square-error difference of Eq. 15.4-4 is equivalent to minimization of $(f_i - s_i)^2$ for all coefficients. Hueckel has performed this minimization, invoking some simplifying approximations and has formulated a set of nonlinear equations expressing the estimated edge parameter set in terms of the expansion coefficients f_i.

Nalwa and Binford (24) have proposed an edge-fitting scheme in which the edge angle is first estimated by a sequential least-squares fit within a 5×5 region. Then, the image data along the edge direction is fit to a hyperbolic tangent function

$$\tanh \rho = \frac{e^\rho - e^{-\rho}}{e^\rho + e^{-\rho}} \qquad (15.4\text{-}6)$$

as shown in Figure 15.4-2.

Edge-fitting methods require substantially more computation than do derivative edge detection methods. Their relative performance is considered in the following section.

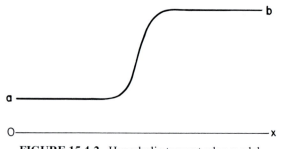

FIGURE 15.4-2. Hyperbolic tangent edge model.

15.5. LUMINANCE EDGE DETECTOR PERFORMANCE

Relatively few comprehensive studies of edge detector performance have been reported in the literature (15,25,26). A performance evaluation is difficult because of the large number of methods proposed, problems in determining the optimum parameters associated with each technique and the lack of definitive performance criteria.

In developing performance criteria for an edge detector, it is wise to distinguish between mandatory and auxiliary information to be obtained from the detector. Obviously, it is essential to determine the pixel location of an edge. Other information of interest includes the height and slope angle of the edge as well as its spatial orientation. Another useful item is a confidence factor associated with the edge decision, for example, the closeness of fit between actual image data and an idealized model. Unfortunately, few edge detectors provide this full gamut of information.

The next sections discuss several performance criteria. No attempt is made to provide a comprehensive comparison of edge detectors.

15.5.1. Edge Detection Probability

The probability of correct edge detection P_D and the probability of false edge detection P_F, as specified by Eq. 15.2-24, are useful measures of edge detector performance. The trade-off between P_D and P_F can be expressed parametrically in terms of the detection threshold. Figure 15.5-1 presents analytically derived plots of P_D versus P_F for several differential operators for vertical and diagonal edges and a signal-to-noise ratio of 1.0 and 10.0 (13). From these curves it is apparent that the Sobel and Prewitt 3×3 operators are superior to the Roberts 2×2 operators. The Prewitt operator is better than the Sobel operator for a vertical edge. But for a diagonal edge, the Sobel operator is superior. In the case of template-matching operators, the Robinson three-level and five-level operators exhibit almost identical performance, which is superior to the Kirsch and Prewitt compass gradient operators. Finally, the Sobel and Prewitt differential operators perform slightly better than the Robinson three- and Robinson five-level operators. It has not been possible to apply this statistical approach to any of the larger operators because of analytic difficulties in evaluating the detection probabilities.

15.5.2. Edge Detection Orientation

An important characteristic of an edge detector is its sensitivity to edge orientation. Abdou and Pratt (15) have analytically determined the gradient response of 3×3 template matching edge detectors and 2×2 and 3×3 orthogonal gradient edge detectors for square-root and magnitude combinations of the orthogonal gradients. Figure 15.5-2 shows plots of the edge gradient as a function of actual edge orientation for a unit-width ramp edge model. The figure clearly shows that magnitude combination of the orthogonal gradients is inferior to square-root combination.

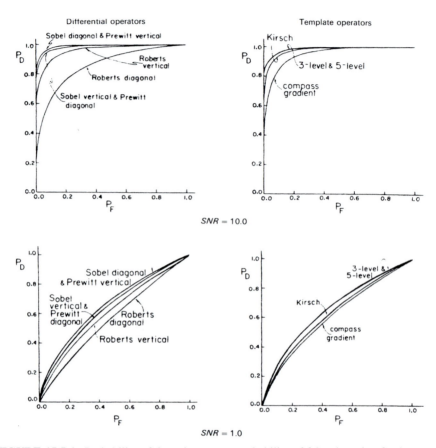

FIGURE 15.5-1. Probability of detection versus probability of false detection for 2×2 and 3×3 operators.

Figure 15.5-3 is a plot of the detected edge angle as a function of the actual orientation of an edge. The Sobel operator provides the most linear response. Laplacian edge detectors are rotationally symmetric operators, and hence are invariant to edge orientation. The edge angle can be determined to within 45° increments during the 3×3 pixel zero-crossing detection process.

15.5.3. Edge Detection Localization

Another important property of an edge detector is its ability to localize an edge. Abdou and Pratt (15) have analyzed the edge localization capability of several first derivative operators for unit width ramp edges. Figure 15.5-4 shows edge models in which the sampled continuous ramp edge is displaced from the center of the operator. Figure 15.5-5 shows plots of the gradient response as a function of edge

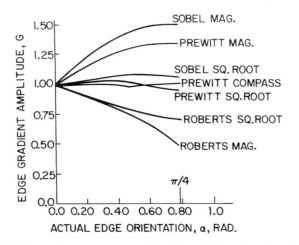

FIGURE 15.5-2. Edge gradient response as a function of edge orientation for 2×2 and 3×3 first derivative operators.

displacement distance for vertical and diagonal edges for 2×2 and 3×3 orthogonal gradient and 3×3 template matching edge detectors. All of the detectors, with the exception of the Kirsch operator, exhibit a desirable monotonically decreasing response as a function of edge displacement. If the edge detection threshold is set at one-half the edge height, or greater, an edge will be properly localized in a noise-free environment for all the operators, with the exception of the Kirsch operator, for which the threshold must be slightly higher. Figure 15.5-6 illustrates the gradient response of boxcar operators as a function of their size (5). A gradient response

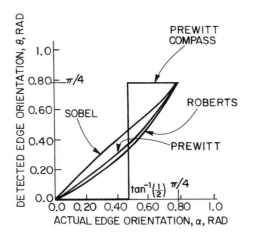

FIGURE 15.5-3. Detected edge orientation as a function of actual edge orientation for 2×2 and 3×3 first derivative operators.

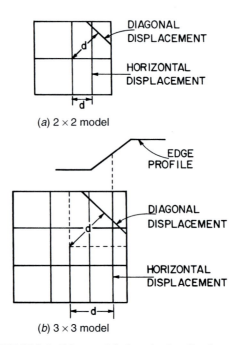

(a) 2 × 2 model

(b) 3 × 3 model

FIGURE 15.5-4. Edge models for edge localization analysis.

comparison of 7×7 orthogonal gradient operators is presented in Figure 15.5-7. For such large operators, the detection threshold must be set relatively high to prevent smeared edge markings. Setting a high threshold will, of course, cause low-amplitude edges to be missed.

Ramp edges of extended width can cause difficulties in edge localization. For first-derivative edge detectors, edges are marked along the edge slope at all points for which the slope exceeds some critical value. Raising the threshold results in the missing of low-amplitude edges. Second derivative edge detection methods are often able to eliminate smeared ramp edge markings. In the case of a unit width ramp edge, a zero crossing will occur only at the midpoint of the edge slope. Extended-width ramp edges will also exhibit a zero crossing at the ramp midpoint provided that the size of the Laplacian operator exceeds the slope width. Figure 15.5-8 illustrates Laplacian of Gaussian (LOG) examples (21).

Berzins (27) has investigated the accuracy to which the LOG zero crossings locate a step edge. Figure 15.5-9 shows the LOG zero crossing in the vicinity of a corner step edge. A zero crossing occurs exactly at the corner point, but the zero-crossing curve deviates from the step edge adjacent to the corner point. The maximum deviation is about $0.3s$, where s is the standard deviation of the Gaussian smoothing function.

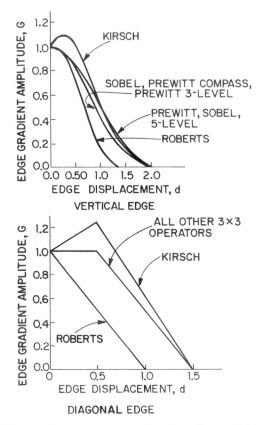

FIGURE 15.5-5. Edge gradient response as a function of edge displacement distance for 2×2 and 3×3 first derivative operators.

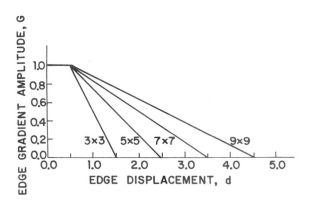

FIGURE 15.5-6. Edge gradient response as a function of edge displacement distance for variable-size boxcar operators.

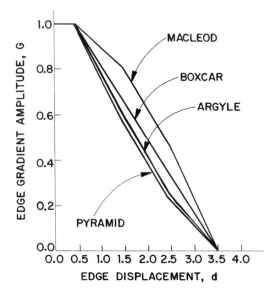

FIGURE 15.5-7 Edge gradient response as a function of edge displacement distance for several 7 × 7 orthogonal gradient operators.

15.5.4. Edge Detector Figure of Merit

There are three major types of error associated with determination of an edge: (1) missing valid edge points, (2) failure to localize edge points, and (3) classification of

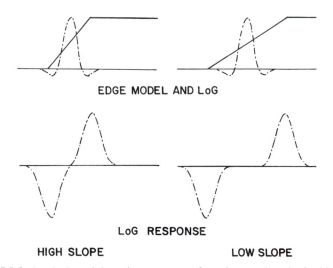

FIGURE 15.5-8. Laplacian of Gaussian response of continuous domain for high- and low-slope ramp edges.

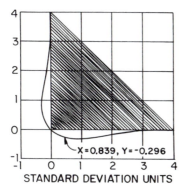

STANDARD DEVIATION UNITS

FIGURE 15.5-9. Locus of zero crossings in vicinity of a corner edge for a continuous Laplacian of Gaussian edge detector.

noise fluctuations as edge points. Figure 15.5-10 illustrates a typical edge segment in a discrete image, an ideal edge representation, and edge representations subject to various types of error.

A common strategy in signal detection problems is to establish some bound on the probability of false detection resulting from noise and then attempt to maximize the probability of true signal detection. Extending this concept to edge detection simply involves setting the edge detection threshold at a level such that the probability of false detection resulting from noise alone does not exceed some desired value. The probability of true edge detection can readily be evaluated by a coincidence comparison of the edge maps of an ideal and an actual edge detector. The penalty for nonlocalized edges is somewhat more difficult to access. Edge detectors that provide a smeared edge location should clearly be penalized; however, credit should be given to edge detectors whose edge locations are localized but biased by a small amount. Pratt (28) has introduced a figure of merit that balances these three types of error. The figure of merit is defined by

$$R = \frac{1}{I_N} \sum_{i=1}^{I_A} \frac{1}{1 + ad^2} \qquad (15.5\text{-}1)$$

where $I_N = \text{MAX}\{I_I, I_A\}$ and I_I and I_A represent the number of ideal and actual edge map points, a is a scaling constant, and d is the separation distance of an actual edge point normal to a line of ideal edge points. The rating factor is normalized so that $R = 1$ for a perfectly detected edge. The scaling factor may be adjusted to penalize edges that are localized but offset from the true position. Normalization by the maximum of the actual and ideal number of edge points ensures a penalty for smeared or fragmented edges. As an example of performance, if $a = \frac{1}{9}$, the rating of

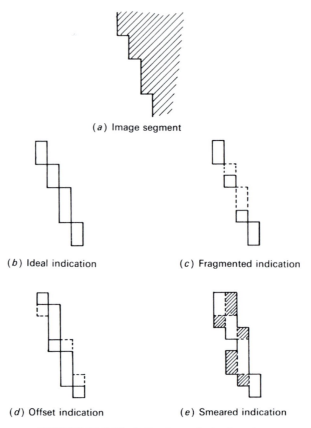

(*a*) Image segment

(*b*) Ideal indication (*c*) Fragmented indication

(*d*) Offset indication (*e*) Smeared indication

FIGURE 15.5-10. Indications of edge location.

a vertical detected edge offset by one pixel becomes $R = 0.90$, and a two-pixel offset gives a rating of $R = 0.69$. With $a = \frac{1}{9}$, a smeared edge of three pixels width centered about the true vertical edge yields a rating of $R = 0.93$, and a five-pixel-wide smeared edge gives $R = 0.84$. A higher rating for a smeared edge than for an offset edge is reasonable because it is possible to thin the smeared edge by morphological postprocessing.

The figure-of-merit criterion described above has been applied to the assessment of some of the edge detectors discussed previously, using a test image consisting of a 64×64 pixel array with a vertically oriented edge of variable contrast and slope placed at its center. Independent Gaussian noise of standard deviation σ_n has been added to the edge image. The signal-to-noise ratio is defined as $\mathrm{SNR} = (h/\sigma_n)^2$, where h is the edge height scaled over the range 0.0 to 1.0. Because the purpose of the testing is to compare various edge detection methods, for fairness it is important that each edge detector be tuned to its best capabilities. Consequently, each edge detector has been permitted to train both on random noise fields without edges and the actual test images before evaluation. For each edge detector, the threshold parameter has been set to achieve the maximum figure of merit subject to the maximum allowable false detection rate.

Figure 15.5-11 shows plots of the figure of merit for a vertical ramp edge as a function of signal-to-noise ratio for several edge detectors (5). The figure of merit is also plotted in Figure 15.5-12 as a function of edge width. The figure of merit curves in the figures follow expected trends: low for wide and noisy edges; and high in the opposite case. Some of the edge detection methods are universally superior to others for all test images. As a check on the subjective validity of the edge location figure of merit, Figures 15.5-13 and 15.5-14 present the edge maps obtained for several high-and low-ranking edge detectors. These figures tend to corroborate the utility of the figure of merit. A high figure of merit generally corresponds to a well-located edge upon visual scrutiny, and vice versa.

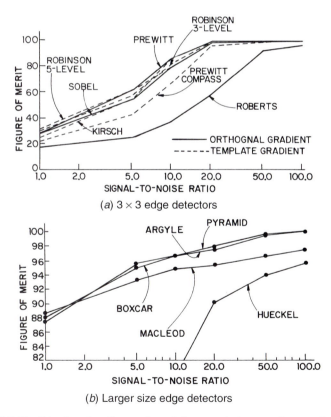

FIGURE 15.5-11. Edge location figure of merit for a vertical ramp edge as a function of signal-to-noise ratio for $h = 0.1$ and $w = 1$.

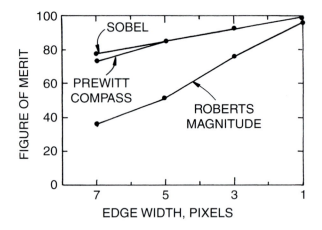

FIGURE 15.5-12. Edge location figure of merit for a vertical ramp edge as a function of signal-to-noise ratio for $h = 0.1$ and SNR = 100.

15.5.5. Subjective Assessment

In many, if not most applications in which edge detection is performed to outline objects in a real scene, the only performance measure of ultimate importance is how well edge detector markings match with the visual perception of object boundaries. A human observer is usually able to discern object boundaries in a scene quite accurately in a perceptual sense. However, most observers have difficulty recording their observations by tracing object boundaries. Nevertheless, in the evaluation of edge detectors, it is useful to assess them in terms of how well they produce outline drawings of a real scene that are meaningful to a human observer.

The peppers image of Figure 15.2-2 has been used for the subjective assessment of edge detectors. The peppers in the image are visually distinguishable objects, but shadows and nonuniform lighting create a challenge to edge detectors, which by definition do not utilize higher-order perceptive intelligence. Figures 15.5-15 and 15.5-16 present edge maps of the peppers image for several edge detectors. The parameters of the various edge detectors have been chosen to produce the best visual delineation of objects.

Heath et al. (26) have performed extensive visual testing of several complex edge detection algorithms, including the Canny and Nalwa–Binford methods, for a number of natural images. The judgment criterion was a numerical rating as to how well the edge map generated by an edge detector allows for easy, quick, and accurate recognition of objects within a test image.

SNR = 100
(*a*) Original   (*b*) Edge map, *R* = 100%

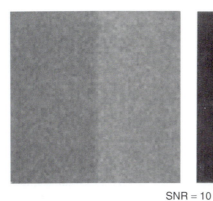

SNR = 10
(*c*) Original   (*d*) Edge map, *R* = 85.1%

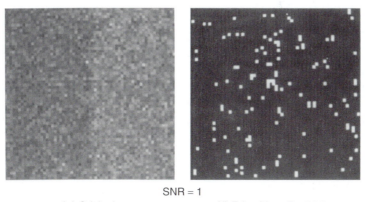

SNR = 1
(*e*) Original   (*f*) Edge Map, *R* = 24.2%

FIGURE 15.5-13. Edge location performance of Sobel edge detector as a function of signal-to-noise ratio, $h = 0.1$, $w = 1$, $a = 1/9$.

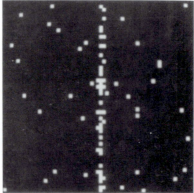

(a) Original   (b) East compass, $R = 66.1\%$

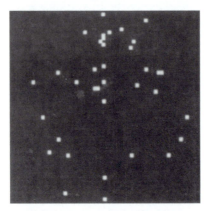
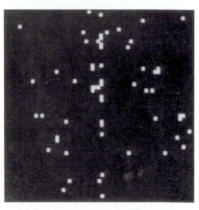

(c) Roberts magnitude, $R = 31.5\%$ (d) Roberts square root, $R = 37.0\%$

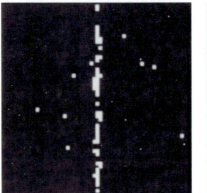
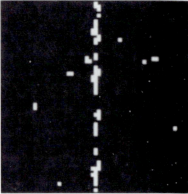

(e) Sobel, $R = 85.1\%$ (f) Kirsch, $R = 80.8\%$

FIGURE 15.5-14. Edge location performance of several edge detectors for SNR = 10, $h = 0.1, w = 1, a = 1/9$.

(a) 2 × 2 Roberts, t = 0.08

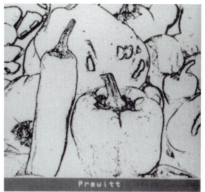
(b) 3 × 3 Prewitt, t = 0.08

(c) 3 × 3 Sobel, t = 0.09

(d) 3 × 3 Robinson five-level

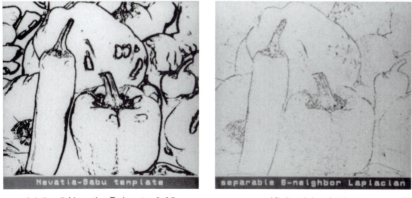
(e) 5 × 5 Nevatia–Babu, t = 0.05

(f) 3 × 3 Laplacian

FIGURE 15.5-15. Edge maps of the peppers_mon image for several small edge detectors.

(a) 7 × 7 boxcar, t = 0.10

(b) 9 × 9 truncated pyramid, t = 0.10

(c) 11 × 11 Argyle, t = 0.05

(d) 11 × 11 Macleod, t = 0.10

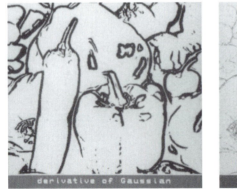

(e) 11 × 11 derivative of Gaussian, t = 0.11

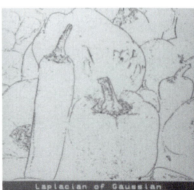

(f) 11 × 11 Laplacian of Gaussian

FIGURE 15.5-16. Edge maps of the `peppers_mon` image for several large edge detectors.

15.6. COLOR EDGE DETECTION

In Chapter 3 it was established that color images may be described quantitatively at each pixel by a set of three tristimulus values T_1, T_2, T_3, which are proportional to the amount of red, green, and blue primary lights required to match the pixel color. The luminance of the color is a weighted sum $Y = a_1 T_1 + a_2 T_2 + a_3 T_3$ of the tristimulus values, where the a_i are constants that depend on the spectral characteristics of the primaries.

Several definitions of a color edge have been proposed (29). An edge in a color image can be said to exist if and only if the luminance field contains an edge. This definition ignores discontinuities in hue and saturation that occur in regions of constant luminance. Another definition is to judge a color edge present if an edge exists in any of its constituent tristimulus components. A third definition is based on forming the sum

$$G(j, k) = G_1(j, k) + G_2(j, k) + G_3(j, k) \tag{15.6-1}$$

of the gradients $G_i(j, k)$ of the three tristimulus values or some linear or nonlinear color components. A color edge exists if the gradient exceeds a threshold. Still another definition is based on the vector sum gradient

$$G(j, k) = [[G_1(j, k)]^2 + [G_2(j, k)]^2 + [G_3(j, k)]^2]^{1/2} \tag{15.6-2}$$

With the tricomponent definitions of color edges, results are dependent on the particular color coordinate system chosen for representation. Figure 15.6-1 is a color photograph of the peppers image and monochrome photographs of its red, green, and blue components. The YIQ and $L*a*b*$ coordinates are shown in Figure 15.6-2. Edge maps of the individual RGB components are shown in Figure 15.6-3 for Sobel edge detection. This figure also shows the logical OR of the RGB edge maps plus the edge maps of the gradient sum and the vector sum. The RGB gradient vector sum edge map provides slightly better visual edge delineation than that provided by the gradient sum edge map; the logical OR edge map tends to produce thick edges and numerous isolated edge points. Sobel edge maps for the YIQ and the $L*a*b*$ color components are presented in Figures 15.6-4 and 15.6-5. The YIQ gradient vector sum edge map gives the best visual edge delineation, but it does not delineate edges quite as well as the RGB vector sum edge map. Edge detection results for the $L*a*b*$ coordinate system are quite poor because the $a*$ component is very noise sensitive.

15.7 LINE AND SPOT DETECTION

A line in an image could be considered to be composed of parallel, closely spaced edges. Similarly, a spot could be considered to be a closed contour of edges. This

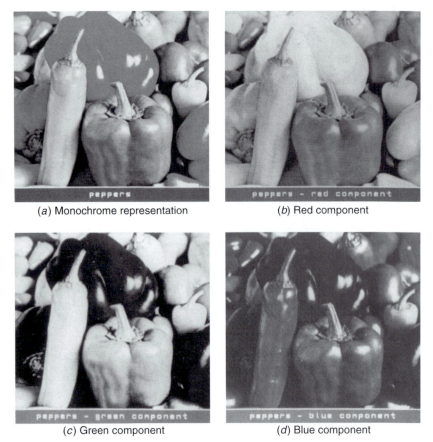

(a) Monochrome representation   (b) Red component

(c) Green component   (d) Blue component

FIGURE 15.6-1. The peppers_gamma color image and its *RGB* color components. See insert for a color representation of this figure.

method of line and spot detection involves the application of scene analysis techniques to spatially relate the constituent edges of the lines and spots. The approach taken in this chapter is to consider only small-scale models of lines and edges and to apply the detection methodology developed previously for edges.

Figure 15.1-4 presents several discrete models of lines. For the unit-width line models, line detection can be accomplished by threshold detecting a line gradient

$$G(j, k) = \underset{m=1}{\overset{4}{\text{MAX}}}\{|F(j, k) \circledast H_m(j, k)|\} \qquad (15.7\text{-}1)$$

(*a*) *Y* component

(*b*) *L** component

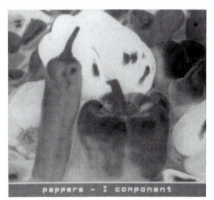
(*c*) *I* component

(*d*) *a** component

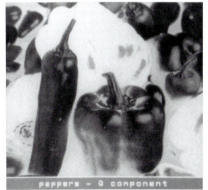
(*e*) *Q* component

(*f*) *b** component

FIGURE 15.6-2. *YIQ* and *L*a*b** color components of the `peppers_gamma` image.

(a) Red edge map

(b) Logical OR of *RGB* edges

(c) Green edge map

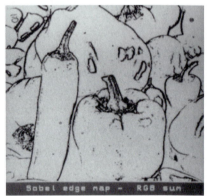

(d) *RGB* sum edge map

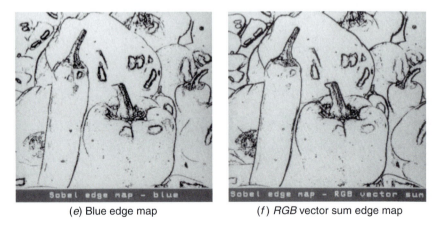

(e) Blue edge map   (f) *RGB* vector sum edge map

FIGURE 15.6-3. Sobel edge maps for edge detection using the *RGB* color components of the peppers_gamma image.

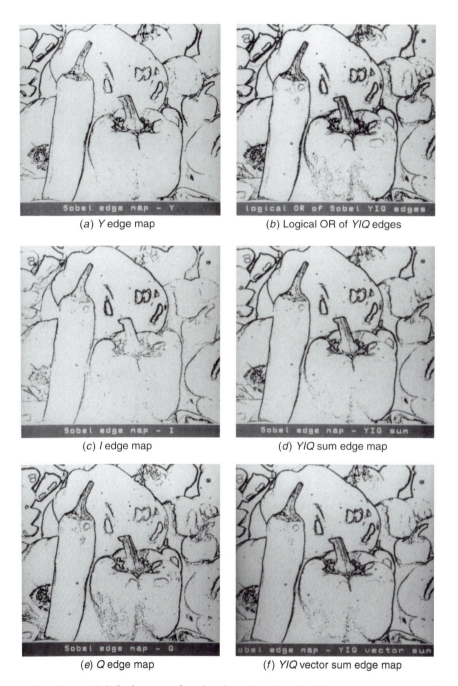

(a) Y edge map

(b) Logical OR of YIQ edges

(c) I edge map

(d) YIQ sum edge map

(e) Q edge map

(f) YIQ vector sum edge map

FIGURE 15.6-4. Sobel edge maps for edge detection using the YIQ color components of the peppers_gamma image.

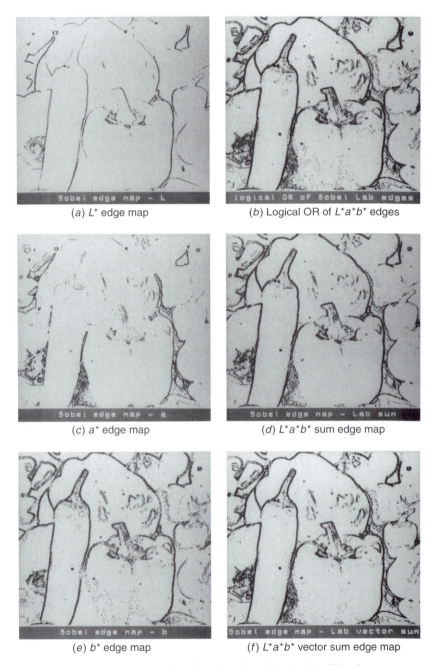

(a) L^* edge map

(b) Logical OR of $L^*a^*b^*$ edges

(c) a^* edge map

(d) $L^*a^*b^*$ sum edge map

(e) b^* edge map

(f) $L^*a^*b^*$ vector sum edge map

FIGURE 15.6-5. Sobel edge maps for edge detection using the $L^*a^*b^*$ color components of the peppers_gamma image.

where $H_m(j, k)$ is a 3×3 line detector impulse response array corresponding to a specific line orientation. Figure 15.7-1 contains two sets of line detector impulse response arrays, weighted and unweighted, which are analogous to the Prewitt and Sobel template matching edge detector impulse response arrays. The detection of ramp lines, as modeled in Figure 15.1-4, requires 5×5 pixel templates.

Unit-width step spots can be detected by thresholding a spot gradient

$$G(j, k) = F(j, k) \circledast H(j, k) \tag{15.7-2}$$

where $H(j, k)$ is an impulse response array chosen to accentuate the gradient of a unit-width spot. One approach is to use one of the three types of 3×3 Laplacian operators defined by Eq. 15.3-5, 15.3-6, or 15.3-8, which are discrete approximations to the sum of the row and column second derivatives of an image. The gradient responses to these impulse response arrays for the unit-width spot model of Figure 15.1-6a are simply a replicas of each array centered at the spot, scaled by the spot height h and zero elsewhere. It should be noted that the Laplacian gradient responses are thresholded for spot detection, whereas the Laplacian responses are examined for sign changes (zero crossings) for edge detection. The disadvantage to using Laplacian operators for spot detection is that they evoke a gradient response for edges, which can lead to false spot detection in a noisy environment. This problem can be alleviated by the use of a 3×3 operator that approximates the continuous

	Unweighted line	Weighted line
H_1	$\begin{bmatrix} -1 & 2 & -1 \\ -1 & 2 & -1 \\ -1 & 2 & -1 \end{bmatrix}$	$\begin{bmatrix} -1 & 2 & -1 \\ -2 & 4 & -2 \\ -1 & 2 & -1 \end{bmatrix}$
H_2	$\begin{bmatrix} -1 & -1 & -1 \\ 2 & 2 & 2 \\ -1 & -1 & -1 \end{bmatrix}$	$\begin{bmatrix} -1 & -2 & -1 \\ 2 & 4 & 2 \\ -1 & -2 & -1 \end{bmatrix}$
H_3	$\begin{bmatrix} -1 & -1 & 2 \\ -1 & 2 & -1 \\ 2 & -1 & -1 \end{bmatrix}$	$\begin{bmatrix} -2 & -1 & 2 \\ -1 & 4 & -1 \\ 2 & -1 & -2 \end{bmatrix}$
H_4	$\begin{bmatrix} 2 & -1 & -1 \\ -1 & 2 & -1 \\ -1 & -1 & 2 \end{bmatrix}$	$\begin{bmatrix} 2 & -1 & -2 \\ -1 & 4 & -1 \\ -2 & -1 & 2 \end{bmatrix}$
Scale factor	$\dfrac{1}{6}$	$\dfrac{1}{8}$

FIGURE 15.7-1. Line detector 3×3 impulse response arrays.

cross second derivative $\partial^2/\partial x^2 \partial y^2$. Prewitt (1, p. 126) has suggested the following discrete approximation:

$$\mathbf{H} = \frac{1}{8} \begin{bmatrix} 1 & -2 & 1 \\ -2 & 4 & -2 \\ 1 & -2 & 1 \end{bmatrix} \qquad (15.7\text{-}3)$$

The advantage of this operator is that it evokes no response for horizontally or vertically oriented edges, however, it does generate a response for diagonally oriented edges. The detection of unit-width spots modeled by the ramp model of Figure 15.1-5 requires a 5×5 impulse response array. The cross second derivative operator of Eq. 15.7-3 and the separable eight-connected Laplacian operator are deceptively similar in appearance; often, they are mistakenly exchanged with one another in the literature. It should be noted that the cross second derivative is identical to within a scale factor with the ninth Chebyshev polynomial impulse response array of Figure 15.3-5.

Cook and Rosenfeld (30) and Zucker et al. (31) have suggested several algorithms for detection of large spots. In one algorithm, an image is first smoothed with a $W \times W$ low-pass filter impulse response array. Then the value of each point in the averaged image is compared to the average value of its north, south, east, and west neighbors spaced W pixels away. A spot is marked if the difference is sufficiently large. A similar approach involves formation of the difference of the average pixel amplitude in a $W \times W$ window and the average amplitude in a surrounding ring region of width W.

Chapter 19 considers the general problem of detecting objects within an image by template matching. Such templates can be developed to detect large spots.

REFERENCES

1. J. M. S. Prewitt, "Object Enhancement and Extraction," in *Picture Processing and Psychopictorics*, B. S. Lipkin and A. Rosenfeld, Eds., Academic Press, New York. 1970.

2. L. G. Roberts, "Machine Perception of Three-Dimensional Solids," in *Optical and Electro-Optical Information Processing*, J. T. Tippett et al., Eds., MIT Press, Cambridge, MA, 1965, 159–197.

3. R. O. Duda and P. E. Hart, *Pattern Classification and Scene Analysis*, Wiley, New York, 1973.

4. W. Frei and C. Chen, "Fast Boundary Detection: A Generalization and a New Algorithm," *IEEE Trans. Computers*, **C-26**, 10, October 1977, 988–998.

5. I. Abdou, "Quantitative Methods of Edge Detection," USCIPI Report 830, Image Processing Institute, University of Southern California, Los Angeles, 1973.

6. E. Argyle, "Techniques for Edge Detection," *Proc. IEEE*, **59**, 2, February 1971, 285–287.

7. I. D. G. Macleod, "On Finding Structure in Pictures," in *Picture Processing and Psychopictorics*, B. S. Lipkin and A. Rosenfeld, Eds., Academic Press, New York, 1970.

8. I. D. G. Macleod, "Comments on Techniques for Edge Detection," *Proc. IEEE*, **60**, 3, March 1972, 344.

9. J. Canny, "A Computational Approach to Edge Detection," *IEEE Trans. Pattern Analysis and Machine Intelligence*, **PAMI-8**, 6, November 1986, 679–698.

10. D. Demigny and T. Kamle, "A Discrete Expression of Canny's Criteria for Step Edge Detector Performances Evaluation," *IEEE Trans. Pattern Analysis and Machine Intelligence*, PAMI-**19**, 11, November 1997, 1199–1211.

11. R. Kirsch, "Computer Determination of the Constituent Structure of Biomedical Images," *Computers and Biomedical Research*, **4**, 3, 1971, 315–328.

12. G. S. Robinson, "Edge Detection by Compass Gradient Masks," *Computer Graphics and Image Processing*, **6**, 5, October 1977, 492–501.

13. R. Nevatia and K. R. Babu, "Linear Feature Extraction and Description," *Computer Graphics and Image Processing*, **13**, 3, July 1980, 257–269.

14. A. P. Paplinski, "Directional Filtering in Edge Detection," *IEEE Trans. Image Processing*, **IP-7**, 4, April 1998, 611–615.

15. I. E. Abdou and W. K. Pratt, "Quantitative Design and Evaluation of Enhancement/Thresholding Edge Detectors," *Proc. IEEE*, **67**, 5, May 1979, 753–763.

16. K. Fukunaga, *Introduction to Statistical Pattern Recognition*, Academic Press, New York, 1972.

17. P. V. Henstock and D. M. Chelberg, "Automatic Gradient Threshold Determination for Edge Detection," *IEEE Trans. Image Processing*, **IP-5**, 5, May 1996, 784–787.

18. V. Torre and T. A. Poggio, "On Edge Detection," *IEEE Trans. Pattern Analysis and Machine Intelligence*, **PAMI-8**, 2, March 1986, 147–163.

19. D. Marr and E. Hildrith, "Theory of Edge Detection," *Proc. Royal Society of London*, **B207**, 1980, 187–217.

20. J. S. Wiejak, H. Buxton, and B. F. Buxton, "Convolution with Separable Masks for Early Image Processing," *Computer Vision, Graphics, and Image Processing*, **32**, 3, December 1985, 279–290.

21. A. Huertas and G. Medioni, "Detection of Intensity Changes Using Laplacian-Gaussian Masks," *IEEE Trans. Pattern Analysis and Machine Intelligence*, **PAMI-8**, 5, September 1986, 651–664.

22. R. M. Haralick, "Digital Step Edges from Zero Crossing of Second Directional Derivatives," *IEEE Trans. Pattern Analysis and Machine Intelligence*, **PAMI-6**, 1, January 1984, 58–68.

23. M. Hueckel, "An Operator Which Locates Edges in Digital Pictures," *J. Association for Computing Machinery*, **18**, 1, January 1971, 113–125.

24. V. S. Nalwa and T. O. Binford, "On Detecting Edges," *IEEE Trans. Pattern Analysis and Machine Intelligence*, **PAMI-6**, November 1986, 699–714.

25. J. R. Fram and E. S. Deutsch, "On the Evaluation of Edge Detection Schemes and Their Comparison with Human Performance," *IEEE Trans. Computers*, **C-24**, 6, June 1975, 616–628.

26. M. D. Heath, et al., "A Robust Visual Method for Assessing the Relative Performance of Edge-Detection Algorithms," *IEEE Trans. Pattern Analysis and Machine Intelligence*, PAMI-**19**, 12, December 1997, 1338–1359.

27. V. Berzins, "Accuracy of Laplacian Edge Detectors," *Computer Vision, Graphics, and Image Processing*, **27**, 2, August 1984, 195–210.

28. W. K. Pratt, *Digital Image Processing*, Wiley-Interscience, New York, 1978, 497–499.

29. G. S. Robinson, "Color Edge Detection," *Proc. SPIE Symposium on Advances in Image Transmission Techniques*, **87**, San Diego, CA, August 1976.

30. C. M. Cook and A. Rosenfeld, "Size Detectors," *Proc. IEEE Letters*, **58**, 12, December 1970, 1956–1957.

31. S. W. Zucker, A. Rosenfeld, and L. S. Davis, "Picture Segmentation by Texture Discrimination," *IEEE Trans. Computers*, **C-24**, 12, December 1975, 1228–1233.

16

IMAGE FEATURE EXTRACTION

An *image feature* is a distinguishing primitive characteristic or attribute of an image. Some features are natural in the sense that such features are defined by the visual appearance of an image, while other, artificial features result from specific manipulations of an image. Natural features include the luminance of a region of pixels and gray scale textural regions. Image amplitude histograms and spatial frequency spectra are examples of artificial features.

Image features are of major importance in the isolation of regions of common property within an image (*image segmentation*) and subsequent identification or labeling of such regions (*image classification*). Image segmentation is discussed in Chapter 16. References 1 to 4 provide information on image classification techniques.

This chapter describes several types of image features that have been proposed for image segmentation and classification. Before introducing them, however, methods of evaluating their performance are discussed.

16.1. IMAGE FEATURE EVALUATION

There are two quantitative approaches to the evaluation of image features: prototype performance and figure of merit. In the prototype performance approach for image classification, a prototype image with regions (segments) that have been independently categorized is classified by a classification procedure using various image features to be evaluated. The classification error is then measured for each feature set. The best set of features is, of course, that which results in the least classification error. The prototype performance approach for image segmentation is similar in nature. A prototype image with independently identified regions is segmented by a segmentation procedure using a test set of features. Then, the detected segments are compared to the known segments, and the segmentation error is evaluated. The problems associated with the prototype performance methods of feature evaluation are the integrity of the prototype data and the fact that the performance indication is dependent not only on the quality of the features but also on the classification or segmentation ability of the classifier or segmenter.

The figure-of-merit approach to feature evaluation involves the establishment of some functional distance measurements between sets of image features such that a large distance implies a low classification error, and vice versa. Faugeras and Pratt (5) have utilized the *Bhattacharyya distance* (3) figure-of-merit for texture feature evaluation. The method should be extensible for other features as well. The Bhattacharyya distance (B-distance for simplicity) is a scalar function of the probability densities of features of a pair of classes defined as

$$B(S_1, S_2) = -\ln\left\{\int [p(\mathbf{x}|S_1)p(\mathbf{x}|S_2)]^{1/2} d\mathbf{x}\right\} \quad (16.1\text{-}1)$$

where $\mathbf{x}$ denotes a vector containing individual image feature measurements with conditional density $p(\mathbf{x}|S_i)$. It can be shown (3) that the B-distance is related monotonically to the *Chernoff bound* for the probability of classification error using a Bayes classifier. The bound on the error probability is

$$P \le [P(S_1)P(S_2)]^{1/2} \exp\{-B(S_1, S_2)\} \quad (16.1\text{-}2)$$

where $P(S_i)$ represents the a priori class probability. For future reference, the Chernoff error bound is tabulated in Table 16.1-1 as a function of B-distance for equally likely feature classes.

For Gaussian densities, the B-distance becomes

$$B(S_1, S_2) = \tfrac{1}{8}(\mathbf{u}_1 - \mathbf{u}_2)^T\left(\frac{\mathbf{\Sigma}_1 + \mathbf{\Sigma}_2}{2}\right)^{-1}(\mathbf{u}_1 - \mathbf{u}_2) + \tfrac{1}{2}\ln\left\{\frac{\tfrac{1}{2}|\mathbf{\Sigma}_1 + \mathbf{\Sigma}_2|}{|\mathbf{\Sigma}_1|^{1/2}|\mathbf{\Sigma}_2|^{1/2}}\right\} \quad (16.1\text{-}3)$$

where $\mathbf{u}_i$ and $\mathbf{\Sigma}_i$ represent the feature mean vector and the feature covariance matrix of the classes, respectively. Calculation of the B-distance for other densities is generally difficult. Consequently, the B-distance figure of merit is applicable only for Gaussian-distributed feature data, which fortunately is the common case. In practice, features to be evaluated by Eq. 16.1-3 are measured in regions whose class has been determined independently. Sufficient feature measurements need be taken so that the feature mean vector and covariance can be estimated accurately.

TABLE 16.1-1 Relationship of Bhattacharyya Distance and Chernoff Error Bound

$B(S_1, S_2)$	Error Bound
1	1.84×10^{-1}
2	6.77×10^{-2}
4	9.16×10^{-3}
6	1.24×10^{-3}
8	1.68×10^{-4}
10	2.27×10^{-5}
12	2.07×10^{-6}

16.2. AMPLITUDE FEATURES

The most basic of all image features is some measure of image amplitude in terms of luminance, tristimulus value, spectral value, or other units. There are many degrees of freedom in establishing image amplitude features. Image variables such as luminance or tristimulus values may be utilized directly, or alternatively, some linear, nonlinear, or perhaps noninvertible transformation can be performed to generate variables in a new amplitude space. Amplitude measurements may be made at specific image points, [e.g., the amplitude $F(j, k)$ at pixel coordinate (j, k), or over a neighborhood centered at (j, k)]. For example, the average or mean image amplitude in a $W \times W$ pixel neighborhood is given by

$$M(j, k) = \frac{1}{W^2} \sum_{m = -w}^{w} \sum_{n = -w}^{w} F(j + m, k + n) \qquad (16.2-1)$$

where $W = 2w + 1$. An advantage of a neighborhood, as opposed to a point measurement, is a diminishing of noise effects because of the averaging process. A disadvantage is that object edges falling within the neighborhood can lead to erroneous measurements.

The median of pixels within a $W \times W$ neighborhood can be used as an alternative amplitude feature to the mean measurement of Eq. 16.2-1, or as an additional feature. The *median* is defined to be that pixel amplitude in the window for which one-half of the pixels are equal or smaller in amplitude, and one-half are equal or greater in amplitude. Another useful image amplitude feature is the neighborhood standard deviation, which can be computed as

$$S(j, k) = \frac{1}{W} \left[\sum_{m = -w}^{w} \sum_{n = -w}^{w} [F(j + m, k + n) - M(j + m, k + n)]^2 \right]^{1/2} \qquad (16.2-2)$$

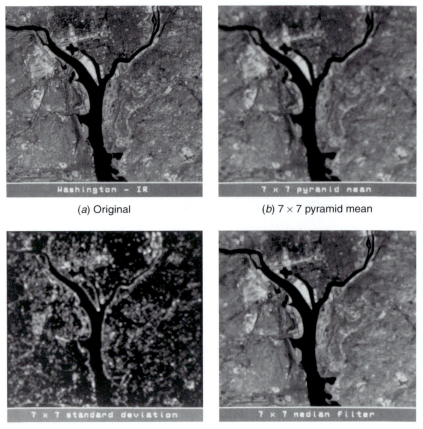

(a) Original

(b) 7 × 7 pyramid mean

(c) 7 × 7 standard deviation

(d) 7 × 7 plus median

FIGURE 16.2-1. Image amplitude features of the `washington_ir` image.

In the literature, the standard deviation image feature is sometimes called the *image dispersion*. Figure 16.2-1 shows an original image and the mean, median, and standard deviation of the image computed over a small neighborhood.

The mean and standard deviation of Eqs. 16.2-1 and 16.2-2 can be computed indirectly in terms of the histogram of image pixels within a neighborhood. This leads to a class of image amplitude *histogram features*. Referring to Section 5.7, the first-order probability distribution of the amplitude of a quantized image may be defined as

$$P(b) = P_R [F(j, k) = r_b] \qquad (16.2-3)$$

where r_b denotes the quantized amplitude level for $0 \le b \le L - 1$. The first-order histogram estimate of $P(b)$ is simply

$$P(b) \approx \frac{N(b)}{M} \tag{16.2-4}$$

where M represents the total number of pixels in a neighborhood window centered about (j, k), and $N(b)$ is the number of pixels of amplitude r_b in the same window.

The shape of an image histogram provides many clues as to the character of the image. For example, a narrowly distributed histogram indicates a low-contrast image. A bimodal histogram often suggests that the image contains an object with a narrow amplitude range against a background of differing amplitude. The following measures have been formulated as quantitative shape descriptions of a first-order histogram (6).

Mean:

$$S_M \equiv \bar{b} = \sum_{b=0}^{L-1} bP(b) \tag{16.2-5}$$

Standard deviation:

$$S_D \equiv \sigma_b = \left[\sum_{b=0}^{L-1} (b - \bar{b})^2 P(b) \right]^{1/2} \tag{16.2-6}$$

Skewness:

$$S_S = \frac{1}{\sigma_b^3} \sum_{b=0}^{L-1} (b - \bar{b})^3 P(b) \tag{16.2-7}$$

Kurtosis:

$$S_K = \frac{1}{\sigma_b^4} \sum_{b=0}^{L-1} (b - \bar{b})^4 P(b) - 3 \tag{16.2-8}$$

Energy:

$$S_N = \sum_{b=0}^{L-1} [P(b)]^2 \tag{16.2-9}$$

Entropy:

$$S_E = - \sum_{b=0}^{L-1} P(b) \log_2 \{P(b)\} \tag{16.2-10}$$

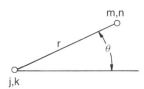

FIGURE 16.2-2. Relationship of pixel pairs.

The factor of 3 inserted in the expression for the Kurtosis measure normalizes S_K to zero for a zero-mean, Gaussian-shaped histogram. Another useful histogram shape measure is the *histogram mode*, which is the pixel amplitude corresponding to the histogram peak (i.e., the most commonly occurring pixel amplitude in the window). If the histogram peak is not unique, the pixel at the peak closest to the mean is usually chosen as the histogram shape descriptor.

Second-order histogram features are based on the definition of the joint probability distribution of pairs of pixels. Consider two pixels $F(j, k)$ and $F(m, n)$ that are located at coordinates (j, k) and (m, n), respectively, and, as shown in Figure 16.2-2, are separated by r radial units at an angle θ with respect to the horizontal axis. The joint distribution of image amplitude values is then expressed as

$$P(a, b) = P_R [F(j, k) = r_a, F(m, n) = r_b] \qquad (16.2\text{-}11)$$

where r_a and r_b represent quantized pixel amplitude values. As a result of the discrete rectilinear representation of an image, the separation parameters (r, θ) may assume only certain discrete values. The histogram estimate of the second-order distribution is

$$P(a, b) \approx \frac{N(a, b)}{M} \qquad (16.2\text{-}12)$$

where M is the total number of pixels in the measurement window and $N(a, b)$ denotes the number of occurrences for which $F(j, k) = r_a$ and $F(m, n) = r_b$.

If the pixel pairs within an image are highly correlated, the entries in $P(a, b)$ will be clustered along the diagonal of the array. Various measures, listed below, have been proposed (6,7) as measures that specify the energy spread about the diagonal of $P(a, b)$.

Autocorrelation:

$$S_A = \sum_{a=0}^{L-1} \sum_{b=0}^{L-1} abP(a, b) \qquad (16.2\text{-}13)$$

Covariance:

$$S_C = \sum_{a=0}^{L-1} \sum_{b=0}^{L-1} (a-\bar{a})(b-\bar{b})P(a,b) \tag{16.2-14a}$$

where

$$\bar{a} = \sum_{a=0}^{L-1} \sum_{b=0}^{L-1} aP(a,b) \tag{16.2-14b}$$

$$\bar{b} = \sum_{a=0}^{L-1} \sum_{b=0}^{L-1} bP(a,b) \tag{16.2-14c}$$

Inertia:

$$S_I = \sum_{a=0}^{L-1} \sum_{b=0}^{L-1} (a-b)^2 P(a,b) \tag{16.2-15}$$

Absolute value:

$$S_V = \sum_{a=0}^{L-1} \sum_{b=0}^{L-1} |a-b|P(a,b) \tag{16.2-16}$$

Inverse difference:

$$S_F = \sum_{a=0}^{L-1} \sum_{b=0}^{L-1} \frac{P(a,b)}{1+(a-b)^2} \tag{16.2-17}$$

Energy:

$$S_G = \sum_{a=0}^{L-1} \sum_{b=0}^{L-1} [P(a,b)]^2 \tag{16.2-18}$$

Entropy:

$$S_T = -\sum_{a=0}^{L-1} \sum_{b=0}^{L-1} P(a,b) \log_2\{P(a,b)\} \tag{16.2-19}$$

The utilization of second-order histogram measures for texture analysis is considered in Section 16.6.

16.3. TRANSFORM COEFFICIENT FEATURES

The coefficients of a two-dimensional transform of a luminance image specify the amplitude of the luminance patterns (two-dimensional *basis functions*) of a transform such that the weighted sum of the luminance patterns is identical to the image. By this characterization of a transform, the coefficients may be considered to indicate the degree of correspondence of a particular luminance pattern with an image field. If a basis pattern is of the same spatial form as a feature to be detected within the image, image detection can be performed simply by monitoring the value of the transform coefficient. The problem, in practice, is that objects to be detected within an image are often of complex shape and luminance distribution, and hence do not correspond closely to the more primitive luminance patterns of most image transforms.

Lendaris and Stanley (8) have investigated the application of the continuous two-dimensional Fourier transform of an image, obtained by a coherent optical processor, as a means of image feature extraction. The optical system produces an electric field radiation pattern proportional to

$$\mathcal{F}(\omega_x, \omega_y) = \int_{-\infty}^{\infty}\int_{-\infty}^{\infty} F(x, y) \exp\{-i(\omega_x x + \omega_y y)\} \, dx \, dy \qquad (16.3\text{-}1)$$

where (ω_x, ω_y) are the image spatial frequencies. An optical sensor produces an output

$$M(\omega_x, \omega_y) = \left| \mathcal{F}(\omega_x, \omega_y) \right|^2 \qquad (16.3\text{-}2)$$

proportional to the intensity of the radiation pattern. It should be observed that $\mathcal{F}(\omega_x, \omega_y)$ and $F(x, y)$ are unique transform pairs, but $M(\omega_x, \omega_y)$ is not uniquely related to $F(x, y)$. For example, $M(\omega_x, \omega_y)$ does not change if the origin of $F(x, y)$ is shifted. In some applications, the translation invariance of $M(\omega_x, \omega_y)$ may be a benefit. Angular integration of $M(\omega_x, \omega_y)$ over the spatial frequency plane produces a spatial frequency feature that is invariant to translation and rotation. Representing $M(\omega_x, \omega_y)$ in polar form, this feature is defined as

$$N(\rho) = \int_0^{2\pi} M(\rho, \theta) \, d\theta \qquad (16.3\text{-}3)$$

where $\theta = \arctan\{\omega_x / \omega_y\}$ and $\rho^2 = \omega_x^2 + \omega_y^2$. Invariance to changes in scale is an attribute of the feature

$$P(\theta) = \int_0^{\infty} M(\rho, \theta) \, d\rho \qquad (16.3\text{-}4)$$

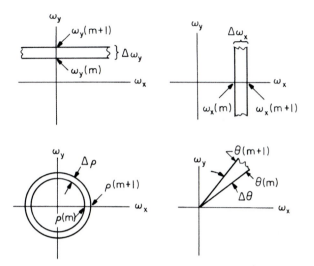

FIGURE 16.3-1. Fourier transform feature masks.

The Fourier domain intensity pattern $\mathcal{M}(\omega_x, \omega_y)$ is normally examined in specific regions to isolate image features. As an example, Figure 16.3-1 defines regions for the following Fourier features:

Horizontal slit:

$$S_1(m) = \int_{-\infty}^{\infty} \int_{\omega_y(m)}^{\omega_y(m+1)} \mathcal{M}(\omega_x, \omega_y) \, d\omega_x \, d\omega_y \qquad (16.3\text{-}5)$$

Vertical slit:

$$S_2(m) = \int_{\omega_x(m)}^{\omega_x(m+1)} \int_{-\infty}^{\infty} \mathcal{M}(\omega_x, \omega_y) \, d\omega_x \, d\omega_y \qquad (16.3\text{-}6)$$

Ring:

$$S_3(m) = \int_{\rho(m)}^{\rho(m+1)} \int_{0}^{2\pi} \mathcal{M}(\rho, \theta) \, d\rho \, d\theta \qquad (16.3\text{-}7)$$

Sector:

$$S_4(m) = \int_{0}^{\infty} \int_{\theta(m)}^{\theta(m+1)} \mathcal{M}(\rho, \theta) \, d\rho \, d\theta \qquad (16.3\text{-}8)$$

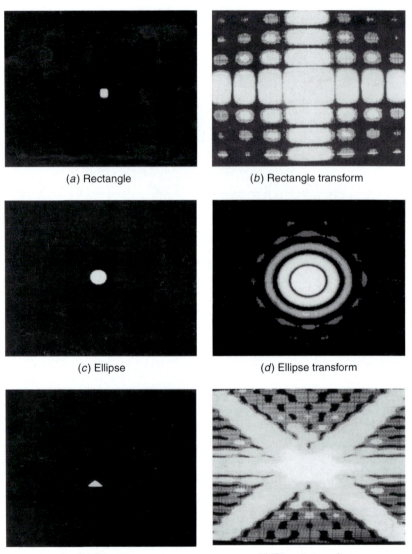

(a) Rectangle

(b) Rectangle transform

(c) Ellipse

(d) Ellipse transform

(e) Triangle

(f) Triangle transform

FIGURE 16.3-2. Discrete Fourier spectra of objects; log magnitude displays.

For a discrete image array $F(j, k)$, the discrete Fourier transform

$$\mathcal{F}(u, v) = \frac{1}{N} \sum_{j=0}^{N-1} \sum_{k=0}^{N-1} F(j, k) \exp\left\{\frac{-2\pi i}{N}(ux + vy)\right\} \qquad (16.3\text{-}9)$$

for $u, v = 0, ..., N-1$ can be examined directly for feature extraction purposes. Horizontal slit, vertical slit, ring, and sector features can be defined analogous to Eqs. 16.3-5 to 16.3-8. This concept can be extended to other unitary transforms, such as the Hadamard and Haar transforms. Figure 16.3-2 presents discrete Fourier transform log magnitude displays of several geometric shapes.

16.4. TEXTURE DEFINITION

Many portions of images of natural scenes are devoid of sharp edges over large areas. In these areas, the scene can often be characterized as exhibiting a consistent structure analogous to the texture of cloth. Image texture measurements can be used to segment an image and classify its segments.

Several authors have attempted qualitatively to define *texture*. Pickett (9) states that "texture is used to describe two dimensional arrays of variations... The elements and rules of spacing or arrangement may be arbitrarily manipulated, provided a characteristic repetitiveness remains." Hawkins (10) has provided a more detailed description of texture: "The notion of texture appears to depend upon three ingredients: (1) some local 'order' is repeated over a region which is large in comparison to the order's size, (2) the order consists in the nonrandom arrangement of elementary parts and (3) the parts are roughly uniform entities having approximately the same dimensions everywhere within the textured region." Although these descriptions of texture seem perceptually reasonably, they do not immediately lead to simple quantitative textural measures in the sense that the description of an edge discontinuity leads to a quantitative description of an edge in terms of its location, slope angle, and height.

Texture is often qualitatively described by its coarseness in the sense that a patch of wool cloth is coarser than a patch of silk cloth under the same viewing conditions. The coarseness index is related to the spatial repetition period of the local structure. A large period implies a coarse texture; a small period implies a fine texture. This perceptual coarseness index is clearly not sufficient as a quantitative texture measure, but can at least be used as a guide for the slope of texture measures; that is, small numerical texture measures should imply fine texture, and large numerical measures should indicate coarse texture. It should be recognized that texture is a neighborhood property of an image point. Therefore, texture measures are inherently dependent on the size of the observation neighborhood. Because texture is a spatial property, measurements should be restricted to regions of relative uniformity. Hence it is necessary to establish the boundary of a uniform textural region by some form of image segmentation before attempting texture measurements.

Texture may be classified as being artificial or natural. Artificial textures consist of arrangements of symbols, such as line segments, dots, and stars placed against a neutral background. Several examples of artificial texture are presented in Figure 16.4-1 (9). As the name implies, natural textures are images of natural scenes containing semirepetitive arrangements of pixels. Examples include photographs of brick walls, terrazzo tile, sand, and grass. Brodatz (11) has published an album of photographs of naturally occurring textures. Figure 16.4-2 shows several natural texture examples obtained by digitizing photographs from the Brodatz album.

FIGURE 16.4-1. Artificial texture.

(*a*) Sand (*b*) Grass

(*c*) Wool (*d*) Raffia

FIGURE 16.4-2. Brodatz texture fields.

16.5. VISUAL TEXTURE DISCRIMINATION

A discrete stochastic field is an array of numbers that are randomly distributed in amplitude and governed by some joint probability density (12). When converted to light intensities, such fields can be made to approximate natural textures surprisingly well by control of the generating probability density. This technique is useful for generating realistic appearing artificial scenes for applications such as airplane flight simulators. Stochastic texture fields are also an extremely useful tool for investigating human perception of texture as a guide to the development of texture feature extraction methods.

In the early 1960s, Julesz (13) attempted to determine the parameters of stochastic texture fields of perceptual importance. This study was extended later by Julesz et al. (14–16). Further extensions of Julesz's work have been made by Pollack (17),

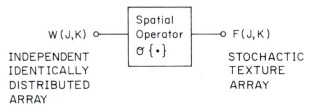

FIGURE 16.5-1. Stochastic texture field generation model.

Purks and Richards (18), and Pratt et al. (19). These studies have provided valuable insight into the mechanism of human visual perception and have led to some useful quantitative texture measurement methods.

Figure 16.5-1 is a model for stochastic texture generation. In this model, an array of independent, identically distributed random variables $W(j, k)$ passes through a linear or nonlinear spatial operator $O\{\cdot\}$ to produce a stochastic texture array $F(j, k)$. By controlling the form of the generating probability density $p(W)$ and the spatial operator, it is possible to create texture fields with specified statistical properties. Consider a continuous amplitude pixel x_0 at some coordinate (j, k) in $F(j, k)$. Let the set $\{z_1, z_2, ..., z_J\}$ denote neighboring pixels but not necessarily nearest geometric neighbors, raster scanned in a conventional top-to-bottom, left-to-right fashion. The conditional probability density of x_0 conditioned on the state of its neighbors is given by

$$p(x_0 | z_1, ..., z_J) = \frac{p(x_0, z_1, ..., z_J)}{p(z_1, ..., z_J)} \tag{16.5-1}$$

The first-order density $p(x_0)$ employs no conditioning, the second-order density $p(x_0 | z_1)$ implies that $J = 1$, the third-order density implies that $J = 2$, and so on.

16.5.1. Julesz Texture Fields

In his pioneering texture discrimination experiments, Julesz utilized Markov process state methods to create stochastic texture arrays independently along rows of the array. The family of Julesz stochastic arrays are defined below.

1. *Notation.* Let $x_n = F(j, k - n)$ denote a row neighbor of pixel x_0 and let $P(m)$, for $m = 1, 2,..., M$, denote a desired probability generating function.

2. *First-order process.* Set $x_0 = m$ for a desired probability function $P(m)$. The resulting pixel probability is

$$P(x_0) = P(x_0 = m) = P(m) \tag{16.5-2}$$

3. *Second-order process.* Set $F(j, 1) = m$ for $P(m) = 1/M$, and set $x_0 = (x_1 + m)\text{MOD}\{M\}$, where the modulus function $p\,\text{MOD}\{q\} \equiv p - [q \times (p \div q)]$ for integers p and q. This gives a first-order probability

$$P(x_0) = \frac{1}{M} \tag{16.5-3a}$$

and a transition probability

$$p(x_0|x_1) = P[x_0 = (x_1 + m)\,\text{MOD}\{M\}] = P(m) \tag{16.5-3b}$$

4. *Third-order process.* Set $F(j, 1) = m$ for $P(m) = 1/M$, and set $F(j, 2) = n$ for $P(n) = 1/M$. Choose x_0 to satisfy $2x_0 = (x_1 + x_2 + m)\,\text{MOD}\{M\}$. The governing probabilities then become

$$P(x_0) = \frac{1}{M} \tag{16.5-4a}$$

$$p(x_0|x_1) = \frac{1}{M} \tag{16.5-4b}$$

$$p(x_0|x_1, x_2) = P[2x_0 = (x_1 + x_2 + m)\,\text{MOD}\,\{M\}] = P(m) \tag{16.5-4c}$$

This process has the interesting property that pixel pairs along a row are independent, and consequently, the process is spatially uncorrelated.

Figure 16.5-2 contains several examples of Julesz texture field discrimination tests performed by Pratt et al. (19). In these tests, the textures were generated according to the presentation format of Figure 16.5-3. In these and subsequent visual texture discrimination tests, the perceptual differences are often small. Proper discrimination testing should be performed using high-quality photographic transparencies, prints, or electronic displays. The following moments were used as simple indicators of differences between generating distributions and densities of the stochastic fields.

$$\eta = E\{x_0\} \tag{16.5-5a}$$

$$\sigma^2 = E\{[x_0 - \eta]^2\} \tag{16.5-5b}$$

$$\alpha = \frac{E\{[x_0 - \eta][x_1 - \eta]\}}{\sigma^2} \tag{16.5-5c}$$

$$\theta = \frac{E\{[x_0 - \eta][x_1 - \eta][x_2 - \eta]\}}{\sigma^3} \tag{16.5-5d}$$

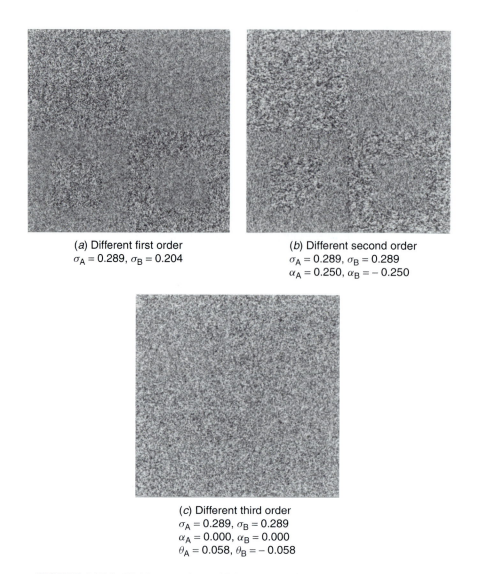

(a) Different first order
$\sigma_A = 0.289$, $\sigma_B = 0.204$

(b) Different second order
$\sigma_A = 0.289$, $\sigma_B = 0.289$
$\alpha_A = 0.250$, $\alpha_B = -0.250$

(c) Different third order
$\sigma_A = 0.289$, $\sigma_B = 0.289$
$\alpha_A = 0.000$, $\alpha_B = 0.000$
$\theta_A = 0.058$, $\theta_B = -0.058$

FIGURE 16.5-2. Field comparison of Julesz stochastic fields; $\eta_A = \eta_B = 0.500$.

The examples of Figure 16.5-2a and b indicate that texture field pairs differing in their first- and second-order distributions can be discriminated. The example of Figure 16.5-2c supports the conjecture, attributed to Julesz, that differences in third-order, and presumably, higher-order distribution texture fields cannot be perceived provided that their first-order and second- distributions are pairwise identical.

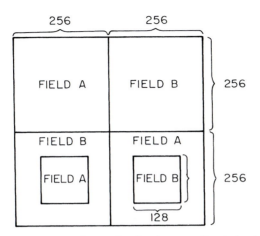

FIGURE 16.5-3. Presentation format for visual texture discrimination experiments.

16.5.2. Pratt, Faugeras, and Gagalowicz Texture Fields

Pratt et al. (19) have extended the work of Julesz et al. (13–16) in an attempt to study the discriminability of spatially correlated stochastic texture fields. A class of Gaussian fields was generated according to the conditional probability density

$$p(x_0 | z_1, \dots, z_J) = \frac{\left[(2\pi)^{J+1} |\mathbf{K}_{J+1}| \right]^{-1/2} \exp\left\{ -\frac{1}{2} (\mathbf{v}_{J+1} - \mathbf{\eta}_{J+1})^T (\mathbf{K}_{J+1})^{-1} (\mathbf{v}_{J+1} - \mathbf{\eta}_{J+1}) \right\}}{\left[(2\pi)^{J} |\mathbf{K}_{J}| \right]^{-1/2} \exp\left\{ -\frac{1}{2} (\mathbf{v}_{J} - \mathbf{\eta}_{J})^T (\mathbf{K}_{J})^{-1} (\mathbf{v}_{J} - \mathbf{\eta}_{J}) \right\}}$$

$$(16.5\text{-}6a)$$

where

$$\mathbf{v}_J = \begin{bmatrix} z_1 \\ \vdots \\ z_J \end{bmatrix}$$

$$(16.5\text{-}6b)$$

$$\mathbf{v}_{J+1} = \begin{bmatrix} x_0 \\ \mathbf{v}_J \end{bmatrix}$$

$$(16.5\text{-}6c)$$

The covariance matrix of Eq. 16.5-6a is of the parametric form

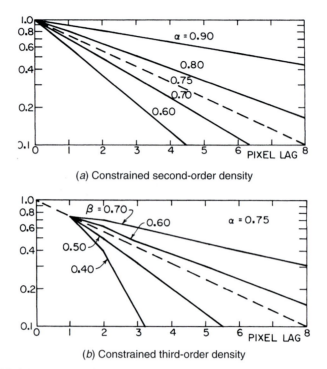

(a) Constrained second-order density

(b) Constrained third-order density

FIGURE 16.5-4. Row correlation factors for stochastic field generation. Dashed line, field A; solid line, field B.

$$\mathbf{K}_{J+1} = \begin{bmatrix} 1 & \alpha & \beta & \gamma & \cdots \\ \alpha & & & & \\ \beta & & \sigma^{-2}\mathbf{K}_J & & \\ \gamma & & & & \\ \vdots & & & & \end{bmatrix} \qquad (16.5\text{-}7)$$

where $\alpha, \beta, \gamma, \ldots$ denote correlation lag terms. Figure 16.5-4 presents an example of the row correlation functions used in the texture field comparison tests described below.

Figures 16.5-5 and 16.5-6 contain examples of Gaussian texture field comparison tests. In Figure 16.5-5, the first-order densities are set equal, but the second-order nearest neighbor conditional densities differ according to the covariance function plot of Figure 16.5-4a. Visual discrimination can be made in Figure 16.5-5, in which the correlation parameter differs by 20%. Visual discrimination has been found to be marginal when the correlation factor differs by less than 10% (19). The first- and second-order densities of each field are fixed in Figure 16.5-6, and the third-order

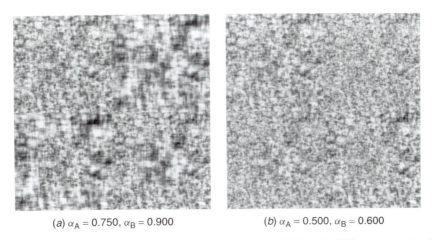

(a) $\alpha_A = 0.750$, $\alpha_B = 0.900$ (b) $\alpha_A = 0.500$, $\alpha_B = 0.600$

FIGURE 16.5-5. Field comparison of Gaussian stochastic fields with different second-order nearest neighbor densities; $\eta_A = \eta_B = 0.500$, $\sigma_A = \sigma_B = 0.167$.

conditional densities differ according to the plan of Figure 16.5-4b. Visual discrimination is possible. The test of Figure 16.5-6 seemingly provides a counterexample to the Julesz conjecture. In this test, $[p^A(x_0) = p^B(x_0)]$ and $p^A(x_0, x_1) = p^B(x_0, x_1)$, but $p^A(x_0, x_1, x_2) \neq p^B(x_0, x_1, x_2)$. However, the general second-order density pairs $p^A(x_0, z_j)$ and $p^B(x_0, z_j)$ are not necessarily equal for an arbitrary neighbor z_j, and therefore the conditions necessary to disprove Julesz's conjecture are violated.

To test the Julesz conjecture for realistically appearing texture fields, it is necessary to generate a pair of fields with identical first-order densities, identical Markovian type second-order densities, and differing third-order densities for every

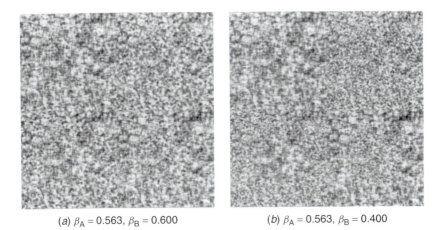

(a) $\beta_A = 0.563$, $\beta_B = 0.600$ (b) $\beta_A = 0.563$, $\beta_B = 0.400$

FIGURE 16.5-6. Field comparison of Gaussian stochastic fields with different third-order nearest neighbor densities; $\eta_A = \eta_B = 0.500$, $\sigma_A = \sigma_B = 0.167$, $\alpha_A = \alpha_B = 0.750$.

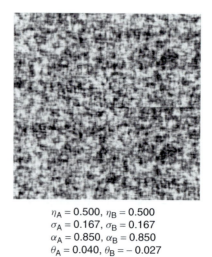

$$\eta_A = 0.500, \ \eta_B = 0.500$$
$$\sigma_A = 0.167, \ \sigma_B = 0.167$$
$$\alpha_A = 0.850, \ \alpha_B = 0.850$$
$$\theta_A = 0.040, \ \theta_B = -0.027$$

FIGURE 16.5-7. Field comparison of correlated Julesz stochastic fields with identical first- and second-order densities, but different third-order densities.

pair of similar observation points in both fields. An example of such a pair of fields is presented in Figure 16.5-7 for a non-Gaussian generating process (19). In this example, the texture appears identical in both fields, thus supporting the Julesz conjecture.

Gagalowicz has succeeded in generating a pair of texture fields that disprove the Julesz conjecture (20). However, the counterexample, shown in Figure 16.5-8, is not very realistic in appearance. Thus, it seems likely that if a statistically based texture measure can be developed, it need not utilize statistics greater than second-order.

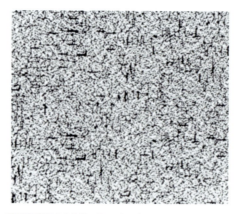

FIGURE 16.5-8. Gagalowicz counterexample.

$$\eta_A = 0.413, \ \eta_B = 0.412$$
$$\sigma_A = 0.078, \ \sigma_B = 0.078$$
$$\alpha_A = 0.915, \ \alpha_B = 0.917$$
$$\theta_A = 1.512, \ \theta_B = 0.006$$

FIGURE 16.5-9. Field comparison of correlated stochastic fields with identical means, variances, and autocorrelation functions, but different nth-order probability densities generated by different processing of the same input field. Input array consists of uniform random variables raised to the 256th power. Moments are computed.

Because a human viewer is sensitive to differences in the mean, variance, and autocorrelation function of the texture pairs, it is reasonable to investigate the sufficiency of these parameters in terms of texture representation. Figure 16.5-9 presents examples of the comparison of texture fields with identical means, variances, and autocorrelation functions, but different nth-order probability densities. Visual discrimination is readily accomplished between the fields. This leads to the conclusion that these low-order moment measurements, by themselves, are not always sufficient to distinguish texture fields.

16.6. TEXTURE FEATURES

As noted in Section 16.4, there is no commonly accepted quantitative definition of visual texture. As a consequence, researchers seeking a quantitative texture measure have been forced to search intuitively for texture features, and then attempt to evaluate their performance by techniques such as those presented in Section 16.1. The following subsections describe several texture features of historical and practical important. References 20 to 22 provide surveys on image texture feature extraction. Randen and Husoy (23) have performed a comprehensive study of many texture feature extraction methods.

16.6.1. Fourier Spectra Methods

Several studies (8,24,25) have considered textural analysis based on the Fourier spectrum of an image region, as discussed in Section 16.2. Because the degree of texture coarseness is proportional to its spatial period, a region of coarse texture should have its Fourier spectral energy concentrated at low spatial frequencies. Conversely, regions of fine texture should exhibit a concentration of spectral energy at high spatial frequencies. Although this correspondence exists to some degree, difficulties often arise because of spatial changes in the period and phase of texture pattern repetitions. Experiments (10) have shown that there is considerable spectral overlap of regions of distinctly different natural texture, such as urban, rural, and woodland regions extracted from aerial photographs. On the other hand, Fourier spectral analysis has proved successful (26,27) in the detection and classification of coal miner's black lung disease, which appears as diffuse textural deviations from the norm.

16.6.2. Edge Detection Methods

Rosenfeld and Troy (28) have proposed a measure of the number of edges in a neighborhood as a textural measure. As a first step in their process, an edge map array $E(j, k)$ is produced by some edge detector such that $E(j, k) = 1$ for a detected edge and $E(j, k) = 0$ otherwise. Usually, the detection threshold is set lower than the normal setting for the isolation of boundary points. This texture measure is defined as

$$T(j, k) = \frac{1}{W^2} \sum_{m = -w}^{w} \sum_{n = -w}^{w} E(j + m, k + n) \tag{16.6-1}$$

where $W = 2w + 1$ is the dimension of the observation window. A variation of this approach is to substitute the edge gradient $G(j, k)$ for the edge map array in Eq. 16.6-1. A generalization of this concept is presented in Section 16.6.4.

16.6.3. Autocorrelation Methods

The autocorrelation function has been suggested as the basis of a texture measure (28). Although it has been demonstrated in the preceding section that it is possible to generate visually different stochastic fields with the same autocorrelation function, this does not necessarily rule out the utility of an autocorrelation feature set for natural images. The *autocorrelation function* is defined as

$$A_F(m, n) = \sum_{j} \sum_{k} F(j, k)F(j - m, k - n) \tag{16.6-2}$$

for computation over a $W \times W$ window with $-T \le m, n \le T$ pixel lags. Presumably, a region of coarse texture will exhibit a higher correlation for a fixed shift (m, n) than will a region of fine texture. Thus, texture coarseness should be proportional to the spread of the autocorrelation function. Faugeras and Pratt (5) have proposed the following set of autocorrelation spread measures:

$$S(u, v) = \sum_{m=0}^{T} \sum_{n=-T}^{T} (m - \eta_m)^u (n - \eta_n)^v A_F(m, n) \tag{16.6-3a}$$

where

$$\eta_m = \sum_{m=0}^{T} \sum_{n=-T}^{T} m A_F(m, n) \tag{16.6-3b}$$

$$\eta_n = \sum_{m=0}^{T} \sum_{n=-T}^{T} n A_F(m, n) \tag{16.6-3c}$$

In Eq. 16.6-3, computation is only over one-half of the autocorrelation function because of its symmetry. Features of potential interest include the profile spreads $S(2, 0)$ and $S(0, 2)$, the cross-relation $S(1, 1)$, and the second-degree spread $S(2, 2)$.

Figure 16.6-1 shows perspective views of the autocorrelation functions of the four Brodatz texture examples (5). Bhattacharyya distance measurements of these texture fields, performed by Faugeras and Pratt (5), are presented in Table 16.6-1. These B-distance measurements indicate that the autocorrelation shape features are marginally adequate for the set of four shape features, but unacceptable for fewer features. Tests by Faugeras and Pratt (5) verify that the B-distances are low for

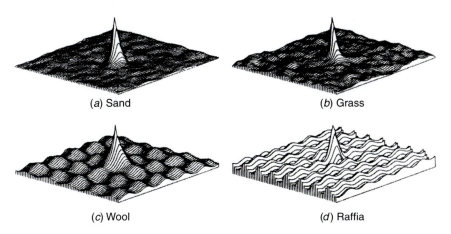

(a) Sand (b) Grass

(c) Wool (d) Raffia

FIGURE 16.6-1. Perspective views of autocorrelation functions of Brodatz texture fields.

TABLE 16.6-1. Bhattacharyya Distance of Texture Feature Sets for Prototype Texture Fields: Autocorrelation Features

Field Pair	Set 1[a]	Set 2[b]	Set 3[c]
Grass – sand	5.05	4.29	2.92
Grass – raffia	7.07	5.32	3.57
Grass – wool	2.37	0.21	0.04
Sand – raffia	1.49	0.58	0.35
Sand – wool	6.55	4.93	3.14
Raffia – wool	8.70	5.96	3.78
Average	5.21	3.55	2.30

[a]1: S(2, 0), S(0, 2), S(1, 1), S(2,2).
[b]2: S(1,1), S(2,2).
[c]3: S(2,2).

the stochastic field pairs of Figure 16.5-9, which have the same autocorrelation functions but are visually distinct.

16.6.4. Decorrelation Methods

Stochastic texture fields generated by the model of Figure 16.5-1 can be described quite compactly by specification of the spatial operator $O\{\cdot\}$ and the stationary first-order probability density $p(W)$ of the independent, identically distributed generating process $W(j, k)$. This observation has led to a texture feature extraction procedure, developed by Faugeras and Pratt (5), in which an attempt has been made to invert the model and estimate its parameters. Figure 16.6-2 is a block diagram of their decorrelation method of texture feature extraction. In the first step of the method, the spatial autocorrelation function $A_F(m, n)$ is measured over a texture field to be analyzed. The autocorrelation function is then used to develop a whitening filter, with an impulse response $H_W(j, k)$, using techniques described in Section 19.2. The *whitening filter* is a special type of *decorrelation operator*. It is used to generate the whitened field

$$\hat{W}(j, k) = F(j, k) \circledast H_W(j, k) \tag{16.6-4}$$

This whitened field, which is spatially uncorrelated, can be utilized as an estimate of the independent generating process $W(j, k)$ by forming its first-order histogram.

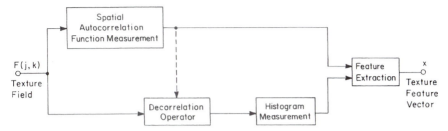

FIGURE 16.6-2. Decorrelation method of texture feature extraction.

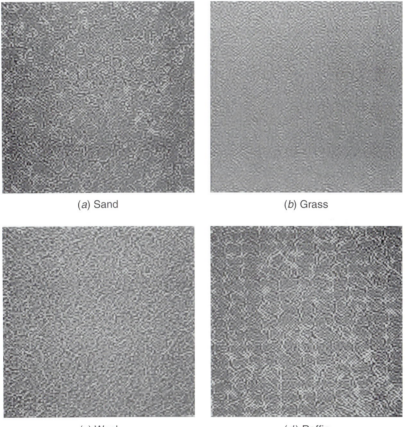

(a) Sand

(b) Grass

(c) Wool

(d) Raffia

FIGURE 16.6-3. Whitened Brodatz texture fields.

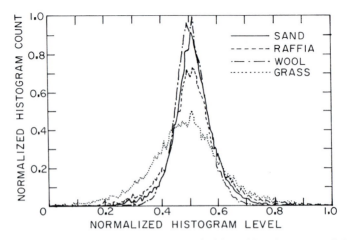

FIGURE 16.6-4. First-order histograms of whitened Brodatz texture fields.

If $W(j, k)$ were known exactly, then, in principle, it could be used to identify $O\{\cdot\}$ from the texture observation $F(j, k)$. But, the whitened field estimate $\hat{W}(j, k)$ can only be used to identify the autocorrelation function, which, of course, is already known. As a consequence, the texture generation model cannot be inverted. However, the shape of the histogram of $\hat{W}(j, k)$ augmented by the shape of the autocorrelation function have proved to be useful texture features.

Figure 16.6-3 shows the whitened texture fields of the Brodatz test images. Figure 16.6-4 provides plots of their histograms. The whitened fields are observed to be visually distinctive; their histograms are also different from one another. Tables 16.6-2 and 16.6-3 list, respectively, the Bhattacharyya distance measurements for histogram shape features alone, and histogram and autocorrelation shape features. The B-distance is relatively low for some of the test textures for histogram-only features. A combination of the autocorrelation shape and histogram shape features provides good results, as noted in Table 16.6-3.

An obvious disadvantage of the decorrelation method of texture measurement, as just described, is the large amount of computation involved in generating the whitening operator. An alternative is to use an approximate decorrelation operator. Two candidates, investigated by Faugeras and Pratt (5), are the Laplacian and Sobel gradients. Figure 16.6-5 shows the resultant decorrelated fields for these operators. The B-distance measurements using the Laplacian and Sobel gradients are presented in Tables 16.6-2 and 16.6-3. These tests indicate that the whitening operator is superior, on average, to the Laplacian operator. But the Sobel operator yields the largest average and largest minimum B-distances.

TABLE 16.6-2 Bhattacharyya Distance of Texture Feature Sets for Prototype Texture Fields: Histogram Features

	Texture Feature											
	Whitening				Laplacian				Sobel			
Field Pair	Set 1[a]	Set 2[b]	Set 3[c]	Set 4[d]	Set 1	Set 2	Set 3	Set 4	Set 1	Set 2	Set 3	Set 4
Grass – sand	4.61	4.52	4.04	0.77	1.29	1.28	0.19	0.66	9.90	7.15	4.41	2.31
Grass – raffia	1.15	1.04	0.51	0.52	3.48	3.38	0.55	1.87	2.20	1.00	0.27	0.02
Grass – wool	1.68	1.59	1.07	0.14	2.23	2.19	1.76	0.13	2.98	1.67	1.01	1.46
Sand – raffia	12.76	12.60	10.93	0.24	2.23	2.14	1.57	0.28	5.09	4.79	3.51	2.30
Sand – wool	12.61	12.55	8.24	2.19	7.73	7.65	7.42	1.40	9.98	5.01	1.67	0.56
Raffia – wool	4.20	3.87	0.39	1.47	4.59	4.43	1.53	3.13	7.73	2.31	0.41	1.41
Average	6.14	6.03	4.20	0.88	3.59	3.51	2.17	1.24	6.31	3.66	1.88	1.35

[a]Set 1: S_M, S_D, S_S, S_K.
[b]Set 2: S_S, S_K.
[c]Set 3: S_S.
[d]Set 4: S_K.

TABLE 16.6-3 Bhattacharyya Distance of Texture Feature Sets for Prototype Texture Fields: Autocorrelation and Histogram Features

Field Pair	Whitening				Texture Feature Laplacian				Sobel			
	Set 1[a]	Set 2[b]	Set 3[c]	Set 4	Set 1	Set 2	Set 3	Set 4	Set 1	Set 2	Set 3	Set 4
Grass – sand	9.80	9.72	8.94	7.48	6.39	6.37	5.61	4.21	15.34	12.34	11.48	10.12
Grass – raffia	8.47	8.34	6.56	4.66	10.61	10.49	8.74	6.95	9.46	8.15	6.33	4.59
Grass – wool	4.17	4.03	1.87	1.70	4.64	4.59	2.48	2.31	5.62	4.05	1.87	1.72
Sand – raffia	15.26	15.08	13.22	12.98	3.85	3.76	2.74	2.49	6.75	6.40	5.39	5.13
Sand – wool	19.14	19.08	17.43	15.72	14.43	14.38	12.72	10.86	18.75	12.3	10.52	8.29
Raffia – wool	13.29	13.14	10.32	7.96	13.93	13.75	10.90	8.47	17.28	11.19	8.24	6.08
Average	11.69	11.57	9.72	8.42	8.98	8.89	7.20	5.88	12.20	9.08	7.31	5.99

[a]Set 1: S_M, S_D, S_S, S_K, $S(2,0)$, $S(0,2)$, $S(1,1)$, $S(2,2)$.
[b]Set 2: S_S, S_K, $S(2,0)$, $S(0,2)$, $S(1,1)$, $S(2,2)$.
[c]Set 3: S_S, S_K, $S(1,1)$, $S(2,2)$.
[d]Set 4: S_S, S_K, $S(2,2)$.

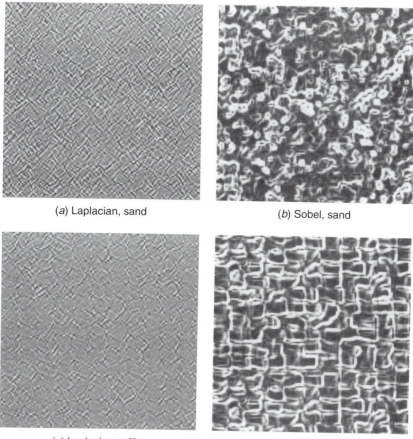

(a) Laplacian, sand

(b) Sobel, sand

(c) Laplacian, raffia

(d) Sobel, raffia

FIGURE 16.6-5. Laplacian and Sobel gradients of Brodatz texture fields.

16.6.5. Dependency Matrix Methods

Haralick et al. (7) have proposed a number of textural features based on the joint amplitude histogram of pairs of pixels. If an image region contains fine texture, the two-dimensional histogram of pixel pairs will tend to be uniform, and for coarse texture, the histogram values will be skewed toward the diagonal of the histogram. Consider the pair of pixels $F(j, k)$ and $F(m, n)$ that are separated by r radial units at an angle θ with respect to the horizontal axis. Let $P(a, b; j, k, r, \theta)$ represent the two-dimensional histogram measurement of an image over some $W \times W$ window where each pixel is quantized over a range $0 \leq a, b \leq L - 1$. The two-dimensional histogram can be considered as an estimate of the joint probability distribution

$$P(a, b; j, k, r, \theta) \approx P_R[F(j, k) = a, F(m, n) = b] \qquad (16.6\text{-}5)$$

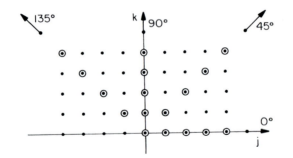

FIGURE 16.6-6. Geometry for measurement of gray scale dependency matrix.

For each member of the parameter set (j, k, r, θ), the two-dimensional histogram may be regarded as a $L \times L$ array of numbers relating the measured statistical dependency of pixel pairs. Such arrays have been called a *gray scale dependency matrix* or a *co-occurrence matrix*. Because a $L \times L$ histogram array must be accumulated for each image point (j, k) and separation set (r, θ) under consideration, it is usually computationally necessary to restrict the angular and radial separation to a limited number of values. Figure 16.6-6 illustrates geometrical relationships of histogram measurements made for four radial separation points and angles of $\theta = 0, \pi/4, \pi/2, 3\pi/4$ radians under the assumption of angular symmetry. To obtain statistical confidence in estimation of the joint probability distribution, the histogram must contain a reasonably large average occupancy level. This can be achieved either by restricting the number of amplitude quantization levels or by utilizing a relatively large measurement window. The former approach results in a loss of accuracy in the measurement of low-amplitude texture, while the latter approach causes errors if the texture changes over the large window. A typical compromise is to use 16 gray levels and a window of about 30 to 50 pixels on each side. Perspective views of joint amplitude histograms of two texture fields are presented in Figure 16.6-7.

For a given separation set (r, θ), the histogram obtained for fine texture tends to be more uniformly dispersed than the histogram for coarse texture. Texture coarseness can be measured in terms of the relative spread of histogram occupancy cells about the main diagonal of the histogram. Haralick et al. (7) have proposed a number of spread indicators for texture measurement. Several of these have been presented in Section 16.2. As an example, the inertia function of Eq. 16.2-15 results in a texture measure of the form

$$T(j, k, r, \theta) = \sum_{a=0}^{L-1} \sum_{b=0}^{L-1} (a - b)^2 P(a, b; j, k, r, \theta) \qquad (16.6\text{-}6)$$

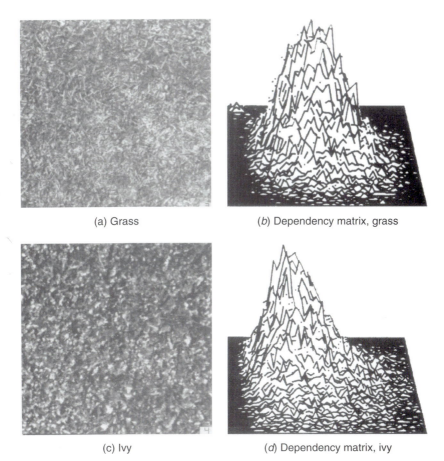

(a) Grass

(b) Dependency matrix, grass

(c) Ivy

(d) Dependency matrix, ivy

FIGURE 16.6-7. Perspective views of gray scale dependency matrices for $r = 4$, $\theta = 0$.

If the textural region of interest is suspected to be angularly invariant, it is reasonable to average over the measurement angles of a particular measure to produce the mean textural measure (20)

$$M_T(j, k, r) = \frac{1}{N_\theta} \sum_\theta T(j, k, r, \theta) \qquad (16.6\text{-}7)$$

where the summation is over the angular measurements, and N_θ represents the number of such measurements. Similarly, an angular-independent texture variance may be defined as

$$V_T(j, k, r) = \frac{1}{N_\theta} \sum_\theta [T(j, k, r, \theta) - M_T(j, k, r)]^2 \qquad (16.6\text{-}8)$$

$$M_i(j, k) = F(j, k) \circledast H_i(j, k)$$

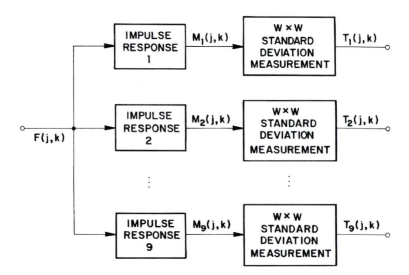

FIGURE 16.6-8. Laws microstructure texture feature extraction method.

Another useful measurement is the angular independent spread defined by

$$S(j, k, r) = \underset{\theta}{\text{MAX}}\{T(j, k, r, \theta)\} - \underset{\theta}{\text{MIN}}\{T(j, k, r, \theta)\} \qquad (16.6\text{-}9)$$

16.6.6. Microstructure Methods

Examination of the whitened, Laplacian, and Sobel gradient texture fields of Figures 16.6-3 and 16.6-5 reveals that they appear to accentuate the microstructure of the texture. This observation was the basis of a texture feature extraction scheme developed by Laws (29), and described in Figure 16.6-8. Laws proposed that the set of nine 3×3 pixel impulse response arrays $H_i(j, k)$ shown in Figure 16.6-9, be convolved with a texture field to accentuate its microstructure. The ith microstructure array is defined as

$$M_i(j, k) = F(j, k) \circledast H_i(j, k) \qquad (16.6\text{-}10)$$

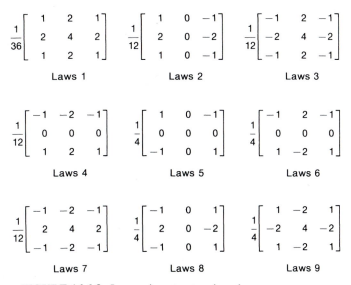

FIGURE 16.6-9. Laws microstructure impulse response arrays.

Then, the energy of these microstructure arrays is measured by forming their moving window standard deviation $T_i(j, k)$ according to Eq. 16.2-2, over a window that contains a few cycles of the repetitive texture.

Figure 16.6-10 shows a mosaic of several Brodatz texture fields that have been used to test the Laws feature extraction method. Note that some of the texture fields appear twice in the mosaic. Figure 16.6-11 illustrates the texture arrays $T_i(j, k)$. In classification tests of the Brodatz textures performed by Laws (29), the correct texture was identified in nearly 90% of the trials.

Many of the microstructure detection operators of Figure 16.6-9 have been encountered previously in this book: the pyramid average, the Sobel horizontal and vertical gradients, the weighted line horizontal and vertical gradients, and the cross second derivative. The nine Laws operators form a basis set that can be generated from all outer product combinations of the three vectors

$$\mathbf{v}_1 = \frac{1}{6} \begin{bmatrix} 1 \\ 2 \\ 1 \end{bmatrix} \qquad (16.6\text{-}11a)$$

$$\mathbf{v}_2 = \frac{1}{2} \begin{bmatrix} 1 \\ 0 \\ -1 \end{bmatrix} \qquad (16.6\text{-}11b)$$

FIGURE 16.6-10. Mosaic of Brodatz texture fields.

$$\mathbf{v}_3 = \frac{1}{2}\begin{bmatrix} 1 \\ -2 \\ 1 \end{bmatrix} \tag{16.6-11c}$$

Alternatively, the 3×3 Chebyshev basis set proposed by Haralick (30) for edge detection, as described in Section 16.3.3, can be used for texture measurement. The first Chebyshev basis vector is $\mathbf{v}_1 = \frac{1}{3}\begin{bmatrix} 1 & 1 & 1 \end{bmatrix}^T$. The other two are identical to Eqs. 16.6-11b and 16.6-11c. The Laws procedure can be extended by using larger size Chebyshev arrays or other types of basis arrays (31).

Ade (32) has suggested a microstructure texture feature extraction procedure similar in nature to the Laws method, which is based on a principal components transformation of a texture sample. In the development of this transformation, pixels within a 3×3 neighborhood are regarded as being column stacked into a 9×1 vector, as shown in Figure 16.6-12a. Then a 9×9 covariance matrix $\mathbf{K}$ that specifies all pairwise covariance relationships of pixels within the stacked vector is estimated from a set of prototype texture fields. Next, a 9×9 transformation matrix $\mathbf{T}$ that diagonalizes the covariance matrix $\mathbf{K}$ is computed, as described in Eq. 5.8-7. The rows of $\mathbf{T}$ are eigenvectors of the principal components transformation. Each eigen

(a) Laws no. 1

(b) Laws no. 2

(c) Laws no. 3

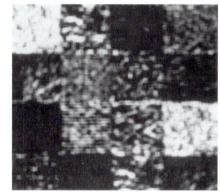

(d) Laws no. 4

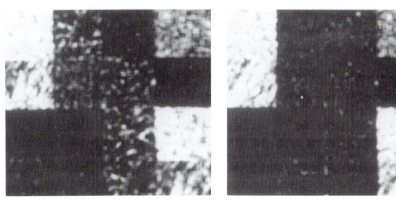

(e) Laws no. 5

(f) Laws no. 6

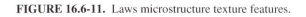

FIGURE 16.6-11. Laws microstructure texture features.

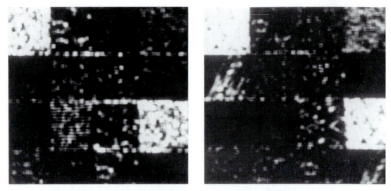

(g) Laws no. 7   (h) Laws no. 8

(i) Laws no. 9

FIGURE 16.6-11. (*continued*) Laws microstructure texture features.

vector is then cast into a 3×3 impulse response array by the destacking operation of Eq. 5.3-4. The resulting nine eigenmatrices are then used in place of the Laws fixed impulse response arrays, as shown in Figure 16.6-8. Ade (32,33) has computed eigenmatrices for a Brodatz texture field and a cloth sample. Interestingly, these eigenmatrices are similar in structure to the Laws arrays.

16.6.7. Gabor Filter Methods

The microstructure method of texture feature extraction is not easily scalable. Microstructure arrays must be derived to match the inherent periodicity of each texture to be characterized. Bovik et al. (34–36) have utilized Gabor filters (37) as an efficient means of scaling the impulse response function arrays of Figure 16.6-8 to the texture periodicity. A two-dimensional *Gabor filter* is a complex field sinuso-idal grating that is modulated by a two-dimensional Gaussian function in the spatial

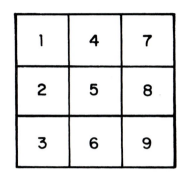

(a) 3 × 3 neighborhood

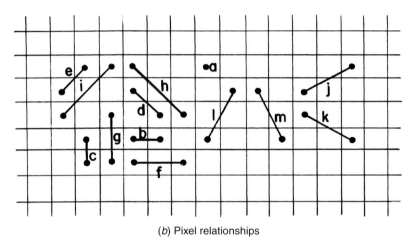

(b) Pixel relationships

FIGURE 16.6-12. Neighborhood covariance relationships.

domain (35). Gabor filters have tunable orientation and radial frequency passbands and tunable center frequencies. A special case of the Gabor filter is the *daisy petal filter*, in which the filter lobes radiate from the origin of the spatial frequency domain. The continuous domain impulse response function of the daisy petal Gabor filter is given by (35)

$$H(x, y) = G(x', y') \exp\{2\pi i F x'\} \tag{16.6-12}$$

where F is a scaling factor and $i = \sqrt{-1}$. The Gaussian component is

$$G(x, y) = \frac{1}{2\pi\lambda\sigma^2} \exp\left\{-\frac{(x/\lambda)^2 + y^2}{2\sigma^2}\right\} \tag{16.6-13}$$

where σ is the Gaussian spread factor and λ is the aspect ratio between the x and y axes. The rotation of coordinates is specified by

$$(x', y') = (x \cos \phi + y \sin \phi, -x \sin \phi + y \cos \phi) \tag{16.6-14}$$

where ϕ is the orientation angle with respect to the x axis. The continuous domain filter transfer function is given by (35)

$$\mathcal{H}(u, v) = \exp\{-2\pi^2\sigma^2[(u' - F)^2 + (v')^2]\} \tag{16.6-15}$$

Figure 16.6-13 shows the relationship between the real and imaginary components of the impulse response array and the magnitude of the transfer function (35).

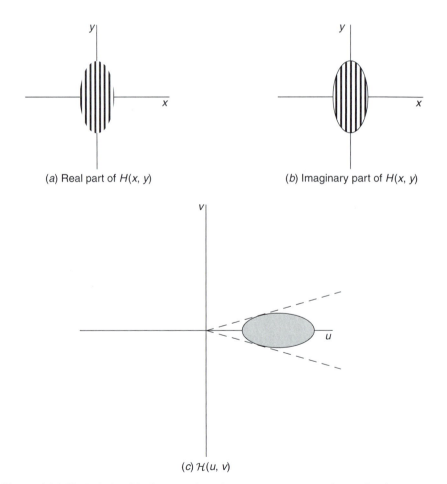

(a) Real part of $H(x, y)$ (b) Imaginary part of $H(x, y)$

(c) $\mathcal{H}(u, v)$

Figure 16.6-13. Relationship between impulse response array and transfer function of a Gabor filter.

The impulse response array is composed of sine-wave gratings within the elliptical region. The half energy profile of the transfer function is shown in gray.

In the comparative study of texture classification methods by Randen and Husoy (23), The Gabor filter method, like many other methods, gave mixed results. It performed well on some texture samples, but poorly on others.

16.6.8. Transform and Wavelet Methods

The Fourier spectra method of texture feature extraction can be generalized to other unitary transforms. The concept is straightforward. A $N \times N$ texture sample is subdivided into $M \times M$ pixel arrays, and a unitary transform is performed for each array yielding a $M^2 \times 1$ feature vector. The window size needs to large enough to contain several cycles of the texture periodicity.

Mallat (38) has used the discrete wavelet transform, based on Haar wavelets (see Section 8.4.2) as a means of generating texture feature vectors. Improved results have been obtained by Unser (39), who has used a complete Haar-based wavelet transform for an 8×8 window. In their comparative study of texture classification, Randen and Husoy (23) used several types of Daubechies transforms up to size 10 (see Section 8.4-4).

The transform and wavelet methods provide reasonably good classification for many texture samples (23). However, the computational requirement is high for large windows.

16.6.9. Singular-Value Decomposition Methods

Ashjari (40) has proposed a texture measurement method based on the singular-value decomposition of a texture sample. In this method, a $N \times N$ texture sample is treated as a $N \times N$ matrix $\mathbf{X}$ and the amplitude-ordered set of singular values $s(n)$ for $n = 1, 2, \ldots, N$ is computed, as described in Appendix A1.2. If the elements of $\mathbf{X}$ are spatially unrelated to one another, the singular values tend to be uniformly distributed in amplitude. On the other hand, if the elements of $\mathbf{X}$ are highly structured, the singular-value distribution tends to be skewed such that the lower-order singular values are much larger than the higher-order ones.

Figure 16.6-14 contains measurements of the singular-value distributions of the four Brodatz textures performed by Ashjari (40). In this experiment, the 512×512 pixel texture originals were first subjected to a statistical rescaling process to produce four normalized texture images whose first-order distributions were Gaussian with identical moments. Next, these normalized texture images were subdivided into 196 nonoverlapping 32×32 pixel blocks, and an SVD transformation was taken of each block. Figure 16.6-14 is a plot of the average value of each singular value. The shape of the singular-value distributions can be quantified by the one-dimensional shape descriptors defined in Section 16.2. Table 16.6-4 lists Bhattacharyya distance measurements obtained by Ashjari (30) for the mean, standard deviation, skewness, and kurtosis shape descriptors. For this experiment, the B-distances are relatively high, and therefore good classification results should be expected.

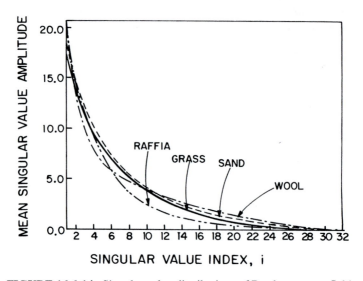

FIGURE 16.6-14. Singular-value distributions of Brodatz texture fields.

TABLE 16.6-4. Bhattacharyya Distance of SVD Texture Feature Sets for Prototype Texture Fields: SVD Features

Field Pair	
Grass – sand	1.25
Grass – raffia	2.42
Grass – wool	3.31
Sand – raffia	6.33
Sand – wool	2.56
Raffia – wool	9.24
Average	4.19

REFERENCES

1. H. C. Andrews, *Introduction to Mathematical Techniques in Pattern Recognition*, Wiley-Interscience, New York, 1972.

2. R. O. Duda, P. E. Hart, and D. G. Stork, *Pattern Classification,* 2nd ed., Wiley-Interscience, New York, 2001.

3. K. Fukunaga, *Introduction to Statistical Pattern Recognition*, 2nd ed., Academic Press, New York, 1990.

4. W. S. Meisel, *Computer-Oriented Approaches to Pattern Recognition*, Academic Press, New York, 1972.

5. O. D. Faugeras and W. K. Pratt, "Decorrelation Methods of Texture Feature Extraction," *IEEE Trans. Pattern Analysis and Machine Intelligence*, **PAMI-2**, 4, July 1980, 323–332.

6. R. O. Duda, "Image Data Extraction," unpublished notes, July 1975.

7. R. M. Haralick, K. Shanmugan, and I. Dinstein, "Texture Features for Image Classification," *IEEE Trans. Systems, Man and Cybernetics*, **SMC-3**, November 1973, 610–621.

8. G. G. Lendaris and G. L. Stanley, "Diffraction Pattern Sampling for Automatic Pattern Recognition," *Proc. IEEE*, **58**, 2, February 1970, 198–216.

9. R. M. Pickett, "Visual Analysis of Texture in the Detection and Recognition of Objects," in *Picture Processing and Psychopictorics*, B. C. Lipkin and A. Rosenfeld, Eds., Academic Press, New York, 1970, 289–308.

10. J. K. Hawkins, "Textural Properties for Pattern Recognition," in *Picture Processing and Psychopictorics*, B. C. Lipkin and A. Rosenfeld, Eds., Academic Press, New York, 1970, 347–370.

11. P. Brodatz, *Texture: A Photograph Album for Artists and Designers*, Dover Publications, New York, 1956.

12. J. W. Woods, "Two-Dimensional Discrete Markov Random Fields," *IEEE Trans. Information Theory*, **IT-18**, 2, March 1972, 232–240.

13. B. Julesz, "Visual Pattern Discrimination," *IRE Trans. Information Theory*, **IT-8**, 1, February 1962, 84–92.

14. B. Julesz et al., "Inability of Humans to Discriminate Between Visual Textures That Agree in Second-Order Statistics Revisited," *Perception*, **2**, 1973, 391–405.

15. B. Julesz, *Foundations of Cyclopean Perception*, University of Chicago Press, Chicago, 1971.

16. B. Julesz, "Experiments in the Visual Perception of Texture," *Scientific American*, **232**, 4, April 1975, 2–11.

17. I. Pollack, *Perceptual Psychophysics*, **13**, 1973, 276–280.

18. S. R. Purks and W. Richards, "Visual Texture Discrimination Using Random-Dot Patterns," *J. Optical Society America*, **67**, 6, June 1977, 765–771.

19. W. K. Pratt, O. D. Faugeras, and A. Gagalowicz, "Visual Discrimination of Stochastic Texture Fields," *IEEE Trans. Systems, Man and Cybernetics*, **SMC-8**, 11, November 1978, 796–804.

20. E. L. Hall et al., "A Survey of Preprocessing and Feature Extraction Techniques for Radiographic Images," *IEEE Trans. Computers*, **C-20**, 9, September 1971, 1032–1044.

21. R. M. Haralick, "Statistical and Structural Approach to Texture," *Proc. IEEE*, **67**, 5, May 1979, 786–804.

22. T. R. Reed and J. M. H. duBuf, "A Review of Recent Texture Segmentation and Feature Extraction Techniques," *CVGIP: Image Understanding*, **57**, May 1993, 358–372.

23. T. Randen and J. H. Husoy, "Filtering for Classification: A Comparative Study," *IEEE Trans. Pattern Analysis and Machine Intelligence*, **PAMI 21**, 4, April 1999, 291–310.

24. A. Rosenfeld, "Automatic Recognition of Basic Terrain Types from Aerial Photographs," *Photogrammic Engineering*, **28**, 1, March 1962, 115–132.

25. J. M. Coggins and A. K. Jain, "A Spatial Filtering Approach to Texture Analysis," *Pattern Recognition Letters*, **3**, 3, 1985, 195–203.

26. R. P. Kruger, W. B. Thompson, and A. F. Turner, "Computer Diagnosis of Pneumoconiosis," *IEEE Trans. Systems, Man and Cybernetics*, **SMC-4**, 1, January 1974, 40–49.

27. R. N. Sutton and E. L. Hall, "Texture Measures for Automatic Classification of Pulmonary Disease," *IEEE Trans. Computers*, **C-21**, July 1972, 667–676.

28. A. Rosenfeld and E. B. Troy, "Visual Texture Analysis," *Proc. UMR–Mervin J. Kelly Communications Conference*, University of Missouri–Rolla, Rolla, MO, October 1970, Sec. 10-1.

29. K. I. Laws, "Textured Image Segmentation," USCIPI Report 940, University of Southern California, Image Processing Institute, Los Angeles, January 1980.

30. R. M. Haralick, "Digital Step Edges from Zero Crossing of Second Directional Derivatives," *IEEE Trans. Pattern Analysis and Machine Intelligence*, **PAMI-6**, 1, January 1984, 58–68.

31. M. Unser and M. Eden, "Multiresolution Feature Extraction and Selection for Texture Segmentation," *IEEE Trans. Pattern Analysis and Machine Intelligence*, **PAMI-11**, 7, July 1989, 717–728.

32. F. Ade, "Characterization of Textures by Eigenfilters," *Signal Processing*, September 1983.

33. F. Ade, "Application of Principal Components Analysis to the Inspection of Industrial Goods," *Proc. SPIE International Technical Conference/Europe*, Geneva, April 1983.

34. M. Clark and A. C. Bovik, "Texture Discrimination Using a Model of Visual Cortex," *Proc. IEEE International Conference on Systems, Man and Cybernetics*, Atlanta, GA, 1986

35. A. C. Bovik, M. Clark, and W. S. Geisler, "Multichannel Texture Analysis Using Localized Spatial Filters," *IEEE Trans. Pattern Analysis and Machine Intelligence*, **PAMI-12**, 1, January 1990, 55–73.

36. A. C. Bovik, "Analysis of Multichannel Narrow-Band Filters for Image Texture Segmentation," *IEEE Trans. Signal Processing,* **39**, 9, September 1991, 2025–2043.

37. D. Gabor, "Theory of Communication," *J. Institute of Electrical Engineers*, **93**, 1946, 429–457.

38. S. G. Mallat, "A Theory for Multiresolution Signal Decomposition: The Wavelet Representation," *IEEE Trans. Pattern Analysis and Machine Intelligence*, **PAMI-11**, 7, July 1989, 674–693.

39. M. Unser, "Texture Classification and Segmentation Using Wavelet Frames," *IEEE Trans. Image Processing*, **IP-4**, 11, November 1995, 1549–1560.

40. B. Ashjari, "Singular Value Decomposition Texture Measurement for Image Classification," Ph.D. dissertation, University of Southern California, Department of Electrical Engineering, Los Angeles February 1982.

17

IMAGE SEGMENTATION

Segmentation of an image entails the division or separation of the image into regions of similar attribute. The most basic attribute for segmentation is image luminance amplitude for a monochrome image and color components for a color image. Image edges and texture are also useful attributes for segmentation.

The definition of segmentation adopted in this chapter is deliberately restrictive; no contextual information is utilized in the segmentation. Furthermore, segmentation does not involve classifying each segment. The segmenter only subdivides an image; it does not attempt to recognize the individual segments or their relationships to one another.

There is no theory of image segmentation. As a consequence, no single standard method of image segmentation has emerged. Rather, there are a collection of ad hoc methods that have received some degree of popularity. Because the methods are ad hoc, it would be useful to have some means of assessing their performance. Haralick and Shapiro (1) have established the following qualitative guideline for a good image segmentation: "Regions of an image segmentation should be uniform and homogeneous with respect to some characteristic such as gray tone or texture. Region interiors should be simple and without many small holes. Adjacent regions of a segmentation should have significantly different values with respect to the characteristic on which they are uniform. Boundaries of each segment should be simple, not ragged, and must be spatially accurate." Unfortunately, no quantitative image segmentation performance metric has been developed.

Several generic methods of image segmentation are described in the following sections. Because of their complexity, it is not feasible to describe all the details of the various algorithms. Surveys of image segmentation methods are given in References 1 to 6.

17.1. AMPLITUDE SEGMENTATION METHODS

This section considers several image segmentation methods based on the thresholding of luminance or color components of an image. An amplitude projection segmentation technique is also discussed.

17.1.1. Bilevel Luminance Thresholding

Many images can be characterized as containing some object of interest of reasonably uniform brightness placed against a background of differing brightness. Typical examples include handwritten and typewritten text, microscope biomedical samples, and airplanes on a runway. For such images, luminance is a distinguishing feature that can be utilized to segment the object from its background. If an object of interest is white against a black background, or vice versa, it is a trivial task to set a midgray threshold to segment the object from the background. Practical problems occur, however, when the observed image is subject to noise and when both the object and background assume some broad range of gray scales. Another frequent difficulty is that the background may be nonuniform.

Figure 17.1-1*a* shows a digitized typewritten text consisting of dark letters against a lighter background. A gray scale histogram of the text is presented in Figure 17.1-1*b*. The expected bimodality of the histogram is masked by the relatively large percentage of background pixels. Figure 17.1-1*c* to *e* are threshold displays in which all pixels brighter than the threshold are mapped to unity display luminance and all the remaining pixels below the threshold are mapped to the zero level of display luminance. The photographs illustrate a common problem associated with image thresholding. If the threshold is set too low, portions of the letters are deleted (the stem of the letter "p" is fragmented). Conversely, if the threshold is set too high, object artifacts result (the loop of the letter "e" is filled in).

Several analytic approaches to the setting of a luminance threshold have been proposed (7,8). One method is to set the gray scale threshold at a level such that the cumulative gray scale count matches an a priori assumption of the gray scale probability distribution (9). For example, it may be known that black characters cover 25% of the area of a typewritten page. Thus, the threshold level on the image might be set such that the quartile of pixels with the lowest luminance are judged to be black. Another approach to luminance threshold selection is to set the threshold at the minimum point of the histogram between its bimodal peaks (10). Determination of the minimum is often difficult because of the jaggedness of the histogram. A solution to this problem is to fit the histogram values between the peaks with some analytic function and then obtain its minimum by differentiation. For example, let *y* and *x* represent the histogram ordinate and abscissa, respectively. Then the quadratic curve

$$y = ax^2 + bx + c \qquad (17.1-1)$$

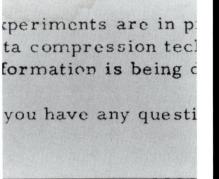

(a) Gray scale text

(b) Histogram

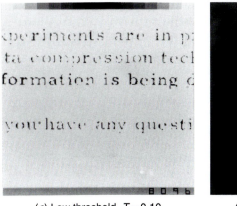

(c) High threshold, T = 0.67

(d) Medium threshold, T = 0.50

(e) Low threshold, T = 0.10

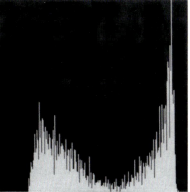

(f) Histogram, Laplacian mask

FIGURE 17.1-1. Luminance thresholding segmentation of typewritten text.

where a, b, and c are constants provides a simple histogram approximation in the vicinity of the histogram valley. The minimum histogram valley occurs for $x = -b/2a$. Papamarkos and Gatos (11) have extended this concept for threshold selection.

Weska et al. (12) have suggested the use of a Laplacian operator to aid in luminance threshold selection. As defined in Eq. 15.3-1, the Laplacian forms the spatial second partial derivative of an image. Consider an image region in the vicinity of an object in which the luminance increases from a low plateau level to a higher plateau level in a smooth ramplike fashion. In the flat regions and along the ramp, the Laplacian is zero. Large positive values of the Laplacian will occur in the transition region from the low plateau to the ramp; large negative values will be produced in the transition from the ramp to the high plateau. A gray scale histogram formed of only those pixels of the original image that lie at coordinates corresponding to very high or low values of the Laplacian tends to be bimodal with a distinctive valley between the peaks. Figure 17.1-1f shows the histogram of the text image of Figure 17.1-1a after the Laplacian mask operation.

If the background of an image is nonuniform, it often is necessary to adapt the luminance threshold to the mean luminance level (13,14). This can be accomplished by subdividing the image into small blocks and determining the best threshold level for each block by the methods discussed previously. Threshold levels for each pixel may then be determined by interpolation between the block centers. Yankowitz and Bruckstein (15) have proposed an adaptive thresholding method in which a threshold surface is obtained by interpolating an image only at points where its gradient is large.

17.1.2. Multilevel Luminance Thresholding

Effective segmentation can be achieved in some classes of images by a recursive *multilevel thresholding* method suggested by Tomita et al. (16). In the first stage of the process, the image is thresholded to separate brighter regions from darker regions by locating a minimum between luminance modes of the histogram. Then histograms are formed of each of the segmented parts. If these histograms are not unimodal, the parts are thresholded again. The process continues until the histogram of a part becomes unimodal. Figures 17.1-2 to 17.1-4 provide an example of this form of amplitude segmentation in which the peppers image is segmented into four gray scale segments.

17.1.3. Multilevel Color Component Thresholding

The multilevel luminance thresholding concept can be extended to the segmentation of color and multispectral images. Ohlander et al. (17, 18) have developed a segmentation scheme for natural color images based on multidimensional thresholding of color images represented by their *RGB* color components, their luma/chroma *YIQ* components, and by a set of nonstandard color components, loosely called intensity,

(a) Original

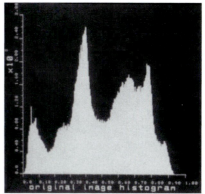

(b) Original histogram

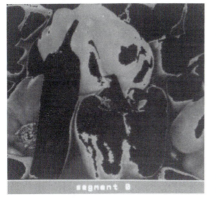

(c) Segment 0

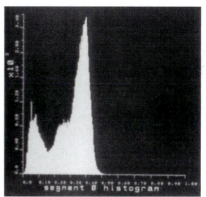

(d) Segment 0 histogram

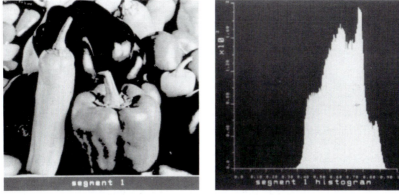

(e) Segment 1

(f) Segment 1 histogram

FIGURE 17.1-2. Multilevel luminance thresholding image segmentation of the peppers_
mon image; first-level segmentation.

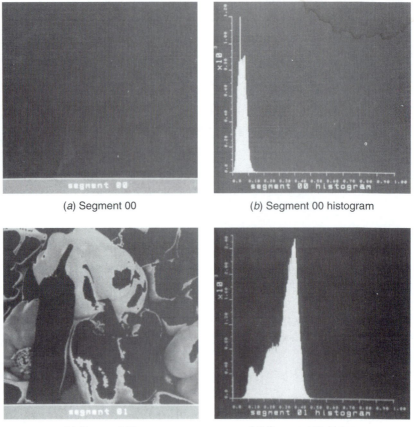

(a) Segment 00

(b) Segment 00 histogram

(c) Segment 01

(d) Segment 01 histogram

FIGURE 17.1-3. Multilevel luminance thresholding image segmentation of the peppers_ mon image; second-level segmentation, 0 branch.

hue, and saturation. Figure 17.1-5 provides an example of the *property histograms* of these nine color components for a scene. The histograms, have been measured over those parts of the original scene that are relatively devoid of texture: the non-busy parts of the scene. This important step of the segmentation process is necessary to avoid false segmentation of homogeneous textured regions into many isolated parts. If the property histograms are not all unimodal, an ad hoc procedure is invoked to determine the best property and the best level for thresholding of that property. The first candidate is image intensity. Other candidates are selected on a priority basis, depending on contrast level and location of the histogram modes. After a threshold level has been determined, the image is subdivided into its segmented parts. The procedure is then repeated on each part until the resulting property histograms become unimodal or the segmentation reaches a reasonable

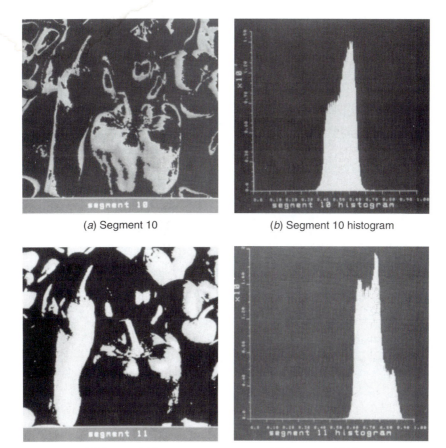

(*a*) Segment 10

(*b*) Segment 10 histogram

(*c*) Segment 11

(*d*) Segment 11 histogram

FIGURE 17.1-4. Multilevel luminance thresholding image segmentation of the peppers_ mon image; second-level segmentation, 1 branch.

stage of separation under manual surveillance. Ohlander's segmentation technique using multidimensional thresholding aided by texture discrimination has proved quite effective in simulation tests. However, a large part of the segmentation control has been performed by a human operator; human judgment, predicated on trial threshold setting results, is required for guidance.

In Ohlander's segmentation method, the nine property values are obviously interdependent. The *YIQ* and intensity components are linear combinations of *RGB*; the hue and saturation measurements are nonlinear functions of *RGB*. This observation raises several questions. What types of linear and nonlinear transformations of *RGB* are best for segmentation? Ohta et al. (19) suggest an approximation to the spectral Karhunen–Loeve transform. How many property values should be used? What is the best form of property thresholding? Perhaps answers to these last two questions may

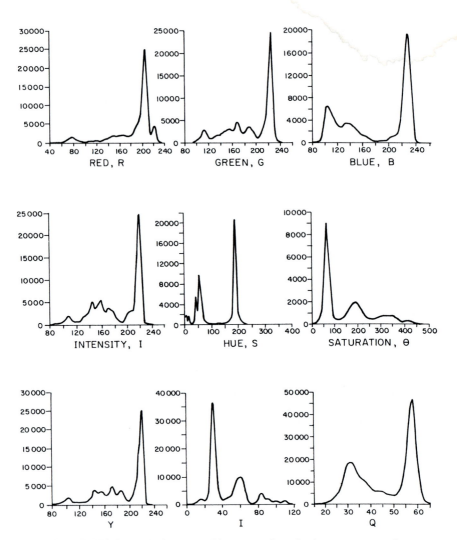

FIGURE 17.1-5. Typical property histograms for color image segmentation.

be forthcoming from a study of clustering techniques in pattern recognition (20). Property value histograms are really the marginal histograms of a joint histogram of property values. Clustering methods can be utilized to specify multidimensional decision boundaries for segmentation. This approach permits utilization of all the property values for segmentation and inherently recognizes their respective cross correlation. The following section discusses clustering methods of image segmentation.

17.1.4. Amplitude Projection

Image segments can sometimes be effectively isolated by forming the average *amplitude projections* of an image along its rows and columns (21,22). The horizontal (row) and vertical (column) projections are defined as

$$H(k) = \frac{1}{J} \sum_{j=1}^{J} F(j, k) \tag{17.1-2}$$

and

$$V(j) = \frac{1}{K} \sum_{k=1}^{K} F(j, k) \tag{17.1-3}$$

Figure 17.1-6 illustrates an application of gray scale projection segmentation of an image. The rectangularly shaped segment can be further delimited by taking projections over oblique angles.

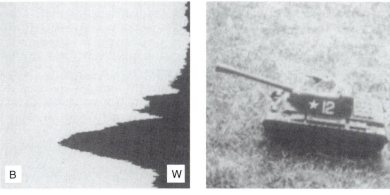

(*a*) Row projection

(*b*) Original

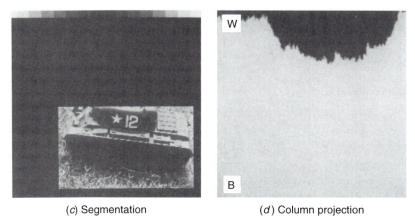

(*c*) Segmentation

(*d*) Column projection

FIGURE 17.1-6. Gray scale projection image segmentation of a toy tank image.

17.2. CLUSTERING SEGMENTATION METHODS

One of the earliest examples of image segmentation, by Haralick and Kelly (23) using data clustering, was the subdivision of multispectral aerial images of agricultural land into regions containing the same type of land cover. The clustering segmentation concept is simple; however, it is usually computationally intensive.

Consider a vector $\mathbf{x} = [x_1, x_2, ..., x_N]^T$ of measurements at each pixel coordinate (j, k) in an image. The measurements could be point multispectral values, point color components, and derived color components, as in the Ohlander approach described previously, or they could be neighborhood feature measurements such as the moving window mean, standard deviation, and mode, as discussed in Section 16.2. If the measurement set is to be effective for image segmentation, data collected at various pixels within a segment of common attribute should be similar. That is, the data should be tightly clustered in an N-dimensional measurement space. If this condition holds, the segmenter design task becomes one of subdividing the N-dimensional measurement space into mutually exclusive compartments, each of which envelopes typical data clusters for each image segment. Figure 17.2-1 illustrates the concept for two features. In the segmentation process, if a measurement vector for a pixel falls within a measurement space compartment, the pixel is assigned the segment name or label of that compartment.

Coleman and Andrews (24) have developed a robust and relatively efficient image segmentation clustering algorithm. Figure 17.2-2 is a flowchart that describes a simplified version of the algorithm for segmentation of monochrome images. The first stage of the algorithm involves feature computation. In one set of experiments, Coleman and Andrews used 12 mode measurements in square windows of size 1, 3, 7, and 15 pixels. The next step in the algorithm is the clustering stage, in which the optimum number of clusters is determined along with the feature space center of each cluster. In the segmenter, a given feature vector is assigned to its closest cluster center.

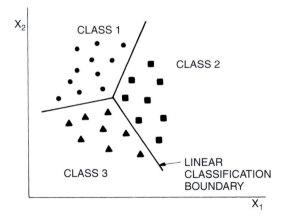

FIGURE 17.2-1. Data clustering for two feature measurements.

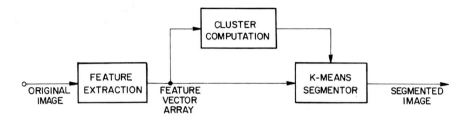

FIGURE 17.2-2. Simplified version of Coleman–Andrews clustering image segmentation method.

The cluster computation algorithm begins by establishing two initial trial cluster centers. All feature vectors of an image are assigned to their closest cluster center. Next, the number of cluster centers is successively increased by one, and a clustering quality factor β is computed at each iteration until the maximum value of β is determined. This establishes the optimum number of clusters. When the number of clusters is incremented by one, the new cluster center becomes the feature vector that is farthest from its closest cluster center. The β factor is defined as

$$\beta = \text{tr}\{\mathbf{S}_W\} \, \text{tr}\{\mathbf{S}_B\} \tag{17.2-1}$$

where $\mathbf{S}_W$ and $\mathbf{S}_B$ are the within- and between-cluster *scatter matrices*, respectively, and $\text{tr}\{\cdot\}$ denotes the trace of a matrix. The within-cluster scatter matrix is computed as

$$\mathbf{S}_W = \frac{1}{K} \sum_{k=1}^{K} \frac{1}{M_k} \sum_{\mathbf{x}_i \in S_k} (\mathbf{x}_i - \mathbf{u}_k)(\mathbf{x}_i - \mathbf{u}_k)^T \tag{17.2-2}$$

where K is the number of clusters, M_k is the number of vector elements in the kth cluster, $\mathbf{x}_i$ is a vector element in the kth cluster, $\mathbf{u}_k$ is the mean of the kth cluster, and S_k is the set of elements in the kth cluster. The between-cluster scatter matrix is defined as

$$\mathbf{S}_B = \frac{1}{K} \sum_{k=1}^{K} (\mathbf{u}_k - \mathbf{u}_0)(\mathbf{u}_k - \mathbf{u}_0)^T \tag{17.2-3}$$

where $\mathbf{u}_0$ is the mean of all of the feature vectors as computed by

$$\mathbf{u}_0 = \frac{1}{M} \sum_{i=1}^{M} \mathbf{x}_i \tag{17.2-4}$$

where M denotes the number of pixels to be clustered. Coleman and Andrews (24) have obtained subjectively good results for their clustering algorithm in the segmentation of monochrome and color images.

17.3. REGION SEGMENTATION METHODS

The amplitude and clustering methods described in the preceding sections are based on point properties of an image. The logical extension, as first suggested by Muerle and Allen (25), is to utilize spatial properties of an image for segmentation.

17.3.1. Region Growing

Region growing is one of the conceptually simplest approaches to image segmentation; neighboring pixels of similar amplitude are grouped together to form a segmented region. However, in practice, constraints, some of which are reasonably complex, must be placed on the growth pattern to achieve acceptable results.

Brice and Fenema (26) have developed a region-growing method based on a set of simple growth rules. In the first stage of the process, pairs of quantized pixels are combined together in groups called *atomic regions* if they are of the same amplitude and are four-connected. Two heuristic rules are next invoked to dissolve weak boundaries between atomic boundaries. Referring to Figure 17.3-1, let R_1 and R_2 be two adjacent regions with perimeters P_1 and P_2, respectively, which have previously been merged. After the initial stages of region growing, a region may contain previously merged subregions of different amplitude values. Also, let C denote the length of the common boundary and let D represent the length of that portion of C for which the amplitude difference Y across the boundary is smaller than a significance factor ε_1. The regions R_1 and R_2 are then merged if

$$\frac{D}{\mathrm{MIN}\{P_1, P_2\}} > \varepsilon_2 \qquad (17.3\text{-}1)$$

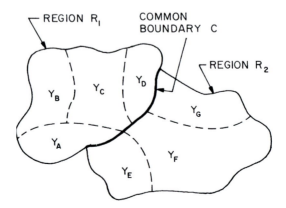

FIGURE 17.3-1. Region-growing geometry.

where ε_2 is a constant typically set at $\varepsilon_2 = \frac{1}{2}$. This heuristic prevents merger of adjacent regions of the same approximate size, but permits smaller regions to be absorbed into larger regions. The second rule merges weak common boundaries remaining after application of the first rule. Adjacent regions are merged if

$$\frac{D}{C} > \varepsilon_3 \qquad\qquad (17.3\text{-}2)$$

where ε_3 is a constant set at about $\varepsilon_3 = \frac{3}{4}$. Application of only the second rule tends to overmerge regions.

The Brice and Fenema region growing method provides reasonably accurate segmentation of simple scenes with few objects and little texture (26, 27) but does not perform well on more complex scenes. Yakimovsky (28) has attempted to improve the region-growing concept by establishing merging constraints based on estimated Bayesian probability densities of feature measurements of each region.

17.3.2. Split and Merge

Split and merge image segmentation techniques (29) are based on a quad tree data representation whereby a square image segment is broken (split) into four quadrants if the original image segment is nonuniform in attribute. If four neighboring squares are found to be uniform, they are replaced (merge) by a single square composed of the four adjacent squares.

In principle, the split and merge process could start at the full image level and initiate split operations. This approach tends to be computationally intensive. Conversely, beginning at the individual pixel level and making initial merges has the drawback that region uniformity measures are limited at the single pixel level. Initializing the split and merge process at an intermediate level enables the use of more powerful uniformity tests without excessive computation.

The simplest uniformity measure is to compute the difference between the largest and smallest pixels of a segment. Fukada (30) has proposed the segment variance as a uniformity measure. Chen and Pavlidis (31) suggest more complex statistical measures of uniformity. The basic split and merge process tends to produce rather blocky segments because of the rule that square blocks are either split or merged. Horowitz and Pavlidis (32) have proposed a modification of the basic process whereby adjacent pairs of regions are merged if they are sufficiently uniform.

17.3.3. Watershed

Topographic and hydrology concepts have proved useful in the development of region segmentation methods (33–36). In this context, a monochrome image is considered to be an altitude surface in which high-amplitude pixels correspond to ridge points, and low-amplitude pixels correspond to valley points. If a drop of water were

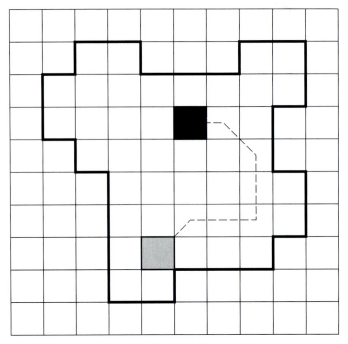

Figure 17.3-2. Rainfall watershed.

to fall on any point of the altitude surface, it would move to a lower altitude until it reached a local altitude minimum. The accumulation of water in the vicinity of a local minimum is called a *catchment basin*. All points that drain into a common catchment basin are part of the same *watershed*. A *valley* is a region that is surrounded by a ridge. A *ridge* is the loci of maximum gradient of the altitude surface. There are two basic algorithmic approaches to the computation of the watershed of an image: rainfall and flooding.

In the *rainfall* approach, local minima are found throughout the image. Each local minima is given a unique tag. Adjacent local minima are combined with a unique tag. Next, a conceptual water drop is placed at each untagged pixel. The drop moves to its lower-amplitude neighbor until it reaches a tagged pixel, at which time it assumes the tag value. Figure 17.3-2 illustrates a section of a digital image encompassing a watershed in which the local minimum pixel is black and the dashed line indicates the path of a water drop to the local minimum.

In the *flooding* approach, conceptual single pixel holes are pierced at each local minima, and the amplitude surface is lowered into a large body of water. The water enters the holes and proceeds to fill each catchment basin. If a basin is about to overflow, a conceptual dam is built on its surrounding ridge line to a height equal to the highest- altitude ridge point. Figure 17.3-3 shows a profile of the filling process of a catchment basin (37). Figure 17.3-4 is an example of watershed segmentation provided by Moga and Gabbouj (38).

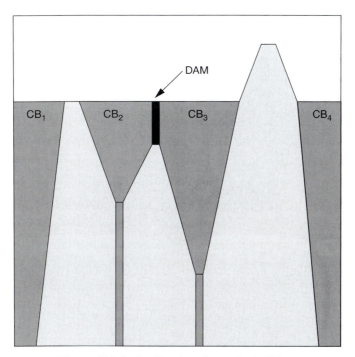

Figure 17.3-3. Profile of catchment basis filling.

(*a*) Original   (*b*) Segmentation

FIGURE 17.3-4. Watershed image segmentation of the `peppers_mon` image. Courtesy of Alina N. Moga and M. Gabbouj, Tampere University of Technology, Finland.

(*a*) Original

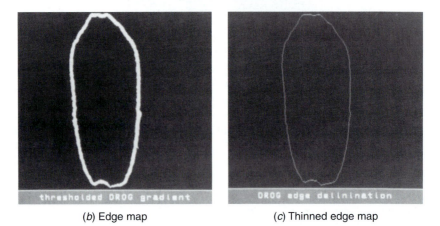

(*b*) Edge map   (*c*) Thinned edge map

FIGURE 17.4-1. Boundary detection image segmentation of the `projectile` image.

Simple watershed algorithms tend to produce results that are oversegmented (39). Najman and Schmitt (37) have applied morphological methods in their watershed algorithm to reduce over segmentation. Wright and Acton (40) have performed watershed segmentation on a pyramid of different spatial resolutions to avoid oversegmentation.

17.4. BOUNDARY DETECTION

It is possible to segment an image into regions of common attribute by detecting the boundary of each region for which there is a significant change in attribute across the boundary. Boundary detection can be accomplished by means of edge detection as described in Chapter 15. Figure 17.4-1 illustrates the segmentation of a projectile from its background. In this example a 11×11 derivative of Gaussian edge detector is used to generate the edge map of Figure 17.4-1*b*. Morphological thinning of this edge map results in Figure 17.4-1*c*. The resulting boundary appears visually to be correct when overlaid on the original image. If an image is noisy or if its region attributes differ by only a small amount between regions, a detected boundary may often be broken. *Edge linking* techniques can be employed to bridge short gaps in such a region boundary.

17.4.1. Curve-Fitting Edge Linking

In some instances, edge map points of a broken segment boundary can be linked together to form a closed contour by curve-fitting methods. If a priori information is available as to the expected shape of a region in an image (e.g., a rectangle or a circle), the fit may be made directly to that closed contour. For more complex-shaped regions, as illustrated in Figure 17.4-2, it is usually necessary to break up the supposed closed contour into chains with broken links. One such chain, shown in Figure 17.4-2 starting at point *A* and ending at point *B*, contains a single broken link. Classical curve-fitting methods (29) such as *Bezier polynomial* or *spline fitting* can be used to fit the broken chain.

   In their book, Duda and Hart (41) credit Forsen as being the developer of a sim-ple piecewise linear curve-fitting procedure called the *iterative endpoint fit*. In the first stage of the algorithm, illustrated in Figure 17.4-3, data endpoints *A* and *B* are connected by a straight line. The point of greatest departure from the straight-line (point *C*) is examined. If the separation of this point is too large, the point becomes an anchor point for two straight-line segments (*A* to *C* and *C* to *B*). The procedure then continues until the data points are well fitted by line segments. The principal advantage of the algorithm is its simplicity; its disadvantage is error caused by incorrect data points. Ramer (42) has used a technique similar to the iterated end-point procedure to determine a polynomial approximation to an arbitrary-shaped closed curve. Pavlidis and Horowitz (43) have developed related algorithms for polygonal curve fitting. The curve-fitting approach is reasonably effective for sim-ply structured objects. Difficulties occur when an image contains many overlapping objects and its corresponding edge map contains branch structures.

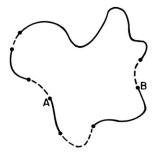

FIGURE 17.4-2. Region boundary with missing links indicated by dashed lines.

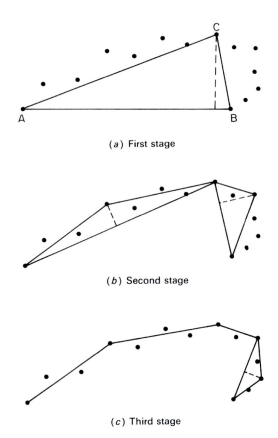

(*a*) First stage

(*b*) Second stage

(*c*) Third stage

FIGURE 17.4-3. Iterative endpoint curve fitting.

17.4.2. Heuristic Edge-Linking Methods

The edge segmentation technique developed by Roberts (44) is typical of the philosophy of many heuristic edge-linking methods. In Roberts' method, edge gradients are examined in 4×4 pixels blocks. The pixel whose magnitude gradient is largest is declared a tentative edge point if its magnitude is greater than a threshold value. Then north-, east-, south-, and west-oriented lines of length 5 are fitted to the gradient data about the tentative edge point. If the ratio of the best fit to the worst fit, measured in terms of the fit correlation, is greater than a second threshold, the tentative edge point is declared valid, and it is assigned the direction of the best fit. Next, straight lines are fitted between pairs of edge points if they are in adjacent 4×4 blocks and if the line direction is within ± 23 degrees of the edge direction of either edge point. Those points failing to meet the linking criteria are discarded. A typical boundary at this stage, shown in Figure 17.4-4*a*, will contain gaps and multiply connected edge points. Small triangles are eliminated by deleting the longest side; small

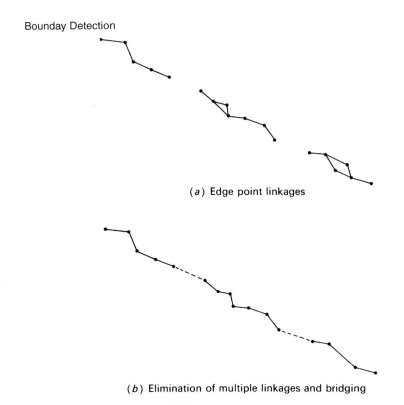

Bounday Detection

(*a*) Edge point linkages

(*b*) Elimination of multiple linkages and bridging

FIGURE 17.4-4. Roberts edge linking.

rectangles are replaced by their longest diagonal, as indicated in Figure 17.4-4*b*. Short spur lines are also deleted. At this stage, short gaps are bridged by straight-line connection. This form of edge linking can be used with a wide variety of edge detectors. Nevatia (45) has used a similar method for edge linking of edges produced by a Heuckel edge detector.

Robinson (46) has suggested a simple but effective edge-linking algorithm in which edge points from an edge detector providing eight edge compass directions are examined in 3×3 blocks as indicated in Figure 17.4-5. The edge point in the center of the block is declared a valid edge if it possesses directional neighbors in the proper orientation. Extensions to larger windows should be beneficial, but the number of potential valid edge connections will grow rapidly with window size.

17.4.3. Hough Transform Edge Linking

The *Hough transform* (47–49) can be used as a means of edge linking. The Hough transform involves the transformation of a line in Cartesian coordinate space to a

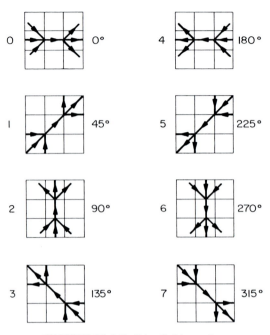

FIGURE 17.4-5. Edge linking rules.

point in polar coordinate space. With reference to Figure 17.4-6a, a straight line can be described parametrically as

$$\rho = x \cos \theta + y \sin \theta \qquad (17.4\text{-}1)$$

where ρ is the normal distance of the line from the origin and θ is the angle of the origin with respect to the x axis. The Hough transform of the line is simply a point at coordinate (ρ, θ) in the polar domain as shown in Figure 17.4-6b. A family of lines passing through a common point, as shown in Figure 17.4-6c, maps into the connected set of ρ–θ points of Figure 17.4-6d. Now consider the three collinear points of Figure 17.4-6e. The Hough transform of the family of curves passing through the three points results in the set of three parametric curves in the ρ–θ space of Figure 17.4-6f. These three curves cross at a single point (ρ_0, θ_0) corresponding to the dashed line passing through the collinear points.

Duda and Hart Version. Duda and Hart (48) have adapted the Hough transform technique for line and curve detection in discrete binary images. Each nonzero data point in the image domain is transformed to a curve in the ρ–θ domain, which is quantized into cells. If an element of a curve falls in a cell, that particular cell is

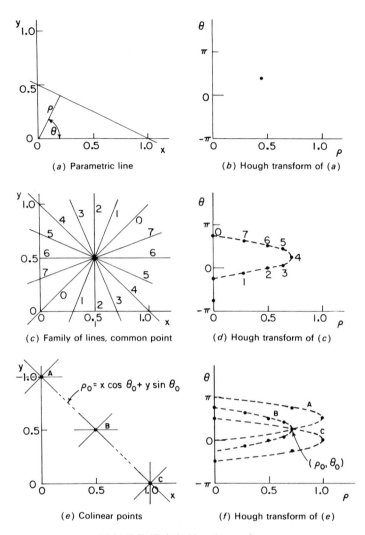

FIGURE 17.4-6. Hough transform.

incremented by one count. After all data points are transformed, the $\rho-\theta$ cells are examined. Large cell counts correspond to colinear data points that may be fitted by a straight line with the appropriate $\rho-\theta$ parameters. Small counts in a cell generally indicate isolated data points that can be deleted.

Figure 17.4-7*a* presents the geometry utilized for the development of an algorithm for the Duda and Hart version of the Hough transform. Following the notation adopted in Section 13.1, the origin of the image is established at the lower left corner of the image. The discrete Cartesian coordinates of the image point (j, k) are

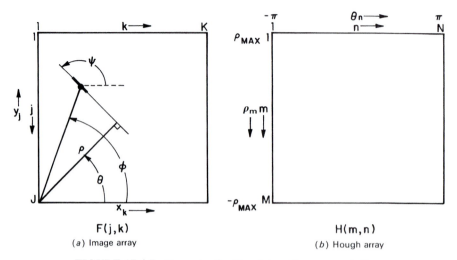

FIGURE 17.4-7. Geometry for Hough transform computation.

$$x_k = k - \tfrac{1}{2} \tag{17.4-2a}$$

$$y_j = J + \tfrac{1}{2} - j \tag{17.4-2b}$$

Consider a line segment in a binary image $F(j, k)$, which contains a point at coordinate (j, k) that is at an angle ϕ with respect to the horizontal reference axis. When the line segment is projected, it intersects a normal line of length ρ emanating from the origin at an angle θ with respect to the horizontal axis. The Hough array $H(m, n)$ consists of cells of the quantized variables ρ_m and θ_n. It can be shown that

$$-\frac{\rho_{max}}{2} \leq \rho_m \leq \rho_{max} \tag{17.4-3a}$$

$$-\frac{\pi}{2} \leq \theta_n \leq \pi \tag{17.4-3b}$$

where

$$\rho_{max} = [(x_K)^2 + (y_1)^2]^{1/2} \tag{17.4-3c}$$

For ease of interpretation, it is convenient to adopt the symmetrical limits of Figure 17.4-7b and to set M and N as odd integers so that the center cell of the Hough array represents $\rho_m = 0$ and $\theta_n = 0$. The Duda and Hart (D & H) Hough transform algorithm follows.

1. Initialize the Hough array to zero.

2. For each (j, k) for which $F(j, k) = 1$, compute

$$\rho(n) = x_k \cos \theta_n + y_j \sin \theta_n \qquad (17.4\text{-}4)$$

where

$$\theta_n = \pi - \frac{2\pi(N-n)}{N-1} \qquad (17.4\text{-}5)$$

is incremented over the range $1 \leq n \leq N$ under the restriction that

$$\phi - \frac{\pi}{2} \leq \theta_n \leq \phi + \frac{\pi}{2} \qquad (17.4\text{-}6)$$

where

$$\phi = \arctan\left\{\frac{y_j}{x_k}\right\} \qquad (17.4\text{-}7)$$

3. Determine the m index of the quantized rho value.

$$m = \left[M - \frac{[\rho_{max} - \rho(n)](M-1)}{2\rho_{max}}\right]_N \qquad (17.4\text{-}8)$$

where $[\cdot]_N$ denotes the nearest integer value of its argument.

4. Increment the Hough array.

$$H(m, n) = H(m, n) + 1 \qquad (17.4\text{-}9)$$

It is important to observe the restriction of Eq. 17.4-6; not all $\rho - \theta$ combinations are legal for a given pixel coordinate (j, k).

Computation of the Hough array requires on the order of N evaluations of Eqs. 17.4-4 to 17.4-9 for each nonzero pixel of $F(j, k)$. The size of the Hough array is not strictly dependent on the size of the image array. However, as the image size increases, the Hough array size should also be increased accordingly to maintain computational accuracy of rho and theta. In most applications, the Hough array size should be set at least one quarter the image size to obtain reasonably accurate results.

Figure 17.4-8 presents several examples of the D & H version of the Hough transform. In these examples, $M = N = 127$ and $J = K = 512$. The Hough arrays

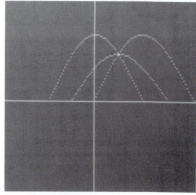

(a) Three dots: upper left, center, lower right   (b) Hough transform of dots

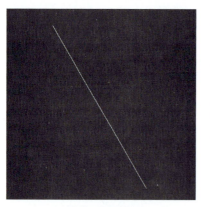

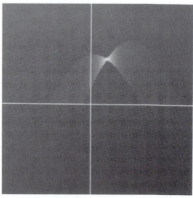

(c) Straight line   (d) Hough transform of line

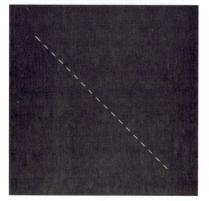

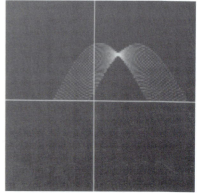

(e) Straight dashed line   (f) Hough transform of dashed line

FIGURE 17.4-8. Duda and Hart version of the Hough transform.

have been flipped bottom to top for display purposes so that the positive rho and positive theta quadrant is in the normal Cartesian first quadrant (i.e., the upper right quadrant).

O 'Gorman and Clowes Version. O' Gorman and Clowes (50) have proposed a modification of the Hough transformation for linking-edge points in an image. In their procedure, the angle θ for entry in $\rho-\theta$ space is obtained from the gradient direction of an edge. The corresponding ρ value is then computed from Eq. 17.4-4 for an edge coordinate (j, k). However, instead of incrementing the (ρ, θ) cell by unity, the cell is incremented by the edge gradient magnitude in order to give greater importance to strong edges than weak edges.

The following is an algorithm for computation of the O' Gorman and Clowes (O & C) version of the Hough transform. Figure 17.4-7a defines the edge angles referenced in the algorithm.

1. Initialize the Hough array to zero.

2. Given a gray scale image $F(j, k)$, generate a first-order derivative edge gradient array $G(j, k)$ and an edge gradient angle array $\gamma(j, k)$ using one of the edge detectors described in Section 15.2.1.

3. For each (j, k) for which $G(j, k) > T$, where T is the edge detector threshold value, compute

$$\rho(j, k) = x_k \cos\{\theta(j, k)\} + y_j \sin\{\theta(j, k)\} \qquad (17.4\text{-}10)$$

where

$$\theta = \begin{cases} \psi + \dfrac{\pi}{2} & \text{for } \psi < \phi \qquad (17.4\text{-}11a) \\[2ex] \psi + \dfrac{\pi}{2} & \text{for } \psi \geq \phi \qquad (17.4\text{-}11b) \end{cases}$$

with

$$\phi = \arctan\left\{\frac{y_j}{x_k}\right\} \qquad (17.4\text{-}12)$$

and

$$\psi = \begin{cases} \gamma + \dfrac{3\pi}{2} & \text{for } -\pi \leq \gamma < -\dfrac{\pi}{2} \qquad (17.4\text{-}13a) \\[2ex] \gamma + \dfrac{\pi}{2} & \text{for } -\dfrac{\pi}{2} \leq \gamma < \dfrac{\pi}{2} \qquad (17.4\text{-}13b) \\[2ex] \gamma - \dfrac{\pi}{2} & \text{for } \dfrac{\pi}{2} \leq \gamma < \pi \qquad (17.4\text{-}13c) \end{cases}$$

4. Determine the m and n indices of the quantized rho and theta values.

$$m = \left[M - \frac{[\rho_{max} - \rho(j, k)](M-1)}{2\rho_{max}} \right]_N \qquad (17.4\text{-}14a)$$

$$n = \left[N - \frac{[\pi - \theta](N-1)}{2\pi} \right]_N \qquad (17.4\text{-}14b)$$

5. Increment the Hough array.

$$H(m, n) = H(m, n) + G(j, k) \qquad (17.4\text{-}15)$$

Figure 17.4-9 gives an example of the O'Gorman and Clowes version of the Hough transform. The original image is 512×512 pixels, and the Hough array is of size 511×511 cells. The Hough array has been flipped bottom to top for display.

Hough Transform Edge Linking. The Hough transform can be used for edge linking in the following manner. Each (ρ, θ) cell whose magnitude is sufficiently large defines a straight line that passes through the original image. If this line is overlaid with the image edge map, it should cover the missing links of straight-line edge segments, and therefore, it can be used as a mask to fill-in the missing links using some heuristic method, such as those described in the preceding section. Another approach, described below, is to use the line mask as a spatial control function for morphological image processing.

Figure 17.4-10 presents an example of Hough transform morphological edge linking. Figure 17.4-10a is an original image of a noisy octagon, and Figure 17.4-10b shows an edge map of the original image obtained by Sobel edge detection followed by morphological thinning, as defined in Section 14.3. Although this form of edge detection performs reasonably well, there are gaps in the contour of the object caused by the image noise. Figure 17.4-10c shows the D & H version of the Hough transform. The eight largest cells in the Hough array have been used to generate the eight Hough lines shown as gray lines overlaid on the original image in Figure 17.4-10d. These Hough lines have been widened to a width of 3 pixels and used as a *region-of-interest* (ROI) mask that controls the edge linking morphological processing such that the processing is performed only on edge map pixels within the ROI. Edge map pixels outside the ROI are left unchanged. The morphological processing consists of three iterations of 3×3 pixel dilation, as shown in Figure 17.4-10e, followed by five iterations of 3×3 pixel thinning. The linked edge map is presented in Figure 17.4-10f.

(*a*) Original

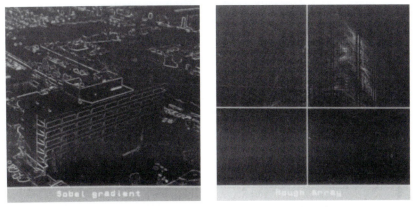

| (*b*) Sobel edge gradient | (*c*) Hough array |

FIGURE 17.4-9. O'Gorman and Clowes version of the Hough transform of the `building` image.

17.4.4. Snakes Boundary Detection

Snakes, developed by Kass et al. (51), is a method of molding a closed contour to the boundary of an object in an image. The snake model is a controlled continuity closed contour that deforms under the influence of internal forces, image forces, and external constraint forces. The internal contour forces provide a piecewise smoothness constraint. The image forces manipulate the contour toward image edges. The external forces are the result of the initial positioning of the contour by some a priori means.

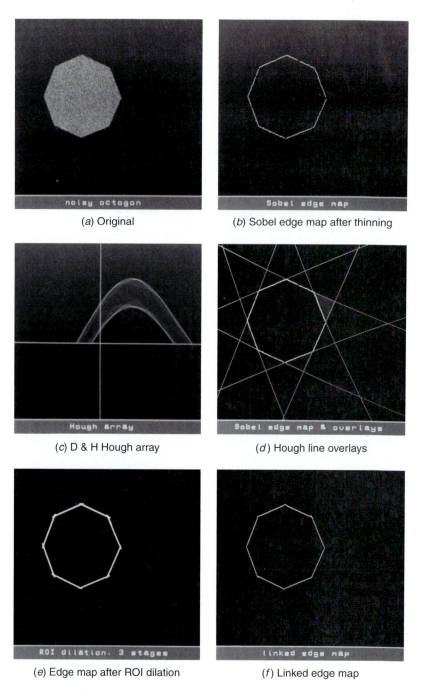

(a) Original

(b) Sobel edge map after thinning

(c) D & H Hough array

(d) Hough line overlays

(e) Edge map after ROI dilation

(f) Linked edge map

FIGURE 17.4-10. Hough transform morphological edge linking.

Let $\mathbf{v}(s) = [x(s), y(s)]$ denote a parametric curve in the continuous domain where s is the arc length of the curve. The continuous domain snake energy is defined as (51)

$$E_S = \int_0^1 E_N\{\mathbf{v}(s)\}\,ds + \int_0^1 E_I\{\mathbf{v}(s)\}\,ds + \int_0^1 E_T\{\mathbf{v}(s)\}\,ds \qquad (17.4\text{-}16)$$

where E_N denotes the internal energy of the contour due to bending or discontinuities, E_I represents the image energy, and E_T is the constraint energy. In the discrete domain, the snake energy is

$$E_S = \sum_{n=1}^{N} E_N\{\mathbf{v}_n\} + \sum_{n=1}^{N} E_I\{\mathbf{v}_n\} + \sum_{n=1}^{N} E_T\{\mathbf{v}_n\} \qquad (17.4\text{-}17)$$

where $\mathbf{v}_n = [x_n, y_n]$ for $n = 0, 1, ..., N$ represents the discrete contour. The location of a snake corresponds to the local minima of the energy functional of Eq. 17.4-17.

Kass et al. (51) have derived a set of N differential equations whose solution minimizes the snake energy. Samadani (52) has investigated the stability of these snake model solutions. The *greedy algorithm* (53,54) expresses the internal snake energy in terms of its continuity energy E_C and curvature energy E_K as

$$E_N = \alpha(n)E_C\{\mathbf{v}_n\} + \beta(n)E_K\{\mathbf{v}_n\} \qquad (17.4\text{-}18)$$

where $\alpha(n)$ and $\beta(n)$ control the elasticity and rigidity of the snake model. The continuity energy is defined as

$$E_C = \frac{d - |\mathbf{v}_n - \mathbf{v}_{n-1}|}{\mathrm{MAX}\{d - |\mathbf{v}_n(j) - \mathbf{v}_{n-1}|\}} \qquad (17.4\text{-}19)$$

and the curvature energy is defined as

$$E_K = \frac{|\mathbf{v}_{n-1} - 2\mathbf{v}_n + \mathbf{v}_{n+1}|^2}{\mathrm{MAX}\{|\mathbf{v}_{n-1} - 2\mathbf{v}_n(j) + \mathbf{v}_{n+1}|^2\}} \qquad (17.4\text{-}19)$$

where d is the average curve length and $\mathbf{v}_n(j)$ represents the eight neighbors of a point $\mathbf{v}_n$ for $j = 1, 2, ..., 8$.

The conventional snake model algorithms suffer from the inability to mold a contour to severe object concavities. Another problem is the generation of false contours due to the creation of unwanted contour loops. Ji and Yan (55) have developed a loop-free snake model segmentation algorithm that overcomes these problems. Figure 17.4-11 illustrates the performance of their algorithm. Figure 17.4-11a shows the initial contour around the pliers object, Figure 17.4-11b is the segmentation

(*a*) Original with initial contour

(*b*) Segmentation with greedy algorithm   (*c*) Segmentation with loop-free algorithm

FIGURE 17.4-11. Snakes image segmentation of the `pliers` image. Courtesy of Lilian Ji and Hong Yan, University of Sydney, Australia.

using the greedy algorithm, and Figure 17.4-11*c* is the result with the loop-free algorithm.

17.5. TEXTURE SEGMENTATION

It has long been recognized that texture should be a valuable feature for image segmentation. Putting this proposition to practice, however, has been hindered by the lack of reliable and computationally efficient means of texture measurement.

One approach to texture segmentation, fostered by Rosenfeld et al. (56–58), is to compute some texture coarseness measure at all image pixels and then detect changes in the coarseness of the texture measure. In effect, the original image is preprocessed to convert texture to an amplitude scale for subsequent amplitude segmentation. A major problem with this approach is that texture is measured over a window area, and therefore, texture measurements in the vicinity of the boundary between texture regions represent some average texture computation. As a result, it becomes difficult to locate a texture boundary accurately.

Another approach to texture segmentation is to detect the transition between regions of differing texture. The basic concept of texture edge detection is identical to that of luminance edge detection; the dissimilarity between textured regions is enhanced over all pixels in an image, and then the enhanced array is thresholded to locate texture discontinuities. Thompson (59) has suggested a means of texture enhancement analogous to the Roberts gradient presented in Section 15.2. Texture measures are computed in each of four adjacent $W \times W$ pixel subregions scanned over the image, and the sum of the cross-difference magnitudes is formed and thresholded to locate significant texture changes. This method can be generalized to include computation in adjacent windows arranged in 3×3 groups. Then, the resulting texture measures of each window can be combined in some linear or nonlinear manner analogous to the 3×3 luminance edge detection methods of Section 15.2.

Zucker et al. (60) have proposed a histogram thresholding method of texture segmentation based on a texture analysis technique developed by Tsuji and Tomita (61). In this method a texture measure is computed at each pixel by forming the spot gradient followed by a dominant neighbor suppression algorithm. Then a histogram is formed over the resultant modified gradient data. If the histogram is multimodal, thresholding of the gradient at the minimum between histogram modes should provide a segmentation of textured regions. The process is repeated on the separate parts until segmentation is complete.

17.6. SEGMENT LABELING

The result of any successful image segmentation is the labeling of each pixel that lies within a specific distinct segment. One means of labeling is to append to each pixel of an image the label number or index of its segment. A more succinct method is to specify the closed contour of each segment. If necessary, contour filling techniques (29) can be used to label each pixel within a contour. The following describes two common techniques of contour following.

The contour following approach to image segment representation is commonly called *bug following*. In the binary image example of Figure 17.6-1, a conceptual bug begins marching from the white background to the black pixel region indicated by the closed contour. When the bug crosses into a black pixel, it makes a left turn

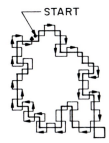

FIGURE 17.6-1. Contour following.

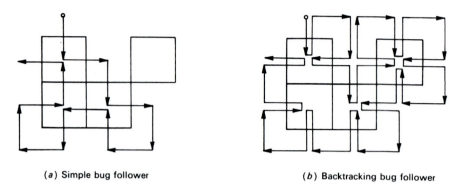

(*a*) Simple bug follower   (*b*) Backtracking bug follower

FIGURE 17.6-2. Comparison of bug follower algorithms.

and proceeds to the next pixel. If that pixel is black, the bug again turns left, and if the pixel is white, the bug turns right. The procedure continues until the bug returns to the starting point. This simple bug follower may miss spur pixels on a boundary. Figure 17.6-2*a* shows the boundary trace for such an example. This problem can be overcome by providing the bug with some memory and intelligence that permit the bug to remember its past steps and backtrack if its present course is erroneous.

Figure 17.6-2*b* illustrates the boundary trace for a *backtracking bug follower*. In this algorithm, if the bug makes a white-to-black pixel transition, it returns to its previous starting point and makes a right turn. The bug makes a right turn whenever it makes a white-to-white transition. Because of the backtracking, this bug follower takes about twice as many steps as does its simpler counterpart.

While the bug is following a contour, it can create a list of the pixel coordinates of each boundary pixel. Alternatively, the coordinates of some reference pixel on the boundary can be recorded, and the boundary can be described by a relative movement code. One such simple code is the *crack code* (62), which is generated for each side p of a pixel on the boundary such that $C(p) = 0, 1, 2, 3$ for movement to the right, down, left, or up, respectively, as shown in Figure 17.6-3. The crack code for the object of Figure 17.6-2 is as follows:

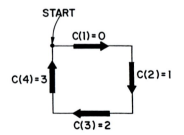

FIGURE 17.6-3. Crack code definition.

p:	1	2	3	4	5	6	7	8	9	10	11	12
$C(p)$:	0	1	0	3	0	1	2	1	2	2	3	3

Upon completion of the boundary trace, the value of the index p is the perimeter of the segment boundary. Section 18.2 describes a method for computing the enclosed area of the segment boundary during the contour following.

Freeman (63, 64) has devised a method of boundary coding, called *chain coding*, in which the path from the centers of connected boundary pixels are represented by an eight-element code. Figure 17.6-4 defines the chain code and provides an example of its use. Freeman has developed formulas for perimeter and area calculation based on the chain code of a closed contour.

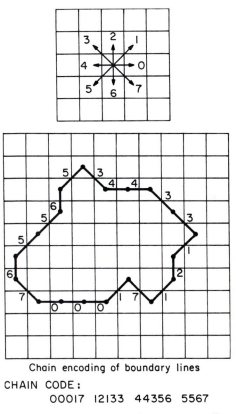

Chain encoding of boundary lines

CHAIN CODE:
00017 12133 44356 5567

FIGURE 17.6-4. Chain coding contour coding.

REFERENCES

1. R. M. Haralick and L. G. Shapiro, "Image Segmentation Techniques," *Computer Vision, Graphics, and Image Processing*, **29**, 1, January 1985, 100–132.

2. S. W. Zucker, "Region Growing: Childhood and Adolescence," *Computer Graphics and Image Processing*, **5**, 3, September 1976, 382–389.

3. E. M. Riseman and M. A. Arbib, "Computational Techniques in the Visual Segmentation of Static Scenes," *Computer Graphics and Image Processing*, **6**, 3, June 1977, 221–276.

4. T. Kanade, "Region Segmentation: Signal vs. Semantics," *Computer Graphics and Image Processing*, **13**, 4, August 1980, 279–297.

5. K. S. Fu and J. K. Mui, "A Survey on Image Segmentation," *Pattern Recognition*, **13**, 1981, 3–16.

6. N. R. Pal and S. K. Pal, "A Review on Image Segmentation Techniques," *Pattern Recognition*, **26**, 9, 1993, 1277–1294.

7. J. S. Weska, "A Survey of Threshold Selection Techniques," *Computer Graphics and Image Processing*, **7**, 2, April 1978, 259–265.

8. B. Sankur, A. T. Abak, and U. Baris, "Assessment of Thresholding Algorithms for Document Processing," *Proc. IEEE International Conference on Image Processing*, Kobe, Japan, October 1999, **1**, 580–584.

9. W. Doyle, "Operations Useful for Similarity-Invariant Pattern Recognition," *J. Association for Computing Machinery*, **9**, 2, April 1962, 259–267.

10 J. M. S. Prewitt and M. L. Mendelsohn, "The Analysis of Cell Images," *Ann. New York Academy of Sci*ence, **128**, 1966, 1036–1053.

11. N. Papamarkos and B. Gatos, "A New Approach for Multilevel Threshold Selection," *CVGIP: Graphical Models and Image Processing*, **56**, 5, September 1994, 357–370.

12. J. S. Weska, R. N. Nagel, and A. Rosenfeld, "A Threshold Selection Technique," *IEEE Trans. Computers*, **C-23**, 12, December 1974, 1322–1326.

13. M. R. Bartz, "The IBM 1975 Optical Page Reader, II: Video Thresholding System," *IBM J. Research and Development*, **12**, September 1968, 354–363.

14. C. K. Chow and T. Kaneko, "Boundary Detection of Radiographic Images by a Threshold Method," in *Frontiers of Pattern Recognition*, S. Watanabe, Ed., Academic Press, New York, 1972.

15. S. D. Yankowitz and A. M. Bruckstein, "A New Method for Image Segmentation," *Computer Vision, Graphics, and Image Processing*, **46**, 1, April 1989, 82–95.

16. F. Tomita, M. Yachida, and S. Tsuji, "Detection of Homogeneous Regions by Structural Analysis," *Proc. International Joint Conference on Artificial Intelligence*, Stanford, CA, August 1973, 564–571.

17. R. B. Ohlander, "Analysis of Natural Scenes," Ph.D. dissertation, Carnegie-Mellon University, Department of Computer Science, Pittsburgh, PA, April 1975.

18. R. B. Ohlander, K. Price, and D. R. Ready, "Picture Segmentation Using a Recursive Region Splitting Method," *Computer Graphics and Image Processing*, **8**, 3, December 1978, 313–333.

19. Y. Ohta, T. Kanade, and T. Saki, "Color Information for Region Segmentation," *Computer Graphics and Image Processing*, **13**, 3, July 1980, 222–241.

20. R. O. Duda, P. E. Hart, and D. G. Stork, *Pattern Classification,* 2nd ed., Wiley-Interscience, New York, 2001.

21. H. C. Becker et al., "Digital Computer Determination of a Medical Diagnostic Index Directly from Chest X-ray Images," *IEEE Trans. Biomedical Engineering.,* **BME-11**, 3, July 1964, 67–72.

22. R. P. Kruger et al., "Radiographic Diagnosis via Feature Extraction and Classification of Cardiac Size and Shape Descriptors," *IEEE Trans. Biomedical Engineering,* **BME-19**, 3, May 1972, 174–186.

23. R. M. Haralick and G. L. Kelly, "Pattern Recognition with Measurement Space and Spatial Clustering for Multiple Images," *Proc. IEEE,* **57**, 4, April 1969, 654–665.

24. G. B. Coleman and H. C. Andrews, "Image Segmentation by Clustering," *Proc. IEEE,* **67**, 5, May 1979, 773–785.

25. J. L. Muerle and D. C. Allen, "Experimental Evaluation of Techniques for Automatic Segmentation of Objects in a Complex Scene," in *Pictorial Pattern Recognition,* G. C. Cheng et al., Eds., Thompson, Washington, DC, 1968, 3–13.

26. C. R. Brice and C. L. Fenema, "Scene Analysis Using Regions," *Artificial Intelligence,* **1**, 1970, 205–226.

27. H. G. Barrow and R. J. Popplestone, "Relational Descriptions in Picture Processing," in *Machine Intelligence,* Vol. 6, B. Meltzer and D. Michie, Eds., University Press, Edinburgh, 1971, 377–396.

28. Y. Yakimovsky, "Scene Analysis Using a Semantic Base for Region Growing," Report AIM-209, Stanford University, Stanford, Calif., 1973.

29. T. Pavlidis, *Algorithms for Graphics and Image Processing,* Computer Science Press, Rockville, MD, 1982.

30. Y. Fukada, "Spatial Clustering Procedures for Region Analysis," *Pattern Recognition,* **12**, 1980, 395–403.

31. P. C. Chen and T. Pavlidis, "Image Segmentation as an Estimation Problem," *Computer Graphics and Image Processing,* **12**, 2, February 1980, 153–172.

32. S. L. Horowitz and T. Pavlidis, "Picture Segmentation by a Tree Transversal Algorithm," *J. Association for Computing Machinery,* **23**, 1976, 368–388.

33. R. M. Haralick, "Ridges and Valleys on Digital Images," *Computer Vision, Graphics and Image Processing,* **22**, 10, April 1983, 28–38.

34. S. Beucher and C. Lantuejoul, "Use of Watersheds in Contour Detection," *Proc. International Workshop on Image Processing, Real Time Edge and Motion Detection/Estimation,* Rennes, France, September 1979.

35. S. Beucher and F. Meyer, "The Morphological Approach to Segmentation: The Watershed Transformation," in *Mathematical Morphology in Image Processing,* E. R. Dougherty, ed., Marcel Dekker, New York, 1993.

36. L. Vincent and P. Soille, "Watersheds in Digital Spaces: An Efficient Algorithm Based on Immersion Simulations," *IEEE Trans. Pattern Analysis and Machine Intelligence,* **PAMI-13**, 6, June 1991, 583–598.

37. L. Najman and M. Schmitt, "Geodesic Saliency of Watershed Contours and Hierarchical Segmentation," *IEEE Trans. Pattern Analysis and Machine Intelligence,* **PAMI-18**, 12, December 1996.

38. A. N. Morga and M. Gabbouj, "Parallel Image Component Labeling with Watershed Transformation," *IEEE Trans. Pattern Analysis and Machine Intelligence*, **PAMI-19**, 5, May 1997, 441–440.

39. A. M. Lopez et al., "Evaluation of Methods for Ridge and Valley Detection," *IEEE Trans. Pattern Analysis and Machine Intelligence*, **PAMI-21**, 4, April 1999, 327–335.

40. A. S. Wright and S. T. Acton, "Watershed Pyramids for Edge Detection, *Proc. 1997 International Conference on Image Processing*, II, Santa Bartara, CA, 1997, 578–581.

41. R. O. Duda and P. E. Hart, *Pattern Classification and Scene Analysis*, Wiley-Interscience, New York, 1973.

42. U. Ramer, "An Iterative Procedure for the Polygonal Approximation of Plane Curves," *Computer Graphics and Image Processing*, **1**, 3, November 1972, 244–256.

43. T. Pavlidis and S. L. Horowitz, "Segmentation of Plane Curves," *IEEE Trans. Computers*, **C-23**, 8, August 1974, 860–870.

44. L. G. Roberts, "Machine Perception of Three Dimensional Solids," in *Optical and Electro-Optical Information Processing*, J. T. Tippett et al., Eds., MIT Press, Cambridge, MA, 1965.

45. R. Nevatia, "Locating Object Boundaries in Textured Environments," *IEEE Trans. Computers*, **C-25**, 11, November 1976, 1170–1175.

46. G. S. Robinson, "Detection and Coding of Edges Using Directional Masks," *Proc. SPIE Conference on Advances in Image Transmission Techniques*, San Diego, CA, August 1976.

47. P. V. C. Hough, "Method and Means for Recognizing Complex patterns," U.S. patent 3,069,654, December 18, 1962.

48. R. O. Duda and P. E. Hart, "Use of the Hough Transformation to Detect Lines and Curves in Pictures," *Communication of the ACM*, **15**, 1, January 1972, 11–15.

49. J. Illingworth and J. Kittler, "A Survey of the Hough Transform," *Computer Vision, Graphics, and Image Processing*, **44**, 1, October 1988, 87–116.

50. F. O'Gorman and M. B. Clowes, "Finding Picture Edges Through Colinearity of Feature Points," *IEEE Trans. Computers*, **C-25**, 4, April 1976, 449–456.

51. M. Kass, A. Witkin, and D. Terzopoulos, "Snakes: Active Contour Models," *International J. Computer Vision*, **1**, 4, 1987, 321–331.

52. R. Samadani, "Adaptive Snakes: Control of Damping and Material Parameters," *Proc. SPIE Conference, on Geometric Methods in Computer Vision*, **1570**, San Diego, CA, 202–213.

53. D. J. Williams and M. Shah, "A Fast Algorithm for Active Contours and Curve Estimation," *CVGIP: Image Understanding*, **55**, 1, 1992, 14–26.

54. K.-H. Lam and H. Yan, "Fast Greedy Algorithm for Active Contours," *Electronic Letters*, **30**, 1, January 1994, 21–23.

55. L. Ji and H. Yan, "Loop-Free Snakes for Image Segmentation," *Proc. 1999 International Conference on Image Processing*, **3**, Kobe, Japan, 1999, 193–197.

56. A. Rosenfeld and M. Thurston, "Edge and Curve Detection for Visual Scene Analysis," *IEEE Trans. Computers*, **C-20**, 5, May 1971, 562–569.

57. A. Rosenfeld, M. Thurston, and Y. H. Lee, "Edge and Curve Detection: Further Experiments," *IEEE Trans. Computers*, **C-21**, 7, July 1972, 677–715.

58. K. C. Hayes, Jr., A. N. Shah, and A. Rosenfeld, "Texture Coarseness: Further Experiments," *IEEE Trans. Systems, Man and Cybernetics* (Correspondence), **SMC-4**, 5, September 1974, 467–472.

59. W. B. Thompson, "Textural Boundary Analysis," Report USCIPI 620, University of Southern California, Image Processing Institute, Los Angeles, September 1975, 124–134.

60. S. W. Zucker, A. Rosenfeld, and L. S. Davis, "Picture Segmentation by Texture Discrimination," *IEEE Trans. Computers,* **C-24**, 12, December 1975, 1228–1233.

61. S. Tsuji and F. Tomita, "A Structural Analyzer for a Class of Textures," *Computer Graphics and Image Processing*, **2**, 3/4, December 1973, 216–231.

62. Z. Kulpa, "Area and Perimeter Measurements of Blobs in Discrete Binary Pictures," *Computer Graphics and Image Processing*, **6**, 4, December 1977, 434–451.

63. H. Freeman, "On the Encoding of Arbitrary Geometric Configurations," *IRE Trans. Electronic Computers*, **EC-10**, 2, June 1961, 260–268.

64. H. Freeman, "Boundary Encoding and Processing," in *Picture Processing and Psychopictorics*, B. S. Lipkin and A. Rosenfeld, Eds., Academic Press, New York, 1970, 241–266.

18

SHAPE ANALYSIS

Several qualitative and quantitative techniques have been developed for characterizing the shape of objects within an image. These techniques are useful for classifying objects in a pattern recognition system and for symbolically describing objects in an image understanding system. Some of the techniques apply only to binary-valued images; others can be extended to gray level images.

18.1. TOPOLOGICAL ATTRIBUTES

Topological shape attributes are properties of a shape that are invariant under *rubber-sheet* transformation (1–3). Such a transformation or mapping can be visualized as the stretching of a rubber sheet containing the image of an object of a given shape to produce some spatially distorted object. Mappings that require cutting of the rubber sheet or connection of one part to another are not permissible. Metric distance is clearly not a topological attribute because distance can be altered by rubber-sheet stretching. Also, the concepts of perpendicularity and parallelism between lines are not topological properties. Connectivity is a topological attribute. Figure 18.1-1a is a binary-valued image containing two connected object components. Figure 18.1-1b is a spatially stretched version of the same image. Clearly, there are no stretching operations that can either increase or decrease the connectivity of the objects in the stretched image. Connected components of an object may contain holes, as illustrated in Figure 18.1-1c. The number of holes is obviously unchanged by a topological mapping.

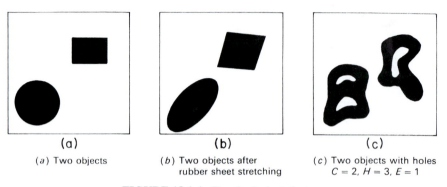

(a) Two objects

(b) Two objects after
rubber sheet stretching

(c) Two objects with holes
$C = 2$, $H = 3$, $E = 1$

FIGURE 18.1-1. Topological attributes.

There is a fundamental relationship between the number of connected object components C and the number of object holes H in an image called the *Euler number*, as defined by

$$E = C - H \qquad (18.1\text{-}1)$$

The Euler number is also a topological property because C and H are topological attributes.

Irregularly shaped objects can be described by their topological constituents. Consider the tubular-shaped object letter R of Figure 18.1-2a, and imagine a rubber band stretched about the object. The region enclosed by the rubber band is called the *convex hull* of the object. The set of points within the convex hull, which are not in the object, form the *convex deficiency* of the object. There are two types of convex deficiencies: regions totally enclosed by the object, called *lakes*; and regions lying between the convex hull perimeter and the object, called *bays*. In some applications it is simpler to describe an object indirectly in terms of its convex hull and convex deficiency. For objects represented over rectilinear grids, the definition of the convex hull must be modified slightly to remain meaningful. Objects such as discretized circles and triangles clearly should be judged as being convex even though their

(a) Object

(b) Convex hull

(c) Bays and lake

FIGURE 18.1-2. Definitions of convex shape descriptors.

boundaries are jagged. This apparent difficulty can be handled by considering a rubber band to be stretched about the discretized object. A pixel lying totally within the rubber band, but not in the object, is a member of the convex deficiency. Sklansky et al. (4,5) have developed practical algorithms for computing the convex attributes of discretized objects.

18.2. DISTANCE, PERIMETER, AND AREA MEASUREMENTS

Distance is a real-valued function $d\{(j_1, k_1), (j_2, k_2)\}$ of two image points (j_1, k_1) and (j_2, k_2) satisfying the following properties (6):

$$d\{(j_1, k_1), (j_2, k_2)\} \geq 0 \tag{18.2-1a}$$

$$d\{(j_1, k_1), (j_2, k_2)\} = d\{(j_2, k_2), (j_1, k_1)\} \tag{18.2-1b}$$

$$d\{(j_1, k_1), (j_2, k_2)\} + d\{(j_2, k_2), (j_3, k_3)\} \geq d\{(j_1, k_1), (j_3, k_3)\} \tag{18.2-1c}$$

There are a number of distance functions that satisfy the defining properties. The most common measures encountered in image analysis are the *Euclidean distance,*

$$d_E = \left[(j_1 - j_2)^2 + (k_1 - k_2)^2 \right]^{1/2} \tag{18.2-2a}$$

the *magnitude distance,*

$$d_M = |j_1 - j_2| + |k_1 - k_2| \tag{18.2-2b}$$

and the *maximum value distance,*

$$d_X = \text{MAX}\{|j_1 - j_2|, |k_1 - k_2|\} \tag{18.2-2c}$$

In discrete images, the coordinate differences $(j_1 - j_2)$ and $(k_1 - k_2)$ are integers, but the Euclidean distance is usually not an integer.

Perimeter and area measurements are meaningful only for binary images. Consider a discrete binary image containing one or more objects, where $F(j, k) = 1$ if a pixel is part of the object and $F(j, k) = 0$ for all nonobject or background pixels.

The perimeter of each object is the count of the number of pixel sides traversed around the boundary of the object starting at an arbitrary initial boundary pixel and returning to the initial pixel. The area of each object within the image is simply the count of the number of pixels in the object for which $F(j, k) = 1$. As an example, for a 2×2 pixel square, the object area is $A_O = 4$ and the object perimeter is $P_O = 8$. An object formed of three diagonally connected pixels possesses $A_O = 3$ and $P_O = 12$.

The enclosed area of an object is defined to be the total number of pixels for which $F(j, k) = 0$ or 1 within the outer perimeter boundary P_E of the object. The enclosed area can be computed during a boundary-following process while the perimeter is being computed (7,8). Assume that the initial pixel in the boundary-following process is the first black pixel encountered in a raster scan of the image. Then, proceeding in a clockwise direction around the boundary, a crack code $C(p)$, as defined in Section 17.6, is generated for each side p of the object perimeter such that $C(p) = 0, 1, 2, 3$ for directional angles 0, 90, 180, 270°, respectively. The enclosed area is

$$A_E = \sum_{p=1}^{P_E} j(p-1)\, \Delta k(p) \tag{18.2-3a}$$

where P_E is the perimeter of the enclosed object and

$$j(p) = \sum_{i=1}^{p} \Delta j(i) \tag{18.2-3b}$$

with $j(0) = 0$. The delta terms are defined by

$$\Delta j(p) = \begin{cases} 1 & \text{if } C(p) = 1 \quad\quad\quad (18.2\text{-}4a) \\ 0 & \text{if } C(p) = 0 \text{ or } 2 \quad (18.2\text{-}4b) \\ -1 & \text{if } C(p) = 3 \quad\quad\quad (18.2\text{-}4c) \end{cases}$$

$$\Delta k(p) = \begin{cases} 1 & \text{if } C(p) = 0 \quad\quad\quad (18.2\text{-}4d) \\ 0 & \text{if } C(p) = 1 \text{ or } 3 \quad (18.2\text{-}4e) \\ -1 & \text{if } C(p) = 2 \quad\quad\quad (18.2\text{-}4f) \end{cases}$$

Table 18.2-1 gives an example of computation of the enclosed area of the following four-pixel object:

TABLE 18.2-1. Example of Perimeter and Area Computation

p	$C(p)$	$\Delta\, j(p)$	$\Delta\, k(p)$	$j(p)$	$A(p)$
1	0	0	1	0	0
2	3	−1	0	−1	0
3	0	0	1	−1	−1
4	1	1	0	0	−1
5	0	0	1	0	−1
6	3	−1	0	−1	−1
7	2	0	−1	−1	0
8	3	−1	0	−2	0
9	2	0	−1	−2	2
10	2	0	−1	−2	4
11	1	1	0	−1	4
12	1	1	0	0	4

$$
\begin{array}{ccccc}
0 & 0 & 0 & 0 & 0 \\
0 & 1 & 0 & 1 & 0 \\
0 & 1 & 1 & 0 & 0 \\
0 & 0 & 0 & 0 & 0
\end{array}
$$

18.2.1. Bit Quads

Gray (9) has devised a systematic method of computing the area and perimeter of binary objects based on matching the logical state of regions of an image to binary patterns. Let $n\{Q\}$ represent the count of the number of matches between image pixels and the pattern Q within the curly brackets. By this definition, the object area is then

$$
A_O = n\{1\} \tag{18.2-5}
$$

If the object is enclosed completely by a border of white pixels, its perimeter is equal to

$$
P_O = 2n\{0\ 1\} + 2n\left\{ \begin{matrix} 0 \\ 1 \end{matrix} \right\} \tag{18.2-6}
$$

Now, consider the following set of 2×2 pixel patterns called *bit quads* defined in Figure 18.2-1. The object area and object perimeter of an image can be expressed in terms of the number of bit quad counts in the image as

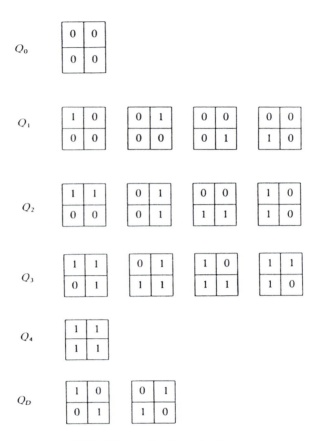

FIGURE 18.2-1. Bit quad patterns.

$$A_O = \tfrac{1}{4}[n\{Q_1\} + 2n\{Q_2\} + 3n\{Q_3\} + 4n\{Q_4\} + 2n\{Q_D\}] \qquad (18.2\text{-}7a)$$

$$P_O = n\{Q_1\} + n\{Q_2\} + n\{Q_3\} + 2n\{Q_D\} \qquad (18.2\text{-}7b)$$

These area and perimeter formulas may be in considerable error if they are utilized to represent the area of a continuous object that has been coarsely discretized. More accurate formulas for such applications have been derived by Duda (10):

$$A_O = \tfrac{1}{4}n\{Q_1\} + \tfrac{1}{2}n\{Q_2\} + \tfrac{7}{8}n\{Q_3\} + n\{Q_4\} + \tfrac{3}{4}n\{Q_D\} \qquad (18.2\text{-}8a)$$

$$P_O = n\{Q_2\} + \tfrac{1}{\sqrt{2}}[n\{Q_1\} + n\{Q_3\} + 2n\{Q_D\}] \qquad (18.2\text{-}8b)$$

Bit quad counting provides a very simple means of determining the Euler number of an image. Gray (9) has determined that under the definition of four-connectivity, the Euler number can be computed as

$$E = \frac{1}{4}[n\{Q_1\} - n\{Q_3\} + 2n\{Q_D\}]$$ (18.2-9a)

and for eight-connectivity

$$E = \frac{1}{4}[n\{Q_1\} - n\{Q_3\} - 2n\{Q_D\}]$$ (18.2-9b)

It should be noted that although it is possible to compute the Euler number E of an image by local neighborhood computation, neither the number of connected components C nor the number of holes H, for which $E = C - H$, can be separately computed by local neighborhood computation.

18.2.2. Geometric Attributes

With the establishment of distance, area, and perimeter measurements, various geometric attributes of objects can be developed. In the following, it is assumed that the number of holes with respect to the number of objects is small (i.e., E is approximately equal to C).

The *circularity* of an object is defined as

$$C_O = \frac{4\pi A_O}{(P_O)^2}$$ (18.2-10)

This attribute is also called the *thinness ratio*. A circle-shaped object has a circularity of unity; oblong-shaped objects possess a circularity of less than 1.

If an image contains many components but few holes, the Euler number can be taken as an approximation of the number of components. Hence, the average area and perimeter of connected components, for $E > 0$, may be expressed as (9)

$$A_A = \frac{A_O}{E}$$ (18.2-11)

$$P_A = \frac{P_O}{E}$$ (18.2-12)

For images containing thin objects, such as typewritten or script characters, the average object length and width can be approximated by

$$L_A = \frac{P_A}{2} \tag{18.2-13}$$

$$W_A = \frac{2A_A}{P_A} \tag{18.2-14}$$

These simple measures are useful for distinguishing gross characteristics of an image. For example, does it contain a multitude of small pointlike objects, or fewer bloblike objects of larger size; are the objects fat or thin? Figure 18.2-2 contains images of playing card symbols. Table 18.2-2 lists the geometric attributes of these objects.

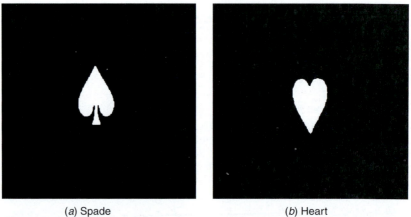

(a) Spade   (b) Heart

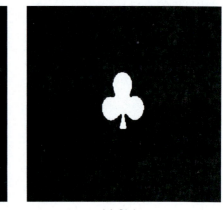

(c) Diamond   (d) Club

FIGURE 18.2-2. Playing card symbol images.

TABLE 18.2-2 Geometric Attributes of Playing Card Symbols

Attribute	Spade	Heart	Diamond	Club
Outer perimeter	652	512	548	668
Enclosed area	8,421	8,681	8.562	8.820
Average area	8,421	8,681	8,562	8,820
Average perimeter	652	512	548	668
Average length	326	256	274	334
Average width	25.8	33.9	31.3	26.4
Circularity	0.25	0.42	0.36	0.25

18.3. SPATIAL MOMENTS

From probability theory, the (m, n)th moment of the joint probability density $p(x, y)$ is defined as

$$M(m, n) = \int_{-\infty}^{\infty} \int_{-\infty}^{\infty} x^m y^n p(x, y) \, dx \, dy \qquad (18.3\text{-}1)$$

The central moment is given by

$$U(m, n) = \int_{-\infty}^{\infty} \int_{-\infty}^{\infty} (x - \eta_x)^m (y - \eta_y)^n p(x, y) \, dx \, dy \qquad (18.3\text{-}2)$$

where η_x and η_y are the marginal means of $p(x, y)$. These classical relationships of probability theory have been applied to shape analysis by Hu (11) and Alt (12). The concept is quite simple. The joint probability density $p(x, y)$ of Eqs. 18.3-1 and 18.3-2 is replaced by the continuous image function $F(x, y)$. Object shape is characterized by a few of the low-order moments. Abu-Mostafa and Psaltis (13,14) have investigated the performance of spatial moments as features for shape analysis.

18.3.1. Discrete Image Spatial Moments

The spatial moment concept can be extended to discrete images by forming spatial summations over a discrete image function $F(j, k)$. The literature (15–17) is notationally inconsistent on the discrete extension because of the differing relationships defined between the continuous and discrete domains. Following the notation established in Chapter 13, the (m, n)th spatial moment is defined as

$$M_U(m, n) = \sum_{j=1}^{J} \sum_{k=1}^{K} (x_k)^m (y_j)^n F(j, k) \qquad (18.3\text{-}3)$$

where, with reference to Figure 13.1-1, the scaled coordinates are

$$x_k = k - \frac{1}{2} \tag{18.3-4a}$$

$$y_j = J + \frac{1}{2} - j \tag{18.3-4b}$$

The origin of the coordinate system is the lower left corner of the image. This formulation results in moments that are extremely scale dependent; the ratio of second-order ($m + n = 2$) to zero-order ($m = n = 0$) moments can vary by several orders of magnitude (18). The spatial moments can be restricted in range by spatially scaling the image array over a unit range in each dimension. The (m, n)th scaled spatial moment is then defined as

$$M(m, n) = \frac{1}{J^n K^m} \sum_{j=1}^{J} \sum_{k=1}^{K} (x_k)^m (y_j)^n F(j, k) \tag{18.3-5}$$

Clearly,

$$M(m, n) = \frac{M_U(m, n)}{J^n K^m} \tag{18.3-6}$$

It is instructive to explicitly identify the lower-order spatial moments. The zero-order moment

$$M(0, 0) = \sum_{j=1}^{J} \sum_{k=1}^{K} F(j, k) \tag{18.3-7}$$

is the sum of the pixel values of an image. It is called the *image surface*. If $F(j, k)$ is a binary image, its surface is equal to its area. The *first-order row moment* is

$$M(1, 0) = \frac{1}{K} \sum_{j=1}^{J} \sum_{k=1}^{K} x_k F(j, k) \tag{18.3-8}$$

and the *first-order column moment* is

$$M(0, 1) = \frac{1}{J} \sum_{j=1}^{J} \sum_{k=1}^{K} y_j F(j, k) \tag{18.3-9}$$

Table 18.3-1 lists the scaled spatial moments of several test images. These images include unit-amplitude gray scale versions of the playing card symbols of Figure 18.2-2, several rotated, minified and magnified versions of these symbols, as shown in Figure 18.3-1, as well as an elliptically shaped gray scale object shown in Figure 18.3-2. The ratios

TABLE 18.3-1. Scaled Spatial Moments of Test Images

Image	$M(0,0)$	$M(1,0)$	$M(0,1)$	$M(2,0)$	$M(1,1)$	$M(0,2)$	$M(3,0)$	$M(2,1)$	$M(1,2)$	$M(0,3)$
Spade	8,219.98	4,013.75	4,281.28	1,976.12	2,089.86	2,263.11	980.81	1,028.31	1,104.36	1,213.73
Rotated spade	8,215.99	4,186.39	3,968.30	2,149.35	2,021.65	1,949.89	1,111.69	1,038.04	993.20	973.53
Heart	8,616.79	4,283.65	4,341.36	2,145.90	2,158.40	2,223.79	1,083.06	1,081.72	1,105.73	1,156.35
Rotated Heart	8,613.79	4,276.28	4,337.90	2,149.18	2,143.52	2,211.15	1,092.92	1,071.95	1,008.05	1,140.43
Magnified heart	34,523.13	17,130.64	17,442.91	8,762.68	8,658.34	9,402.25	4,608.05	4,442.37	4,669.42	5,318.58
Minified heart	2,104.97	1,047.38	1,059.44	522.14	527.16	535.38	260.78	262.82	266.41	271.61
Diamond	8,561.82	4,349.00	4,704.71	2,222.43	2,390.10	2,627.42	1,142.44	1,221.53	1,334.97	1,490.26
Rotated diamond	8,562.82	4,294.89	4,324.09	2,196.40	2,168.00	2,196.97	1,143.83	1,108.30	1,101.11	1,122.93
Club	8,781.71	4,323.54	4,500.10	2,150.47	2,215.32	2,344.02	1,080.29	1,101.21	1,153.76	1,241.04
Rotated club	8,787.71	4,363.23	4,220.96	2,196.08	2,103.88	2,057.66	1,120.12	1,062.39	1,028.90	1,017.60
Ellipse	8,721.74	4,326.93	4,377.78	2,175.86	2,189.76	2,226.61	1,108.47	1,109.92	1,122.62	1,146.97

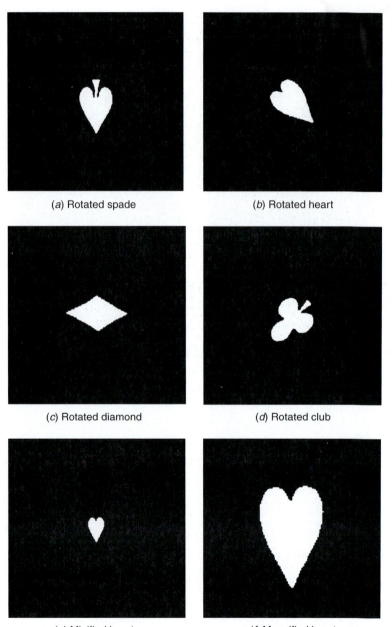

(a) Rotated spade   (b) Rotated heart

(c) Rotated diamond   (d) Rotated club

(e) Minified heart   (f) Magnified heart

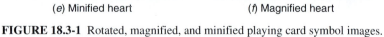

FIGURE 18.3-1 Rotated, magnified, and minified playing card symbol images.

FIGURE 18.3-2 Eliptically shaped object image.

$$\bar{x}_k = \frac{M(1, 0)}{M(0, 0)} \qquad (18.3\text{-}10a)$$

$$\bar{y}_j = \frac{M(0, 1)}{M(0, 0)} \qquad (18.3\text{-}10b)$$

of first- to zero-order spatial moments define the *image centroid*. The centroid, called the *center of gravity*, is the balance point of the image function $F(j, k)$ such that the mass of $F(j, k)$ left and right of $\bar{x}_k$ and above and below $\bar{y}_j$ is equal.

With the centroid established, it is possible to define the scaled spatial central moments of a discrete image, in correspondence with Eq. 18.3-2, as

$$U(m, n) = \frac{1}{J^n K^m} \sum_{j=1}^{J} \sum_{k=1}^{K} (x_k - \bar{x}_k)^m (y_j - \bar{y}_j)^n F(j, k) \qquad (18.3\text{-}11)$$

For future reference, the (m, n)th unscaled spatial central moment is defined as

$$U_U(m, n) = \sum_{j=1}^{J} \sum_{k=1}^{K} (x_k - \tilde{x}_k)^m (y_j - \tilde{y}_j)^n F(j, k) \qquad (18.3\text{-}12)$$

where

$$\tilde{x}_k = \frac{M_U(1, 0)}{M_U(0, 0)} \qquad (18.3\text{-}13a)$$

$$\tilde{y}_j = \frac{M_U(0, 1)}{M_U(0, 0)} \qquad (18.3\text{-}13b)$$

It is easily shown that

$$U(m, n) = \frac{U_U(m, n)}{J^n K^m} \qquad (18.3\text{-}14)$$

The three second-order scaled central moments are the *row moment of inertia,*

$$U(2, 0) = \frac{1}{K^2} \sum_{j=1}^{J} \sum_{k=1}^{K} (x_k - \tilde{x}_k)^2 F(j, k) \qquad (18.3\text{-}15)$$

the *column moment of inertia,*

$$U(0, 2) = \frac{1}{J^n} \sum_{j=1}^{J} \sum_{k=1}^{K} (y_j - \tilde{y}_j)^2 F(j, k) \qquad (18.3\text{-}16)$$

and the *row–column cross moment of inertia,*

$$U(1, 1) = \frac{1}{JK} \sum_{j=1}^{J} \sum_{k=1}^{K} (x_k - \tilde{x}_k)(y_j - \tilde{y}_j) F(j, k) \qquad (18.3\text{-}17)$$

The central moments of order 3 can be computed directly from Eq. 18.3-11 for $m + n = 3$, or indirectly according to the following relations:

$$U(3, 0) = M(3, 0) - 3\tilde{y}_j M(2, 0) + 2(\tilde{y}_j)^2 M(1, 0) \qquad (18.3\text{-}18a)$$

$$U(2, 1) = M(2, 1) - 2\tilde{y}_j M(1, 1) - \tilde{x}_k M(2, 0) + 2(\tilde{y}_j)^2 M(0, 1) \qquad (18.3\text{-}18b)$$

$$U(1, 2) = M(1, 2) - 2\bar{x}_k M(1, 1) - \bar{y}_j M(0, 2) + 2(\bar{x}_k)^2 M(1, 0) \qquad (18.3\text{-}18c)$$

$$U(0, 3) = M(0, 3) - 3\bar{x}_k M(0, 2) + 2(\bar{x}_k)^2 M(0, 1) \qquad (18.3\text{-}18d)$$

Table 18.3-2 presents the horizontal and vertical centers of gravity and the scaled central spatial moments of the test images.

The three second-order moments of inertia defined by Eqs. 18.3-15, 18.3-16, and 18.3-17 can be used to create the moment of inertia covariance matrix,

$$\mathbf{U} = \begin{bmatrix} U(2, 0) & U(1, 1) \\ U(1, 1) & U(0, 2) \end{bmatrix} \qquad (18.3\text{-}19)$$

Performing a singular-value decomposition of the covariance matrix results in the diagonal matrix

$$\mathbf{E}^T \mathbf{U} \mathbf{E} = \mathbf{\Lambda} \qquad (18.3\text{-}20)$$

where the columns of

$$\mathbf{E} = \begin{bmatrix} e_{11} & e_{12} \\ e_{21} & e_{22} \end{bmatrix} \qquad (18.3\text{-}21)$$

are the eigenvectors of $\mathbf{U}$ and

$$\mathbf{\Lambda} = \begin{bmatrix} \lambda_1 & 0 \\ 0 & \lambda_2 \end{bmatrix} \qquad (18.3\text{-}22)$$

contains the eigenvalues of $\mathbf{U}$. Expressions for the eigenvalues can be derived explicitly. They are

$$\lambda_1 = \tfrac{1}{2}[U(2, 0) + U(0, 2)] + \tfrac{1}{2}[U(2, 0)^2 + U(0, 2)^2 - 2U(2, 0)U(0, 2) + 4U(1, 1)^2]^{1/2}$$

$$(18.3\text{-}23a)$$

$$\lambda_2 = \tfrac{1}{2}[U(2, 0) + U(0, 2)] - \tfrac{1}{2}[U(2, 0)^2 + U(0, 2)^2 - 2U(2, 0)U(0, 2) + 4U(1, 1)^2]^{1/2}$$

$$(18.3\text{-}23b)$$

TABLE 18.3-2 Centers of Gravity and Scaled Spatial Central Moments of Test Images

Image	Horizontal COG	Vertical COG	$U(2,0)$	$U(1,1)$	$U(0,2)$	$U(3,0)$	$U(2,1)$	$U(1,2)$	$U(0,3)$
Spade	0.488	0.521	16.240	-0.653	33.261	0.026	-0.285	-0.017	0.363
Rotated spade	0.510	0.483	16.207	-0.366	33.215	-0.013	0.284	-0.002	-0.357
Heart	0.497	0.504	16.380	0.194	36.506	-0.012	0.371	0.027	-0.831
Rotated heart	0.496	0.504	26.237	-10.009	26.584	-0.077	-0.438	0.411	0.122
Magnified heart	0.496	0.505	262.321	3.037	589.162	0.383	11.991	0.886	-27.284
Minified heart	0.498	0.503	0.984	0.013	2.165	0.000	0.011	0.000	-0.025
Diamond	0.508	0.549	13.337	0.324	42.186	-0.002	-0.026	0.005	0.136
Rotated diamond	0.502	0.505	42.198	-0.853	13.366	-0.158	0.009	0.029	-0.005
Club	0.492	0.512	21.834	-0.239	37.979	0.037	-0.545	-0.039	0.950
Rotated club	0.497	0.480	29.675	8.116	30.228	0.268	-0.505	-0.557	0.216
Ellipse	0.496	0.502	29.236	17.913	29.236	0.000	0.000	0.000	0.000

Let $\lambda_M = \text{MAX}\{\lambda_1, \lambda_2\}$ and $\lambda_N = \text{MIN}\{\lambda_1, \lambda_2\}$, and let the orientation angle θ be defined as

$$\theta = \begin{cases} \arctan\left\{\dfrac{e_{21}}{e_{11}}\right\} & \text{if } \lambda_M = \lambda_1 \qquad (18.3\text{-}24a) \\[4ex] \arctan\left\{\dfrac{e_{22}}{e_{12}}\right\} & \text{if } \lambda_M = \lambda_2 \qquad (18.3\text{-}24b) \end{cases}$$

The orientation angle can be expressed explicitly as

$$\theta = \arctan\left\{\frac{\lambda_M - U(0, 2)}{U(1, 1)}\right\} \qquad (18.3\text{-}24c)$$

The eigenvalues λ_M and λ_N and the orientation angle θ define an ellipse, as shown in Figure 18.3-2, whose major axis is λ_M and whose minor axis is λ_N. The major axis of the ellipse is rotated by the angle θ with respect to the horizontal axis. This elliptically shaped object has the same moments of inertia along the horizontal and vertical axes and the same moments of inertia along the principal axes as does an actual object in an image. The ratio

$$R_A = \frac{\lambda_N}{\lambda_M} \qquad (18.3\text{-}25)$$

of the minor-to-major axes is a useful shape feature.

Table 18.3-3 provides moment of inertia data for the test images. It should be noted that the orientation angle can only be determined to within plus or minus $\pi/2$ radians.

TABLE 18.3-3 Moment of Intertia Data of Test Images

Image	Largest Eigenvalue	Smallest Eigenvalue	Orientation (radians)	Eigenvalue Ratio
Spade	33.286	16.215	−0.153	0.487
Rotated spade	33.223	16.200	−1.549	0.488
Heart	36.508	16.376	1.561	0.449
Rotated heart	36.421	16.400	−0.794	0.450
Magnified heart	589.190	262.290	1.562	0.445
Minified heart	2.165	0.984	1.560	0.454
Diamond	42.189	13.334	1.560	0.316
Rotated diamond	42.223	13.341	−0.030	0.316
Club	37.982	21.831	−1.556	0.575
Rotated club	38.073	21.831	0.802	0.573
Ellipse	47.149	11.324	0.785	0.240

Hu (11) has proposed a normalization of the unscaled central moments, defined by Eq. 18.3-12, according to the relation

$$V(m, n) = \frac{U_U(m, n)}{[M(0, 0)]^\alpha}$$

(18.3-26a)

where

$$\alpha = \frac{m + n}{2} + 1$$

(18.3-26b)

for $m + n = 2, 3,...$ These normalized central moments have been used by Hu to develop a set of seven compound spatial moments that are invariant in the continuous image domain to translation, rotation, and scale change. The *Hu invariant moments* are defined below.

$$h_1 = V(2, 0) + V(0, 2)$$

(18.3-27a)

$$h_2 = [V(2, 0) - V(0, 2)]^2 + 4[V(1, 1)]^2$$

(18.3-27b)

$$h_3 = [V(3, 0) - 3V(1, 2)]^2 + [V(0, 3) - 3V(2, 1)]^2$$

(18.3-27c)

$$h_4 = [V(3, 0) + V(1, 2)]^2 + [V(0, 3) - V(2, 1)]^2$$

(18.3-27d)

$$h_5 = [V(3, 0) - 3V(1, 2)][V(3, 0) + V(1, 2)][[V(3, 0) + V(1, 2)]^2 - 3[V(0, 3) + V(2, 1)]^2]$$

$$+ [3V(2, 1) - V(0, 3)][V(0, 3) + V(2, 1)][3[V(3, 0) + V(1, 2)]^2$$

$$- [V(0, 3) + V(2, 1)]^2]$$

(18.3-27e)

$$h_6 = [V(2, 0) - V(0, 2)][[V(3, 0) + V(1, 2)]^2 - [V(0, 3) + V(2, 1)]^2]$$

$$+ 4V(1, 1)[V(3, 0) + V(1, 2)][V(0, 3) + V(2, 1)]$$

(18.3-27f)

$$h_7 = [3V(2, 1) - V(0, 3)][V(3, 0) + V(1, 2)][[V(3, 0) + V(1, 2)]^2 - 3[V(0, 3) + V(2, 1)]^2]$$

$$+ [3V(1, 2) - V(3, 0)][V(0, 3) + V(2, 1)][3[V(3, 0) + V(1, 2)]^2$$

$$- [V(0, 3) + V(2, 1)]^2]$$

(18.3-27g)

Table 18.3-4 lists the moment invariants of the test images. As desired, these moment invariants are in reasonably close agreement for the geometrically modified versions of the same object, but differ between objects. The relatively small degree of variability of the moment invariants for the same object is due to the spatial discretization of the objects.

TABLE 18.3-4 Invariant Moments of Test Images

Image	$h_1 \times 10^1$	$h_2 \times 10^3$	$h_3 \times 10^3$	$h_4 \times 10^5$	$h_5 \times 10^9$	$h_6 \times 10^6$	$h_7 \times 10^1$
Spade	1.920	4.387	0.715	0.295	0.123	0.185	−14.159
Rotated spade	1.919	4.371	0.704	0.270	0.097	0.162	−11.102
Heart	1.867	5.052	1.435	8.052	27.340	5.702	−15.483
Rotated heart	1.866	5.004	1.434	8.010	27.126	5.650	−14.788
Magnified heart	1.873	5.710	1.473	8.600	30.575	6.162	0.559
Minified heart	1.863	4.887	1.443	8.019	27.241	5.583	0.658
Diamond	1.986	10.648	0.018	0.475	0.004	0.490	0.004
Rotated diamond	1.987	10.663	0.024	0.656	0.082	0.678	−0.020
Club	2.033	3.014	2.313	5.641	20.353	3.096	10.226
Rotated club	2.033	3.040	2.323	5.749	20.968	3.167	13.487
Ellipse	2.015	15.242	0.000	0.000	0.000	0.000	0.000

The terms of Eq. 18.3-27 contain differences of relatively large quantities, and therefore, are sometimes subject to significant roundoff error. Liao and Pawlak (19) have investigated the numerical accuracy of moment measures.

18.4. SHAPE ORIENTATION DESCRIPTORS

The spatial orientation of an object with respect to a horizontal reference axis is the basis of a set of orientation descriptors developed at the Stanford Research Institute (20). These descriptors, defined below, are described in Figure 18.4-1.

1. *Image-oriented bounding box:* the smallest rectangle oriented along the rows of the image that encompasses the object

2. *Image-oriented box height:* dimension of box height for image-oriented box

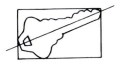
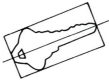
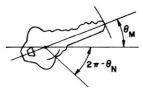

(a) Image-oriented bounding box   (b) Object-oriented bounding box   (c) Radial measures

FIGURE 18.4-1. Shape orientation descriptors.

3. *Image-oriented box width:* dimension of box width for image-oriented box

4. *Image-oriented box area:* area of image-oriented bounding box

5. *Image oriented box ratio:* ratio of box area to enclosed area of an object for an image-oriented box

6. *Object-oriented bounding box:* the smallest rectangle oriented along the major axis of the object that encompasses the object

7. *Object-oriented box height:* dimension of box height for object-oriented box

8. *Object-oriented box width:* dimension of box width for object-oriented box

9. *Object-oriented box area:* area of object-oriented bounding box

10. *Object-oriented box ratio:* ratio of box area to enclosed area of an object for an object-oriented box

11. *Minimum radius:* the minimum distance between the centroid and a perimeter pixel

12. *Maximum radius:* the maximum distance between the centroid and a perimeter pixel

13. *Minimum radius angle:* the angle of the minimum radius vector with respect to the horizontal axis

14. *Maximum radius angle:* the angle of the maximum radius vector with respect to the horizontal axis

15. *Radius ratio:* ratio of minimum radius angle to maximum radius angle

Table 18.4-1 lists the orientation descriptors of some of the playing card symbols.

TABLE 18.4-1 Shape Orientation Descriptors of the Playing Card Symbols

Descriptor	Spade	Rotated Heart	Rotated Diamond	Rotated Club
Row-bounding box height	155	122	99	123
Row-bounding box width	95	125	175	121
Row-bounding box area	14,725	15,250	17,325	14,883
Row-bounding box ratio	1.75	1.76	2.02	1.69
Object-bounding box height	94	147	99	148
Object-bounding box width	154	93	175	112
Object-bounding box area	14,476	13,671	17,325	16,576
Object-bounding box ratio	1.72	1.57	2.02	1.88
Minimum radius	11.18	38.28	38.95	26.00
Maximum radius	92.05	84.17	88.02	82.22
Minimum radius angle	−1.11	0.35	1.06	0.00
Maximum radius angle	−1.54	−0.76	0.02	0.85

18.5. FOURIER DESCRIPTORS

The perimeter of an arbitrary closed curve can be represented by its instantaneous curvature at each perimeter point. Consider the continuous closed curve drawn on the complex plane of Figure 18.5-1, in which a point on the perimeter is measured by its polar position $z(s)$ as a function of arc length s. The complex function $z(s)$ may be expressed in terms of its real part $x(s)$ and imaginary part $y(s)$ as

$$z(s) = x(s) + iy(s) \tag{18.5-1}$$

The tangent angle defined in Figure 18.5-1 is given by

$$\Phi(s) = \arctan\left\{\frac{dy(s)/ds}{dx(s)/ds}\right\} \tag{18.5-2}$$

and the curvature is the real function

$$k(s) = \frac{d\Phi(s)}{ds} \tag{18.5-3}$$

The coordinate points $x(s), y(s)$ can be obtained from the curvature function by the reconstruction formulas

$$x(s) = x(0) + \int_0^s k(\alpha)\cos\{\Phi(\alpha)\}\, d\alpha \tag{18.5-4a}$$

$$y(s) = y(0) + \int_0^s k(\alpha)\sin\{\Phi(\alpha)\}\, d\alpha \tag{18.5-4b}$$

where $x(0)$ and $y(0)$ are the starting point coordinates.

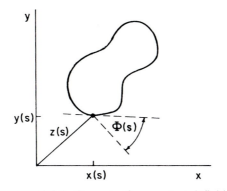

FIGURE 18.5-1. Geometry for curvature definition.

Because the curvature function is periodic over the perimeter length P, it can be expanded in a Fourier series as

$$k(s) = \sum_{n=-\infty}^{\infty} c_n \exp\left\{\frac{2\pi ins}{P}\right\} \tag{18.5-5a}$$

where the coefficients c_n are obtained from

$$c_n = \frac{1}{P}\int_0^P k(s)\exp\left\{-\frac{2\pi in}{P}\right\} ds \tag{18.5-5b}$$

This result is the basis of an analysis technique developed by Cosgriff (21) and Brill (22) in which the Fourier expansion of a shape is truncated to a few terms to produce a set of Fourier descriptors. These Fourier descriptors are then utilized as a symbolic representation of shape for subsequent recognition.

If an object has sharp discontinuities (e.g., a rectangle), the curvature function is undefined at these points. This analytic difficulty can be overcome by the utilization of a cumulative shape function

$$\theta(s) = \int_0^s k(\alpha)\, d\alpha - \frac{2\pi s}{P} \tag{18.5-6}$$

proposed by Zahn and Roskies (23). This function is also periodic over P and can therefore be expanded in a Fourier series for a shape description.

Bennett and MacDonald (24) have analyzed the discretization error associated with the curvature function defined on discrete image arrays for a variety of connectivity algorithms. The discrete definition of curvature is given by

$$z(s_j) = x(s_j) + iy(s_j) \tag{18.5-7a}$$

$$\Phi(s_j) = \arctan\left\{\frac{y(s_j)-y(s_{j-1})}{x(s_j)-x(s_{j-1})}\right\} \tag{18.5-7b}$$

$$k(s_j) = \Phi(s_j) - \Phi(s_{j-1}) \tag{18.5-7c}$$

where s_j represents the jth step of arc position. Figure 18.5-2 contains results of the Fourier expansion of the discrete curvature function.

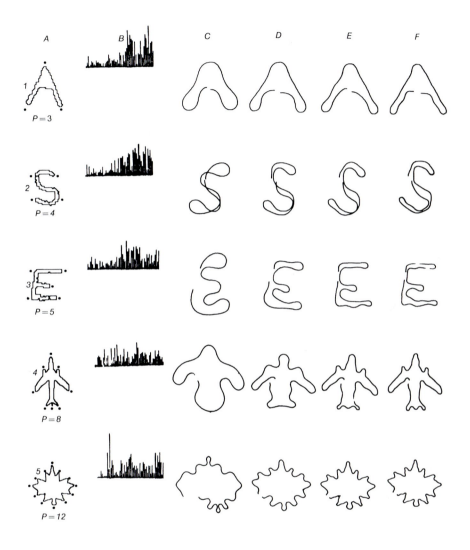

FIGURE 18.5-2. Fourier expansions of curvature function.

REFERENCES

1. R. O. Duda and P. E. Hart, *Pattern Classification and Scene Analysis*, Wiley-Interscience, New York, 1973.

2. E. C. Greanis et al., "The Recognition of Handwritten Numerals by Contour Analysis," *IBM J. Research and Development*, **7**, 1, January 1963, 14–21.

3. M. A. Fischler, "Machine Perception and Description of Pictorial Data," *Proc. International Joint Conference on Artificial Intelligence*, D. E. Walker and L. M. Norton, Eds., May 1969, 629–639.

4. J. Sklansky, "Recognizing Convex Blobs," *Proc. International Joint Conference on Artificial Intelligence*, D. E. Walker and L. M. Norton, Eds., May 1969, 107–116.

5. J. Sklansky, L. P. Cordella, and S. Levialdi, "Parallel Detection of Concavities in Cellular Blobs," *IEEE Trans. Computers*, **C-25**, 2, February 1976, 187–196.

6. A. Rosenfeld and J. L. Pflatz, "Distance Functions on Digital Pictures," *Pattern Recognition*, **1**, July 1968, 33–62.

7. Z. Kulpa, "Area and Perimeter Measurements of Blobs in Discrete Binary Pictures," *Computer Graphics and Image Processing*, **6**, 5, October 1977, 434–451.

8. G. Y. Tang, "A Discrete Version of Green's Theorem," *IEEE Trans. Pattern Analysis and Machine Intelligence*, **PAMI-7**, 3, May 1985, 338–344.

9. S. B. Gray, "Local Properties of Binary Images in Two Dimensions," *IEEE Trans. Computers*, **C-20**, 5, May 1971, 551–561.

10. R. O. Duda, "Image Segmentation and Description," unpublished notes, 1975.

11. M. K. Hu, "Visual Pattern Recognition by Moment Invariants," *IRE Trans. Information Theory*, **IT-8**, 2, February 1962, 179–187.

12. F. L. Alt, "Digital Pattern Recognition by Moments," *J. Association for Computing Machinery*, **9**, 2, April 1962, 240–258.

13. Y. S. Abu-Mostafa and D. Psaltis, "Recognition Aspects of Moment Invariants," *IEEE Trans. Pattern Analysis and Machine Intelligence*, **PAMI-6**, 6, November 1984, 698–706.

14. Y. S. Abu-Mostafa and D. Psaltis, "Image Normalization by Complex Moments," *IEEE Trans. Pattern Analysis and Machine Intelligence*, **PAMI-7**, 6, January 1985, 46–55.

15. S. A. Dudani et al., "Aircraft Identification by Moment Invariants," *IEEE Trans. Computers*, **C-26**, February 1962, 179–187.

16. F. W. Smith and M. H. Wright, "Automatic Ship Interpretation by the Method of Moments," *IEEE Trans. Computers*, **C-20**, 1971, 1089–1094.

17. R. Wong and E. Hall, "Scene Matching with Moment Invariants," *Computer Graphics and Image Processing*, **8**, 1, August 1978, 16–24.

18. A. Goshtasby, "Template Matching in Rotated Images," *IEEE Trans. Pattern Analysis and Machine Intelligence*, **PAMI-7**, 3, May 1985, 338–344.

19. S. X. Liao and M. Pawlak, "On Image Analysis by Moments,"*IEEE Trans. Pattern Analysis and Machine Intelligence*, **PAMI-18**, 3, March 1996, 254–266.

20. Stanford Research Institute, unpublished notes.

21. R. L. Cosgriff, "Identification of Shape," Report 820-11, ASTIA AD 254 792, Ohio State University Research Foundation, Columbus, OH, December 1960.

22. E. L. Brill, "Character Recognition via Fourier Descriptors," *WESCON Convention Record*, Paper 25/3, Los Angeles, 1968.

23. C. T. Zahn and R. Z. Roskies, "Fourier Descriptors for Plane Closed Curves," *IEEE Trans. Computers*, **C-21**, 3, March 1972, 269–281.

24. J. R. Bennett and J. S. MacDonald, "On the Measurement of Curvature in a Quantized Environment," *IEEE Trans. Computers*, **C-25**, 8, August 1975, 803–820.

19

IMAGE DETECTION AND REGISTRATION

This chapter covers two related image analysis tasks: detection and registration. Image detection is concerned with the determination of the presence or absence of objects suspected of being in an image. Image registration involves the spatial alignment of a pair of views of a scene.

19.1. TEMPLATE MATCHING

One of the most fundamental means of object detection within an image field is by *template matching*, in which a replica of an object of interest is compared to all unknown objects in the image field (1–4). If the template match between an unknown object and the template is sufficiently close, the unknown object is labeled as the template object.

As a simple example of the template-matching process, consider the set of binary black line figures against a white background as shown in Figure 19.1-1a. In this example, the objective is to detect the presence and location of right triangles in the image field. Figure 19.1-1b contains a simple template for localization of right triangles that possesses unit value in the triangular region and zero elsewhere. The width of the legs of the triangle template is chosen as a compromise between localization accuracy and size invariance of the template. In operation, the template is sequentially scanned over the image field and the common region between the template and image field is compared for similarity.

A template match is rarely ever exact because of image noise, spatial and amplitude quantization effects, and a priori uncertainty as to the exact shape and structure of an object to be detected. Consequently, a common procedure is to produce a difference measure $D(m, n)$ between the template and the image field at all points of

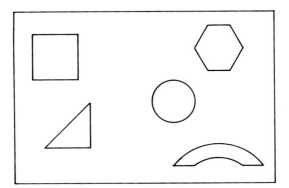

(*a*) Array of objects

(*b*) Triangle template

FIGURE 19.1-1. Template-matching example.

the image field where $-M \leq m \leq M$ and $-N \leq n \leq N$ denote the trial offset. An object is deemed to be matched wherever the difference is smaller than some established level $L_D(m, n)$. Normally, the threshold level is constant over the image field. The usual difference measure is the mean-square difference or error as defined by

$$D(m, n) = \sum_j \sum_k [F(j, k) - T(j - m, k - n)]^2 \qquad (19.1\text{-}1)$$

where $F(j, k)$ denotes the image field to be searched and $T(j, k)$ is the template. The search, of course, is restricted to the overlap region between the translated template and the image field. A template match is then said to exist at coordinate (m, n) if

$$D(m, n) < L_D(m, n) \qquad (19.1\text{-}2)$$

Now, let Eq. 19.1-1 be expanded to yield

$$D(m, n) = D_1(m, n) - 2D_2(m, n) + D_3(m, n) \qquad (19.1\text{-}3)$$

where

$$D_1(m, n) = \sum_j \sum_k [F(j, k)]^2 \qquad (19.1\text{-}4a)$$

$$D_2(m, n) = \sum_j \sum_k [F(j, k)T(j - m, k - n)] \qquad (19.1\text{-}4b)$$

$$D_3(m, n) = \sum_j \sum_k [T(j - m, k - n)]^2 \qquad (19.1\text{-}4c)$$

The term $D_3(m, n)$ represents a summation of the template energy. It is constant valued and independent of the coordinate (m, n). The image energy over the window area represented by the first term $D_1(m, n)$ generally varies rather slowly over the image field. The second term should be recognized as the cross correlation $R_{FT}(m, n)$ between the image field and the template. At the coordinate location of a template match, the cross correlation should become large to yield a small difference. However, the magnitude of the cross correlation is not always an adequate measure of the template difference because the image energy term $D_1(m, n)$ is position variant. For example, the cross correlation can become large, even under a condition of template mismatch, if the image amplitude over the template region is high about a particular coordinate (m, n). This difficulty can be avoided by comparison of the normalized cross correlation

$$\tilde{R}_{FT}(m, n) = \frac{D_2(m, n)}{D_1(m, n)} = \frac{\sum_j \sum_k [F(j, k)T(j - m, k - n)]}{\sum_j \sum_k [F(j, k)]^2} \qquad (19.1\text{-}5)$$

to a threshold level $L_R(m, n)$. A template match is said to exist if

$$\tilde{R}_{FT}(m, n) > L_R(m, n) \qquad (19.1\text{-}6)$$

The normalized cross correlation has a maximum value of unity that occurs if and only if the image function under the template exactly matches the template.

One of the major limitations of template matching is that an enormous number of templates must often be test matched against an image field to account for changes in rotation and magnification of template objects. For this reason, template matching is usually limited to smaller local features, which are more invariant to size and shape variations of an object. Such features, for example, include edges joined in a Y or T arrangement.

19.2. MATCHED FILTERING OF CONTINUOUS IMAGES

Matched filtering, implemented by electrical circuits, is widely used in one-dimensional signal detection applications such as radar and digital communication (5–7). It is also possible to detect objects within images by a two-dimensional version of the matched filter (8–12).

In the context of image processing, the *matched filter* is a spatial filter that provides an output measure of the spatial correlation between an input image and a reference image. This correlation measure may then be utilized, for example, to determine the presence or absence of a given input image, or to assist in the spatial registration of two images. This section considers matched filtering of deterministic and stochastic images.

19.2.1. Matched Filtering of Deterministic Continuous Images

As an introduction to the concept of the matched filter, consider the problem of detecting the presence or absence of a known continuous, deterministic signal or reference image $F(x, y)$ in an unknown or input image $F_U(x, y)$ corrupted by additive stationary noise $N(x, y)$ independent of $F(x, y)$. Thus, $F_U(x, y)$ is composed of the signal image plus noise,

$$F_U(x, y) = F(x, y) + N(x, y) \qquad (19.2\text{-}1a)$$

or noise alone,

$$F_U(x, y) = N(x, y) \qquad (19.2\text{-}1b)$$

The unknown image is spatially filtered by a matched filter with impulse response $H(x, y)$ and transfer function $\mathcal{H}(\omega_x, \omega_y)$ to produce an output

$$F_O(x, y) = F_U(x, y) \circledast H(x, y) \qquad (19.2\text{-}2)$$

The matched filter is designed so that the ratio of the signal image energy to the noise field energy at some point (ε, η) in the filter output plane is maximized.

The instantaneous signal image energy at point (ε, η) of the filter output in the absence of noise is given by

$$|S(\varepsilon, \eta)|^2 = |F(x, y) \circledast H(x, y)|^2 \qquad (19.2\text{-}3)$$

with $x = \varepsilon$ and $y = \eta$. By the convolution theorem,

$$|S(\varepsilon, \eta)|^2 = \left| \int_{-\infty}^{\infty} \int_{-\infty}^{\infty} \mathcal{F}(\omega_x, \omega_y) \mathcal{H}(\omega_x, \omega_y) \exp\{i(\omega_x \varepsilon + \omega_y \eta)\} \, d\omega_x \, d\omega_y \right|^2 \quad (19.2\text{-}4)$$

where $\mathcal{F}(\omega_x, \omega_y)$ is the Fourier transform of $F(x, y)$. The additive input noise component $N(x, y)$ is assumed to be stationary, independent of the signal image, and described by its noise power-spectral density $\mathcal{W}_N(\omega_x, \omega_y)$. From Eq. 1.4-27, the total noise power at the filter output is

$$N = \int_{-\infty}^{\infty} \int_{-\infty}^{\infty} \mathcal{W}_N(\omega_x, \omega_y) |\mathcal{H}(\omega_x, \omega_y)|^2 \, d\omega_x \, d\omega_y \quad (19.2\text{-}5)$$

Then, forming the signal-to-noise ratio, one obtains

$$\frac{|S(\varepsilon, \eta)|^2}{N} = \frac{\left| \int_{-\infty}^{\infty} \int_{-\infty}^{\infty} \mathcal{F}(\omega_x, \omega_y) \mathcal{H}(\omega_x, \omega_y) \exp\{i(\omega_x \varepsilon + \omega_y \eta)\} \, d\omega_x \, d\omega_y \right|^2}{\int_{-\infty}^{\infty} \int_{-\infty}^{\infty} \mathcal{W}_N(\omega_x, \omega_y) |\mathcal{H}(\omega_x, \omega_y)|^2 d\omega_x d\omega_y} \quad (19.2\text{-}6)$$

This ratio is found to be maximized when the filter transfer function is of the form (5,8)

$$\mathcal{H}(\omega_x, \omega_y) = \frac{\mathcal{F}^*(\omega_x, \omega_y) \exp\{-i(\omega_x \varepsilon + \omega_y \eta)\}}{\mathcal{W}_N(\omega_x, \omega_y)} \quad (19.2\text{-}7)$$

If the input noise power-spectral density is white with a flat spectrum, $\mathcal{W}_N(\omega_x, \omega_y) = n_w/2$, the matched filter transfer function reduces to

$$\mathcal{H}(\omega_x, \omega_y) = \frac{2}{n_w} \mathcal{F}^*(\omega_x, \omega_y) \exp\{-i(\omega_x \varepsilon + \omega_y \eta)\} \quad (19.2\text{-}8)$$

and the corresponding filter impulse response becomes

$$H(x, y) = \frac{2}{n_w} F^*(\varepsilon - x, \eta - y) \quad (19.2\text{-}9)$$

In this case, the matched filter impulse response is an amplitude scaled version of the complex conjugate of the signal image rotated by $180°$.

For the case of white noise, the filter output can be written as

$$F_O(x, y) = \frac{2}{n_w} F_U(x, y) \circledast F^*(\varepsilon - x, \eta - y) \quad (19.2\text{-}10a)$$

or

$$F_O(x, y) = \frac{2}{n_w} \int_{-\infty}^{\infty}\int_{-\infty}^{\infty} F_U(\alpha, \beta)F^*(\alpha + \varepsilon - x, \beta + \eta - y)\, d\alpha\, d\beta \quad (19.2\text{-}10b)$$

If the matched filter offset (ε, η) is chosen to be zero, the filter output

$$F_O(x, y) = \frac{2}{n_w} \int_{-\infty}^{\infty}\int_{-\infty}^{\infty} F_U(\alpha, \beta)F^*(\alpha - x, \beta - y)\, d\alpha\, d\beta \quad (19.2\text{-}11)$$

is then seen to be proportional to the mathematical correlation between the input image and the complex conjugate of the signal image. Ordinarily, the parameters (ε, η) of the matched filter transfer function are set to be zero so that the origin of the output plane becomes the point of no translational offset between $F_U(x, y)$ and $F(x, y)$.

If the unknown image $F_U(x, y)$ consists of the signal image translated by distances $(\Delta x, \Delta y)$ plus additive noise as defined by

$$F_U(x, y) = F(x + \Delta x, y + \Delta y) + N(x, y) \quad (19.2\text{-}12)$$

the matched filter output for $\varepsilon = 0$, $\eta = 0$ will be

$$F_O(x, y) = \frac{2}{n_w} \int_{-\infty}^{\infty}\int_{-\infty}^{\infty} [F(\alpha + \Delta x, \beta + \Delta y) + N(x, y)]F^*(\alpha - x, \beta - y)\, d\alpha\, d\beta \quad (19.2\text{-}13)$$

A correlation peak will occur at $x = \Delta x$, $y = \Delta y$ in the output plane, thus indicating the translation of the input image relative to the reference image. Hence the matched filter is translation invariant. It is, however, not invariant to rotation of the image to be detected.

It is possible to implement the general matched filter of Eq. 19.2-7 as a two-stage linear filter with transfer function

$$\mathcal{H}(\omega_x, \omega_y) = \mathcal{H}_A(\omega_x, \omega_y)\mathcal{H}_B(\omega_x, \omega_y) \quad (19.2\text{-}14)$$

The first stage, called a *whitening filter*, has a transfer function chosen such that noise $N(x, y)$ with a power spectrum $W_N(\omega_x, \omega_y)$ at its input results in unit energy white noise at its output. Thus

$$W_N(\omega_x, \omega_y)|\mathcal{H}_A(\omega_x, \omega_y)|^2 = 1 \quad (19.2\text{-}15)$$

The transfer function of the whitening filter may be determined by a spectral factorization of the input noise power-spectral density into the product (7)

$$W_N(\omega_x, \omega_y) = W_N^+(\omega_x, \omega_y) W_N^-(\omega_x, \omega_y) \qquad (19.2\text{-}16)$$

such that the following conditions hold:

$$W_N^+(\omega_x, \omega_y) = [W_N^-(\omega_x, \omega_y)]^* \qquad (19.2\text{-}17a)$$

$$W_N^-(\omega_x, \omega_y) = [W_N^+(\omega_x, \omega_y)]^* \qquad (19.2\text{-}17b)$$

$$W_N(\omega_x, \omega_y) = |W_N^+(\omega_x, \omega_y)|^2 = |W_N^-(\omega_x, \omega_y)|^2 \qquad (19.2\text{-}17c)$$

The simplest type of factorization is the spatially noncausal factorization

$$W_N^+(\omega_x, \omega_y) = \sqrt{W_N(\omega_x, \omega_y)} \exp\{i\theta(\omega_x, \omega_y)\} \qquad (19.2\text{-}18)$$

where $\theta(\omega_x, \omega_y)$ represents an arbitrary phase angle. Causal factorization of the input noise power-spectral density may be difficult if the spectrum does not factor into separable products. For a given factorization, the whitening filter transfer function may be set to

$$\mathcal{H}_A(\omega_x, \omega_y) = \frac{1}{W_N^+(\omega_x, \omega_y)} \qquad (19.2\text{-}19)$$

The resultant input to the second-stage filter is $F_1(x, y) + N_W(x, y)$, where $N_W(x, y)$ represents unit energy white noise and

$$F_1(x, y) = F(x, y) \circledast H_A(x, y) \qquad (19.2\text{-}20)$$

is a modified image signal with a spectrum

$$\mathcal{F}_1(\omega_x, \omega_y) = \mathcal{F}(\omega_x, \omega_y)\mathcal{H}_A(\omega_x, \omega_y) = \frac{\mathcal{F}(\omega_x, \omega_y)}{W_N^+(\omega_x, \omega_y)} \qquad (19.2\text{-}21)$$

From Eq. 19.2-8, for the white noise condition, the optimum transfer function of the second-stage filter is found to be

$$H_B(\omega_x, \omega_y) = \frac{\mathcal{F}^*(\omega_x, \omega_y)}{W_N^-(\omega_x, \omega_y)} \exp\{-i(\omega_x\varepsilon + \omega_y\eta)\} \qquad (19.2\text{-}22)$$

Calculation of the product $H_A(\omega_x, \omega_y)H_B(\omega_x, \omega_y)$ shows that the optimum filter expression of Eq. 19.2-7 can be obtained by the whitening filter implementation.

The basic limitation of the normal matched filter, as defined by Eq. 19.2-7, is that the correlation output between an unknown image and an image signal to be detected is primarily dependent on the energy of the images rather than their spatial structure. For example, consider a signal image in the form of a bright hexagonally shaped object against a black background. If the unknown image field contains a circular disk of the same brightness and area as the hexagonal object, the correlation function resulting will be very similar to the correlation function produced by a perfect match. In general, the normal matched filter provides relatively poor discrimination between objects of different shape but of similar size or energy content. This drawback of the normal matched filter is overcome somewhat with the *derivative matched filter* (8), which makes use of the edge structure of an object to be detected. The transfer function of the *p*th-order derivative matched filter is given by

$$H_p(\omega_x, \omega_y) = \frac{(\omega_x^2 + \omega_y^2)^p \mathcal{F}^*(\omega_x, \omega_y)\exp\{-i(\omega_x\varepsilon + \omega_y\eta)\}}{W_N(\omega_x, \omega_y)} \qquad (19.2\text{-}23)$$

where p is an integer. If $p = 0$, the normal matched filter

$$H_0(\omega_x, \omega_y) = \frac{\mathcal{F}^*(\omega_x, \omega_y)\exp\{-i(\omega_x\varepsilon + \omega_y\eta)\}}{W_N(\omega_x, \omega_y)} \qquad (19.2\text{-}24)$$

is obtained. With $p = 1$, the resulting filter

$$H_p(\omega_x, \omega_y) = (\omega_x^2 + \omega_y^2)H_0(\omega_x, \omega_y) \qquad (19.2\text{-}25)$$

is called the *Laplacian matched filter*. Its impulse response function is

$$H_1(x, y) = \left(\frac{\partial}{\partial x^2} + \frac{\partial}{\partial y^2}\right) \circledast H_0(x, y) \qquad (19.2\text{-}26)$$

The *p*th-order derivative matched filter transfer function is

$$H_p(\omega_x, \omega_y) = (\omega_x^2 + \omega_y^2)^p H_0(\omega_x, \omega_y) \qquad (19.2\text{-}27)$$

Hence the derivative matched filter may be implemented by cascaded operations consisting of a generalized derivative operator whose function is to enhance the edges of an image, followed by a normal matched filter.

19.2.2. Matched Filtering of Stochastic Continuous Images

In the preceding section, the ideal image $F(x, y)$ to be detected in the presence of additive noise was assumed deterministic. If the state of $F(x, y)$ is not known exactly, but only statistically, the matched filtering concept can be extended to the detection of a stochastic image in the presence of noise (13). Even if $F(x, y)$ is known deterministically, it is often useful to consider it as a random field with a mean $E\{F(x, y)\} = F(x, y)$. Such a formulation provides a mechanism for incorporating a priori knowledge of the spatial correlation of an image in its detection. Conventional matched filtering, as defined by Eq. 19.2-7, completely ignores the spatial relationships between the pixels of an observed image.

For purposes of analysis, let the observed unknown field

$$F_U(x, y) = F(x, y) + N(x, y) \qquad (19.2\text{-}28a)$$

or noise alone

$$F_U(x, y) = N(x, y) \qquad (19.2\text{-}28b)$$

be composed of an ideal image $F(x, y)$, which is a sample of a two-dimensional stochastic process with known moments, plus noise $N(x, y)$ independent of the image, or be composed of noise alone. The unknown field is convolved with the matched filter impulse response $H(x, y)$ to produce an output modeled as

$$F_O(x, y) = F_U(x, y) \circledast H(x, y) \qquad (19.2\text{-}29)$$

The stochastic matched filter is designed so that it maximizes the ratio of the average squared signal energy without noise to the variance of the filter output. This is simply a generalization of the conventional signal-to-noise ratio of Eq. 19.2-6. In the absence of noise, the expected signal energy at some point (ε, η) in the output field is

$$|S(\varepsilon, \eta)|^2 = |E\{F(x, y)\} \circledast H(x, y)|^2 \qquad (19.2\text{-}30)$$

By the convolution theorem and linearity of the expectation operator,

$$|S(\varepsilon, \eta)|^2 = \left| \int_{-\infty}^{\infty} \int_{-\infty}^{\infty} E\{\mathcal{F}(\omega_x, \omega_y)\} \mathcal{H}(\omega_x, \omega_y) \exp\{i(\omega_x \varepsilon + \omega_y \eta)\} \, d\omega_x \, d\omega_y \right|^2 \qquad (19.2\text{-}31)$$

The variance of the matched filter output, under the assumption of stationarity and signal and noise independence, is

$$N = \int_{-\infty}^{\infty} \int_{-\infty}^{\infty} [\mathcal{W}_F(\omega_x, \omega_y) + \mathcal{W}_N(\omega_x, \omega_y)] |\mathcal{H}(\omega_x, \omega_y)|^2 \, d\omega_x \, d\omega_y \qquad (19.2\text{-}32)$$

where $\mathcal{W}_F(\omega_x, \omega_y)$ and $\mathcal{W}_N(\omega_x, \omega_y)$ are the image signal and noise power spectral densities, respectively. The generalized signal-to-noise ratio of the two equations above, which is of similar form to the specialized case of Eq. 19.2-6, is maximized when

$$\mathcal{H}(\omega_x, \omega_y) = \frac{E\{\mathcal{F}^*(\omega_x, \omega_y)\} \exp\{-i(\omega_x \varepsilon + \omega_y \eta)\}}{\mathcal{W}_F(\omega_x, \omega_y) + \mathcal{W}_N(\omega_x, \omega_y)} \qquad (19.2\text{-}33)$$

Note that when $F(x, y)$ is deterministic, Eq. 19.2-33 reduces to the matched filter transfer function of Eq. 19.2-7.

The stochastic matched filter is often modified by replacement of the mean of the ideal image to be detected by a replica of the image itself. In this case, for $\varepsilon = \eta = 0$,

$$\mathcal{H}(\omega_x, \omega_y) = \frac{\mathcal{F}^*(\omega_x, \omega_y)}{\mathcal{W}_F(\omega_x, \omega_y) + \mathcal{W}_N(\omega_x, \omega_y)} \qquad (19.2\text{-}34)$$

A special case of common interest occurs when the noise is white, $\mathcal{W}_N(\omega_x, \omega_y) = n_W/2$, and the ideal image is regarded as a first-order nonseparable Markov process, as defined by Eq. 1.4-17, with power spectrum

$$\mathcal{W}_F(\omega_x, \omega_y) = \frac{2}{\alpha^2 + \omega_x^2 + \omega_y^2} \qquad (19.2\text{-}35)$$

where $\exp\{-\alpha\}$ is the adjacent pixel correlation. For such processes, the resultant modified matched filter transfer function becomes

$$\mathcal{H}(\omega_x, \omega_y) = \frac{2(\alpha^2 + \omega_x^2 + \omega_y^2)\mathcal{F}^*(\omega_x, \omega_y)}{4 + n_W(\alpha^2 + \omega_x^2 + \omega_y^2)} \qquad (19.2\text{-}36)$$

At high spatial frequencies and low noise levels, the modified matched filter defined by Eq. 19.2-36 becomes equivalent to the Laplacian matched filter of Eq. 19.2-25.

19.3. MATCHED FILTERING OF DISCRETE IMAGES

A matched filter for object detection can be defined for discrete as well as continuous images. One approach is to perform discrete linear filtering using a discretized version of the matched filter transfer function of Eq. 19.2-7 following the techniques outlined in Section 9.4. Alternatively, the discrete matched filter can be developed by a vector-space formulation (13,14). The latter approach, presented in this section, is advantageous because it permits a concise analysis for nonstationary image and noise arrays. Also, image boundary effects can be dealt with accurately. Consider an observed image vector

$$\mathbf{f}_U = \mathbf{f} + \mathbf{n} \qquad (19.3\text{-}1a)$$

or

$$\mathbf{f}_U = \mathbf{n} \qquad (19.3\text{-}1b)$$

composed of a deterministic image vector $\mathbf{f}$ plus a noise vector $\mathbf{n}$, or noise alone. The discrete matched filtering operation is implemented by forming the inner product of $\mathbf{f}_U$ with a matched filter vector $\mathbf{m}$ to produce the scalar output

$$f_O = \mathbf{m}^T \mathbf{f}_U \qquad (19.3\text{-}2)$$

Vector $\mathbf{m}$ is chosen to maximize the signal-to-noise ratio. The signal power in the absence of noise is simply

$$S = [\mathbf{m}^T \mathbf{f}]^2 \qquad (19.3\text{-}3)$$

and the noise power is

$$N = E\{[\mathbf{m}^T \mathbf{n}][\mathbf{m}^T \mathbf{n}]^T\} = \mathbf{m}^T \mathbf{K_n} \mathbf{m} \qquad (19.3\text{-}4)$$

where $\mathbf{K_n}$ is the noise covariance matrix. Hence the signal-to-noise ratio is

$$\frac{S}{N} = \frac{[\mathbf{m}^T \mathbf{f}]^2}{\mathbf{m}^T \mathbf{K_n} \mathbf{m}} \qquad (19.3\text{-}5)$$

The optimal choice of $\mathbf{m}$ can be determined by differentiating the signal-to-noise ratio of Eq. 19.3-5 with respect to $\mathbf{m}$ and setting the result to zero. These operations lead directly to the relation

$$m = \left[\frac{m^T K_n m}{m^T f} \right] K_n^{-1} f \qquad (19.3\text{-}6)$$

where the term in brackets is a scalar, which may be normalized to unity. The matched filter output

$$f_O = f^T K_n^{-1} f_U \qquad (19.3\text{-}7)$$

reduces to simple vector correlation for white noise. In the general case, the noise covariance matrix may be spectrally factored into the matrix product

$$K_n = KK^T \qquad (19.3\text{-}8)$$

with $K = E\Lambda_n^{-1/2}$, where E is a matrix composed of the eigenvectors of K_n and Λ_n is a diagonal matrix of the corresponding eigenvalues (14). The resulting matched filter output

$$f_O = [K^{-1}f_U]^T [K^{-1}f_U] \qquad (19.3\text{-}9)$$

can be regarded as vector correlation after the unknown vector f_U has been *whitened* by premultiplication by K^{-1}.

Extensions of the previous derivation for the detection of stochastic image vectors are straightforward. The signal energy of Eq. 19.3-3 becomes

$$S = [m^T \eta_f]^2 \qquad (19.3\text{-}10)$$

where η_f is the mean vector of f and the variance of the matched filter output is

$$N = m^T K_f m + m^T K_n m \qquad (19.3\text{-}11)$$

under the assumption of independence of f and n. The resulting signal-to-noise ratio is maximized when

$$m = [K_f + K_n]^{-1} \eta_f \qquad (19.3\text{-}12)$$

Vector correlation of m and f_U to form the matched filter output can be performed directly using Eq. 19.3-2 or alternatively, according to Eq. 19.3-9, where $K = E\Lambda^{-1/2}$ and E and Λ denote the matrices of eigenvectors and eigenvalues of

$[\mathbf{K}_f + \mathbf{K}_n]$, respectively (14). In the special but common case of white noise and a separable, first-order Markovian covariance matrix, the whitening operations can be performed using an efficient Fourier domain processing algorithm developed for Wiener filtering (15).

19.4. IMAGE REGISTRATION

In many image processing applications, it is necessary to form a pixel-by-pixel comparison of two images of the same object field obtained from different sensors, or of two images of an object field taken from the same sensor at different times. To form this comparison, it is necessary to spatially register the images, and thereby, to correct for relative translation shifts, rotational differences, scale differences and even perspective view differences. Often, it is possible to eliminate or minimize many of these sources of misregistration by proper static calibration of an image sensor. However, in many cases, a posteriori misregistration detection and subsequent correction must be performed. Chapter 13 considered the task of spatially warping an image to compensate for physical spatial distortion mechanisms. This section considers means of detecting the parameters of misregistration.

Consideration is given first to the common problem of detecting the translational misregistration of two images. Techniques developed for the solution to this problem are then extended to other forms of misregistration.

19.4.1. Translational Misregistration Detection

The classical technique for registering a pair of images subject to unknown translational differences is to (1) form the normalized cross correlation function between the image pair, (2) determine the translational offset coordinates of the correlation function peak, and (3) translate one of the images with respect to the other by the offset coordinates (16,17). This subsection considers the generation of the basic cross correlation function and several of its derivatives as means of detecting the translational differences between a pair of images.

Basic Correlation Function. Let $F_1(j, k)$ and $F_2(j, k)$, for $1 \leq j \leq J$ and $1 \leq k \leq K$, represent two discrete images to be registered. $F_1(j, k)$ is considered to be the reference image, and

$$F_2(j, k) = F_1(j - j_o, k - k_o) \tag{19.4-1}$$

is a translated version of $F_1(j, k)$ where (j_o, k_o) are the offset coordinates of the translation. The normalized cross correlation between the image pair is defined as

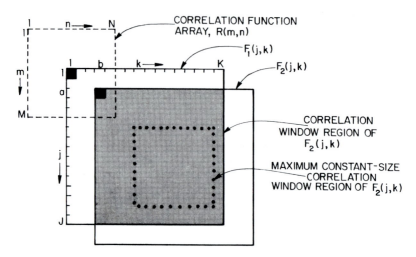

FIGURE 19.4-1. Geometrical relationships between arrays for the cross correlation of an image pair.

$$R(m, n) = \frac{\displaystyle\sum_j \sum_k F_1(j, k) F_2(j - m + (M+1)/2, k - n + (N+1)/2)}{\left[\displaystyle\sum_j \sum_k [F_1(j, k)]^2\right]^{\frac{1}{2}} \left[\displaystyle\sum_j \sum_k [F_2(j - m + (M+1)/2, k - n + (N+1)/2)]^2\right]^{\frac{1}{2}}}$$

$$(19.4\text{-}2)$$

for $m = 1, 2, \ldots, M$ and $n = 1, 2, \ldots, N$, where M and N are odd integers. This formulation, which is a generalization of the template matching cross correlation expression, as defined by Eq. 19.1-5, utilizes an upper left corner–justified definition for all of the arrays. The dashed-line rectangle of Figure 19.4-1 specifies the bounds of the correlation function region over which the upper left corner of $F_2(j, k)$ moves in space with respect to $F_1(j, k)$. The bounds of the summations of Eq. 19.4-2 are

$$\text{MAX}\{1, m - (M-1)/2\} \le j \le \text{MIN}\{J, J + m - (M+1)/2\} \qquad (19.4\text{-}3a)$$

$$\text{MAX}\{1, n - (N-1)/2\} \le k \le \text{MIN}\{K, K + n - (N+1)/2\} \qquad (19.4\text{-}3b)$$

These bounds are indicated by the shaded region in Figure 19.4-1 for the trial offset (a, b). This region is called the *window region* of the correlation function computation. The computation of Eq. 19.4-2 is often restricted to a constant-size window area less than the overlap of the image pair in order to reduce the number of calculations. This $P \times Q$ constant-size window region, called a *template region*, is defined by the summation bounds

$$m \leq j \leq m + J - M \tag{19.4-4a}$$

$$n \leq k \leq n + K - N \tag{19.4-4b}$$

The dotted lines in Figure 19.4-1 specify the maximum constant-size template region, which lies at the center of $F_2(j, k)$. The sizes of the $M \times N$ correlation function array, the $J \times K$ search region, and the $P \times Q$ template region are related by

$$M = J - P + 1 \tag{19.4-5a}$$

$$N = K - Q + 1 \tag{19.4-5b}$$

For the special case in which the correlation window is of constant size, the correlation function of Eq. 19.4-2 can be reformulated as a template search process. Let $S(u, v)$ denote a $U \times V$ search area within $F_1(j, k)$ whose upper left corner is at the offset coordinate (j_s, k_s). Let $T(p, q)$ denote a $P \times Q$ template region extracted from $F_2(j, k)$ whose upper left corner is at the offset coordinate (j_t, k_t). Figure 19.4-2 relates the template region to the search area. Clearly, $U > P$ and $V > Q$. The normalized cross correlation function can then be expressed as

$$R(m, n) = \frac{\sum_u \sum_v S(u, v) T(u - m + (M+1)/2, v - n + (N+1)/2)}{\left[\sum_u \sum_v [S(u, v)]^2\right]^{\frac{1}{2}} \left[\sum_u \sum_v [T(u - m + (M+1)/2, v - n + (N+1)/2)]^2\right]^{\frac{1}{2}}} \tag{19.4-6}$$

for $m = 1, 2, \ldots, M$ and $n = 1, 2, \ldots, N$ where

$$M = U - P + 1 \tag{19.4-7a}$$

$$N = V - Q + 1 \tag{19.4-7b}$$

The summation limits of Eq. 19.4-6 are

$$m \leq u \leq m + P - 1 \tag{19.4-8a}$$

$$n \leq v \leq n + Q - 1 \tag{19.4-8b}$$

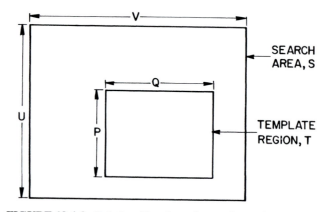

FIGURE 19.4-2. Relationship of template region and search area.

Computation of the numerator of Eq. 19.4-6 is equivalent to raster scanning the template $T(p, q)$ over the search area $S(u, v)$ such that the template always resides within $S(u, v)$, and then forming the sum of the products of the template and the search area under the template. The left-hand denominator term is the square root of the sum of the terms $[S(u, v)]^2$ within the search area defined by the template position. The right-hand denominator term is simply the square root of the sum of the template terms $[T(p, q)]^2$ independent of (m, n). It should be recognized that the numerator of Eq. 19.4-6 can be computed by convolution of $S(u, v)$ with an impulse response function consisting of the template $T(p, q)$ spatially rotated by 180°. Similarly, the left-hand term of the denominator can be implemented by convolving the square of $S(u, v)$ with a $P \times Q$ uniform impulse response function. For large templates, it may be more computationally efficient to perform the convolutions indirectly by Fourier domain filtering.

Statistical Correlation Function. There are two problems associated with the basic correlation function of Eq. 19.4-2. First, the correlation function may be rather broad, making detection of its peak difficult. Second, image noise may mask the peak correlation. Both problems can be alleviated by extending the correlation function definition to consider the statistical properties of the pair of image arrays.

The statistical correlation function (14) is defined as

$$R_S(m, n) = \frac{\displaystyle\sum_j \sum_k G_1(j, k) G_2(j - m + (M+1)/2, k - n + (N+1)/2)}{\left[\displaystyle\sum_j \sum_k [G_1(j, k)]^2\right]^{1/2} \left[\displaystyle\sum_j \sum_k [G_2(j - m + (M+1)/2, k - n + (N+1)/2)]^2\right]^{1/2}}$$

$$(19.4\text{-}9)$$

The arrays $G_i(j, k)$ are obtained by the convolution operation

$$G_i(j, k) = [F_i(j, k) - \bar{F}_i(j, k)] \circledast D_i(j, k) \tag{19.4-10}$$

where $\bar{F}_i(j, k)$ is the spatial average of $F_i(j, k)$ over the correlation window. The impulse response functions $D_i(j, k)$ are chosen to maximize the peak correlation when the pair of images is in best register. The design problem can be solved by recourse to the theory of matched filtering of discrete arrays developed in the preceding section. Accordingly, let $\mathbf{f}_1$ denote the vector of column-scanned elements of $F_1(j, k)$ in the window area and let $\mathbf{f}_2(m, n)$ represent the elements of $F_2(j, k)$ over the window area for a given registration shift (m, n) in the search area. There are a total of $M \cdot N$ vectors $\mathbf{f}_2(m, n)$. The elements within $\mathbf{f}_1$ and $\mathbf{f}_2(m, n)$ are usually highly correlated spatially. Hence, following the techniques of stochastic method filtering, the first processing step should be to whiten each vector by premultiplication with whitening filter matrices $\mathbf{H}_1$ and $\mathbf{H}_2$ according to the relations

$$\mathbf{g}_1 = [\mathbf{H}_1]^{-1} \mathbf{f}_1 \tag{19.4-11a}$$

$$\mathbf{g}_2(m, n) = [\mathbf{H}_2]^{-1} \mathbf{f}_2(m, n) \tag{19.4-11b}$$

where $\mathbf{H}_1$ and $\mathbf{H}_2$ are obtained by factorization of the image covariance matrices

$$\mathbf{K}_1 = \mathbf{H}_1 \mathbf{H}_1^T \tag{19.4-12a}$$

$$\mathbf{K}_2 = \mathbf{H}_2 \mathbf{H}_2^T \tag{19.4-12b}$$

The factorization matrices may be expressed as

$$\mathbf{H}_1 = \mathbf{E}_1 [\mathbf{\Lambda}_1]^{1/2} \tag{19.4-13a}$$

$$\mathbf{H}_2 = \mathbf{E}_2 [\mathbf{\Lambda}_2]^{1/2} \tag{19.4-13b}$$

where $\mathbf{E}_1$ and $\mathbf{E}_2$ contain eigenvectors of $\mathbf{K}_1$ and $\mathbf{K}_2$, respectively, and $\mathbf{\Lambda}_1$ and $\mathbf{\Lambda}_2$ are diagonal matrices of the corresponding eigenvalues of the covariance matrices.

The statistical correlation function can then be obtained by the normalized inner-product computation

$$R_S(m, n) = \frac{\mathbf{g}_1^T \mathbf{g}_2(m, n)}{[\mathbf{g}_1^T \mathbf{g}_1]^{1/2} [\mathbf{g}_2^T(m, n)\mathbf{g}_2(m, n)]^{1/2}} \qquad (19.4\text{-}14)$$

Computation of the statistical correlation function requires calculation of two sets of eigenvectors and eigenvalues of the covariance matrices of the two images to be registered. If the window area contains $P \cdot Q$ pixels, the covariance matrices $\mathbf{K}_1$ and $\mathbf{K}_2$ will each be $(P \cdot Q) \times (P \cdot Q)$ matrices. For example, if $P = Q = 16$, the covariance matrices $\mathbf{K}_1$ and $\mathbf{K}_2$ are each of dimension 256×256. Computation of the eigenvectors and eigenvalues of such large matrices is numerically difficult. However, in special cases, the computation can be simplified appreciably (14). For example, if the images are modeled as separable Markov process sources and there is no observation noise, the convolution operators of Eq. 19.5-9 reduce to the statistical mask operator

$$\mathbf{D}_i = \frac{1}{(1 + \rho^2)^2} \begin{bmatrix} \rho^2 & -\rho(1 + \rho^2) & \rho^2 \\ -\rho(1 + \rho^2) & (1 + \rho^2)^2 & -\rho(1 + \rho^2) \\ \rho^2 & -\rho(1 + \rho^2) & \rho^2 \end{bmatrix} \qquad (19.4\text{-}15)$$

where ρ denotes the adjacent pixel correlation (18). If the images are spatially uncorrelated, then $\rho = 0$, and the correlation operation is not required. At the other extreme, if $\rho = 1$, then

$$\mathbf{D}_i = \frac{1}{4} \begin{bmatrix} 1 & -2 & 1 \\ -2 & 4 & -2 \\ 1 & -2 & 1 \end{bmatrix} \qquad (19.4\text{-}16)$$

This operator is an orthonormally scaled version of the cross second derivative spot detection operator of Eq. 15.7-3. In general, when an image is highly spatially correlated, the statistical correlation operators $\mathbf{D}_i$ produce outputs that are large in magnitude only in regions of an image for which its amplitude changes significantly in both coordinate directions simultaneously.

Figure 19.4-3 provides computer simulation results of the performance of the statistical correlation measure for registration of the toy tank image of Figure 17.1-6b. In the simulation, the reference image $F_1(j, k)$ has been spatially offset horizontally by three pixels and vertically by four pixels to produce the translated image $F_2(j, k)$. The pair of images has then been correlated in a window area of 16×16 pixels over a search area of 32×32 pixels. The curves in Figure 19.4-3 represent the normalized statistical correlation measure taken through the peak of the correlation

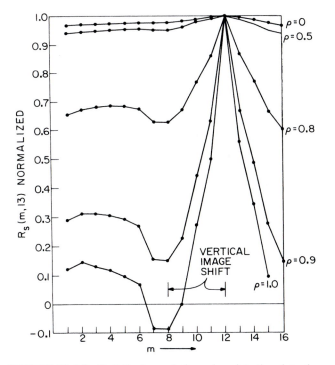

FIGURE 19.4-3. Statistical correlation misregistration detection.

function. It should be noted that for $\rho = 0$, corresponding to the basic correlation measure, it is relatively difficult to distinguish the peak of $R_S(m, n)$. For $\rho = 0.9$ or greater, $R(m, n)$ peaks sharply at the correct point.

The correlation function methods of translation offset detection defined by Eqs. 19.4-2 and 19.4-9 are capable of estimating any translation offset to an accuracy of $\pm\frac{1}{2}$ pixel. It is possible to improve the accuracy of these methods to subpixel levels by interpolation techniques (19). One approach (20) is to spatially interpolate the correlation function and then search for the peak of the interpolated correlation function. Another approach is to spatially interpolate each of the pair of images and then correlate the higher-resolution pair.

A common criticism of the correlation function method of image registration is the great amount of computation that must be performed if the template region and the search areas are large. Several computational methods that attempt to overcome this problem are presented next.

Two-State Methods. Rosenfeld and Vandenburg (21,22) have proposed two efficient two-stage methods of translation offset detection. In one of the methods, called *coarse–fine matching*, each of the pair of images is reduced in resolution by conventional techniques (low-pass filtering followed by subsampling) to produce coarse representations of the images. Then the coarse images are correlated and the resulting correlation peak is determined. The correlation peak provides a rough estimate of the translation offset, which is then used to define a spatially restricted search area for correlation at the fine resolution of the original image pair. The other method, suggested by Vandenburg and Rosenfeld (22), is to use a subset of the pixels within the window area to compute the correlation function in the first stage of the two-stage process. This can be accomplished by restricting the size of the window area or by performing subsampling of the images within the window area. Goshtasby et al. (23) have proposed random rather than deterministic subsampling. The second stage of the process is the same as that of the coarse–fine method; correlation is performed over the full window at fine resolution. Two-stage methods can provide a significant reduction in computation, but they can produce false results.

Sequential Search Method. With the correlation measure techniques, no decision can be made until the correlation array is computed for all (m, n) elements. Furthermore, the amount of computation of the correlation array is the same for all degrees of misregistration. These deficiencies of the standard correlation measures have led to the search for efficient sequential search algorithms.

An efficient sequential search method has been proposed by Barnea and Silverman (24). The basic form of this algorithm is deceptively simple. The absolute value difference error

$$\mathcal{E}_S = \sum_j \sum_k |F_1(j, k) - F_2(j - m, k - n)| \qquad (19.4\text{-}17)$$

is accumulated for pixel values in a window area. If the error exceeds a predetermined threshold value before all $P \cdot Q$ pixels in the window area are examined, it is assumed that the test has failed for the particular offset (m, n), and a new offset is checked. If the error grows slowly, the number of pixels examined when the threshold is finally exceeded is recorded as a rating of the test offset. Eventually, when all test offsets have been examined, the offset with the largest rating is assumed to be the proper misregistration offset.

Phase Correlation Method. Consider a pair of continuous domain images

$$F_2(x, y) = F_1(x - x_o, y - y_o) \qquad (19.4\text{-}18)$$

that are translated by an offset (x_o, y_o) with respect to one another. By the Fourier transform shift property of Eq. 1.3-13a, the Fourier transforms of the images are related by

$$\mathcal{F}_2(\omega_x, \omega_y) = \mathcal{F}_1(\omega_x, \omega_y) \exp\{-i(\omega_x x_o + \omega_y y_o)\} \qquad (19.4\text{-}19)$$

The exponential phase shift factor can be computed by the *cross-power spectrum* (25) of the two images as given by

$$G(\omega_x, \omega_y) \equiv \frac{\mathcal{F}_1(\omega_x, \omega_y) \mathcal{F}_2^*(\omega_x, \omega_y)}{|\mathcal{F}_1(\omega_x, \omega_y) \mathcal{F}_2(\omega_x, \omega_y)|} = \exp\{i(\omega_x x_o + \omega_y y_o)\} \qquad (19.4\text{-}20)$$

Taking the inverse Fourier transform of Eq. 19.4-20 yields the spatial offset

$$G(x, y) = \delta(x - x_o, y - y_o) \qquad (19.4\text{-}21)$$

in the space domain.

The cross-power spectrum approach can be applied to discrete images by utilizing discrete Fourier transforms in place of the continuous Fourier transforms in Eq. 19.4-20. However, care must be taken to prevent wraparound error. Figure 19.4-4 presents an example of translational misregistration detection using the phase correlation method. Figure 19.4-4a and b show translated portions of a scene embedded in a zero background. The scene in Figure 19.4-4a was obtained by extracting the first 480 rows and columns of the 500×500 washington_ir source image. The scene in Figure 19.4-4b consists of the last 480 rows and columns of the source image. Figure 19.4-4c and d are the logarithm magnitudes of the Fourier transforms of the two images, and Figure 19.4-4e is inverse Fourier transform of the cross-power spectrum of the pair of images. The bright pixel in the upper left corner of Figure 19.4-4e, located at coordinate (20,20), is the correlation peak.

19.4.2. Scale and Rotation Misregistration Detection

The phase correlation method for translational misregistration detection has been extended to scale and rotation misregistration detection (25,26). Consider a a pair of images in which a second image is translated by an offset (x_o, y_o) and rotated by an angle θ_o with respect to the first image. Then

$$F_2(x, y) = F_1(x \cos \theta_o + y \sin \theta_o - x_o, -x \sin \theta_o + y \cos \theta_o - y_o) \qquad (19.4\text{-}22)$$

Taking Fourier transforms of both sides of Eq. 19.4-22, one obtains the relationship (25)

$$\mathcal{F}_2(\omega_x, \omega_y) = \mathcal{F}_1(\omega_x \cos \theta_o + \omega_y \sin \theta_o, -\omega_x \sin \theta_o + \omega_y \cos \theta_o) \exp\{-i(\omega_x x_o + \omega_y y_o)\}$$

$$(19.4\text{-}23)$$

(a) Embedded image 1   (b) Embedded image 2

(c) Log magnitude of Fourier   (d) Log magnitude of Fourier
transform of image 1   transform of image 1

(e) Phase correlation spatial array

FIGURE 19.4-4. Translational misregistration detection on the `washington_ir1` and `washington_ir2` images using the phase correlation method. See white pixel in upper left corner of (e).

The rotation component can be isolated by taking the magnitudes $\mathcal{M}_1(\omega_x, \omega_y)$ and $\mathcal{M}_2(\omega_x, \omega_y)$ of both sides of Eq. 19.4-19. By representing the frequency variables in polar form,

$$\mathcal{M}_2(\rho, \theta) = \mathcal{M}_1(\rho, \theta - \theta_o) \tag{19.4-24}$$

the phase correlation method can be used to determine the rotation angle θ_o.

If a second image is a size-scaled version of a first image with scale factors (a, b), then from the Fourier transform scaling property of Eq. 1.3-12,

$$\mathcal{F}_2(\omega_x, \omega_y) = \frac{1}{|ab|} \mathcal{F}_1\left(\frac{\omega_x}{a}, \frac{\omega_y}{b}\right) \tag{19.4-25}$$

By converting the frequency variables to a logarithmic scale, scaling can be converted to a translational movement. Then

$$\mathcal{F}_2(\log \omega_x, \log \omega_y) = \frac{1}{|ab|} \mathcal{F}_1(\log \omega_x - \log a, \log \omega_y - \log b) \tag{19.4-26}$$

Now, the phase correlation method can be applied to determine the unknown scale factors (a, b).

19.4.3. Generalized Misregistration Detection

The basic correlation concept for translational misregistration detection can be generalized, in principle, to accommodate rotation and size scaling. As an illustrative example, consider an observed image $F_2(j, k)$ that is an exact replica of a reference image $F_1(j, k)$ except that it is rotated by an unknown angle θ measured in a clockwise direction about the common center of both images. Figure 19.4-5 illustrates the geometry of the example. Now suppose that $F_2(j, k)$ is rotated by a trial angle θ_r measured in a counterclockwise direction and that it is resampled with appropriate interpolation. Let $F_2(j, k; \theta_r)$ denote the trial rotated version of $F_2(j, k)$. This procedure is then repeated for a set of angles $\theta_1 \leq \theta \leq \theta_R$ expected to span the unknown angle θ in the reverse direction. The normalized correlation function can then be expressed as

$$R(r) = \frac{\displaystyle\sum_j \sum_k F_1(j, k) F_2(j, k; r)}{\left[\displaystyle\sum_j \sum_k [F_1(j, k)]^2\right]^{1/2} \left[\displaystyle\sum_j \sum_k [F_2(j, k; r)]^2\right]^{1/2}} \tag{19.4-27}$$

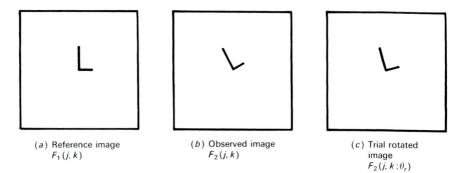

(*a*) Reference image
$F_1(j, k)$

(*b*) Observed image
$F_2(j, k)$

(*c*) Trial rotated
image
$F_2(j, k; \theta_r)$

FIGURE 19.4-5 Rotational misregistration detection.

for $r = 1, 2, \ldots, R$. Searching for the peak of $R(r)$ leads to an estimate of the unknown rotation angle θ. The procedure does, of course, require a significant amount of computation because of the need to resample $F_2(j, k)$ for each trial rotation angle θ_r.

The rotational misregistration example of Figure 19.4-5 is based on the simplifying assumption that the center of rotation is known. If it is not, then to extend the correlation function concept, it is necessary to translate $F_2(j, k)$ to a trial translation coordinate (j_p, k_q), rotate that image by a trial angle θ_r, and translate that image to the translation coordinate $(-j_p, -k_q)$. This results in a trial image $F_2(j, k; j_p, k_q, \theta_r)$, which is used to compute one term of a three-dimensional correlation function $R(p, q, r)$, the peak of which leads to an estimate of the unknown translation and rotation. Clearly, this procedure is computationally intensive.

It is possible to apply the correlation concept to determine unknown row and column size scaling factors between a pair of images. The straightforward extension requires the computation of a two-dimensional correlation function. If all five misregistration parameters are unknown, then again, in principle, a five-dimensional correlation function can be computed to determine an estimate of the unknown parameters. This formidable computational task is further complicated by the fact that, as noted in Section 13.1, the order of the geometric manipulations is important.

The complexity and computational load of the correlation function method of misregistration detection for combined translation, rotation, and size scaling can be reduced significantly by a procedure in which the misregistration of only a few chosen common points between a pair of images is determined. This procedure, called *control point detection*, can be applied to the general rubber-sheet warping problem. A few pixels that represent unique points on objects within the pair of images are identified, and their coordinates are recorded to be used in the spatial warping mapping process described in Eq. 13.2-3. The trick, of course, is to accurately identify and measure the control points. It is desirable to locate object features that are reasonably invariant to small-scale geometric transformations. One such set of features are Hu's (27) seven invariant moments defined by Eqs. 18.3-27. Wong and Hall (28) have investigated the use of invariant moment features for matching optical and radar images of the same scene. Goshtasby (29) has applied invariant moment features for registering visible and infrared weather satellite images.

The control point detection procedure begins with the establishment of a small feature template window, typically 8×8 pixels, in the reference image that is sufficiently large to contain a single control point feature of interest. Next, a search window area is established such that it envelops all possible translates of the center of the template window between the pair of images to be registered. It should be noted that the control point feature may be rotated, minified or magnified to a limited extent, as well as being translated. Then the seven Hu moment invariants h_{i1} for i = 1, 2,..., 7 are computed in the reference image. Similarly, the seven moments $h_{i2}(m, n)$ are computed in the second image for each translate pair (m, n) within the search area. Following this computation, the invariant moment correlation function is formed as

$$R(r) = \frac{\sum\limits_{i=1}^{7} h_{i1} h_{i2}(m, n)}{\left[\sum\limits_{i=1}^{7} (h_{i1})^2\right]^{1/2} \left[\sum\limits_{i=1}^{7} [h_{i2}(m, n)]^2\right]^{1/2}} \qquad (19.4\text{-}28)$$

Its peak is found to determine the coordinates of the control point feature in each image of the image pair. The process is then repeated on other control point features until the number of control points is sufficient to perform the rubber-sheet warping of $F_2(j, k)$ onto the space of $F_1(j, k)$.

REFERENCES

1. R. O. Duda and P. E. Hart, *Pattern Classification and Scene Analysis*, Wiley-Interscience, New York, 1973.

2. W. H. Highleyman, "An Analog Method for Character Recognition," *IRE Trans. Electronic Computers*, EC-**10**, 3, September 1961, 502–510.

3. L. N. Kanal and N. C. Randall, "Recognition System Design by Statistical Analysis," *Proc. ACM National Conference*, 1964.

4. J. H. Munson, "Experiments in the Recognition of Hand-Printed Text, I. Character Recognition," *Proc. Fall Joint Computer Conference*, December 1968, 1125–1138.

5. G. L. Turin, "An Introduction to Matched Filters," *IRE Trans. Information Theory*, **IT-6**, 3, June 1960, 311–329.

6. C. E. Cook and M. Bernfeld, *Radar Signals*, Academic Press, New York, 1965.

7. J. B. Thomas, *An Introduction to Statistical Communication Theory*, Wiley, New York, 1965, 187–218.

8. H. C. Andrews, *Computer Techniques in Image Processing*, Academic Press, New York, 1970, 55–71.

9. L. J. Cutrona, E. N. Leith, C. J. Palermo, and L. J. Porcello, "Optical Data Processing and Filtering Systems," *IRE Trans. Information Theory*, **IT-6**, 3, June 1960, 386–400.

10. A. Vander Lugt, F. B. Rotz, and A. Kloester, Jr., "Character-Reading by Optical Spatial Filtering," in *Optical and Electro-Optical Information Processing*, J. Tippett et al., Eds., MIT Press, Cambridge, MA, 1965, 125–141.

11. A. Vander Lugt, "Signal Detection by Complex Spatial Filtering," *IEEE Trans. Information Theory*, **IT-10**, 2, April 1964, 139–145.

12. A. Kozma and D. L. Kelly, "Spatial Filtering for Detection of Signals Submerged in Noise," *Applied Optics*, **4**, 4, April 1965, 387–392.

13. A. Arcese, P. H. Mengert, and E. W. Trombini, "Image Detection Through Bipolar Correlation," *IEEE Trans. Information Theory*, **IT-16**, 5, September 1970, 534–541.

14. W. K. Pratt, "Correlation Techniques of Image Registration," *IEEE Trans. Aerospace and Electronic Systems*, **AES-10**, 3, May 1974, 353–358.

15. W. K. Pratt and F. Davarian, "Fast Computational Techniques for Pseudoinverse and Wiener Image Restoration," *IEEE Trans. Computers*, **C-26**, 6 June, 1977, 571–580.

16. W. Meyer-Eppler and G. Darius, "Two-Dimensional Photographic Autocorrelation of Pictures and Alphabet Letters," *Proc. 3rd London Symposium on Information Theory*, C. Cherry, Ed., Academic Press, New York, 1956, 34–36.

17. P. F. Anuta, "Digital Registration of Multispectral Video Imagery," *SPIE J.*, **7**, September 1969, 168–178.

18. J. M. S. Prewitt, "Object Enhancement and Extraction," in *Picture Processing and Psychopictorics*, B. S. Lipkin and A. Rosenfeld, Eds., Academic Press, New York, 1970.

19. Q. Tian and M. N. Huhns, "Algorithms for Subpixel Registration," *Computer Graphics, Vision, and Image Processing*, **35**, 2, August 1986, 220–233.

20. P. F. Anuta, "Spatial Registration of Multispectral and Multitemporal Imagery Using Fast Fourier Transform Techniques," *IEEE Trans. Geoscience and Electronics*, **GE-8**, 1970, 353–368.

21. A. Rosenfeld and G. J. Vandenburg, "Coarse–Fine Template Matching," *IEEE Trans. Systems, Man and Cybernetics*, **SMC-2**, February 1977, 104–107.

22. G. J. Vandenburg and A. Rosenfeld, "Two-Stage Template Matching," *IEEE Trans. Computers*, **C-26**, 4, April 1977, 384–393.

23. A. Goshtasby, S. H. Gage, and J. F. Bartolic, "A Two-Stage Cross-Correlation Approach to Template Matching," *IEEE Trans. Pattern Analysis and Machine Intelligence*, **PAMI-6**, 3, May 1984, 374–378.

24. D. I. Barnea and H. F. Silverman, "A Class of Algorithms for Fast Image Registration" *IEEE Trans. Computers*, **C-21**, 2, February 1972, 179–186.

25. B. S. Reddy and B. N. Chatterji, "An FFT-Based Technique for Translation, Rotation, and Scale-Invariant Image Registration," *IEEE Trans. Image Processing*, **IP-5**, 8, August 1996, 1266–1271.

26. E. De Castro and C. Morandi, "Registration of Translated and Rotated Images Using Finite Fourier Transforms," *IEEE Trans. Pattern Analysis and Machine Intelligence*, **PAMI-9**, 5, September 1987, 700–703.

27. M. K. Hu, "Visual Pattern Recognition by Moment Invariants," *IRE Trans. Information Theory*, **IT-8**, 2, February 1962, 179–187.

28. R. Y. Wong and E. L. Hall, "Scene Matching with Invariant Moments," *Computer Graphics and Image Processing*, **8**, 1, August 1978, 16–24.

29. A. Goshtasby, "Template Matching in Rotated Images," *IEEE Trans. Pattern Analysis and Machine Intelligence*, **PAMI-7**, 3, May 1985, 338–344.

PART 6

IMAGE PROCESSING SOFTWARE

Digital image processing applications typically are implemented by software calls to an image processing library of functional operators. Many libraries are limited to primitive functions such as lookup table manipulation, convolution, and histogram generation. Sophisticated libraries perform more complex functions such as unsharp masking, edge detection, and spatial moment shape analysis. The interface between an application and a library is an application program interface (API) which defines the semantics and syntax of an operation.

Chapter 20 describes the architecture of a full featured image processing API called the Programmer's Imaging Kernel System (PIKS). PIKS is an international standard developed under the auspices of the International Organization for Standardization (ISO) and the International Electrotechnical Commission (IEC). The PIKS description in Chapter 20 serves two purposes. It explains the architecture and elements of a well designed image processing API. It provides an introduction to PIKS usage to implement the programming exercises in Chapter 21.

20

PIKS IMAGE PROCESSING SOFTWARE

PIKS contains a rich set of operators that perform manipulations of multidimensional images or of data objects extracted from images in order to enhance, restore, or assist in the extraction of information from images. This chapter presents a functional overview of the PIKS standard and a more detailed definition of a functional subset of the standard called *PIKS Core*.

20.1. PIKS FUNCTIONAL OVERVIEW

This section provides a brief functional overview of PIKS. References 1 to 6 provide further information. The PIKS documentation utilizes British spelling conventions, which differ from American spelling conventions for some words (e.g., *colour* instead of *color*). For consistency with the PIKS standard, the British spelling convention has been adopted for this chapter.

20.1.1. PIKS Imaging Model

Figure 20.1-1 describes the PIKS imaging model. The solid lines indicate data flow, and the dashed lines indicate control flow. The PIKS application program interface consists of four major parts:

1. Data objects
2. Operators, tools, and utilities
3. System mechanisms
4. Import and export

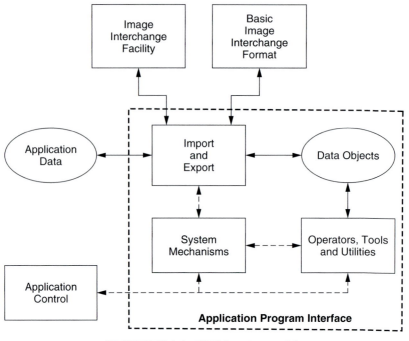

FIGURE 20.1-1. PIKS imaging model.

The PIKS data objects include both image and image-related, non-image data objects. The operators, tools, and utilities are functional elements that are used to process images or data objects extracted from images. The system mechanisms manage and control the processing. PIKS receives information from the application to invoke its system mechanisms, operators, tools, and utilities, and returns certain status and error information to the application. The import and export facility provides the means of accepting images and image-related data objects from an application, and for returning processed images and image-related data objects to the application. PIKS can transmit its internal data objects to an external facility through the ISO/IEC standards Image Interchange Facility (IIF) or the Basic Image Interchange Format (BIIF). Also, PIKS can receive data objects in its internal format, which have been supplied by the IIF or the BIIF. References 7 to 9 provide information and specifications of the IIF and BIIF.

20.1.2. PIKS Data Objects

PIKS supports two types of data objects: image data objects and image-related, non-image data objects.

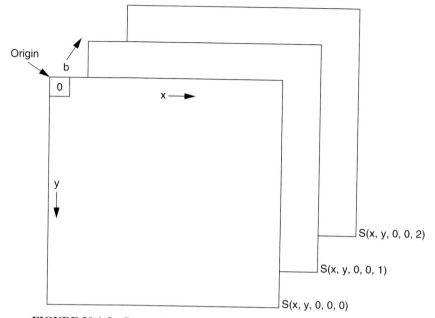

FIGURE 20.1-2. Geometrical representation of a PIKS colour image array.

A PIKS image data object is a five-dimensional collection of pixels whose structure is:

x Horizontal space index, $0 \le x \le X - 1$

y Vertical space index, $0 \le y \le Y - 1$

z Depth space index, $0 \le z \le Z - 1$

t Temporal index, $0 \le t \le T - 1$

b Colour or spectral band index, $0 \le b \le B - 1$

Some of the image dimensions may be unpopulated. For example, as shown in Figure 20.1-2, for a colour image, $Z = T = 1$. PIKS gives semantic meaning to certain dimensional subsets of the five-dimensional image object. These are listed in Table 20.1-1.

PIKS utilizes the following pixel data types:

1. Boolean

2. Non-negative integer

3. Signed integer

4. Real arithmetic

5. Complex arithmetic

TABLE 20.1-1. PIKS Image Objects

Semantic Description	Image Indices
Monochrome	x, y, 0, 0, 0
Volume	x, y, z, 0, 0
Temporal	x, y, 0, t, 0
Colour	x, y, 0, 0, b
Spectral	x, y, 0, 0, b
Volume–temporal	x, y, z, t, 0
Volume–colour	x, y, z, 0, b
Volume–spectral	x, y, z, 0, b
Temporal–colour	x, y, 0, t, b
Temporal–spectral	x, y, 0, t, b
Volume–temporal–colour	x, y, z, t, b
Volume–temporal–spectral	x, y, z, t, b
Generic	x, y, z, t, b

The precision and data storage format of pixel data is implementation dependent.

PIKS supports several image related, non-image data objects. These include:

1. *Chain:* an identifier of a sequence of operators

2. *Composite identifier:* an identifier of a structure of image arrays, lists, and records

3. *Histogram:* a construction of the counts of pixels with some particular amplitude value

4. *Lookup table:* a structure that contains pairs of entries in which the first entry is an input value to be matched and the second is an output value

5. *Matrix:* a two-dimensional array of elements that is used in vector algebra operations

6. *Neighbourhood array:* a multi-dimensional moving window associated with each pixel of an image (e.g., a convolution impulse response function array)

7. *Pixel record:* a sequence of across-band pixel values

8. *Region-of-interest:* a general mechanism for pixel-by-pixel processing selection

9. *Static array:* an identifier of the same dimension as an image to which it is related (e.g., a Fourier filter transfer function)

10. *Tuple:* a collection of data values of the same elementary data type (e.g., image size 5-tuple).

11. *Value bounds collection:* a collection of pairs of elements in which the first element is a pixel coordinate and the second element is an image measurement (e.g., pixel amplitude)

12. *Virtual register:* an identifier of a storage location for numerical values returned from operators in a chain

20.1.3. PIKS Operators, Tools, Utilities, and Mechanisms

PIKS operators are elements that manipulate images or manipulate data objects extracted from images in order to enhance or restore images, or to assist in the extraction of information from images. Exhibit 20.1-1 is a list of PIKS operators categorized by functionality.

PIKS tools are elements that create data objects to be used by PIKS operators. Exhibit 20.1-2 presents a list of PIKS tools functionally classified. PIKS utilities are elements that perform basic mechanical image manipulation tasks. A classification of PIKS utilities is shown in Exhibit 20.1-3. This list contains several file access and display utilities that are defined in a proposed amendment to PIKS. PIKS mechanisms are elements that perform control and management tasks. Exhibit 20.1-4 provides a functional listing of PIKS mechanisms. In Exhibits 20.1-1 to 20.1-4, the elements in PIKS Core are identified by an asterisk.

EXHIBIT 20.1-1. PIKS Operators Classification

Analysis: image-to-non-image operators that extract numerical information from an image

  * Accumulator
  Difference measures
  * Extrema
  * Histogram, one-dimensional
  Histogram, two-dimensional
  Hough transform
  * Line profile
  * Moments
  * Value bounds

Classification: image-to-image operators that classify each pixel of a multispectral image into one of a specified number of classes based on the amplitudes of pixels across image bands

  Classifier, Bayes
  Classifier, nearest neighbour

Colour: image-to-image operators that convert a colour image from one colour space to another

  * Colour conversion, linear
  * Colour conversion, nonlinear
  * Colour conversion, subtractive
  Colour lookup, interpolated
  * Luminance generation

Complex image: image-to-image operators that perform basic manipulations of
images in real and imaginary or magnitude and phase form

   *Complex composition
   *Complex conjugate
   *Complex decomposition
   *Complex magnitude

Correlation: image-to-non-image operators that compute a correlation array of a
pair of images

   Cross-correlation
   Template match

Edge detection: image-to-image operators that detect the edge boundary of objects
within an image

   Edge detection, orthogonal gradient
   Edge detection, second derivative
   Edge detection, template gradient

Enhancement: image-to-image operators that improve the visual appearance of an
image or that convert an image to a form better suited for analysis by
a human or a machine

   Adaptive histogram equalization
   False colour
   Histogram modification
   Outlier removal
   Pseudocolour
   Unsharp mask
   Wallis statistical differencing

Ensemble: image-to-image operators that perform arithmetic, extremal, and logical
combinations of pixels

   *Alpha blend, constant
   Alpha blend, variable
   *Dyadic, arithmetic
   *Dyadic, complex
   *Dyadic, logical
   *Dyadic, predicate
   *Split image
   Z merge

Feature extraction: image-to-image operators that compute a set of image features at each pixel of an image

   Label objects
   Laws texture features
   Window statistics

Filtering: image-to-image operators that perform neighbourhood combinations of pixels directly or by Fourier transform domain processing

   Convolve, five-dimensional
  *Convolve, two-dimensional
   Filtering, homomorphic
  *Filtering, linear

Geometric: image-to-image and ROI-to-ROI operators that perform geometric modifications

   Cartesian to polar
  *Flip, spin, transpose
   Polar to cartesian
  *Rescale
  *Resize
  *Rotate
  *Subsample
  *Translate
   Warp, control point
  *Warp, lookup table
  *Warp, polynomial
  *Zoom

Histogram shape: non-image to non-image operators that generate shape measurements of a pixel amplitude histogram of an image

   Histogram shape, one-dimensional
   Histogram shape, two-dimensional

Morphological: image-to-image operators that perform morphological operations on boolean and grey scale images

  *Erosion or dilation, Boolean
  *Erosion or dilation, grey
  *Fill region
   Hit or miss transformation
  *Morphic processor

Morphology
Neighbour count
Open and close

Pixel modification: image-to-image operators that modify an image by pixel draw-
ing or painting

Draw pixels
Paint pixels

Point: image-to-image operators that perform point manipulation on a pixel-by-
pixel basis

   *Bit shift
   * Complement
   Error function scaling
   *Gamma correction
   Histogram scaling
   Level slice
   *Lookup Lookup, interpolated
   *Monadic, arithmetic
   *Monadic, complex
   *Monadic, logical
   Noise combination
   *Power law scaling
   Rubber band scaling
   *Threshold
   *Unary, integer
   *Unary, real
   *Window-level

Presentation: image-to-image operators that prepare an image for display

   *Diffuse
   *Dither

Shape: Image-to-non-image operators that label objects and perform measurements
of the shape of objects within an image

Perimeter code generator
Shape metrics
Spatial moments, invariant
Spatial moments, scaled

Unitary transform: image-to-image operators that perform multi-dimensional forward and inverse unitary transforms of an image

Transform, cosine
*Transform, Fourier
Transform, Hadamard
Transform, Hartley

3D Specific: image-to-image operators that perform manipulations of three-dimensional image data

Sequence average
Sequence Karhunen-Loeve transform
Sequence running measures
3D slice

EXHIBIT 20.1-2 PIKS Tools Classification

Image generation: Tools that create test images

Image, bar chart
*Image, constant
Image, Gaussian image
Image, grey scale image
Image, random number image

Impulse response function array generation: Tools that create impulse response function neighbourhood array data objects

Impulse, boxcar
*Impulse, derivative of Gaussian
Impulse, difference of Gaussians
*Impulse, elliptical
*Impulse, Gaussian
*Impulse, Laplacian of Gaussian
Impulse, pyramid
*Impulse, rectangular
Impulse, sinc function

Lookup table generation: Tools that create entries of a lookup table data object

* Array to LUT

Matrix generation: tools that create matrix data objects

*Colour conversion matrix

Region-of-interest generation: tools that create region-of-interest data objects from a mathematical description of the region-of-interest

  *ROI, coordinate
  *ROI, elliptical
  *ROI, polygon
  *ROI, rectangular

Static array generation: tools that create filter transfer function, power spectrum, and windowing function static array data objects

  *Filter, Butterworth
  *Filter, Gaussian
   Filter, inverse
   Filter, matched
   Filter, Wiener
   Filter, zonal
   Markov process power spectrum
   Windowing function

EXHIBIT 20.1-3. PIKS Utilities Classification

Display: utilities that perform image display functions

  *Boolean display
  *Close window
  *Colour display
  *Event display
  *Monochrome display
  *Open titled window
  *Open window
  *Pseudocolour display

Export From Piks: Utilities that export image and non-image data objects from PIKS to an application or to the IIF or BIIF

  *Export histogram
  *Export image
  *Export LUT
  *Export matrix
  *Export neighbourhood array
  *Export ROI array
  *Export static array
  *Export tuple

* Export value bounds
* Get colour pixel
* Get pixel
* Get pixel array
  Get pixel record
* Output image file
  Output object

Import to PIKS: utilities that import image and non-image data objects to PIKS
from an application or from the IIF or the BIIF

* Import histogram
* Import image
* Import LUT
* Import matrix
* Import neighbourhood array
* Import ROI array
* Import static array
* Import tuple
* Import value bounds
  Input object
* Input image file
* Input PhotoCD
* Put colour pixel
* Put pixel
* Put pixel array
  Put pixel record

Inquiry: utilities that return information to the application regarding PIKS data
objects, status and implementation

  Inquire chain environment
  Inquire chain status
* Inquire elements
* Inquire image
  Inquire index assignment
* Inquire non-image object
* Inquire PIKS implementation
* Inquire PIKS status
* Inquire repository
* Inquire resampling

Internal: utilities that perform manipulation and conversion of PIKS internal image
and non-image data objects

* Constant predicate

*Convert array to image
*Convert image data type
*Convert image to array
*Convert image to ROI
*Convert ROI to image
*Copy window
*Create tuple
*Equal predicate
*Extract pixel plane
*Insert pixel plane

EXHIBITS 20.1-4 PIKS Mechanisms Classification

Chaining: mechanisms that manage execution of PIKS elements inserted in chains

Chain abort
Chain begin
Chain delete
Chain end
Chain execute
Chain reload

Composite identifier management: mechanisms that perform manipulation of image identifiers inserted in arrays, lists, and records

Composite identifier array equal
Composite identifier array get
Composite identifier array put
Composite identifier list empty
Composite identifier list equal
Composite identifier list get
Composite identifier list insert
Composite identifier list remove
Composite identifier record equal
Composite identifier record get
Composite identifier record put

Control: mechanisms that control the basic operational functionality of PIKS

Abort asynchronous execution
*Close PIKS
*Close PIKS, emergency
*Open PIKS
Synchronize

Error: mechanisms that provide means of reporting operational errors

   *Error handler
   *Error logger
   *Error test

System management: mechanisms that allocate, deallocate, bind, and set attributes
                       of data objects and set global variables

   Allocate chain
   Allocate composite identifier array
   Allocate composite identifier list
   Allocate composite identifier record
   *Allocate display image
   *Allocate histogram
   *Allocate image
   *Allocate lookup table
   *Allocate matrix
   *Allocate neighbourhood array
   Allocate pixel record
   *Allocate ROI
   *Allocate static array
   *Allocate tuple
   *Allocate value bounds collection
   Allocate virtual register
   Bind match point
   *Bind ROI
   *Deallocate data object
   *Define sub image
   *Return repository identifier
   *Set globals
   *Set image attributes
   Set index assignment

Virtual register: mechanisms that manage the use of virtual registers

   Vreg alter
   Vreg clear
   Vreg conditional
   Vreg copy
   Vreg create
   Vreg delete
   Vreg get
   Vreg set
   Vreg wait

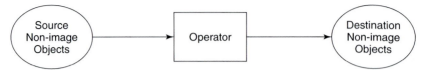

FIGURE 20.1-3. PIKS operator model: non-image to non-image operators.

20.1.4. PIKS Operator Model

The PIKS operator model provides three possible transformations of PIKS data objects by a PIKS operator:

1. Non-image to non-image
2. Image to non-image
3. Image to image

Figure 20.1-3 shows the PIKS operator model for the transformation of non-image data objects to produce destination non-image data objects. An example of such a transformation is the generation of shape features from an image histogram. The operator model for the transformation of image data objects by an operator to produce non-image data objects is shown in Figure 20.1-4. An example of such a transformation is the computation of the least-squares error between a pair of images. In this operator model, processing is subject to two control mechanisms: region-of-interest (ROI) source selection and source match point translation. These control mechanisms are defined later. The dashed line in Figure 20.1-4 indicates the transfer of control information. The dotted line indicates the binding of source ROI objects to source image objects. Figure 20.1-5 shows the PIKS operator model for

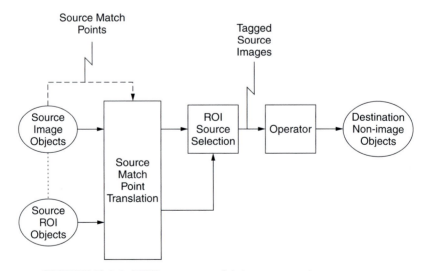

FIGURE 20.1-4. PIKS operator model: image to non-image operators.

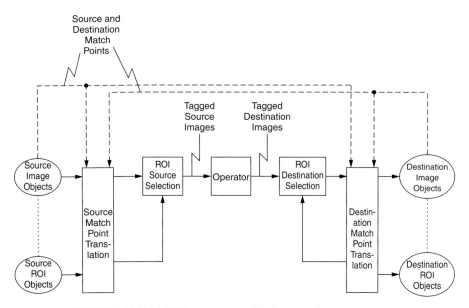

FIGURE 20.1-5. PIKS operator model: image to image operators.

the transformation of image data objects by an operator to produce other image data objects. An example of such an operator is the unsharp masking operator, which enhances detail within an image. In this operator model, processing is subject to four control mechanisms: source match point translation, destination match point translation, ROI source selection, and ROI destination selection.

Index Assignment. Some PIKS image to non-image and image to image operators have the capability of assigning operator indices to image indices. This capability permits operators that are inherently Nth order, where $N < 5$, to be applied to five-dimensional images in a flexible manner. For example, a two-dimensional Fourier transform can be taken of each column slice of a volumetric image using index assignment.

ROI Control. A region-of-interest (ROI) data object can be used to control which pixels within a source image will be processed by an operator and to specify which pixels processed by an operator will be recorded in a destination image. Conceptually, a ROI consists of an array of Boolean value pixels of up to five dimensions. Figure 20.1-6 presents an example of a two-dimensional rectangular ROI. In this example, if the pixels in the cross-hatched region are logically TRUE, the remaining pixels are logically FALSE. Otherwise, if the cross-hatched pixels are set FALSE, the others are TRUE.

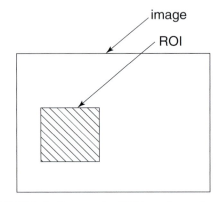

FIGURE 20.1-6. Rectangular ROI bound to an image array.

The size of a ROI need not be the same as the size of an image to which it is associated. When a ROI is to be associated with an image, a binding process occurs in which a ROI control object is generated. If the ROI data object is larger in spatial extent than the image to which it is to be bound, it is clipped to the image size to form the ROI control object. In the opposite case, if the ROI data object is smaller than the image, the ROI control object is set to the FALSE state in the non-overlap region.

Figure 20.1-7 illustrates three cases of ROI functionality for point processing of a monochrome image. In case 1, the destination ROI control object is logically TRUE over the full image extent, and the source ROI control object is TRUE over a cross-hatched rectangular region smaller than the full image. In this case, the destination image consists of the existing destination image with an insert of processed source pixels. For case 2, the source ROI is of full extent, and the destination ROI is of a smaller cross-hatched rectangular extent. The resultant destination image consists of processed pixels inserted into the existing destination image. Functionally, the result is the same as for case 1. The third case shows the destination image when the source and destination ROIs are overlapping rectangles smaller than the image extent. In this case, the processed pixels are recorded only in the overlap area of the source and destination ROIs.

The ROI concept applies to multiple destination images. Each destination image has a separately bound ROI control object which independently controls recording of pixels in the corresponding destination image. The ROI concept also applies to neighbourhood as well as point operators. Each neighbourhood processing element, such as an impulse response array, has a pre-defined key pixel. If the key pixel lies within a source control ROI, the output pixel is formed by the neighbourhood operator even if any or all neighbourhood elements lie outside the ROI.

PIKS provides tools for generating ROI data objects from higher level specifications. Such supported specifications include:

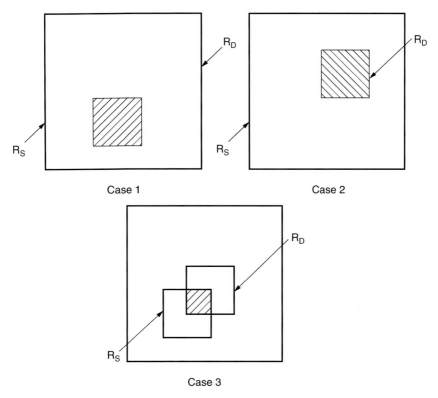

FIGURE 20.1-7. ROI operation.

1. Coordinate list
2. Ellipse
3. Polygon
4. Rectangle

These tools, together with the ROI binding tool, provide the capability to conceptually generate five-dimensional ROI control objects from lower dimensional descriptions by pixel plane extensions. For example, with the elliptical ROI generation tool, it is possible to generate a circular disk ROI in a spatial pixel plane, and then cause the disk to be replicated over the other pixel planes of a volumetric image to obtain a cylinder-shaped ROI.

Match Point Control. Each PIKS image object has an associated match point coordinate set (x, y, z, t, b) which some PIKS operators utilize to control multi-dimensional translations of images prior to processing by an operator. The generic effect of match point control for an operator that creates multiple destination images from

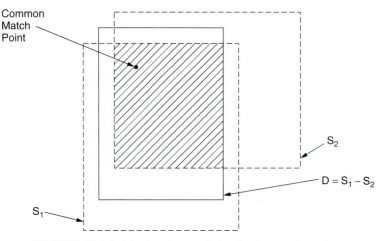

FIGURE 20.1-8. Match point translation for image subtraction.

multiple source images is to translate each source image and each destination image, other than the first source image, such that the match points of these images are aligned with the match point of the first source image prior to processing. Processing then occurs on the spatial intersection of all images. Figure 20.1-8 an example of image subtraction subject to match point control. In the example, the difference image is shown cross-hatched.

Other Features. PIKS provides a number of other features to control processing. These include:

1. Processing of ROI objects in concert with image objects
2. Global setting of image and ROI resampling options
3. Global engagement of ROI control and ROI processing
4. Global engagement of index assignment
5. Global engagement of match point control
6. Global engagement of synchronous or asynchronous operation
7. Heterogeneous bands of dissimilar data types
8. Operator chaining
9. Virtual registers to store intermediate numerical results of an operator chain
10. Composite image management of image and non-image objects

The PIKS Functional Specification (2) provides rigorous specifications of these features. PIKS also contains a data object repository of commonly used impulse response arrays, dither arrays, and colour conversion matrices.

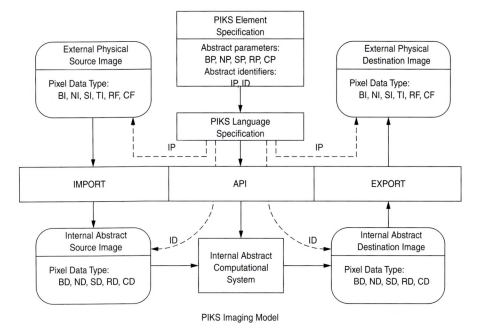

PIKS Imaging Model

FIGURE 20.1-9. PIKS application interface.

20.1.5. PIKS Application Interface

Figure 20.1-9 describes the PIKS application interface for data interchange for an implementation-specific data pathway. PIKS supports a limited number of physical data types that may exist within an application domain or within the PIKS domain. Such data types represent both input and output parameters of PIKS elements and image and non-image data that are interchanged between PIKS and the application.

PIKS provides notational differentiation between most of the elementary abstract data types used entirely within the PIKS domain (PIKS internal), those that are used to convey parameter data between PIKS and the application (PIKS parameter), and those that are used to convey pixel data between PIKS and the application (external physical image). Table 20.1-2 lists the codes for the PIKS abstract data types. The abstract data types are defined in ISO/IEC 12087-1. PIKS internal and parameter data types are of the same class if they refer to the same basic data type. For example, RP and RD data types are of the same class, but RP and SD data types are of different classes. The external physical data types supported by PIKS for the import and export of image data are also listed in Table 20.1-2. PIKS internal pixel data types and external pixel data types are of the same class if they refer to the same basic data type. For example, ND and NI data types are of the same class, but SI and ND data types are of different classes.

TABLE 20.1-2 PIKS Datatype Codes

Data Type	PIKS Internal Code	PIKS Parameter Code	Physical Code
Boolean	BD	BP	BI
Non-negative integer	ND	NP	NI
Signed integer	SD	SP	SI
Fixed-point integer	—	—	TI
Real arithmetic	RD	RP	RF
Complex arithmetic	CD	CP	CF
Character string	CS	CS	—
Data object identifier	ID	IP	—
Enumerated	NA	EP	—
Null	NULL	NULL	—

20.1.6. PIKS Conformance Profiles

Because image processing requirements vary considerably across various applications, PIKS functionality has been subdivided into the following five nested sets of functionality called *conformance profiles*:

1. *PIKS Foundation*: basic image processing functionality for monochrome and colour images whose pixels are represented as Boolean values or as non-negative or signed integers.

2. *PIKS Core*: intermediate image processing functionality for monochrome and colour images whose pixels are represented as Boolean values, non-negative or signed integers, real arithmetic values, and complex arithmetic values. PIKS Core is a superset of PIKS Foundation.

3. *PIKS Technical*: expanded image processing functionality for monochrome, colour, volume, temporal, and spectral images for all pixel data types.

4. *PIKS Scientific*: complete set of image processing functionality for all image structures and pixel data types. PIKS Scientific is a superset of PIKS Technical functionality.

5. *PIKS Full*: complete set of image processing functionality for all image structures and pixel data types plus the capability to chain together PIKS processing elements and to operate asynchronously. PIKS Full is a superset of PIKS Scientific functionality.

Each PIKS profile may include the capability to interface with the IIF, the BIIF, and to include display and input/output functionality, as specified by PIKS Amendment 1.

20.2. PIKS CORE OVERVIEW

The PIKS Core profile provides an intermediate level of functionality designed to service the majority of image processing applications. It supports all pixel data types, but only monochrome and colour images of the full five-dimensional PIKS image data object. It supports the following processing features:

1. Nearest neighbour, bilinear, and cubic convolution global resampling image interpolation

2. Nearest neighbour global resampling ROI interpolation

3. All ROIs

4. Data object repository

The following sections provide details of the data structures for PIKS Core non-image and image data objects.

20.2.1. PIKS Core Non-image Data Objects

PIKS Core supports the non-image data objects listed below. The list contains the PIKS Functional Specific object name code and the definition of each object.

HIST	Histogram
LUT	Look-up table
MATRIX	Matrix
NBHOOD_ARRAY	Neighbourhood array
ROI	Region-of-interest
STATIC_ARRAY	Static array
TUPLE	Tuple
VALUE_BOUNDS	Value bounds collection

The tuple object is defined first because it is used to define other non-image and image data objects. Tuples are also widely used in PIKS to specify operator and tool parameters (e.g., the size of a magnified image). Figure 20.2-1 contains the tree structure of a tuple object. It consists of the tuple size, tuple data type, and a private identifier to the tuple data values. The tuple size is an unsigned integer that specifies the number of tuple data values. The tuple datatype option is a signed integer from 1 to 6 that specifies one of the six options. The identifier to the tuple data array is private in the sense that it is not available to an application; only the tuple data object itself has a public identifier.

A PIKS histogram data object is a one-dimensional array of unsigned integers that stores the histogram of an image plus histogram object attributes. Figure 20.2-2 shows the tree structure of a histogram data object. The histogram array size is an unsigned integer that specifies the number of histogram bins. The lower and upper

Tuple Object
    Tuple data size
        number of tuple data values, e.g. 5
    Tuple datatype option
        choice of BD, ND, SD, RD, CD or CS
    Tuple data array
        private identifier

FIGURE 20.2-1. Tuple object tree structure.

Histogram Object
    Histogram array size
        number of histogram bins, e.g. 512
    Lower amplitude value
        lower amplitude value of histogram range, e.g. 0.1
    Upper amplitude value
        upper amplitude value of histogram range, e.g. 0.9
    Histogram data array
        private identifier

FIGURE 20.2-2. Histogram object tree structure.

amplitude values are real numbers that specify the pixel amplitude range of the histogram.

A PIKS look-up table data object, as shown in Figure 20.2-3, is a two-dimensional array that stores the look-up table data plus a collection of look-up table attributes. The two-dimensional array has the general form following:

$$
\begin{array}{ccccc}
T(0,0) & \cdots & T(b,0) & \cdots & T(B-1,0) \\
\vdots & & \vdots & & \vdots \\
T(0,e) & \cdots & T(b,e) & \cdots & T(B-1,e) \\
\vdots & & \vdots & & \vdots \\
T(0,E-1) & \cdots & T(b,E-1) & \cdots & T(B-1,E-1)
\end{array}
$$

A positive integer e is the input row index to the table. It is derived from a source image by the relationship

$$e = S(x, y, z, t, b) \qquad (20.2\text{-}1)$$

The LUT output is a one-dimensional array

$$a(e) = [T(0,e) \cdots T(b,e) \cdots T(B-1,e)] \qquad (20.2\text{-}2)$$

Lookup Table Object
    Table entries
        number of table entries, e.g. 512
    Table bands
        number of table bands, e.g. 3
    Table input data type option
        choice of ND or SD
    Table output data type option
        choice of BD, ND, SD, RD OR CD
    Lookup table data array
        private identifier

FIGURE 20.2-3. Look-up table object tree structure.

There are two types of usage for PIKS Core: (1) the source and destination images are of the same band dimension, or (2) the source image is monochrome and the destination image is colour. In the former case,

$$D(x, y, 0, 0, b) = T(0, S(x, y, z, t, b))$$
(20.2-3)

In the latter case,

$$D(x, y, 0, 0, b) = T(b, S(x, y, z, t, 0))$$
(20.2-4)

Figure 20.2-4 shows the tree structure of a matrix data object. The matrix is specified by its number of rows R and columns C and the data type of its constituent terms. The matrix is addressed as follows:

$$
\mathbf{M} = \begin{bmatrix}
M(1, 1) & \cdots & M(1, c) & \cdots & M(1, C) \\
\vdots & & \vdots & & \vdots \\
M(r, 1) & \cdots & M(r, c) & \cdots & M(r, C) \\
\vdots & & \vdots & & \vdots \\
M(R, 1) & \cdots & M(R, c) & \cdots & M(R, C)
\end{bmatrix}
$$
(20.2-5)

In PIKS, matrices are used primarily for colour space conversion.

A PIKS Core neighbourhood array is a two-dimensional array and associated attributes as shown in Figure 20.2-5. The array has J columns and K rows. As shown below, it is indexed in the same manner as a two-dimensional image.

Matrix Object
    Column size
        number of matrix columns, e.g. 4
    Row size
        number of matrix rows, e.g. 3
    Matrix data type option
        choice of ND, SD, RD or CD
    Matrix data array
        private identifier

FIGURE 20.2-4. Matrix object tree structure.

Neighbourhood Array Object
    Neighbourhood size
        5-tuple public identifier specification of J, K, 1, 1, 1
    Key pixel
        5-tuple public identifier specification of j_K, k_K, 0, 0, 0
    Scale factor
        integer value
    Semantic label option
        choice of GL, DL, IL, ML, SL
    Neighbourhood data array
        private identifier

FIGURE 20.2-5. Neighbourhood object tree structure.

$$H(j, k) = \frac{1}{S} \begin{bmatrix} H(0, 0) & \dots & H(j, 0) & \dots & H(J-1, 0) \\ \vdots & & \vdots & & \vdots \\ H(0, k) & \dots & H(j, k) & \dots & H(J-1, k) \\ \vdots & & \vdots & & \vdots \\ H(0, K-1) & \dots & H(j, K-1) & \dots & H(J-1, K-1) \end{bmatrix} \quad (20.2\text{-}6)$$

In Eq. 20.2-6, the scale factor S is unity except for the signed integer data. For signed integers, the scale factor can be used to realize fractional elements. The key pixel (j_K, k_K) defines the origin of the neighbourhood array. It need not be with the confines of the array. There are five types of neighbourhood arrays, specified by the following structure codes:

GL   Generic array

DL   Dither array

IL   Impulse response array

ML   Mask array

SL   Structuring element array

```
Region-of-interest Object
        ROI virtual array size
                5-tuple public identifier specification of X_R, Y_R, 1, 1, 1
        ROI structure option
                choice of AR, CR, ER, GR, PR, RR
        Polarity option
                choice of TRUE or FALSE
        Conceptual ROI data array
                private identifier
```

FIGURE 20.2-6. Region-of-interest object tree structure.

Figure 20.2-6 shows the tree structure of a region-of-interest ROI data object. Conceptually, a PIKS Core ROI data object is a two-dimensional array of Boolean value pixels of width X_R and height Y_R. The actual storage method is implementation dependent. The ROI can be constructed by one of the following representations:

AR	ROI array
CR	ROI coordinate list
ER	ROI elliptical
GR	ROI generic
RR	ROI rectangular

The ROI can be defined to be TRUE or FALSE in its interior.

A PIKS Core static array is a two-dimensional array of width X_S and height Y_S as shown in Figure 20.2-7. Following is a list of the types of static arrays supported by PIKS:

GS	Generic static array
PS	Power spectrum
TS	Transfer function
WS	Windowing function

```
Static Array Object
        Static array size
                5-tuple public identifier specification of X_S, Y_S, 1, 1, 1
        Semantic label option
                choice of GS, PS, TS, WS
        Datatype option
                choice of BD, ND, SD, RD or CD
        Static array data array
                private identifier
```

FIGURE 20.2-7. Static array object tree structure.

Value Bounds Collection Object
Collection size
number of collection members
Lower amplitude bound
value of lower amplitude bound
Upper amplitude bound
value of upper amplitude bound
Pixel data type option
choice of NP, SP, RP
Value bounds collection data array
private identifier

FIGURE 20.2-8. Value bounds collection object tree structure.

A value bounds collection is a storage mechanism containing the pixel coordinate and pixel values of all pixels whose amplitudes lie within a lower and an upper bound. Figure 20.2-8 is the tree structure of the value bounds collection data object.

20.2.2. PIKS Core Image Data Object

A PIKS image object is a tree structure of image attributes, processing control attributes, and private identifiers to an image data array of pixels and an associated ROI. Figure 20.2-9 illustrates the tree structure of an image object. The image attributes are created when an image object is allocated. When an image is allocated, there will be no private identifier to the image array data. The private identifier is established automatically when raw image data are imported to a PIKS image object or when a destination image is created by an operator. The processing control attributes are created when a ROI is bound to an image. It should be noted that for PIKS Core, all bands must be of the same datatype and pixel precision. The pixel precision specification must be in accord with the choices provided by a particular PIKS implementation.

20.2.3. PIKS Core C Language Binding

The PIKS Functional Specification document (2) establishes the semantic usage of PIKS. The PIKS C language binding document (10) defines the PIKS syntactical usage for the C programming language. At present, there are no other language bindings. Reader familiarity with the C programming language is assumed.

The PIKS C binding has adopted the *Hungarian prototype naming convention*, in which the datatypes of all entities are specified by prefix codes. Table 20.2-1 lists the datatype prefix codes. The entities in courier font are binding names. Table 20.2-2 gives the relationship between the PIKS Core C binding designators and the PIKS Functional Specification datatypes and data objects. The general structure of the C language binding element prototype is

Image Object
    Image attributes
        Representation
            Size
                5-tuple public identifier specification of X, Y, Z, T, B
            Band datatype
             B-tuple public identifier specification of BD, ND, SD, RD or CD datatype
            Image structure option
                MON or COLR
        Channel
            Band precision
             B-tuple public identifier specification of pixel precision per band
        Colour
            White point
             specification of X_0, Y_0, Z_0
            Colour space option
             29 choices, e.g. CIE L*a*b* or CMYK
    Control
        ROI
            private identifier
        ROI offset
            5-tuple public identifier specification of x_0, y_0, z_0, t_0, b_0
    Image data array
        private identifier

FIGURE 20.2-9. Image object tree structure.

```
void IvElementName
```

or

```
I(prefix)ReturnName  I(prefix)ElementName
```

As an example, the following is the element C binding prototype for two-dimensional convolution of a source image into a destination image:

```
Idnimage InConvolve2D(      /* OUT destination image identifier */
    Idnimage   nSourceImage, /* source image identifier           */
    Idnimage   nDestImage,   /* destination image identifier      */
    Idnnbhood  nImpulse,     /* impulse response array identifier */
    Ipint      iOption       /* convolution 2D option             */
);
```

In this example, the first two components of the prototype are the identifiers to the source and destination images. Next is the identifier to the impulse response neighbourhood array. The last component is the integer option parameter for the convolution boundary option. The following #define convolution options are provided in the piks.h header file:

TABLE 20.2-1 PIKS Datatype Prefix Codes

Prefix	Definition
a	Array
b	Boolean
c	Character
d	Internal data type
e	Enumerated data type
f	Function
i	Integer
m	External image data type
n	Identifier
p	Parameter type
r	Real
s	Structure
t	Pointer
u	Unsigned integer
v	Void
z	Zero terminated string
st	Structure or union pointer
tba	Pointer to Boolean array
tia	Pointer to integer array
tf	Pointer to function
tra	Pointer to real array
tua	Pointer to unsigned integer array

```
ICONVOLVE_UPPER_LEFT      1   /* upper left corner justified    */
ICONVOLVE_ENCLOSED        2   /* enclosed array                 */
ICONVOLVE_KEY_ZERO        3   /* key pixel, zero exterior       */
ICONVOLVE_KEY_REFLECTED   4   /* key pixel, reflected exterior  */
```

As an example, let nSrc and nDst be the identifier names assigned to a source and destination images, respectively, and let nImpulse be the identifier of an impulse response array. In an application program, the two-dimensional convolution operator can be invoked as

```
InConvolve2D(nSrc, nDst, nImpulse, ICONVOLVE_ENCLOSED);
```

or by

```
nDst = InConvolve2D(nSrc, nDst, nImpulse, ICONVOLVE_EN CLOSED);
```

TABLE 20.2-2 PIKS Core C Binding Designators and Functional Specification Datatypes and Data Objects

Binding	Functional Specification	Description
Imbool	BI	External Boolean datatype
Imuint	NI	External non-negative integer datatype
Imint	SI	External signed integer datatype
Imfixed	TI	External fixed point integer datatype
Imfloat	RF	External floating point datatype
Ipbool	BP	Parameter Boolean datatype
Ipuint	NP	Parameter non-negative integer datatype
Ipint	SP	Parameter signed integer datatype
Ipfloat	RP	Parameter real arithmetic datatype
Idnimage	SRC, DST	Image data object
Idnhist	HIST	Histogram data object
Idnlut	LUT	Lookup table data object
Idnmatrix	MATRIX	Matrix data object
Idnnbhood	NBHOOD_ARRAY	Neighbourhood array data object
Idnroi	ROI	Region-of-interest data object
idnstatic	STATIC_ARRAY	Static array data object
Idntuple	TUPLE	Tuple data object
Idnbounds	VALUE_BOUNDS	Value bounds collection data object
Idnrepository	IP	External repository identifier
Ipnerror	IP	External error file identifier
Ipsparameter_basic	IP	External tuple data array pointer union
Ipsparameter_numeric	IP	External matrix data array pointer union
Ipsparameter_pixel	IP	External LUT, neighbourhood, pixel data array pointer union
Ipspiks_pixel_types	IP	External image data array pointer union

where ICONVOLVE_ENCLOSED is a boundary convolution option. The second formulation is useful for nesting of operator calls.

The PIKS C binding provides a number of standardized convenience functions, which are shortcuts for creating tuples, ROIs, and monochrome and colour images.

Reference 5 is a complete C programmer's guide for the PIKS Foundation profile. The compact disk contains a PDF file of a PIKS Core programmer's reference manual. This manual contains program snippets for each of the PIKS elements that explain their use.

REFERENCES

1. "Information Technology, Computer Graphics and Image Processing, Image Processing and Interchange, Functional Specification, Part 1: Common Architecture for Imaging," ISO/IEC 12087-1:1995(E).

2. "Information Technology, Computer Graphics and Image Processing, Image Processing and Interchange, Functional Specification, Part 2: Programmer's Imaging Kernel System Application Program Interface," ISO/IEC 12087-2:1994(E).

3. A. F. Clark, "Image Processing and Interchange: The Image Model," *Proc. SPIE/IS&T Conference on Image Processing and Interchange: Implementation and Systems*, San Jose, CA, February 1992, **1659**, SPIE Press, Bellingham, WA, 106–116.

4. W. K. Pratt, "An Overview of the ISO/IEC Programmer's Imaging Kernel System Application Program Interface," *Proc. SPIE/IS&T Conference on Image Processing and Interchange: Implementation and Systems*, San Jose, CA, February 1992, **1659**, SPIE Press, Bellingham, WA, 117–129.

5. W. K. Pratt, *PIKS Foundation C Programmer's Guide*, Manning Publications, Prentice Hall, Upper Saddle River, NJ, 1995.

6. W. K. Pratt, "Overview of the ISO/IEC Image Processing and Interchange Standard," *in Standards for Electronic Imaging Technologies, Devices, and Systems,* M. C. Nier, Ed., San Jose, CA, February 1996, **CR61**, SPIE Press, Bellingham, WA, 29–53.

7. "Information Technology, Computer Graphics and Image Processing, Image Processing and Interchange, Functional Specification, Part 3: Image Interchange Facility," ISO/IEC 12087-3:1995(E).

8. C. Blum and G. R. Hoffman, "ISO/IEC's Image Interchange Facility," *Proc. SPIE/IS&T Conf. on Image Processing and Interchange: Implementation and Systems*, San Jose, CA, February 1992, **1659**, SPIE Press, Bellingham, WA, 130–141.

9. "Information Technology, Computer Graphics and Image Processing, Image Processing and Interchange, Functional Specification, Part 5: Basic Image Interchange Format," ISO/IEC 12087-5:1998(E).

10. "Information Technology, Computer Graphics and Image Processing, Image Processing and Interchange, Application Program Interface Language Bindings, *Part 4: C*," ISO/IEC 12088-4:1995(E).

21

PIKS IMAGE PROCESSING PROGRAMMING EXERCISES

Digital image processing is best learned by writing and executing software programs that implement image processing algorithms. Toward this end, the compact disk affixed to the back cover of this book provides executable versions of the PIKS Core Application Program Interface C programming language library, which can be used to implement exercises described in this chapter.

The compact disk contains the following items:

A Solaris operating system executable version of the PIKS Core API.

A Windows 2000 and Windows NT operating system executable version of the PIKS Core API.

A Windows 2000 and Windows NT operating system executable version of PIKSTool, a graphical user interface method of executing many of the PIKS Core operators without program compilation.

A PDF file format version of the PIKS Core C Programmer's Reference Manual.

PDF file format and Word versions of the PIKSTool User's Manual.

A PDF file format version of the image database directory.

A digital image database of most of the source images used in the book plus many others widely used in the literature. The images are provided in the PIKS file format. A utility program is provided for conversion from the PIKS file format to the TIFF file format.

Digital images of many of the book photographic figures. The images are provided in the TIFF file format. A utility program is provided for conversion from the TIFF file format to the PIKS file format.

C program source demonstration programs.

C program executable programs of the programming exercises.

To install the CD on a Windows computer, insert the CD into the CD drive and follow the screen instructions. To install the CD on a Solaris computer, create a subdirectory called PIKSrelease, and make that your current working directory by executing:

```
mkdir PIKSrelease
cd PIKSrelease
```

Insert the PIKS CD in the CD drive and type:

```
/cdrom/piks_core_1_6/install.sh
```

See the README text file in the PIKSrelease directory for further installation information.

For further information about the PIKS software, please refer to the PixelSoft,

Inc. web site:

or send email to:

pixelsoft@pixelsoft.com

The following sections contain descriptions of programming exercises. All of them can be implemented using the PIKS API. Some can be more easily implemented using PIKSTool. It is, of course, possible to implement the exercises with other APIs or tools that match the functionality of PIKS Core.

21.1 PROGRAM GENERATION EXERCISES

1.1   Develop a program that:

  (a) Opens a program session.

  (b) Reads file parameters of a source image stored in a file.

  (c) Allocates unsigned integer, monochrome source and destination images.

  (d) Reads an unsigned integer, 8-bit, monochrome source image from a file.

  (e) Opens an image display window and displays the source image.

  (f) Creates a destination image, which is the complement of the source image.

(g) Opens a second display window and displays the destination image.

(h) Closes the program session.

The executable `example_complement_monochrome_ND` performs this exercise. The utility source program `DisplayMonochromeND.c` provides a PIKS template for this exercise. Refer to the *input_image_file* manual page of the PIKS Programmer's Reference Manual for file reading information.

1.2   Develop a program that:

(a) Creates, in application space, an unsigned integer, 8-bit, 512 × 512 pixel array of a source ramp image whose amplitude increases from left-to-right from 0 to 255.

(b) Imports the source image for display.

(c) Creates a destination image by adding value 100 to each pixel

(d) Displays the destination image

What is the visual effect of the display in step (d)? The *monadic_arithmetic* operator can be used for the pixel addition. The executable `example_import_ramp` performs this exercise. See the *monadic_arithmetic*, and *import_image* manual pages.

21.2 IMAGE MANIPULATION EXERCISES

2.1   Develop a program that passes a monochrome image through the log part of the monochrome vision model of Figure 2.4-4. Steps:

(a) Convert an unsigned integer, 8-bit, monochrome source image to floating point datatype.

(b) Scale the source image over the range 1.0 to 100.0.

(c) Compute the source image logarithmic lightness function of Eq. 6.3-4.

(d) Scale the log source image for display.

The executable `example_monochrome_vision` performs this exercise. Refer to the *window-level* manual page for image scaling. See the *unary_real* and *monadic_arithmetic* manual pages for computation of the logarithmic lightness function.

2.2   Develop a program that passes an unsigned integer, monochrome image through a lookup table with a square root function. Steps:

(a) Read an unsigned integer, 8-bit, monochrome source image from a file.

(b) Display the source image.

(c) Allocate a 256 level lookup table.

(d) Load the lookup table with a square root function.

(e) Pass the source image through the lookup table.

(f) Display the destination image.

The executable `example_lookup_monochrome_ND` performs this exercise. See the *allocate_lookup_table*, *import_lut,* and *lookup* manual pages.

2.3   Develop a program that passes a signed integer, monochrome image through a lookup table with a square root function. Steps:

(a) Read a signed integer, 16-bit, monochrome source image from a file.

(b) Linearly scale the source image over its maximum range and display it.

(c) Allocate a 32,768 level lookup table.

(d) Load the lookup table with a square root function over the source image maximum range.

(e) Pass the source image through the lookup table.

(f) Linearly scale the destination image over its maximum range and display it.

The executable `example_lookup_monochrome_SD` performs this exercise. See the extrema, window_level, *allocate_lookup_table*, *import_lut*, and *lookup* manual pages.

21.3 COLOUR SPACE EXERCISES

3.1   Develop a program that converts a linear *RGB* unsigned integer, 8-bit, colour image to the *XYZ* colour space and converts the *XYZ* colour image back to the *RGB* colour space. Steps:

(a) Display the *RGB* source linear colour image.

(b) Display the *R, G* and *B* components as monochrome images.

(c) Convert the source image to unit range.

(d) Convert the *RGB* source image to *XYZ* colour space.

(e) Display the *X, Y* and *Z* components as monochrome images.

(f) Convert the *XYZ* destination image to *RGB* colour space.

(g) Display the *RGB* destination image.

The executable `example_colour_conversion_RGB_XYZ` performs this exercise. See the *extract_pixel_plane, convert_image_datatype, monadic_ arithmetic,* and *colour_conversion_linear* manual pages.

3.2 Develop a program that converts a linear *RGB* colour image to the *L*a*b** colour space and converts the *L*a*b** colour image back to the *RGB* colour space. Steps:

(a) Display the *RGB* source linear colour image.

(b) Display the *R, G* and *B* components as monochrome images.

(c) Convert the source image to unit range.

(d) Convert the *RGB* source image to *L*a*b** colour space.

(e) Display the *L*, a** and *b** components as monochrome images.

(f) Convert the *L*a*b** destination image to *RGB* colour space.

(g) Display the *RGB* destination image.

The executable `example_colour_conversion_RGB_Lab` performs this exercise. See the *extract_pixel_plane, convert_image_datatype, monadic_ arithmetic,* and *colour_conversion_linear* manual pages.

3.3 Develop a program that converts a linear *RGB* colour image to a gamma   corrected *RGB* colour image and converts the gamma colour image back to the linear *RGB* colour space. Steps:

(a) Display the *RGB* source linear colour image.

(b) Display the *R, G* and *B* components as monochrome images.

(c) Convert the source image to unit range.

(d) Perform gamma correction on the linear *RGB* source image.

(e) Display the gamma corrected *RGB* destination image.

(f) Display the *R, G* and *B* gamma corrected components as monochrome images.

(g) Convert the gamma corrected destination image to linear *RGB* colour space.

(h) Display the linear *RGB* destination image.

The executable `example_colour_gamma_correction` performs this exercise. See the *extract_pixel_plane, convert_image_datatype, monadic_arithmetic,* and *gamma_correction* manual pages.

3.4 Develop a program that converts a gamma *RGB* colour image to the *YCbCr* colour space and converts the *YCbCr* colour image back to the gamma *RGB* colour space. Steps:

(a) Display the *RGB* source gamma colour image.

(b) Display the *R, G* and *B* components as monochrome images.

(c) Convert the source image to unit range.

(d) Convert the *RGB* source image to *YCbCr* colour space.

(e) Display the *Y, Cb* and *Cr* components as monochrome images.

(f) Convert the *YCbCr* destination image to gamma *RGB* colour space.

(g) Display the gamma *RGB* destination image.

The executable `example_colour_conversion_RGB_YCbCr` performs this exercise. See the *extract_pixel_plane, convert_image_datatype, monadic_ arithmetic,* and *colour_conversion_linear* manual pages.

3.5   Develop a program that converts a gamma *RGB* colour image to the *IHS* colour space and converts the *IHS* colour image back to the gamma *RGB* colour space. Steps:

(a) Display the *RGB* source gamma colour image.

(b) Display the *R, G* and *B* components as monochrome images.

(c) Convert the source image to unit range.

(d) Convert the *RGB* source image to *IHS* colour space.

(e) Display the *I, H* and *S* components as monochrome images.

(f) Convert the *IHS* destination image to gamma *RGB* colour space.

(g) Display the gamma *RGB* destination image.

The executable `example_colour_conversion_RGB_IHS` performs this exercise. See the *extract_pixel_plane, convert_image_datatype, monadic_ arithmetic,* and *colour_conversion_linear* manual pages.

21.4 REGION-OF-INTEREST EXERCISES

4.1   Develop a program that forms the complement of an unsigned integer, 8-bit, 512×512, monochrome, image under region-of-interest control.

Case 1: Full source and destination ROIs.

Case 2: Rectangular source ROI, upper left corner at (50, 100), lower right corner at (300, 350) and full destination ROI.

Case 3: Full source ROI and rectangular destination ROI, upper left corner at (150, 200), lower right corner at (400, 450).

Case 4: Rectangular source ROI, upper left corner at (50, 100), lower right corner at (300, 350) and rectangular destination ROI, upper left corner at (150, 200), lower right corner at (400, 450).

Steps:

  (a) Display the source monochrome image.

  (b) Create a constant destination image of value 150.

  (c) Complement the source image into the destination image.

  (d) Display the destination image.

  (e) Create a constant destination image of value 150.

  (f) Bind the source ROI to the source image.

  (g) Complement the source image into the destination image.

  (h) Display the destination image.

  (i) Create a constant destination image of value 150.

  (j) Bind the destination ROI to the destination image.

  (k) Complement the source image into the destination image.

  (l) Display the destination image.

  (m) Create a constant destination image of value 150.

  (n) Bind the source ROI to the source image and bind the destination ROI to the destination image.

  (o) Complement the source image into the destination image.

  (p) Display the destination image.

The executable `example_complement_monochrome_roi` performs this exercise. See the *image_constant, generate_2d_roi_rectangular, bind_roi,* and *complement* manual pages.

21.5 IMAGE MEASUREMENT EXERCISES

5.1 Develop a program that computes the extrema of the *RGB* components of an unsigned integer, 8-bit, colour image. Steps:

  (a) Display the source colour image.

  (b) Compute extrema of the colour image and print results for all bands.

The executable `example_extrema_colour` performs this exercise. See the *extrema* manual page.

5.2 Develop a program that computes the mean and standard deviation of an unsigned integer, 8-bit, monochrome image. Steps:

(a) Display the source monochrome image.

(b) Compute moments of the monochrome image and print results.

The executable `example_moments_monochrome` performs this exercise. See the *moments* manual page.

5.3   Develop a program that computes the first-order histogram of an unsigned integer, 8-bit, monochrome image with 16 amplitude bins. Steps:

(a) Display the source monochrome image.

(b) Allocate the histogram.

(c) Compute the histogram of the source image.

(d) Export the histogram and print its contents.

The executable `example_histogram_monochrome` performs this exercise. See the *allocate_histogram, histogram_1d,* and *export_histogram* manual pages.

21.6 QUANTIZATION EXERCISES

6.1   Develop a program that re-quantizes an unsigned integer, 8-bit, monochrome image linearly to three bits per pixel and reconstructs it to eight bits per pixel. Steps:

(a) Display the source image.

(b) Perform a right overflow shift by three bits on the source image.

(c) Perform a left overflow shift by three bits on the right bit-shifted source image.

(d) Scale the reconstruction levels to 3-bit values.

(e) Display the destination image.

The executable `example_linear_quantizer` executes this example. See the *bit_shift, extrema,* and *window_level* manual pages.

6.2   Develop a program that quantizes an unsigned integer, 8-bit, monochrome image according to the cube root lightness function of Eq. 6.3-4 and reconstructs it to eight bits per pixel. Steps:

(a) Display the source image.

(b) Scale the source image to unit range.

(c) Perform the cube root lightness transformation.

(d) Scale the lightness function image to 0 to 255.

(e) Perform a right overflow shift by three bits on the source image.

(f) Perform a left overflow shift by three bits on the right bit-shifted source image.

(g) Scale the reconstruction levels to 3-bit values.

(h) Scale the reconstruction image to the lightness function range.

(i) Perform the inverse lightness function.

(j) Scale the inverse lightness function to the display range.

(k) Display the destination image.

The executable `example_lightness_quantizer` executes this example. See the *monadic_arithmetic*, *unary_integer, window_level,* and *bit_shift* manual pages.

21.7 CONVOLUTION EXERCISES

7.1 Develop a program that convolves a test image with a 3×3 uniform impulse response array for three convolution boundary conditions. Steps:

(a) Create a 101×101 pixel, real datatype test image consisting of a 2×2 cluster of amplitude 1.0 pixels in the upper left corner and a single pixel of amplitude 1.0 in the image center. Set all other pixels to 0.0.

(b) Create a 3×3 uniform impulse response array.

(c) Convolve the source image with the impulse response array for the following three boundary conditions: enclosed array, zero exterior, reflected exterior.

(d) Print a 5×5 pixel image array about the upper left corner and image center for each boundary condition and explain the results.

The executable `example_convolve_boundary` executes this example. See the *allocate_neighbourhood_array, impulse_rectangular, image_constant, put_pixel, get_pixel,* and *convolve_2d* manual pages.

7.2 Develop a program that convolves an unsigned integer, 8-bit, colour image with a 5×5 uniform impulse response array acquired from the data object repository. Steps:

(a) Display the source colour image.

(b) Allocate the impulse response array.

(c) Fetch the impulse response array from the data object repository.

(d) Convolve the source image with the impulse response array.

(e) Display the destination image.

The executable `example_repository_convolve_colour` executes this example. See the *allocate_neighbourhood_array, return_repository_id,* and *convolve_2d* manual pages.

21.8 UNITARY TRANSFORM EXERCISES

8.1   Develop a program that generates the Fourier transform log magnitude ordered display of Figure 8.2-4d for the `smpte_girl_luma` image. Steps:

(a) Display the source monochrome image.

(b) Scale the source image to unit amplitude.

(c) Perform a two-dimensional Fourier transform on the unit amplitude source image with the ordered display option.

(d) Scale the log magnitude according to Eq. 8.2-9 where $a = 1.0$ and $b = 100.0$.

(e) Display the Fourier transformed image.

The executable `example_fourier_transform_spectrum` executes this example. See the *convert_image_datatype, monadic_arithmetic, image_constant, complex_composition, transform_fourier, complex_magnitude, window_level,* and *unary_real* manual pages.

8.2   Develop a program that generates the Hartley transform log magnitude ordered display of Figure 8.3-2c for the `smpte_girl_luma` image by manipulation of the Fourier transform coefficients of the image. Steps:

(a) Display the source monochrome image.

(b) Scale the source image to unit amplitude.

(c) Perform a two-dimensional Fourier transform on the unit amplitude source image with the dc term at the origin option.

(d) Extract the Hartley components from the Fourier components.

(e) Scale the log magnitude according to Eq. 8.2-9 where $a = 1.0$ and $b = 100.0$.

(f) Display the Hartley transformed image.

The executable `example_transform_hartley` executes this example. See the *convert_image_datatype, monadic_arithmetic, image_constant, complex_composition, transform_fourier, complex_decomposition,   dyadic_arithmetic, complex_magnitude, window_level,* and *unary_real* manual pages.

21.9 LINEAR PROCESSING EXERCISES

9.1   Develop a program that performs fast Fourier transform convolution following the steps of Section 9.3. Execute this program using an 11×11 uniform impulse response array on an unsigned integer, 8-bit, 512×512 monochrome image without zero padding. Steps:

(a) Display the source monochrome image.

(b) Scale the source image to unit range.

(c) Perform a two-dimensional Fourier transform of the source image.

(d) Display the clipped magnitude of the source Fourier transform.

(e) Allocate an 11×11 impulse response array.

(f) Create an 11×11 uniform impulse response array.

(g) Convert the impulse response array to an image and embed it in a 512×512 zero background image.

(h) Perform a two-dimensional Fourier transform of the embedded impulse image.

(i) Display the clipped magnitude of the embedded impulse Fourier transform.

(j) Multiply the source and embedded impulse Fourier transforms.

(k) Perform a two-dimensional inverse Fourier transform of the product image.

(l) Display the destination image.

(m) Printout the erroneous pixels along a mid image row.

The executable `example_fourier_filtering` executes this example. See the *monadic_arithmetic, image_constant, complex_composition, transform_fourier, complex_magnitude, allocate_neighbourhood_array, impulse_rectangular, convert_array_to_image, dyadic_complex,* and *complex_decomposition* manual pages.

21.10 IMAGE ENHANCEMENT EXERCISES

10.1 Develop a program that displays the Q component of a *YIQ* colour image over its full dynamic range. Steps:

(a) Display the source monochrome *RGB* image.

(b) Scale the *RGB* image to unit range and convert it to the *YIQ* space.

(c) Extract the Q component image.

(d) Compute the amplitude extrema.

(e) Use the *window_level* conversion function to display the Q component.

The executable `example_Q_display` executes this example. See the *monadic_arithmetic, colour_conversion_linear, extrema, extract_pixel_plane,* and *window_level* manual pages.

10.2 Develop a program to histogram equalize an unsigned integer, 8-bit, monochrome image. Steps:

(a) Display the source monochrome image.

(b) Compute the image histogram.

(c) Compute the image cumulative histogram.

(d) Load the image cumulative histogram into a lookup table.

(e) Pass the image through the lookup table.

(f) Display the enhanced destination image.

The executable `example_histogram_equalization` executes this example. See the *allocate_histogram, histogram_1d, export_histogram, allocate_lookup_table, export_lut,* and *lookup_table* manual pages.

10.3 Develop a program to perform outlier noise cleaning of the unsigned integer, 8-bit, monochrome image `peppers_replacement_noise` following the algorithm of Figure 10.3-9. Steps:

(a) Display the source monochrome image.

(b) Compute a 3×3 neighborhood average image.

(c) Display the neighbourhood image.

(d) Create a magnitude of the difference image between the source image and the neighbourhood image.

(e) Create a Boolean mask image which is TRUE if the magnitude difference image is greater than a specified error tolerance, e.g. 15%.

(f) Convert the mask image to a ROI and use it to generate the outlier destination image.

(g) Display the destination image.

The executable `example_outlier` executes this example. See the *return_repository_id, convolve_2d, dyadic_predicate, allocate_roi, convert_image_to_roi, bind_roi,* and *convert_image_datatype* manual pages.

10.4 Develop a program that performs linear edge crispening of an unsigned integer, 8-bit, colour image by convolution. Steps:

(a) Display the source colour image.

(b) Import the Mask 3 impulse response array defined by Eq.10.3-1c.

(c) Convert the ND source image to SD datatype.

(d) Convolve the colour image with the impulse response array.

(e) Clip the convolved image over the dynamic range of the source image to avoid amplitude undershoot and overshoot.

(f) Display the clipped destination image.

The executable `example_edge_crispening` executes this example. See the *allocate_neighbourhood_array, import_neighbourhood_array, convolve_2d, extrema,* and *window_level* manual pages.

10.5 Develop a program that performs 7×7 plus-shape median filtering of the unsigned integer, 8-bit, monochrome image `peppers_replacement _noise`. Steps:

(a) Display the source monochrome image.

(b) Create a 7×7 Boolean mask array.

(c) Perform median filtering.

(d) Display the destination image.

The executable `example_filtering_median_plus7` executes this example. See the *allocate_neighbourhood_array, import_neighbourhood_array,* and *filtering _median* manual pages.

21.11 IMAGE RESTORATION MODELS EXERCISES

11.1 Develop a program that creates an unsigned integer, 8-bit, monochrome image with zero mean, additive, uniform noise with a signal-to-noise ratio of 10.0. The program should execute for arbitrary size source images. Steps:

(a) Display the source monochrome image.

(b) In application space, create a unit range noise image array using the C `math.h` function `rand`.

(c) Import the noise image array.

(d) Display the noise image array.

(e) Scale the noise image array to produce a noise image array with zero mean and a SNR of 10.0.

(f) Compute the mean and standard deviation of the noise image.

(g) Read an unsigned integer, 8-bit monochrome image source image file and normalize it to unit range.

(h) Add the noise image to the source image and clip to unit range.

(i) Display the noisy source image.

The executable `example_additive_noise` executes this example. See the *monadic_arithmetic, import_image, moments, window_level,* and *dyadic_arithmetic* manual pages.

11.2 Develop a program that creates an unsigned integer, 8-bit, monochrome image with replacement impulse noise. The program should execute for arbitrary size source images. Steps:

(a) Display the source monochrome image.

(b) In application space, create a unit range noise image array using the C `math.h` function `rand`.

(c) Import the noise image array.

(d) Read a source image file and normalize to unit range.

(e) Replace each source image pixel with 0.0 if the noise pixel is less than 1.0%, and replace each source image pixel with 1.0 if the noise pixel is greater than 99%. The replacement operation can be implemented by image copying under ROI control.

(f) Display the noisy source image.

The executable `example_replacement_noise` executes this example. See the *monadic_arithmetic, import_image, dyadic_predicate, allocate_roi, bind_roi, convert_image_datatype,* and *dyadic_arithmetic* manual pages.

21.12 IMAGE RESTORATION EXERCISES

12.1 Develop a program that computes a 512×512 Wiener filter transfer function for the blur impulse response array of Eq. 10.3-2c and white noise with a SNR of 10.0. Steps:

(a) Fetch the impulse response array from the repository.

(b) Convert the impulse response array to an image and embed it in a 512×512 zero background array.

(c) Compute the two-dimensional Fourier transform of the embedded impulse response array.

(d) Form the Wiener filter transfer function according to Eq. 12.2-23.

(e) Display the magnitude of the Wiener filter transfer function.

The executable `example_wiener` executes this example. See the *return_repository_id, transform_fourier, image_constant, complex_conjugate, dyadic_arithmetic,* and *complex_magnitude* manual pages.

21.13 GEOMETRICAL IMAGE MODIFICATION EXERCISES

13.1 Develop a program that minifies an unsigned integer, 8-bit, monochrome image by a factor of two and rotates the minified image by 45 degrees about its center using bilinear interpolation. Display the geometrically modified image. Steps:

(a) Display the source monochrome image.

(b) Set the global interpolation mode to bilinear.

(c) Set the first work image to zero.

(d) Minify the source image into the first work image.

(e) Set the second work image to zero.

(f) Translate the first work image into the center of the second work image.

(g) Set the destination image to zero.

(h) Rotate the second work image about its center into the destination image.

(i) Display the destination image.

The executable `example_minify_rotate` executes this example. See the *image_constant, resize, translate, rotate,* and *set_globals* manual pages.

13.2 Develop a program that performs shearing of the rows of an unsigned integer, 8-bit, monochrome image using the *warp_lut* operator such that the last image row is shifted 10% of the row width and all other rows are shifted proportionally. Steps:

(a) Display the source monochrome image.

(b) Set the global interpolation mode to bilinear.

(c) Set the warp polynomial coefficients.

(d) Perform polynomial warping.

(e) Display the destination image.

The executable `example_shear` executes this example. See the *set_globals, image_constant,* and *warp_lut* manual pages.

21.14 MORPHOLOGICAL IMAGE PROCESSING EXERCISES

14.1 Develop a program that reads the 64 × 64, Boolean test image `boolean_test` and dilates it by one and two iterations with a 3 × 3 structuring element. Steps:

(a) Read the source image and zoom it by a factor of 8:1.

(b) Create a 3 × 3 structuring element array.

(c) Dilate the source image with one iteration.

(d) Display the zoomed destination image.

(e) Dilate the source image with two iterations.

(f) Display the zoomed destination image.

The executable `example_boolean_dilation` executes this example. See the *allocate_neighbourhood_array, import_neighbourhood_array, erosion_dilation_ boolean, zoom,* and *boolean_display* manual pages.

14.2 Develop a program that reads the 64 × 64, Boolean test image `boolean_ test` and erodes it by one and two iterations with a 3 × 3 structuring element. Steps:

(a) Read the source image and zoom it by a factor of 8:1.

(b) Create a 3 × 3 structuring element array.

(c) Erode the source image with one iteration.

(d) Display the zoomed destination image.

(e) Erode the source image with two iterations.

(f) Display the zoomed destination image.

The executable `example_boolean_erosion` executes this example. See the *allocate_neighbourhood_array, import_neighbourhood_array, erosion_dilation _boolean, zoom,* and *boolean_display* manual pages.

14.3 Develop a program that performs gray scale dilation on an unsigned integer, 8-bit, monochrome image with a 5 × 5 zero-value structuring element and a 5 × 5 TRUE state mask. Steps:

(a) Display the source image.

(b) Create a 5 × 5 Boolean mask.

(c) Perform grey scale dilation on the source image.

(d) Display the destination image.

The executable `example_dilation_grey_ND` executes this example. See the *allocate_neighbourhood_array, import_neighbourhood_array,* and *erosion_dilation _grey* manual pages.

14.4 Develop a program that performs gray scale erosion on an unsigned integer, 8-bit, monochrome image with a 5 × 5 zero-value structuring element and a 5 × 5 TRUE state mask. Steps:

(a) Display the source image.

(b) Create a 5 × 5 Boolean mask.

(c) Perform grey scale erosion on the source image.

(d) Display the destination image.

The executable `example_erosion_gray_ND` executes this example. See the *allocate_neighbourhood_array, import_neighbourhood_array*, and *erosion_dilation _gray* manual pages.

21.15 EDGE DETECTION EXERCISES

15.1 Develop a program that generates the Sobel edge gradient according to Figure 15.2-1 using a square root sum of squares gradient combination. Steps:

(a) Display the source image.

(b) Allocate the horizontal and vertical Sobel impulse response arrays.

(c) Fetch the horizontal and vertical Sobel impulse response arrays from the repository.

(d) Convolve the source image with the horizontal Sobel.

(e) Display the Sobel horizontal gradient.

(f) Convolve the source image with the vertical Sobel.

(g) Display the Sobel vertical gradient.

(h) Form the square root sum of squares of the gradients.

(i) Display the Sobel gradient.

The executable `example_sobel_gradient` executes this example. See the *allocate_neighbourhood_array, return_repository_id, convolve_2d, unary_real*, and *dyadic_arithmetic* manual pages.

15.2 Develop a program that generates the Laplacian of Gaussian gradient for a 11 × 11 impulse response array and a standard deviation of 2.0. Steps:

(a) Display the source image.

(b) Allocate the Laplacian of Gaussian impulse response array.

(c) Generate the Laplacian of Gaussian impulse response array.

(d) Convolve the source image with the Laplacian of Gaussian impulse response array.

(e) Display the Laplacian of Gaussian gradient.

The executable `example_LoG_gradient` executes this example. See the *allocate_neighbourhood_array, impulse_laplacian_of_gaussian,* and *convolve_2d* manual pages.

21.16 IMAGE FEATURE EXTRACTION EXERCISES

16.1 Develop a program that generates the 7×7 moving window mean and standard deviation features of an unsigned integer, 8-bit, monochrome image. Steps:

(a) Display the source image.

(b) Scale the source image to unit range.

(c) Create a 7×7 uniform impulse response array.

(d) Compute the moving window mean with the uniform impulse response array.

(e) Display the moving window mean image.

(f) Compute the moving window standard deviation with the uniform impulse response array.

(g) Display the moving window standard deviation image.

The executable `example_amplitude_features` executes this example. See the *allocate_neighbourhood_array, impulse_rectangular, convolve_2d, dyadic_arithmetic,* and *unary_real* manual pages.

16.2 Develop a program that computes the mean, standard deviation, skewness, kurtosis, energy, and entropy first-order histogram features of an unsigned integer, 8-bit, monochrome image. Steps:

(a) Display the source image.

(b) Compute the histogram of the source image.

(c) Export the histogram and compute the histogram features.

The executable `example_histogram_features` executes this example. See the *allocate_histogram, histogram_1d,* and *export_histogram* manual pages.

16.3 Develop a program that computes the nine Laws texture features of an unsigned integer, 8-bit, monochrome image. Use a 7×7 moving window to compute the standard deviation. Steps:

(a) Display the source image.

(b) Allocate nine 3×3 impulse response arrays.

(c) Fetch the nine Laws impulse response arrays from the repository.

(d) For each Laws array:

> convolve the source image with the Laws array.
>
> compute the moving window mean of the Laws convolution.
>
> compute the moving window standard deviation of the Laws convolution image.
>
> display the Laws texture features.

The executable `example_laws_features` executes this example. See the *allocate_neighbourhood_array, impulse_rectangular, return_repository_id, convolve_2d, dyadic_arithmetic,* and *unary_real* manual pages.

21.17 IMAGE SEGMENTATION EXERCISES

17.1 Develop a program that thresholds the monochrome image `parts` and displays the thresholded image. Determine the threshold value that provides the best visual segmentation. Steps:

(a) Display the source image.

(b) Threshold the source image into a Boolean destination image.

(c) Display the destination image.

The executable `example_threshold` executes this example. See the *threshold* and *boolean_display* manual pages.

17.2 Develop a program that locates and tags the watershed segmentation local minima in the monochrome image `segmentation_test`. Steps:

(a) Display the source image.

(b) Generate a 3×3 Boolean mask.

(c) Erode the source image into a work image with the Boolean mask.

(d) Compute the local minima of the work image.

(e) Display the local minima image.

The executable `example_watershed` executes this example. See the *erosion_dilation_grey,* and *dyadic_predicate* manual pages.

21.18 SHAPE ANALYSIS EXERCISES

18.1 Develop a program that computes the scaled second-order central moments of the monochrome image `ellipse`. Steps:

(a) Display the source image.

(b) Normalize the source image to unit range.

(c) Export the source image and perform the computation in application space in double precision.

The executable `example_spatial_moments` executes this example. See the *monadic_arithmetic,* and *export_image* manual pages.

21.19 IMAGE DETECTION AND REGISTRATION EXERCISES

19.1 Develop a program that performs normalized cross-correlation template matching of the monochrome source image `L_source` and the monochrome template image `L_template` using the convolution operator as a means of correlation array computation. Steps:

(a) Display the source image.

(b) Display the template image.

(c) Rotate the template image 180 degrees and convert it to an impulse response array.

(d) Convolve the source image with the impulse response array to form the numerator of the cross-correlation array.

(e) Display the numerator image.

(f) Square the source image and compute its moving window average energy by convolution with a rectangular impulse response array to form the denominator of the cross-correlation array.

(g) Display the denominator image.

(h) Form the cross-correlation array image.

(i) Display the cross-correlation array image.

Note, it is necessary to properly scale the source and template images to obtain valid results. The executable `example_template` executes this example. See the *allocate_neighbourhood_array, flip_spin_transpose, convert_image_to_array, impulse_rectangular, convolve_2d,* and *monadic_arithmetic* manual pages.

APPENDIX 1

VECTOR-SPACE ALGEBRA CONCEPTS

This appendix contains reference material on vector-space algebra concepts used in the book.

A1.1. VECTOR ALGEBRA

This section provides a summary of vector and matrix algebraic manipulation procedures utilized in the book. References 1 to 5 may be consulted for formal derivations and proofs of the statements of definition presented here.

Vector. An $N \times 1$ column vector $\mathbf{f}$ is a one-dimensional vertical arrangement,

$$
\mathbf{f} = \begin{bmatrix} f(1) \\ f(2) \\ \vdots \\ f(n) \\ \vdots \\ f(N) \end{bmatrix}
\tag{A1.1-1}
$$

of the elements $f(n)$, where $n = 1, 2,..., N$. An $1 \times N$ row vector $\mathbf{h}$ is a one-dimensional horizontal arrangement

$$\mathbf{h} = \begin{bmatrix} h(1) & h(2) & ... & h(n) & ... & h(N) \end{bmatrix} \tag{A1.1-2}$$

of the elements $h(n)$, where $n = 1, 2,..., N$. In this book, unless otherwise indicated, all boldface lowercase letters denote column vectors. Row vectors are indicated by the transpose relation

$$\mathbf{f}^T = \begin{bmatrix} f(1) & f(2) & \cdots & f(n) & \cdots & f(N) \end{bmatrix} \tag{A1.1-3}$$

Matrix.   An $M \times N$ matrix $\mathbf{F}$ is a two-dimensional arrangement

$$\mathbf{F} = \begin{bmatrix} F(1, 1) & F(1, 2) & \cdots & F(1, N) \\ F(2, 1) & F(2, 2) & \cdots & F(2, N) \\ \vdots & \vdots & & \vdots \\ F(M, 1) & F(M, 2) & \cdots & F(M, N) \end{bmatrix} \tag{A1.1-4}$$

of the elements $F(m, n)$ into rows and columns, where $m = 1, 2,..., M$ and $n = 1, 2,..., N$. The symbol $\mathbf{0}$ indicates a null matrix whose terms are all zeros. A diagonal matrix is a square matrix, $M = N$, for which all off-diagonal terms are zero; that is, $F(m, n) = 0$ if $m \neq n$. An identity matrix denoted by $\mathbf{I}$ is a diagonal matrix whose diagonal terms are unity. The identity symbol is often subscripted to indicate its dimension: $\mathbf{I}_N$ is an $N \times N$ identity matrix. A submatrix $\mathbf{F}_{pq}$ is a matrix partition of a larger matrix $\mathbf{F}$ of the form

$$\mathbf{F} = \begin{bmatrix} \mathbf{F}_{1, 1} & \mathbf{F}_{1, 2} & \cdots & \mathbf{F}_{1, Q} \\ \vdots & \vdots & & \vdots \\ \mathbf{F}_{P, 1} & \mathbf{F}_{P, 1} & \cdots & \mathbf{F}_{P, Q} \end{bmatrix} \tag{A1.1-5}$$

Matrix Addition.   The sum $\mathbf{C} = \mathbf{A} + \mathbf{B}$ of two matrices is defined only for matrices of the same size. The sum matrix $\mathbf{C}$ is an $M \times N$ matrix whose elements are $C(m, n) = A(m, n) + B(m, n)$.

Matrix Multiplication.   The product $\mathbf{C} = \mathbf{AB}$ of two matrices is defined only when the number of columns of $\mathbf{A}$ equals the number of rows of $\mathbf{B}$. The $M \times N$ product matrix $\mathbf{C}$ of the matrix $\mathbf{A}$ and the $P \times N$ matrix $\mathbf{B}$ is a matrix whose general element is given by

$$C(m, n) = \sum_{p=1}^{P} A(m, p)B(p, n) \qquad \text{(A1.1-6)}$$

Matrix Inverse. The matrix inverse, denoted by $\mathbf{A}^{-1}$, of a square matrix $\mathbf{A}$ has the property that $\mathbf{A}\mathbf{A}^{-1} = \mathbf{I}$ and $\mathbf{A}^{-1}\mathbf{A} = \mathbf{I}$. If such a matrix $\mathbf{A}^{-1}$ exists, the matrix $\mathbf{A}$ is said to be nonsingular; otherwise, $\mathbf{A}$ is singular. If a matrix possesses an inverse, the inverse is unique. The matrix inverse of a matrix inverse is the original matrix. Thus

$$[\mathbf{A}^{-1}]^{-1} = \mathbf{A} \qquad \text{(A1.1-7)}$$

If matrices $\mathbf{A}$ and $\mathbf{B}$ are nonsingular,

$$[\mathbf{A}\mathbf{B}]^{-1} = \mathbf{B}^{-1}\mathbf{A}^{-1} \qquad \text{(A1.1-8)}$$

If matrix $\mathbf{A}$ is nonsingular, and the scalar $k \neq 0$, then

$$[k\mathbf{A}] = \frac{1}{k}\mathbf{A}^{-1} \qquad \text{(A1.1-9)}$$

Inverse operators of singular square matrices and of nonsquare matrices are considered in Section A1.3. The inverse of the partitioned square matrix

$$\mathbf{F} = \begin{bmatrix} \mathbf{F}_{11} & \mathbf{F}_{12} \\ \mathbf{F}_{21} & \mathbf{F}_{22} \end{bmatrix} \qquad \text{(A1.1-10)}$$

may be expressed as

$$\mathbf{F}^{-1} = \begin{bmatrix} [\mathbf{F}_{11} - \mathbf{F}_{12}\mathbf{F}_{22}^{-1}\mathbf{F}_{21}]^{-1} & -\mathbf{F}_{11}^{-1}\mathbf{F}_{12}[\mathbf{F}_{22} - \mathbf{F}_{21}\mathbf{F}_{11}^{-1}\mathbf{F}_{12}]^{-1} \\ -\mathbf{F}_{22}^{-1}\mathbf{F}_{21}[\mathbf{F}_{11} - \mathbf{F}_{12}\mathbf{F}_{22}^{-1}\mathbf{F}_{21}]^{-1} & [\mathbf{F}_{22} - \mathbf{F}_{21}\mathbf{F}_{11}^{-1}\mathbf{F}_{12}]^{-1} \end{bmatrix} \qquad \text{(A1.1-11)}$$

provided that $\mathbf{F}_{11}$ and $\mathbf{F}_{22}$ are nonsingular.

Matrix Transpose. The transpose of an $M \times N$ matrix $\mathbf{A}$ is a $N \times M$ matrix denoted by $\mathbf{A}^{T}$, whose rows are the columns of $\mathbf{A}$ and whose columns are the rows of $\mathbf{A}$. For any matrix $\mathbf{A}$,

$$[\mathbf{A}^{T}]^{T} = \mathbf{A} \qquad \text{(A1.1-12)}$$

If $\mathbf{A} = \mathbf{A}^T$, then $\mathbf{A}$ is said to be *symmetric*. The matrix products $\mathbf{A}\mathbf{A}^T$ and $\mathbf{A}^T\mathbf{A}$ are symmetric. For any matrices $\mathbf{A}$ and $\mathbf{B}$,

$$[\mathbf{A}\mathbf{B}]^T = \mathbf{B}^T\mathbf{A}^T \tag{A1.1-13}$$

If $\mathbf{A}$ is nonsingular, then $\mathbf{A}^T$ is nonsingular and

$$[\mathbf{A}^T]^{-1} = [\mathbf{A}^{-1}]^T \tag{A1.1-14}$$

Matrix Direct Product. The left direct product of a $P \times Q$ matrix $\mathbf{A}$ and an $M \times N$ matrix $\mathbf{B}$ is a $PM \times QN$ matrix defined by

$$\mathbf{C} = \mathbf{A} \otimes \mathbf{B} = \begin{bmatrix} B(1,1)\mathbf{A} & B(1,2)\mathbf{A} & \cdots & B(1,N)\mathbf{A} \\ B(2,1)\mathbf{A} & B(2,2)\mathbf{A} & \cdots & B(2,N)\mathbf{A} \\ \vdots & \vdots & & \vdots \\ B(M,1)\mathbf{A} & \cdots & \cdots & B(M,N)\mathbf{A} \end{bmatrix} \tag{A1.1-15}$$

A right direct product can also be defined in a complementary manner. In this book, only the left direct product will be employed. The direct products $\mathbf{A} \otimes \mathbf{B}$ and $\mathbf{B} \otimes \mathbf{A}$ are not necessarily equal. The product, sum, transpose, and inverse relations are:

$$[\mathbf{A} \otimes \mathbf{B}][\mathbf{C} \otimes \mathbf{D}] = [\mathbf{A}\mathbf{C}] \otimes [\mathbf{B}\mathbf{D}] \tag{A1.1-16}$$

$$[\mathbf{A} + \mathbf{B}] \otimes \mathbf{C} = \mathbf{A} \otimes \mathbf{C} + \mathbf{B} \otimes \mathbf{C} \tag{A1.1-17}$$

$$[\mathbf{A} \otimes \mathbf{B}]^T = \mathbf{A}^T \otimes \mathbf{B}^T \tag{A1.1-18}$$

$$[\mathbf{A} \otimes \mathbf{B}]^{-1} = [\mathbf{A}^{-1} \otimes \mathbf{B}^{-1}] \tag{A1.1-19}$$

Matrix Trace. The trace of an $N \times N$ square matrix $\mathbf{F}$ is the sum of its diagonal elements denoted as

$$\text{tr}\{\mathbf{F}\} = \sum_{n=1}^{N} F(n,n) \tag{A1.1-20}$$

If $\mathbf{A}$ and $\mathbf{B}$ are square matrices,

$$\text{tr}\{\mathbf{A}\mathbf{B}\} = \text{tr}\{\mathbf{B}\mathbf{A}\} \tag{A1.1-21}$$

The trace of the direct product of two matrices equals

$$\text{tr}\{\mathbf{A} \otimes \mathbf{B}\} = \text{tr}\{\mathbf{A}\}\text{tr}\{\mathbf{B}\} \qquad (A1.1\text{-}22)$$

Vector Norm. The Euclidean vector norm of the $N \times 1$ vector $\mathbf{f}$ is a scalar defined as

$$\|\mathbf{f}\| = \mathbf{f}^T\mathbf{f} \qquad (A1.1\text{-}23)$$

Matrix Norm. The Euclidean matrix norm of the $M \times N$ matrix $\mathbf{F}$ is a scalar defined as

$$\|\mathbf{F}\| = \text{tr}[\mathbf{F}^T\mathbf{F}] \qquad (A1.1\text{-}24)$$

Matrix Rank. An $N \times N$ matrix $\mathbf{A}$ is a rank R matrix if the largest nonsingular square submatrix of $\mathbf{A}$ is an $R \times R$ matrix. The rank of a matrix is utilized in the inversion of matrices. If matrices $\mathbf{A}$ and $\mathbf{B}$ are nonsingular, and $\mathbf{C}$ is an arbitrary matrix, then

$$\text{rank}\{\mathbf{C}\} = \text{rank}\{\mathbf{AC}\} = \text{rank}\{\mathbf{CA}\} = \text{rank}\{\mathbf{ACB}\} \qquad (A1.1\text{-}25)$$

The rank of the product of matrices $\mathbf{A}$ and $\mathbf{B}$ satisfies the relations

$$\text{rank}\{\mathbf{AB}\} \le \text{rank}\{\mathbf{A}\} \qquad (A1.1\text{-}26a)$$

$$\text{rank}\{\mathbf{AB}\} \le \text{rank}\{\mathbf{B}\} \qquad (A1.1\text{-}26b)$$

The rank of the sum of matrices $\mathbf{A}$ and $\mathbf{B}$ satisfies the relations

$$\text{rank}\{\mathbf{A} + \mathbf{B}\} \le \text{rank}\{\mathbf{A}\} + \text{rank}\{\mathbf{B}\} \qquad (A1.1\text{-}27)$$

Vector Inner Product. The inner product of the $N \times 1$ vectors $\mathbf{f}$ and $\mathbf{g}$ is a scalar

$$k = \mathbf{g}^T\mathbf{f} \qquad (A1.1\text{-}28)$$

where

$$k = \sum_{n=1}^{N} g(n)f(n) \qquad (A1.1\text{-}29)$$

Vector Outer Product. The outer product of the $M \times 1$ vector $\mathbf{g}$ and the $N \times 1$ vector $\mathbf{f}$ is a matrix

$$\mathbf{A} = \mathbf{g}\mathbf{f}^{T} \tag{A1.1-30}$$

where $A(m, n) = g(m)f(n)$.

Quadratic Form. The quadratic form of an $N \times 1$ vector $\mathbf{f}$ is a scalar

$$k = \mathbf{f}^{T}\mathbf{A}\mathbf{f} \tag{A1.1-31}$$

where $\mathbf{A}$ is an $N \times N$ matrix. Often, the matrix $\mathbf{A}$ is selected to be symmetric.

Vector Differentiation. For a symmetric matrix $\mathbf{A}$, the derivative of the quadratic form $\mathbf{x}^{T}\mathbf{A}\mathbf{x}$ with respect to $\mathbf{x}$ is

$$\frac{\partial[\mathbf{x}^{T}\mathbf{A}\mathbf{x}]}{\partial \mathbf{x}} = 2\mathbf{A}\mathbf{x} \tag{A1.1-32}$$

A1.2. SINGULAR-VALUE MATRIX DECOMPOSITION

Any arbitrary $M \times N$ matrix $\mathbf{F}$ of rank R can be decomposed into the sum of a weighted set of unit rank $M \times N$ matrices by a singular-value decomposition (SVD) (6–8).

According to the SVD matrix decomposition, there exist an $M \times M$ unitary matrix $\mathbf{U}$ and an $N \times N$ unitary matrix $\mathbf{V}$ for which

$$\mathbf{U}^{T}\mathbf{F}\mathbf{V} = \mathbf{\Lambda}^{1/2} \tag{A1.2-1}$$

where

$$\mathbf{\Lambda}^{1/2} = \begin{bmatrix} \lambda^{1/2}(1) & \cdots & & 0 \\ \vdots & \ddots & & \\ & & \lambda^{1/2}(1) & \vdots \\ 0 & \cdots & & 0 \end{bmatrix} \tag{A1.2-2}$$

is an $M \times N$ matrix with a general diagonal entry $\lambda^{1/2}(j)$ called a *singular value* of F. Because U and V are unitary matrices, $\mathbf{U}\mathbf{U}^T = \mathbf{I}_M$ and $\mathbf{V}\mathbf{V}^T = \mathbf{I}_N$. Consequently,

$$\mathbf{F} = \mathbf{U}\mathbf{\Lambda}^{1/2}\mathbf{V}^T \qquad (A1.2\text{-}3)$$

The columns of the unitary matrix U are composed of the eigenvectors $\mathbf{u}_m$ of the symmetric matrix $\mathbf{FF}^T$. The defining relation is

$$\mathbf{U}^T[\mathbf{FF}^T]\mathbf{U} = \begin{bmatrix} \lambda(1) & \cdots & & 0 \\ \vdots & \ddots & & \vdots \\ & & \lambda(R) & \\ 0 & \cdots & & 0 \end{bmatrix} \qquad (A1.2\text{-}4)$$

where $\lambda(j)$ are the nonzero eigenvalues of $\mathbf{FF}^T$. Similarly, the columns of V are the eigenvectors $\mathbf{v}_n$ of the symmetric matrix $\mathbf{F}^T\mathbf{F}$ as defined by

$$\mathbf{V}^T[\mathbf{F}^T\mathbf{F}]\mathbf{V} = \begin{bmatrix} \lambda(1) & \cdots & & 0 \\ \vdots & \ddots & & \vdots \\ & & \lambda(R) & \\ 0 & \cdots & & 0 \end{bmatrix} \qquad (A1.2\text{-}5)$$

where the $\lambda(j)$ are the corresponding nonzero eigenvalues of $\mathbf{F}^T\mathbf{F}$. Consistency is easily established between Eqs. A1.2-3 to A1.2-5. It is possible to express the matrix decomposition of Eq. A1.2-3 in the series form

$$\mathbf{F} = \sum_{j=1}^{R} \lambda^{1/2}(j)\mathbf{u}_j\mathbf{v}_j^T \qquad (A1.2\text{-}6)$$

The outer products $\mathbf{u}_j\mathbf{v}_j^T$ of the eigenvectors form a set of unit rank matrices each of which is scaled by a corresponding singular value of F. The consistency of Eq. A1.2-6 with the previously stated relations can be shown by its substitution into Eq. A1.2-1, which yields

$$\mathbf{\Lambda}^{1/2} = \mathbf{U}^T\mathbf{FV} = \sum_{j=1}^{R} \lambda^{1/2}(j)\mathbf{U}^T\mathbf{u}_j\mathbf{v}_j^T\mathbf{V} \qquad (A1.2\text{-}7)$$

It should be observed that the vector product $\mathbf{U}^T\mathbf{u}_j$ is a column vector with unity in its jth elements and zeros elsewhere. The row vector resulting from the product $\mathbf{v}_j^T\mathbf{V}$ is of similar form. Hence, upon final expansion, the right-hand side of Eq. A1.2-7 reduces to a diagonal matrix containing the singular values of $\mathbf{F}$.

The SVD matrix decomposition of Eq. A1.2-3 and the equivalent series representation of Eq. A1.2-6 apply for any arbitrary matrix. Thus the SVD expansion can be applied directly to discrete images represented as matrices. Another application is the decomposition of linear operators that perform superposition, convolution, or general transformation of images in vector form.

A1.3. PSEUDOINVERSE OPERATORS

A common task in linear signal processing is to invert the transformation equation

$$\mathbf{p} = \mathbf{Tf} \tag{A1.3-1}$$

to obtain the value of the $Q \times 1$ input data vector $\mathbf{f}$, or some estimate $\hat{\mathbf{f}}$ of the data vector, in terms of the $P \times 1$ output vector $\mathbf{p}$. If $\mathbf{T}$ is a square matrix, obviously

$$\hat{\mathbf{f}} = [\mathbf{T}]^{-1}\mathbf{p} \tag{A1.3-2}$$

provided that the matrix inverse exists. If $\mathbf{T}$ is not square, a $Q \times P$ matrix pseudoinverse operator $\mathbf{T}^+$ may be used to determine a solution by the operation

$$\hat{\mathbf{f}} = \mathbf{T}^+\mathbf{p} \tag{A1.3-3}$$

If a unique solution does indeed exist, the proper pseudoinverse operator will provide a perfect estimate in the sense that $\hat{\mathbf{f}} = \mathbf{f}$. That is, it will be possible to extract the vector $\mathbf{f}$ from the observation $\mathbf{p}$ without error. If multiple solutions exist, a pseudoinverse operator may be utilized to determine a minimum norm choice of solution. Finally, if there are no exact solutions, a pseudoinverse operator can provide a best approximate solution. This subject is explored further in the following sections. References 5, 6, and 9 provide background and proofs of many of the following statements regarding pseudoinverse operators.

The first type of pseudoinverse operator to be introduced is the generalized inverse $\mathbf{T}^-$, which satisfies the following relations:

$$\mathbf{T}\mathbf{T}^- = [\mathbf{T}\mathbf{T}^-]^T \tag{A1.3-4a}$$

$$\mathbf{T}^-\mathbf{T} = [\mathbf{T}^-\mathbf{T}]^T \tag{A1.3-4b}$$

$$\mathbf{T}\mathbf{T}^-\mathbf{T} = \mathbf{T} \tag{A1.3-4c}$$

$$\mathbf{T}^-\mathbf{T}\mathbf{T}^- = \mathbf{T}^- \tag{A1.3-4d}$$

The generalized inverse is unique. It may be expressed explicitly under certain circumstances. If $P > Q$, the system of equations of Eq. A1.3-1 is said to be *overdetermined*; that is, there are more observations $\mathbf{p}$ than points $\mathbf{f}$ to be estimated. In this case, if $\mathbf{T}$ is of rank Q, the generalized inverse may be expressed as

$$\mathbf{T}^- = [\mathbf{T}^T\mathbf{T}]^{-1}\mathbf{T}^T \tag{A1.3-5}$$

At the other extreme, if $P < Q$, Eq. A1.3-1 is said to be *underdetermined*. In this case, if $\mathbf{T}$ is of rank P, the generalized inverse is equal to

$$\mathbf{T}^- = \mathbf{T}^T[\mathbf{T}^T\mathbf{T}]^{-1} \tag{A1.3-6}$$

It can easily be shown that Eqs. A1.3-5 and A1.3-6 satisfy the defining relations of Eq. A1.3-4. A special case of the generalized inverse operator of computational interest occurs when $\mathbf{T}$ is direct product separable. Under this condition

$$\mathbf{T}^- = \mathbf{T}_C^- \otimes \mathbf{T}_R^- \tag{A1.3-7}$$

where $\mathbf{T}_R^-$ and $\mathbf{T}_C^-$ are the generalized inverses of the row and column linear operators.

Another type of pseudoinverse operator is the least-squares inverse $\mathbf{T}^\$$, which satisfies the defining relations

$$\mathbf{T}\mathbf{T}^\$\mathbf{T} = \mathbf{T} \tag{A1.3-8a}$$

$$\mathbf{T}\mathbf{T}^\$ = [\mathbf{T}\mathbf{T}^\$]^T \tag{A1.3-8b}$$

Finally, a conditional inverse $\mathbf{T}^\#$ is defined by the relation

$$\mathbf{T}\mathbf{T}^\#\mathbf{T} = \mathbf{T} \tag{A1.3-9}$$

Examination of the defining relations for the three types of pseudoinverse operators reveals that the generalized inverse is also a least-squares inverse, which in turn is also a conditional inverse. Least-squares and conditional inverses exist for a given linear operator **T**; however, they may not be unique. Furthermore, it is usually not possible to explicitly express these operators in closed form.

The following is a list of useful relationships for the generalized inverse operator of a $P \times Q$ matrix **T**.

Generalized inverse of matrix transpose:

$$[\mathbf{T}^T]^- = [\mathbf{T}^-]^T \qquad \text{(A1.3-10)}$$

Generalized inverse of generalized inverse:

$$[\mathbf{T}^-]^- = \mathbf{T} \qquad \text{(A1.3-11)}$$

Rank:

$$\text{rank}\{\mathbf{T}^-\} = \text{rank}\{\mathbf{T}\} \qquad \text{(A1.3-12)}$$

Generalized inverse of matrix product:

$$[\mathbf{T}^T\mathbf{T}]^- = [\mathbf{T}]^-[\mathbf{T}^T]^- \qquad \text{(A1.3-13)}$$

Generalized inverse of orthogonal matrix product:

$$[\mathbf{ATB}]^- = \mathbf{B}^T\mathbf{T}^-\mathbf{A}^T \qquad \text{(A1.3-14)}$$

where **A** is a $P \times P$ orthogonal matrix and **B** is a $Q \times Q$ orthogonal matrix.

A1.4. SOLUTIONS TO LINEAR SYSTEMS

The general system of linear equations specified by

$$\mathbf{p} = \mathbf{Tf} \qquad \text{(A1.4-1)}$$

where **T** is a $P \times Q$ matrix may be considered to represent a system of P equations in Q unknowns. Three possibilities exist:

1. The system of equations has a unique solution $\hat{\mathbf{f}}$ for which $\mathbf{T}\hat{\mathbf{f}} = \mathbf{p}$.
2. The system of equations is satisfied by multiple solutions.
3. The system of equations does not possess an exact solution.

If the system of equations possesses at least one solution, the system is called *consistent*; otherwise, it is *inconsistent*. The lack of a solution to the set of equations often occurs in physical systems in which the vector **p** represents a sequence of physical measurements of observations that are assumed to be generated by some nonobservable driving force represented by the vector **f**. The matrix **T** is formed by mathematically modeling the physical system whose output is **p**. For image restoration, **f** often denotes an ideal image vector, **p** is a blurred image vector and **T** models the discrete superposition effect causing the blur. Because the modeling process is subject to uncertainty, it is possible that the vector observations **p** may not correspond to any possible driving function **f**. Thus, whenever Eq. A1.4-1 is stated, either explicitly or implicitly, its validity should be tested.

Consideration is now given to the existence of solutions to the set of equations **p** = **Tf**. It is clear from the formation of the set of equations that a solution will exist if and only if the vector **p** can be formed by a linear combination of the columns of **T**. In this case, **p** is said to be in the column space of **T**. A more systematic condition for the existence of a solution is given by (5):

A solution to **p** = **Tf** exists if and only if there is a conditional inverse $\mathbf{T}^{\#}$ of **T** for which $\mathbf{T}\mathbf{T}^{\#}\mathbf{p} = \mathbf{p}$.

This condition simply states that the conditional inverse mapping $\mathbf{T}^{\#}$ from observation to image space, followed by the reverse mapping **T** from image to observation space, must yield the same observation vector **p** for a solution to exist. In the case of an underdetermined set of equations $(P < Q)$, when **T** is of full row rank P, a solution exists; in all other cases, including the overdetermined system, the existence of a solution must be tested.

A1.4.1. Solutions to Consistent Linear Systems

On establishment of the existence of a solution of the set of equations

$$\mathbf{p} = \mathbf{Tf} \tag{A1.4-2}$$

investigation should be directed toward the character of the solution. Is the solution unique? Are there multiple solutions? What is the form of the solution? The latter question is answered by the following fundamental theorem of linear equations (5).

If a solution to the set of equations **p** = **Tf** exists, it is of the general form

$$\hat{\mathbf{f}} = \mathbf{T}^{\#}\mathbf{p} + [\mathbf{I} - \mathbf{T}^{\#}\mathbf{T}]\mathbf{v} \tag{A1.4-3}$$

where $\mathbf{T}^{\#}$ is the conditional inverse of **T** and **v** is an arbitrary $Q \times 1$ vector.

Because the generalized inverse $\mathbf{T}^{-}$ and the least-squares inverse $\mathbf{T}^{\$}$ are also conditional inverses, the general solution may also be stated as

$$\hat{\mathbf{f}} = \mathbf{T}^{\$}\mathbf{p} + [\mathbf{I} - \mathbf{T}^{\$}\mathbf{T}]\mathbf{v} \qquad (A1.4\text{-}4a)$$

$$\hat{\mathbf{f}} = \mathbf{T}^{-}\mathbf{p} + [\mathbf{I} - \mathbf{T}^{-}\mathbf{T}]\mathbf{v} \qquad (A1.4\text{-}4b)$$

Clearly, the solution will be unique if $\mathbf{T}^{\#}\mathbf{T} = \mathbf{I}$. In all such cases, $\mathbf{T}^{-}\mathbf{T} = \mathbf{I}$. By examination of the rank of $\mathbf{T}^{-}\mathbf{T}$, it is found that (1):

If a solution to $\mathbf{p} = \mathbf{T}\mathbf{f}$ exists, the solution is unique if and only if the rank of the $P \times Q$ matrix $\mathbf{T}$ is equal to Q.

As a result, it can be immediately deduced that if a solution exists to an underdetermined set of equations, the solution is of multiple form. Furthermore, the only solution that can exist for an overdetermined set of equations is a unique solution. If Eq. A1.4-2 is satisfied exactly, the resulting pseudoinverse estimate

$$\hat{\mathbf{f}} = \mathbf{T}^{+}\mathbf{p} = \mathbf{T}^{+}\mathbf{T}\mathbf{f} \qquad (A1.4\text{-}5)$$

where $\mathbf{T}^{+}$ represents one of the pseudoinverses of $\mathbf{T}$, may not necessarily be perfect because the matrix product $\mathbf{T}^{+}\mathbf{T}$ may not equate to an identity matrix. The residual estimation error between $\mathbf{f}$ and $\hat{\mathbf{f}}$ is commonly expressed as the least-squares difference of the vectors written as

$$\mathcal{E}_E = [\mathbf{f} - \hat{\mathbf{f}}]^{T}[\mathbf{f} - \hat{\mathbf{f}}] \qquad (A1.4\text{-}6a)$$

or equivalently,

$$\mathcal{E}_E = \mathrm{tr}\{[\mathbf{f} - \hat{\mathbf{f}}][\mathbf{f} - \hat{\mathbf{f}}]^{T}\} \qquad (A1.4\text{-}6b)$$

Substitution of Eq. A1.4-5 into Eq. A1.4-6a yields

$$\mathcal{E}_E = \mathbf{f}^{T}[\mathbf{I} - (\mathbf{T}^{+}\mathbf{T})^{T}][\mathbf{I} - (\mathbf{T}^{+}\mathbf{T})]\mathbf{f} \qquad (A1.4\text{-}7)$$

The choice of $\mathbf{T}^{+}$ that minimizes the estimation error of Eq. A1.4-6 can be determined by setting the derivative of $\mathcal{E}_E$, with respect to $\mathbf{f}$, to zero. From Eq. A1.1-32

$$\frac{\partial \mathcal{E}_E}{\partial \mathbf{f}} = \mathbf{0} = 2[\mathbf{I} - (\mathbf{T}^{+}\mathbf{T})^{T}][\mathbf{I} - (\mathbf{T}^{+}\mathbf{T})]\mathbf{f} \qquad (A1.4\text{-}8)$$

Equation A1.4-8 is satisfied if $\mathbf{T}^{+} = \mathbf{T}^{-}$ is the generalized inverse of $\mathbf{T}$. Under this condition, the residual least-squares estimation error reduces to

$$\mathcal{E}_E = \mathbf{f}^T[\mathbf{I} - (\mathbf{T}^-\mathbf{T})]\mathbf{f} \qquad \qquad \text{(A1.4-9a)}$$

or

$$\mathcal{E}_E = \text{tr}\{\mathbf{f}\mathbf{f}^T[\mathbf{I} - (\mathbf{T}^-\mathbf{T})]\} \qquad \qquad \text{(A1.4-9b)}$$

The estimation error becomes zero, as expected, if $\mathbf{T}^-\mathbf{T} = \mathbf{I}$. This will occur, for example, if $\mathbf{T}^-$ is a rank Q generalized inverse as defined in Eq. A1.3-5.

A1.4.2. Approximate Solution to Inconsistent Linear Systems

Inconsistency of the system of equations $\mathbf{p} = \mathbf{Tf}$ means simply that the set of equations does not form an equality for any potential estimate $\mathbf{f} = \hat{\mathbf{f}}$. In such cases, the system of equations can be reformulated as

$$\mathbf{p} = \mathbf{Tf} + \mathbf{e}(\mathbf{f}) \qquad \qquad \text{(A1.4-10)}$$

where $\mathbf{e}(\mathbf{f})$ is an error vector dependent on $\mathbf{f}$. Now, consideration turns toward the determination of an estimate $\hat{\mathbf{f}}$ that minimizes the least-squares modeling error expressed in the equivalent forms

$$\mathcal{E}_M = [\mathbf{e}(\hat{\mathbf{f}})]^T[\mathbf{e}(\hat{\mathbf{f}})] = [\mathbf{p} - \mathbf{T}\hat{\mathbf{f}}]^T[\mathbf{p} - \mathbf{T}\hat{\mathbf{f}}] \qquad \qquad \text{(A1.4-11a)}$$

or

$$\mathcal{E}_M = \text{tr}\{[\mathbf{e}(\hat{\mathbf{f}})][\mathbf{e}(\hat{\mathbf{f}})]^T\} = \text{tr}\{[\mathbf{p} - \mathbf{T}\hat{\mathbf{f}}][\mathbf{p} - \mathbf{T}\hat{\mathbf{f}}]^T\} \qquad \text{(A1.4-11b)}$$

Let the matrix $\mathbf{T}^+$ denote the pseudoinverse that gives the estimate

$$\hat{\mathbf{f}} = \mathbf{T}^+\mathbf{p} \qquad \qquad \text{(A1.4-12)}$$

Then, adding and subtracting the quantity $\mathbf{TT}^+\mathbf{p}$ inside the brackets of Eq. A1.4-11a yields

$$\mathcal{E}_M = [(\mathbf{I} - \mathbf{TT}^+)\mathbf{p} + \mathbf{T}(\mathbf{T}^+\mathbf{p} - \hat{\mathbf{f}})]^T [(\mathbf{I} - \mathbf{TT}^+)\mathbf{p} + \mathbf{T}(\mathbf{T}^+\mathbf{p} - \hat{\mathbf{f}})] \quad \text{(A1.4-13)}$$

Expansion then gives

$$\mathcal{E}_M = [(\mathbf{I} - \mathbf{TT^+})\mathbf{p}]^T[(\mathbf{I} - \mathbf{TT^+})\mathbf{p}] + [\mathbf{T}(\mathbf{T^+p} - \hat{\mathbf{f}})]^T[\mathbf{T}(\mathbf{T^+p} - \hat{\mathbf{f}})]$$

$$+ [(\mathbf{I} - \mathbf{TT^+})\mathbf{p}]^T[\mathbf{T}(\mathbf{T^+p} - \hat{\mathbf{f}})] + [\mathbf{T}(\mathbf{T^+p} - \hat{\mathbf{f}})]^T[(\mathbf{I} - \mathbf{TT^+})\mathbf{p}] \quad (A1.4\text{-}14)$$

The two cross-product terms will be equal zero if $\mathbf{TT^+T} = \mathbf{T}$ and $\mathbf{TT^+} = [\mathbf{TT^+}]^T$. These are the defining conditions for $\mathbf{T^+}$ to be a least-squares inverse of $\mathbf{T}$, (i.e., $\mathbf{T^+} = \mathbf{T^\$}$). Under these circumstances, the residual error becomes equal to the sum of two positive terms:

$$\mathcal{E}_M = [(\mathbf{I} - \mathbf{TT^\$})\mathbf{p}]^T[(\mathbf{I} - \mathbf{TT^\$})\mathbf{p}] + [\mathbf{T}(\mathbf{T^\$p} - \hat{\mathbf{f}})]^T[\mathbf{T}(\mathbf{T^\$p} - \hat{\mathbf{f}})] \quad (A1.4\text{-}15)$$

The second term of Eq. A1.4-15 goes to zero when $\hat{\mathbf{f}}$ equals the least-squares pseudoinverse estimate, $\hat{\mathbf{f}} = \mathbf{T^\$p}$, and the residual error reduces to

$$\mathcal{E}_M = \mathbf{p}^T[\mathbf{I} - \mathbf{TT^\$}]\mathbf{p} \qquad (A1.4\text{-}16)$$

If $\mathbf{TT^\$} = \mathbf{I}$, the residual error goes to zero, as expected.

The least-squares pseudoinverse solution is not necessarily unique. If the pseudoinverse is further restricted such that $\mathbf{T^+TT^+} = \mathbf{T}$ and $\mathbf{T^+T} = [\mathbf{T^+T}]^T$ so that $\mathbf{T^+}$ is a generalized inverse, (i.e. $\mathbf{T^+} = \mathbf{T^-}$), it can be shown that the generalized inverse estimate, $\hat{\mathbf{f}} = \mathbf{T^-p}$, is a minimum norm solution in the sense that

$$\hat{\mathbf{f}}^T\hat{\mathbf{f}} \le \tilde{\mathbf{f}}^T\tilde{\mathbf{f}} \qquad (A1.4\text{-}17)$$

for any least-squares estimate $\tilde{\mathbf{f}}$. That is, the sum of the squares of the elements of the estimate is a minimum for all possible least-squares estimates. If $\mathbf{T^-}$ is a rank-Q generalized inverse, as defined in Eq. A1.3-5, $\mathbf{TT^-}$ is not necessarily an identity matrix, and the least-squares modeling error can be evaluated by Eq. A1.4-16. In the case for which $\mathbf{T^-}$ is a rank-P generalized inverse, as defined in Eq. A1.4-15, $\mathbf{TT^-} = \mathbf{I}$, and the least-squares modeling error is zero.

REFERENCES

1.   F. Ayres, Jr., *Schaum's Outline of Theory and Problems of Matrices,* McGraw-Hill, New York, 1962.

2.   R. E. Bellman, *Introduction to Matrix Analysis*, McGraw-Hill, New York, 1970.

3. H. G. Campbell, *An Introduction to Matrices, Vectors, and Linear Programming*, Appleton, New York, 1965.

4. C. G. Cullen, *Matrices and Linear Transformations*, Addison-Wesley, Reading, MA, 1966.

5. F. A. Graybill, *Introduction to Matrices with Applications in Statistics*, Wadsworth, Belmont, CA, 1969.

6. C. R. Rao and S. K. Mitra, *Generalized Inverse of Matrices and Its Applications*, Wiley, New York, 1971.

7. G. H. Golub and C. Reinsch, "Singular Value Decomposition and Least Squares Solutions," *Numerische Mathematik*, **14**, 1970, 403–420.

8. H. C. Andrews and C. L. Patterson, "Outer Product Expansions and Their Uses in Digital Image Processing," *American Mathematical Monthly*, **1**, 82, January 1975, 1–13.

9. A. Albert, *Regression and the Moore–Penrose Pseudoinverse*, Academic Press, New York, 1972.

APPENDIX 2

COLOR COORDINATE CONVERSION

There are two basic methods of specifying a color in a three primary color system: by its three tristimulus values (T_1, T_2, T_3), and by its chromaticity (t_1, t_2) and its luminance (Y). Given either one of these representations, it is possible to convert from one primary system to another.

CASE 1. TRISTIMULUS TO TRISTIMULUS CONVERSION

Let (T_1, T_2, T_3) represent the tristimulus values in the original coordinate system and $(\tilde{T}_1, \tilde{T}_2, \tilde{T}_3)$ the tristimulus values in a new coordinate system. The conversion between systems is given by

$$\tilde{T}_1 = m_{11}T_1 + m_{12}T_2 + m_{13}T_3 \tag{A2-1}$$

$$\tilde{T}_2 = m_{21}T_1 + m_{22}T_2 + m_{23}T_3 \tag{A2-2}$$

$$\tilde{T}_3 = m_{31}T_1 + m_{32}T_2 + m_{33}T_3 \tag{A2-3}$$

where the m_{ij} are the coordinate conversion constants.

CASE 2. TRISTIMULUS TO LUMINANCE/CHROMINANCE CONVERSION

Let

$$t_1 = \frac{T_1}{T_1 + T_2 + T_3} \tag{A2-4}$$

$$t_2 = \frac{T_2}{T_1 + T_2 + T_3} \tag{A2-5}$$

and

$$\tilde{t}_1 = \frac{\tilde{T}_1}{\tilde{T}_1 + \tilde{T}_2 + \tilde{T}_3} \tag{A2-6}$$

$$\tilde{t}_2 = \frac{\tilde{T}_2}{\tilde{T}_1 + \tilde{T}_2 + \tilde{T}_3} \tag{A2-7}$$

represent the chromaticity coordinates in the original and new coordinate systems, respectively. Then, from Eqs. A2-1 to A2-3,

$$\tilde{t}_1 = \frac{\beta_1 T_1 + \beta_2 T_2 + \beta_3 T_3}{\beta_4 T_1 + \beta_5 T_2 + \beta_6 T_3} \tag{A2-8}$$

$$\tilde{t}_1 = \frac{\beta_7 T_1 + \beta_8 T_2 + \beta_9 T_3}{\beta_4 T_1 + \beta_5 T_2 + \beta_6 T_3} \tag{A2-9}$$

where

$$\beta_1 = m_{11} \tag{A2-10a}$$

$$\beta_2 = m_{12} \tag{A2-10b}$$

$$\beta_3 = m_{13} \tag{A2-10c}$$

$$\beta_4 = m_{11} + m_{21} + m_{31} \tag{A2-10d}$$

$$\beta_5 = m_{12} + m_{22} + m_{32} \tag{A2-10e}$$

$$\beta_6 = m_{13} + m_{23} + m_{33} \tag{A2-10f}$$

$$\beta_7 = m_{21} \tag{A2-10g}$$

$$\beta_8 = m_{22} \tag{A2-10h}$$

$$\beta_9 = m_{23} \tag{A2-10i}$$

and m_{ij} are conversion matrix elements from the (T_1, T_2, T_3) to the $(\tilde{T}_1, \tilde{T}_2, \tilde{T}_3)$ coordinate system. The luminance signal is related to the original tristimulus values by

$$Y = w_{21}T_1 + w_{22}T_2 + w_{23}T_3 \tag{A2-11}$$

where the w_{ij} are conversion elements from the (T_1, T_2, T_3) to the (X, Y, Z) coordinate systems in correspondence with Eq. A2-2.

CASE 3. LUMINANCE/CHROMINANCE TO LUMINANCE CHROMINANCE CONVERSION

Substitution of

$$T_1 = t_1(T_1 + T_2 + T_3) \tag{A2-12}$$

$$T_2 = t_2(T_1 + T_2 + T_3) \tag{A2-13}$$

$$T_3 = (1 - t_1 - t_2)(T_1 + T_2 + T_3) \tag{A2-14}$$

into Eqs. A2-8 and A2-9 gives

$$\tilde{t}_1 = \frac{\alpha_1 t_1 + \alpha_2 t_2 + \alpha_3}{\alpha_4 t_1 + \alpha_5 t_2 + \alpha_6} \tag{A2-15}$$

$$\tilde{t}_2 = \frac{\alpha_7 t_1 + \alpha_8 t_2 + \alpha_9}{\alpha_4 t_1 + \alpha_5 t_2 + \alpha_6} \tag{A2-16}$$

where

$$\alpha_1 = m_{11} - m_{13} \tag{A2-17a}$$

$$\alpha_2 = m_{12} - m_{13} \tag{A2-17b}$$

$$\alpha_1 = m_{13} \tag{A2-17c}$$

$$\alpha_4 = m_{11} + m_{21} + m_{31} - m_{13} - m_{23} - m_{33} \tag{A2-17d}$$

$$\alpha_5 = m_{12} + m_{22} + m_{32} - m_{13} - m_{23} - m_{33} \tag{A2-17e}$$

$$\alpha_6 = m_{13} + m_{23} + m_{33} \tag{A2-17f}$$

$$\alpha_7 = m_{21} - m_{23} \tag{A2-17g}$$

$$\alpha_8 = m_{22} - m_{23} \tag{A2-17h}$$

$$\alpha_9 = m_{23} \tag{A2-17i}$$

and the m_{ij} are conversion matrix elements from the (T_1, T_2, T_3) to the $(\tilde{T}_1, \tilde{T}_2, \tilde{T}_3)$ coordinate system.

CASE 4. LUMINANCE/CHROMINANCE TO TRISTIMULUS CONVERSION

In the general situation in which the original chromaticity coordinates are not the CIE x–y coordinates, the conversion is made in a two-stage process. From Eqs. A2-1 to A2-3,

$$\tilde{T}_1 = n_{11}X + n_{12}Y + n_{13}Z \tag{A2-18}$$

$$\tilde{T}_2 = n_{21}X + n_{22}Y + n_{23}Z \tag{A2-19}$$

$$\tilde{T}_3 = n_{31}X + n_{32}Y + n_{33}Z \tag{A2-20}$$

where the m_{ij} are the constants for a conversion from (X, Y, Z) tristimulus values to $(\tilde{T}_1, \tilde{T}_2, \tilde{T}_3)$ tristimulus values. The X and Z tristimulus values needed for substitution into Eqs. As-18 to A2-20 are related to the source chromaticity coordinates by

$$X = \frac{\alpha_1 t_1 + \alpha_2 t_2 + \alpha_3}{\alpha_7 t_1 + \alpha_8 t_2 + \alpha_9} \qquad \text{(A2-21)}$$

$$Z = \frac{(\alpha_4 - \alpha_1 - \alpha_7)t_1 + (\alpha_5 - \alpha_2 - \alpha_8)t_2 + (\alpha_6 - \alpha_3 - \alpha_9)}{\alpha_7 t_1 + \alpha_8 t_2 - \alpha_9} Y \qquad \text{(A2-22)}$$

where the α_{ij} are constants for a transformation from (t_1, t_2) chromaticity coordinates to (x, y) chromaticity coordinates.

APPENDIX 3

IMAGE ERROR MEASURES

In the development of image enhancement, restoration, and coding techniques, it is useful to have some measure of the difference between a pair of similar images. The most common difference measure is the mean-square error. The mean-square error measure is popular because it correlates reasonable with subjective visual quality tests and it is mathematically tractable.

Consider a discrete $F(j, k)$ for $j = 1, 2, ..., J$ and $k = 1, 2, ... , K$, which is regarded as a reference image, and consider a second image $\hat{F}(j, k)$ of the same spatial dimensions as $F(j, k)$ that is to be compared to the reference image. Under the assumption that $F(j, k)$ and $\hat{F}(j, k)$ represent samples of a stochastic process, the mean-square error between the image pair is defined as

$$\xi_{MSE} = E\{|F(j, k) - \hat{F}(j, k)|^2\} \tag{A3-1}$$

where $E\{\cdot\}$ is the expectation operator. The normalized mean-square error is

$$\xi_{NMSE} = \frac{E\{|F(j, k) - \hat{F}(j, k)|^2\}}{E\{|F(j, k)|^2\}} \tag{A3-2}$$

Error measures analogous to Eqs. A3-1 and A3-2 have been developed for deterministic image arrays. The least-squares error for a pair of deterministic arrays is defined as

$$\xi_{LSE} = \frac{1}{JK} \sum_{j=1}^{J} \sum_{k=1}^{K} |F(j, k) - \hat{F}(j, k)|^2 \tag{A3-3}$$

and the normalized least-squares error is

$$
\xi_{NLSE} = \frac{\displaystyle\sum_{j=1}^{J} \sum_{k=1}^{K} |F(j, k) - \hat{F}(j, k)|^2}{\displaystyle\sum_{j=1}^{J} \sum_{k=1}^{K} |F(j, k)|^2} \tag{A3-4}
$$

Another common form of error normalization is to divide Eq. A3-3 by the squared peak value of $F(j, k)$. This peak least-squares error measure is defined as

$$
\xi_{PLSE} = \frac{\displaystyle\sum_{j=1}^{J} \sum_{k=1}^{K} |F(j, k) - \hat{F}(j, k)|^2}{[MAX\{F(j, k)\}]^2} \tag{A3-5}
$$

In the literature, the least-squares error expressions of Eqs. A3-3 to A3-5 are sometimes called mean-square error measures even though they are computed from deterministic arrays. Image error measures are often expressed in terms of a signal-to-noise ratio (SNR) in decibel units, which is defined as

$$
SNR = -10 \log_{10}\{\xi\} \tag{A3-6}
$$

A common criticism of mean-square error and least-squares error measures is that they do not always correlate well with human subjective testing. In an attempt to improve this situation, a logical extension of the measurements is to substitute processed versions of the pair of images to be compared into the error expressions. The processing is chosen to map the original images into some perceptual space in which just noticeable differences are equally perceptible. One approach is to perform a transformation on each image according to a human visual system model such as that presented in Chapter 2.

BIBLIOGRAPHY

J. K. Aggarwal, R. O. Duda, and A. Rosenfeld, Eds., *Computer Methods in Image Analysis*, IEEE Press, New York, 1977.

N. Ahmed and K. R. Rao, *Orthogonal Transforms for Digital Signal Processing*, Springer-Verlag, New York, 1975.

J. P. Allebach, *Digital Halftoning*, Vol. MS154, SPIE Press, Bellingham, WA, 1999.

H. C. Andrews, with W. K. Pratt and K. Caspari (Contributors), *Computer Techniques in Image Processing*, Academic Press, New York, 1970.

H. C. Andrews and B. R. Hunt, *Digital Image Restoration*, Prentice Hall, Englewood Cliffs, NJ, 1977.

H. C. Andrews, Ed., *Digital Image Processing*, IEEE Press, New York 1978.

D. H. Ballard and C. M. Brown, *Computer Vision*, Prentice Hall, Englewood Cliffs, NJ, 1982.

I. Bankman, Ed., *Handbook of Medical Imaging*, Academic Press, New York, 2000.

G. A. Baxes, *Digital Image Processing: Principles and Applications*, Wiley, New York, 1994.

R. Bernstein, Ed., *Digital Image Processing for Remote Sensing*, IEEE Press, New York, 1978.

J. C. Bezdek, Ed., *Fuzzy Models and Algorithms for Pattern Recognition and Image Processing*, Kluwer, Norwell, MA, 1999.

H. Bischof and W. Kropatsc, *Digital Image Analysis*, Springer-Verlag, New York, 2000.

A. Bovik, Ed., *Handbook of Image and Video Processing*, Academic Press, New York, 2000.

R. N. Bracewell, *Two-Dimensional Imaging*, Prentice Hall, Englewood Cliffs, NJ, 1995.

J. M. Brady, Ed., *Computer Vision*, North-Holland, Amsterdam, 1981.

H. E. Burdick, *Digital Imaging: Theory and Applications*, Wiley, New York, 1997.

K. R. Castleman, *Digital Image Processing*, Prentice Hall, Englewood Cliffs, NJ, 1979.

R. Chellappa and A. A. Sawchuk, *Digital Image Processing and Analysis, Vol. 1, Digital Image Processing*, IEEE Press, New York, 1985.

E. R. Davies, *Machine Vision: Theory, Algorithms, Practicalities*, 2nd ed., Academic Press, New York, 1996.

C. Demant, B. Streicher-Abel, and P. Waszlewitz, *Industrial Image Processing*, Springer-Verlag, New York, 1999.

G. G. Dodd and L. Rossol, Eds., *Computer Vision and Sensor-Based Robots*, Plenum Press, New York, 1979.

E. R. Dougherty and C. R. Giardina, *Image Processing Continuous to Discrete*, Vol. 1, *Geometric, Transform, and Statistical Methods*, Prentice Hall, Englewood Cliffs, NJ, 1987.

E. R. Dougherty and C. R. Giardina, *Matrix-Structured Image Processing*, Prentice Hall, Englewood Cliffs, NJ, 1987.

E. R. Dougherty, *Introduction to Morphological Image Processing*, Vol. TT09, SPIE Press, Bellingham, WA, 1992.

E. R. Dougherty, *Morphological Image Processing*, Marcel Dekker, New York, 1993.

E. R. Dougherty, *Random Processes for Image and Signal Processing*, Vol. PM44, SPIE Press, Bellingham, WA, 1998.

E. R. Dougherty, Ed., *Electronic Imaging Technology*, Vol. PM60, SPIE Press, Bellingham, WA, 1999.

E. R. Dougherty and J. T. Astola, Eds., *Nonlinear Filters for Image Processing*, Vol. PM59, SPIE Press, Bellingham, WA, 1999.

R. O. Duda and P. E. Hart, *Pattern Classification and Scene Analysis*, Wiley-Interscience, New York, 1973.

R. O. Duda, P. E. Hart, and D. G. Stork, *Pattern Classification,* 2nd ed., Wiley, New York, 2001.

M. J. B. Duff, Ed., *Computing Structures for Image Processing*, Academic Press, London, 1983.

M. P. Ekstrom, Ed., *Digital Image Processing Techniques*, Academic Press, New York, 1984.

H. Elias and E. R. Weibel, *Quantitative Methods in Morphology*, Springer-Verlag, Berlin, 1967.

O. W. E. Gardner, Ed., *Machine-Aided Image Analysis*, Institute of Physics, Bristol and London, 1979.

R. C. Gonzalez and P. Wintz, *Digital Image Processing*, 2nd ed., Addison-Wesley, Reading, MA, 1987.

R. C. Gonzalez, R. E. Woods (Contributor), and R. C. Gonzalez, *Digital Image Processing*, 3rd ed., Addison-Wesley, Reading, MA, 1992.

J. Goutsias and L. M. Vincent, Eds., *Mathematical Morphology and Its Applications to Image and Signal Processing*, Kluwer, Norwell, MA, 2000.

A. Grasselli, *Automatic Interpretation and Classification of Images*, Academic Press, New York, 1969.

E. L. Hall, *Computer Image Processing and Recognition*, Academic Press, New York, 1979.

A. R. Hanson and E. M. Riseman, Eds., *Computer Vision Systems*, Academic Press, New York, 1978.

R. M. Haralick and L. G. Shapiro (Contributor), *Computer and Robot Vision*, Addison-Wesley, Reading, MA, 1992.

R. M. Haralick, *Mathematical Morphology: Theory and Hardware*, Oxford Press, Oxford, 1998.

G. C. Holst, *Sampling, Aliasing and Data Fidelity*, Vol. PM55, SPIE Press, Bellingham, WA.

B. K. P. Horn, *Robot Vision*, MIT Press, Cambridge, MA, 1986.

T. S. Huang, Ed., *Topics in Applied Physics: Picture Processing and Digital Filtering*, Vol. 6, Springer-Verlag, New York, 1975.

T. S. Huang, *Image Sequence Processing and Dynamic Scene Analysis*, Springer-Verlag, New York, 1983.

B. Jahne, *Practical Handbook on Image Processing for Scientific Applications*, CRC Press, Boca Raton, FL, 1997.

B. Jahne and B. Jahne, *Digital Image Processing: Concepts, Algorithms, and Scientific Applications*, 4th ed., Springer-Verlag, Berlin, 1997.

B. Jahne et al., Eds., *Handbook of Computer Vision and Applications*, package ed., Academic Press, London, 1999.

B. Jahne and H. Haubecker, *Computer Vision and Applications*, Academic Press, New York, 2000.

A. K. Jain, *Fundamentals of Digital Image Processing*, Prentice Hall, Englewood Cliffs, NJ, 1989.

J. R. Jensen, *Introductory Digital Image Processing: A Remote Sensing Perspective*, Prentice Hall, Englewood Cliffs, NJ, 1985.

I. Kabir, *High Performance Computer Imaging*, Prentice Hall, Englewood Cliffs, NJ, 1996.

S. Kaneff, Ed., *Picture Language Machines*, Academic Press, New York, 1970.

H. R. Kang, *Color Technology for Electronic Imaging Devices*, Vol. PM28, SPIE Press, Bellingham, WA, 1997.

H. R. Kang, *Digital Color Halftoning*, Vol. PM68, SPIE Press, Bellingham, WA, 1999.

F. Klette et al., *Computer Vision*, Springer-Verlag, New York, 1998.

A. C. Kokaram, *Motion Picture Restoration*, Springer-Verlag, New York, 1998.

P. A. Laplante and A. D. Stoyenko, *Real-Time Imaging: Theory, Techniques, and Applications*, Vol. PM36, SPIE Press, Bellingham, WA, 1996.

C. T. Leondes, Ed., *Image Processing and Pattern Recognition*, Academic Press, New York, 1997.

M. D. Levine, *Vision in Man and Machine*, McGraw-Hill, New York, 1985.

J. S. Lim, *Two-Dimensional Signal and Image Processing*, Prentice Hall, Englewood Cliffs, NJ, 1989.

C. A. Lindley, *Practical Image Processing in C*, Wiley, New York, 1991.

B. S. Lipkin and A. Rosenfeld, Eds., *Picture Processing and Psychopictorics*, Academic Press, New York, 1970.

R. P. Loce and E. R. Dougherty, *Enhancement and Restoration of Digital Documents: Statistical Design of Nonlinear Algorithms*, Vol. PM 29, SPIE Press, Bellingham, WA, 1997.

D. A. Lyon and D. A. Lyons, *Image Processing in Java*, Prentice Hall, Englewood Cliffs, NJ, 1999.

A. Macovski, *Medical Imaging Systems*, Prentice Hall, Englewood Cliffs, NJ, 1983.

Y. Mahdavieh and R. C. Gonzalez, *Advances in Image Analysis*, Vol. PM08, SPIE Press, Bellingham, WA, 1992.

S. Marchand-Maillet and Y. M. Sharaiha, *Binary Digital Image Processing*, Academic Press, New York, 1999.

D. Marr, *Vision*, W.H. Freeman, San Francisco, 1982.

S. Mitra and G. Sicuranza, Eds., *Nonlinear Image Processing*, Academic Press, New York, 2000.

H. R. Myler, *Fundamentals of Machine Vision*, Vol. TT33, 1998.

R. Nevatia, *Structure Descriptions of Complex Curved Objects for Recognition and Visual Memory*, Springer-Verlag, New York, 1977.

R. Nevatia, *Machine Perception*, Prentice Hall, Englewood Cliffs, NJ, 1982.

W. Niblack, *An Introduction to Digital Image Processing*, Prentice Hall, Englewood Cliffs, NJ, 1986.

J. R. Parker, *Algorithms for Image Processing and Computer Vision*, Wiley, New York, 1996.

T. Pavlidis, *Algorithms for Graphics and Image Processing*, Computer Science Press, Rockville, MD, 1982.

I. Pitas, *Digital Image Processing Algorithms,* Prentice Hall, Englewood Cliffs, NJ, 1993.

I. Pitas, *Digital Image Processing Algorithms and Applications,* Wiley, New York, 2000.

C. A. Poynton, *A Technical Introduction to Digital Video*, Wiley, New York, 1996.

W. K. Pratt, *Digital Image Processing*, Wiley-Interscience, New York, 1978.

W. K. Pratt, *Digital Image Processing,* 2nd ed., Wiley-Interscience, New York, 1991.

W. K. Pratt, *PIKS Foundation C Programmer's Guide*, Manning Publications, Greenwich, CT, 1995.

W. K. Pratt, *Developing Visual Applications, XIL: An Imaging Foundation Library*, Sun Microsystems Press, Mountain View, CA, 1997.

K. Preston, Jr. and L. Uhr, *Multicomputers and Image Processing, Algorithms and Programs*, Academic Press, New York, 1982.

K. Preston, Jr. and M. J. B. Duff, *Modern Cellular Automata: Theory and Applications*, Plenum Press, New York, 1984.

G. X. Ritter and J. N. Wilson (Contributor), *Handbook of Computer Vision Algorithms in Image Algebra*, Lewis Publications, New York, 1996.

G. X. Ritter, *Handbook of Computer Vision Algorithms in Image Algebra*, 2nd ed., Lewis Publications, New York, 2000.

A. Rosenfeld, *Picture Processing by Computer*, Academic Press, New York, 1969.

A. Rosenfeld; Ed., *Digital Picture Analysis*, Springer-Verlag, New York, 1976.

A. Rosenfeld and A. C. Kak, *Digital Image Processing*, Academic Press, New York, 1976.

A. Rosenfeld and A. C. Kak, *Digital Picture Processing*, 2nd ed., Academic Press, San Diego, CA, 1986.

J. C. Russ, *The Image Processing Handbook*, 3rd ed., CRC Press, Boca Raton, FL, 1999.

S. J. Sangwine and R. E. N. Horne, Eds., *The Colour Image Processing Handbook*, Kluwer, Norwell, MA, 1998.

R. J. Schalkoff, *Digital Image Processing and Computer Vision*, Wiley, New York, 1989.

R. A. Schowengerdt, *Remote Sensing: Models and Methods for Image Processing*, Academic Press, New York, 1997.

J. Serra, *Image Analysis and Mathematical Morphology*, Academic Press, London, 1982.

M. I. Sezan, Ed., *Digital Image Restoration*, Vol. MS47, SPIE Press, Bellingham, WA, 1992.

D. Sinha and E. R. Dougherty, *Introduction to Computer-Based Imaging Systems*, Vol. TT23, SPIE Press, Bellingham, WA, 1997.

G. Stockman and L. G. Shapiro, *Computer Vision*, Prentice Hall, Englewood Cliffs, NJ, 2000.

P. Stucki, Ed., *Advances in Digital Image Processing: Theory, Application, Implementation*, Plenum Press, New York, 1979.

T. Szoplik, Ed., *Morphological Image Processing: Principles and Optoelectronic Implementations*, Vol. MS127, SPIE Press, Bellingham, WA, 1996.

S. Tanamoto and A. Klinger, Eds., *Structured Computer Vision: Machine Perception Through Hierarchical Computation Structures*, Academic Press, New York, 1980.

A. M. Telkap, *Digital Video Processing*, Prentice Hall, Englewwod Cliffs, NJ, 1995.

J. T. Tippett et al., Eds., *Optical and Electro-Optical Information Processing*, MIT Press, Cambridge, MA, 1965.

M. M. Trivedi, *Digital Image Processing*, Vol. MS17, SPIE Press, Bellingham, WA, 1990.

R. Ulichney, *Digital Halftoning*, MIT Press, Cambridge, MA, 1987.

S. Ullman and W. Richards, Eds., *Image Understanding 1984*, Ablex Publishing, Norwood, NJ, 1984.

S. E. Umbaugh, *Computer Vision and Image Processing*, Prentice Hall, Englewood Cliffs, NJ, 1997.

A. Venetsanoupoulos and K. N. Plataniotis, *Color Image Processing and Applications*, Springer-Verlag, New York, 2000.

A. R. Weeks, Jr., *Fundamentals of Electronic Image Processing*, Vol. PM32, SPIE Press, Bellingham, WA, 1996.

P. F. Whelan and D. Molloy, *Machine Vision Algorithms in Java*, Springer-Verlag, New York,

G. Wolberg, *Digital Image Warping*, IEEE Computer Society Press, New York, 1990.

T. Y. Young and K. S. Fu, Eds., *Handbook of Pattern Recognition and Image Processing*, Academic Press, San Diego, CA, 1986.

INDEX